Proceedings
of the
Steklov Institute of Mathematics

edited by
S. M. Nikol'skiĭ

Number 120 (1974)

Selected problems of weighted approximation and spectral analysis

by

N. K. Nikol'skiĭ

Providence, Rhode Island
AMERICAN MATHEMATICAL SOCIETY
1976

Академия Наук
Союза Советских Социалистических Республик

ТРУДЫ

МАТЕМАТИЧЕСКОГО ИНСТИТУТА имени В. А. СТЕКЛОВА

CXX

Ответственный редактор академик АН СССР С. М. НИКОЛЬСКИЙ

ИЗБРАННЫЕ ЗАДАЧИ ВЕСОВОЙ АППРОКСИМАЦИИ И СПЕКТРАЛЬНОГО АНАЛИЗА

Н. К. НИКОЛЬСКИЙ

Издательство "Наука"
Ленинград 1974

Translated by F. A. CEZUS

AMS (MOS) subject classifications (1970). Primary 30A82, 31A05, 47A10; Secondary 30A04, 30A18, 30A98, 46E15, 47B05.

ABSTRACT. In this monograph we consider problems of spectral analysis and spectral synthesis for various classes of operators which act in function spaces or abstract Banach spaces. These problems originate in classical harmonic analysis (Bohr, Wiener, and others) and approximation theory in complex domains (Keldyš, Beurling, and others) and they also come from problems of describing invariant subspaces for linear operators. In this work we consider simultaneously problems on spectral synthesis and completeness of polynomials in some weighted spaces of analytic functions, multiplicity of the spectrum for a shift operator and simultaneous approximation, and spectral singularities of operators with a unitary spectrum. An important step in solving these problems is the verification of precise one-sided and modulus estimates for harmonic and analytic functions in the first chapter.

Library of Congress Cataloging in Publication Data CIP

Nikol'skiĭ, Nikolaĭ Konstantinovich.
Selected problems of weighted approximation and spectral analysis.
(Proceedings of the Steklov Institute of Mathematics ; no. 120 (1974))
Translation of Izbrannye zadachi vesovoĭ approksima-t͡sii i spektral'nogo analiza.
Bibliography: p.
1. Analytic functions. 2. Harmonic functions. 3. Approximation theory. 4. Spectral theory (Mathematics) 5. Linear operators. I. Title. II. Series: Akademii͡a nauk SSSR. Matematicheskiĭ institut. Proceedings ; no. 120.
QA1.A413 No. 120 [QA331] 510'.8s [515'.7] 76-46375
ISBN 0-8218-3020-1

Translation authorized by Meždunarodnaja Kniga, Moscow

PROCEEDINGS OF THE STEKLOV INSTITUTE OF MATHEMATICS IN THE ACADEMY OF SCIENCES OF THE USSR

(Труды математического института им. В. А. Стеклова, т. CXX, 1974)

TABLE OF CONTENTS

Page

FOREWORD

The concept of "analysis-synthesis"—decomposition of an object into simple components or harmonics, followed by study of the components and then recovery of the object from the components—is deeply rooted in various areas of mathematics. Even within the confines of mathematical analysis, realizations of the analysis-synthesis problem can be very diverse, and at first glance their common features may not be recognized even by specialists. In order to underline this diversity, we mention the classical analysis of Fourier series and the contemporary harmonic analysis on groups, the theory of partial differential operators, and weighted approximation by means of polynomials, in which many problems can be given a unified treatment by means of the analysis-synthesis concept. In the present monograph we present results along two directions. I. The analysis-synthesis problem for the operator of multiplication by the independent variable in spaces of analytic functions and the related problem of weighted polynomial approximation. II. The analysis-synthesis problem for some abstract and specialized classes of linear operators acting in general Banach or Hilbert spaces. There is a close interconnection between problems I and II, but the corresponding parts of this monograph (Chapters 1 and 2 and Chapter 3 respectively) can be read independently. Chapter 0 contains a summary of the basic results.

At present the intensity of the investigations into the general analysis-synthesis problem and problems I and II does not correspond to the potential effect which these methods can have on a wide range of questions in the theory of functions of a complex variable, the theory of linear operators, and differential equations. One goal of the present work is to attract attention to the problems of spectral analysis-synthesis as an interesting and deep area of research.

The author owes the idea of writing this monograph to V. P. Il'in, the director of the "Approximate Calculations" laboratory of the Leningrad Branch of the Steklov Institute of Mathematics. Individual results were gleaned from many useful discussions with S. A. Vinogradov, A. S. Markus, V. I. Macaev, B. S. Pavlov and V. P. Havin. The author extends his deep thanks to all of these people.

The author also thanks V. P. Havin for reading the entire manuscript, and S. V. Hruščev and N. A. Širokov for reading individual chapters and making many valuable comments. He thanks A. P. Hus, scientific secretary at the Leningrad Branch of the Steklov Institute, for excellent organization of the technical details of the manuscript. And finally he thanks his family for supporting his mathematical ambitions.

INTRODUCTION

Harmonic analysis and weighted approximation by polynomials, multiplicity of the spectrum and unicellularity of linear operators—these and other questions considered in this monograph are a part of the general problem of spectral "analysis-synthesis". This problem arises in mathematical analysis with the first results in the theory of Fourier series, where it is known as the problem of summability of a formal Fourier series to the (generalized) function which generates it. At the same time a trigonometric system which is invariant with respect to rotations of a circle loses its distinguished position and takes its place among other systems of eigenfunctions of linear operators. The problem of spectral synthesis in this situation reduces to recovery of elements of a space from their formal Fourier series with respect to a given system of eigenfunctions or root functions. Other important areas where such problems arise are mean-periodic functions and convolution equations, the Wiener Tauberian theorem, and theorems on weighted polynomial approximation.

First we give a general formulation of the problem of spectral analysis-synthesis. Let X be a linear (topological) space and let T be a linear operator (or family of such operators) from X into itself which has a complete set of eigenvectors or root vectors. The following questions form the content of the analysis-synthesis problem:

(A) Does each subspace of X which is invariant([1]) with respect to T contain at least one eigenvector ("analysis")?

(B) Is each such subspace generated by the eigenvectors (or root vectors) which it contains ("synthesis")?

In our investigation we consider two realizations of this general problem. In this first we deal with classes of functions which are defined and analytic in the unit disk $\mathbf{D}$ of the complex plane $\mathbf{C}$; that is, in $\mathbf{D} = \{\zeta\colon \zeta \in \mathbf{C}, |\zeta| < 1\}$. We investigate the possibility of arbitrarily precise approximations of an arbitrary function in a given class by products of the form pf, where p is a polynomial and f is a given (fixed) function. Let z be the identity mapping of $\mathbf{D}$ onto itself, $z(\zeta) = \zeta$, $\zeta \in \mathbf{D}$, and let S be the shift operator $Sf = zf$. Under mild assumptions on the original space the eigenfunctions of

([1])An invariant subspace of T is a closed linear manifold E such that $x \in E \Rightarrow Tx \in E$.

the operator S^* which is conjugate to S are the functionals $\Phi_\zeta: f \longrightarrow f(\zeta)$, where $|\zeta| < 1$, and the root vectors are the functionals $\Phi_{k,\zeta}: f \longrightarrow f^{(k)}(\zeta)$, where $k \geqslant 0$ and $|\zeta| < 1$. Therefore the "analysis" problem for S^* leads to the aforementioned polynomial approximation with a "weight" function f satisfying the condition $f(\zeta) \neq 0, \zeta \in \mathbf{D}$. Theorems on solvability of this approximation problem are contained in Chapter 2.

Another set of questions which we consider in this monograph pertains to the possibility of analysis-synthesis for so-called complete operators. This class is a natural extension of the class of finite-dimensional operators, for which the analysis-synthesis problem was completely solved by means of linear algebra. The fundamental theorem of algebra asserts that each finite-dimensional operator has an eigenvector, and this class is closed with respect to taking restrictions to invariant subspaces and with respect to factorizations. Therefore finite-dimensionality of an operator implies the possibility of spectral analysis and synthesis.

A fundamental conclusion which we can draw from the results in the second part of this monograph (Chapter 3) is the fact that the possibility of spectral synthesis can only be guaranteed by very restrictive assumptions on the analytic properties of an operator which has a complete set of eigenvectors, and a small deviation from these assumptions usually leads to very pathological situations.

Chapter 1 occupies a special position. On the one hand it is preliminary to the fundamental results in Chapter 2, and on the other hand it contains estimates for harmonic and analytic functions which are of interest in themselves. The role of these estimates in weighted approximation problems is discussed in Chapter 1 and also in the appropriate sections of Chapter 2.

We turn to a more detailed presentation of the results.

1°. *Estimates for harmonic and analytic functions.* In Chapter 1 we obtain two types of estimates for harmonic or analytic functions in the unit disk $\mathbf{D}$ in the complex plane. Inequalities of the first type estimate the speed of growth of the maximum modulus $M_{\tilde{u}}$ of the harmonic conjugate $\tilde{u}$ of u,

$$M_{\tilde{u}}(r) = \max_{|\zeta|=r} |\tilde{u}(\zeta)|, \quad 0 \leqslant r < 1,$$

as $r \longrightarrow 1-0$ as compared to the growth of $M_u(r)$. The inequalities which we prove admit the following qualitative description: $M_{\tilde{u}}$ grows no faster than M_u if the order of growth of M_u is not less than a power ($M_u(r) \geqslant 1/(1-r)^\epsilon, \epsilon > 0$); $M_{\tilde{u}}$ grows no faster than $M_u(r)\alpha(\log(1/(1-r)))$ if the order of growth of M_u is less than a power. Here α is some nondecreasing function on $[0, +\infty)$ with $\alpha(x) \leqslant \text{const} \cdot x$, and α increases faster the slower M_u increases; $\alpha(x) = x$ beginning with logarithmic growth of M_u ($M_u(r) \leqslant (\log(1/(1-r)))^\epsilon, \epsilon > 0$). We show that the estimates which we obtain are precise in the class of all harmonic functions u with a given majorant for M_u. In the case $M_u(r) \leqslant 1/(1-r)^\epsilon, \epsilon > 0$, these estimates become the Hardy-Littlewood theorem in [136]; see also [39].

The second type of inequality is connected with lower estimates for analytic (in

D) functions f, $f(\zeta) \neq 0$, $|\zeta| < 1$. Here we show(2) that $m_f(r) \overset{\text{def}}{=} \min_{|\zeta|=r} |f(\zeta)|$ decreases no faster than $1/M_f^{\text{const}}$ as $r \to 1-0$ if $M_f(r) \geqslant \exp(1/(1-r)^\epsilon)$ for some $\epsilon > 1$, and no faster than

$$\exp\left\{-\frac{\text{const}}{1-r}\left(\int_0^r \sqrt{\frac{\log M_f(t)}{1-t}}\,dt\right)^2\right\}, \quad 0 \leqslant r < 1,$$

if the power order of the growth of $\log M_f$ is not greater than unity. These estimates are also sharp in the class of all analytic functions with a given majorant for M_f. For exponential majorants $\exp(1/(1-r)^\epsilon)$, $\epsilon > 0$, these estimates coincide with results of Cartwright in [124]. Some other special cases were considered by Linden in [153] and [154]. We mention also the estimate of Macaev [67], [68] for functions having exponential order greater than unity; he obtained a lower bound of the form

$$\frac{1}{\lambda^{\text{const}}(\alpha r + (1-\alpha))}, \quad \text{where } \lambda = M_f,\ 0 < \alpha < 1,\ \alpha = \alpha(\lambda).$$

Detailed comments on the results in Chapter 1 are contained in §1.0 and in the introductory parts of §§1.1–1.4. The method of proof for the lower estimates is based on an inequality of Carathéodory and a formula of Carleman [53], which connects the growth of a function inside a domain with the growth on the boundary, followed by an application of estimates on the harmonic measure of curvilinear trapezoids. These estimates depend on asymptotic formulas of Warschawski for conformal mappings [15] and refinements of these formulas which we develop.

2°. *General remarks on the problem of weighted approximation.* The problem of spectral analysis for the operator S^* adjoint to the shift S admits the following dual formulation. Let X be a space of analytic functions in **D** which is invariant with respect to multiplication by z; that is, $f \in X \Rightarrow zf \in X$. We consider subspaces $E \subset X$ which are also invariant with respect to a shift. Such a subspace is said to be fixed (localized) if there exists a point $\zeta \in \mathbf{D}$ such that $f(\zeta) = 0$ for all $f \in E$. The familiar problem of localization of invariant subspaces (ideals, in algebraic terminology) consists of finding a description of spaces X where all S-invariant subspaces are fixed. It is clear that for such spaces X, and only for such spaces, the operator S^* admits (weak) spectral analysis. If this is so, we arrive at a precise analogue of the Tauberian theorem of Wiener. For the largest space $A(\mathbf{D})$ of all analytic functions in **D** with the topology of compact convergence this theorem was essentially proved in the work of Schilling [168]. On the other hand, Keldyš [45] and Beurling [117] showed that in the spaces $H^2(l_2)$ and H^2, which are determined by the conditions

$$\iint_{\mathbf{D}} |f(\zeta)|^2\,dx\,dy < +\infty, \qquad \sup_{0 \leqslant r < 1} \int_{|\zeta|=1} |f(r\zeta)|^2\,|d\zeta| < +\infty,$$

(2)In this Introduction we are only giving qualitative descriptions of the results; precise formulations will be given in the text.

respectively, not every invariant subspace is fixed. Later Beurling [118] investigated the space of all analytic functions in **D** whose Taylor coefficients $\hat{f}(n)$, $n \geqslant 0$, grow no faster than some power of a given sequence $\{d_n\}_{n \geqslant 0}$: $|\hat{f}(n)| = O(d_n^k)$, $n \geqslant 0$, for some $k > 0$. He showed that under mild restrictions on the "regularity" of growth a necessary and sufficient condition for the localization of all principal ideals is divergence of the series

$$\sum_{n \geqslant 1} \frac{\log d_n}{n^{3/2}}.$$

We also mention the works of Šamojan [112] and Macaev [69] which investigate algebras of functions of power or exponential order of growth and describe all of their ideals. In the same language we can interpret the profound investigations of Šilov [115], Rudin [164], Taylor and Williams [176], and Korenbljum [51], [52] on closed ideals in algebras of functions which are smooth in the closure $\overline{\mathbf{D}}$ of **D**. We should also mention that after the work of Ehrenpreis [129] and Schwartz [169] the problem of spectral analysis-synthesis was actively investigated for domains Ω in **C** which were not equal to **D**, based on the case $\Omega = \mathbf{C}$ (see, for example, [48] and [49]); this remark also applies to the n-dimensional case $\mathbf{C}^n$. The nature of these investigations is different from the ideas in the present work.

Returning to the problem of spectral analysis for spaces of analytic functions in **D**, we first mention a feature which is common to all of the works which we have listed; namely, spaces with "mild" topologies (for instance, topologies defined by a family of seminorms with large "clearances" between them) admit localizations of invariant subspaces, while "severe" (norm) topologies produce the opposite situation. This last observation is connected with the fact that functions in normed spaces have specific "boundary zeros" and with the possibility of distinguishing S-invariant subspaces by such singularities in the behavior on the boundary. But in the general case the assumption of existence of nonfixed invariant subspaces in normed spaces of analytic functions can only be treated as a hypothesis.

In Chapter 2 we solve an individual version of the problem of fixed subspaces, similar to a problem considered by Beurling [118]. In describing this problem we first consider its connection with weighted polynomial approximations. If in a normed space X each S-invariant subspace is fixed, then the system of functions

$$\{z^n f\}_{n \geqslant 0} \tag{1}$$

is complete in X if $f(\zeta) \neq 0$, $\zeta \in \mathbf{D}$. Later we will assume that the set $\mathcal{P}_A$ of all polynomials in z is dense in X; therefore completeness of the system (1) is equivalent to the possibility of achieving the approximation

$$\inf_{p \in \mathcal{P}_A} \| pf - \mathbf{1} \|_X = 0, \tag{2}$$

where $\mathbf{1}(\zeta) \equiv 1$, $\zeta \in \mathbf{D}$. In other words, if $f(\zeta) \neq 0$, $\zeta \in \mathbf{D}$, then the set of polynomials $\mathcal{P}_A$ is dense in the space $X_f = \{g\colon gf \in X\}$ with the weighted norm $\|g\|_f = \|gf\|_X$. In specific situations the question of completeness of the polynomials in such a space with an "analytic" weight has been investigated by Keldyš [145], Markuševič [64], Šaginjan [110], Džrbašjan [31], [32], and others (see the survey [70]).

We consider the possible methods for achieving the approximation $\lim_n \|p_n f - 1\|_X = 0$. The most natural method of pushing the function f towards **1** is by using the Abel means f_r of f: $f_r(\zeta) = f(r\zeta)$, $0 \leqslant r < 1$, $\zeta \in \mathbf{D}$. In fact $\lim_{r\to 1-0} f(\zeta)/f_r(\zeta) = 1$ for all $\zeta \in \mathbf{D}$, and the function $1/f_r$ can easily be approximated by polynomials in the necessary sense. But even in simple cases ($X = H^2$) it can turn out that (see [167]) $\sup_{0\leqslant r<1} \|f/f_r\|_X = +\infty$, despite the approximation (2), so that this method is unsuitable. We note in passing that in the general case of weighted completeness of polynomials (without assuming "analyticity" of the weight) the use of the means f_r, presented by Džrbašjan in [31], turns out to be useful (see also [70]). Another method of pushing f towards **1** (within the closed linear hull $\mathfrak{L}(z^n f\colon n \geqslant 0) \stackrel{\text{def}}{=} E(f)$ of the set $\{z^n f\colon n \geqslant 0\}$) consists of considering the powers f^t, $0 \leqslant t \leqslant 1$. Under certain assumptions on the function f and the space X we can perform the division f/f^ϵ, $\epsilon > 0$, without leaving the subspace $E(f)$, and after a finite number of steps we reach the desired result, namely, the inclusion $\mathbf{1} \in E(f)$. This method was used by Lebedev [100] and Shapiro [170], [171], [113]; and this method underlies §2.1 in the present work. This method shares a deficiency with the first method, namely, too big a "piece" of the function f splits off from it after each step; the division f/f^ϵ becomes impossible in $E(f)$ if the maxima and minima of $|f|$ on the circles $|\zeta| = r$, $0 \leqslant r < 1$, are too far from each other (for details, see §2.1).

It turns out to be more acceptable to push f towards **1** by means of logarithms or in a similar fashion by means of derivatives (Beurling [118]). The first approach calls for approximating the functions $(\log f)^n f$, $n \geqslant 0$, and then approximating **1**. The Beurling method is based on the use of the functions $D^n f$, $n \geqslant 0$, where $Df = zf'$. The applicability of these methods to the question of weighted approximations depends upon their connection with the problem of quasi-analyticity of classes of functions which are analytic in a domain and are smooth up to the boundary. We consider, for example, pushing by means of logarithms (for details, see §2.4). Let e_f be the vector-valued function defined in $\overline{\mathbf{D}}$ by the formula $e_f(t) = f^t$, $t \in \overline{\mathbf{D}}$. Under appropriate conditions e_f is infinitely differentiable in $\overline{\mathbf{D}}$ (as an X-valued function), and $e_f^{(n)}(1) = (\log f)^n f$, $n \geqslant 0$. If the scalar functions $t \longrightarrow \Phi(e_f(t))$, $t \in \overline{\mathbf{D}}$, where Φ is a linear functional on X, are quasi-analytic, then the inclusion $(\log f)^n f \in E(f)$ and the Hahn-Banach theorem [43] imply that $\mathbf{1} \in E(f)$. Thus the original weighted approximation problem splits into two easier problems. This approach is used in the main part of Chapter 2 (§§2.4–2.7).

Later, after a more detailed description of the results in Chapters 2 and 3, a function f for which the system (1) is complete will be said to be weakly invertible (in the space X). A vector x in X is said to be a cyclic (weakly invertible, reproducing) vector of the operator T if $\mathfrak{L}(T^n x\colon n \geqslant 0) = X$.

3°. *Criteria for weak invertibility, based on pushing by means of powers.* These sufficient conditions for weighted completeness of polynomials in spaces of the type X_f, $f \in X$, use the aforementioned pushing by means of the powers $\{f^t\}_{0\leqslant t<1}$. They are suitable only if $\max_{|\zeta|=r} |f(\zeta)|$ does not grow too fast (with respect to X) as $r \to 1-0$, and $\min_{|\zeta|=r} |f(\zeta)|$ does not decrease too fast. But these elementary criteria enjoy a wide generality; they are valid in all spaces of analytic functions which satisfy a few explicit conditions which are usually easy to verify. A typical criterion in simplified form reads as follows: if X contains together with the function f, $f(\zeta) \neq 0$, $\zeta \in \mathbf{D}$, each analytic function g (for some $\epsilon > 0$) such that

$$|g(\zeta)| \leqslant |f(\zeta)| \left(\min_{|\xi|=|\zeta|} |f(\xi)|\right)^{-\varepsilon}, \quad \zeta \in \mathbf{D},$$

then $E(f) = X$ and f is weakly invertible. In §2.1, where we prove this and other elementary criteria for weak invertibility, we pay particular attention to weak invertibility of inner and outer (in the sense of Beurling) functions. We recall that for functions of a certain kind, namely

$$\sup_{0 \leqslant r < 1} \int_{|\zeta|=1} \log^+ |f(r\zeta)| \, |d\zeta| < +\infty,$$

we have the Nevanlinna-Riesz factorization theorem (see [23] and [95]):

$$f = \frac{V_1}{V_2} F,$$

where V_1 and V_2 are bounded regular functions in $\mathbf{D}$ whose moduli on $\partial\mathbf{D}$ equal 1 almost everywhere (V_1 and V_2 are inner functions), and F is a function with no zeros in $\mathbf{D}$, whose logarithm can be recovered by means of the Schwarz integral from the values of $\log|F|$ on $\partial\mathbf{D}$ (outer function). It is known (see [117]) that in the Hardy spaces H^p, $1 \leqslant p < \infty$ (see [23], [95], or §§2.1 and 2.2 below; the definition of H^2 is given below) a function f is weakly invertible if and only if f is an outer function (that is, $V_1 = V_2 = 1$ in the canonical factorization). In §2.1 we show that an outer function is weakly invertible in any space which satisfies the axioms in this section. In an arbitrary space X weakly invertible functions are analogues of the outer functions of Beurling (this analogy is discussed in §2.1). We investigate weak invertibility in the nonseparable space $A_\infty(\lambda)$, which consists of all functions $f \in A(\mathbf{D})$ with $f(\zeta) = O(\lambda(|\zeta|))$, $|\zeta| \to 1$, where λ is a given majorant, and we study its connection with simultaneous approximations in the disk $\mathbf{D}$ and on the circumference $\partial\mathbf{D}$.

Nonseparability of the space $A_\infty(\lambda)$ forces us to abandon approximations of functions by means of polynomials in the norm $\|g\|_f = \|gf\|_{A_\infty(\lambda)}$. Here it is natural to go to a weaker type of convergence as closely connected as possible with the original space. In §2.2 we consider the case $\lambda \equiv \mathbf{1}$. Then $A_\infty(\lambda) = H^\infty$ is the collection of all bounded analytic functions in $\mathbf{D}$ with the sup-norm $\|\cdot\|_\infty$. It is known that H^∞ is isometrically imbedded in the space L^∞ of all (classes of) bounded measurable functions on $\partial\mathbf{D}$. A natural convergence in which to consider weighted completeness of the polynomials is $(*)$-convergence; by definition $(*)\text{-}\lim_n f_n = f$ if $\sup_n \|f_n\|_\infty < \infty$ and $f_n(\zeta) \to f(\zeta)$ for almost all $\zeta \in \partial\mathbf{D}$. Questions of weighted completeness with respect to this form of convergence are treated in §2.2. It turns out that $E(f) = H^\infty$ (here $E(f)$ is defined in a natural way) if and only if f is an outer function whose modulus has a certain degree of "smoothness". We give a complete description of the subspaces of $E(f)$ in the general case, where $E(f) \neq H^\infty$.

4°. *Estimates on Taylor coefficients, boundary spaces, and Bernšteĭn inequalities.* The method of weak invertibility discussed in 2° makes essential use of imbedding theorems of the type

$$l_A^1(\|z^n\|_X) \subset X \subset l_A^\infty(\|z^n\|_X), \tag{3}$$

where $l_A^p(w_n)$ is the set of all analytic functions in $\mathbf{D}$ with a finite norm

$$\|f\|_{p,w} = \Big(\sum_{n\geq 1} |\hat{f}(n)|^p w_n^p\Big)^{1/p}.$$

The inclusion on the left side of (3) is obviously valid for an arbitrary Banach space X which contains the polynomials, but the inclusion $X \subset l_A^\infty(\|z^n\|_X)$ is less trivial, and its validity requires special restrictions. This inclusion is equivalent to the following system of estimates on Taylor coefficients:

$$|\hat{g}(n)| \leqslant \text{const}\,\|g\|_X \|z^n\|_X^{-1}, \quad n \geqslant 0, \quad g \in X. \tag{4}$$

It is not hard to see (§2.3) that, for instance, the assumption of invariance of the norm $\|\cdot\|_X$ with respect to rotations ($f \to f_\alpha$, $f_\alpha(\zeta) = f(\alpha\zeta)$, $|\alpha| = 1$, $\zeta \in \mathbf{D}$) leads to the required estimates (3) and (4). In order to extend the class of spaces which admit the estimates (3) and (4), in §2.3 we insert some "nonhomogeneity" into invariant norms by means of "analytic" weights. More precisely, beginning with a given class $\mathfrak{S}$ for which (3) and (4) hold, we describe all spaces $X = \mathfrak{S}_f$, $f \in \mathfrak{S}$, in which analogous estimates hold. For example, for all classes $\mathfrak{S}$ which are determined by the conditions

$$\iint_{\mathbf{D}} |g(r\zeta)|^p \, dl_1(\zeta)\, d\mathcal{M}(r) < +\infty, \tag{5}$$

where $\mathcal{M}$ is a finite measure on $[0, 1)$ and $1 \leqslant p < \infty$, the admissible weight functions f are given by the condition $f \in H^p$, $1/f \in H^{p'}$, where $1/p + 1/p' = 1$ and H^p and $H^{p'}$ are Hardy classes; here l_1 is normalized Lebesgue measure on $\partial\mathbf{D}$.

In the general case the weighted classes $X = \mathfrak{S}_f$ with estimates (3) and (4) are distinguished from other such classes by the fact that f and $1/f$ belong to the boundary spaces $\partial\mathfrak{S}$ and $\partial^*\mathfrak{K}\mathfrak{S}^*$ which are defined in §2.3. The space $\partial\mathfrak{S}$ consists of all functions $f \in \mathfrak{S}$ such that

$$\|f\|_{\partial\mathfrak{S}} \overset{\text{def}}{=} \varlimsup_n \frac{\|z^n f\|_{\mathfrak{S}}}{\|z^n\|_{\mathfrak{S}}} < +\infty.$$

The class $\partial^*\mathfrak{K}\mathfrak{S}^*$ has a dual definition in terms of the conjugate space $\mathfrak{S}^*$. In §2.3 we carry out an explicit calculation of the boundary norms $\|\cdot\|_{\partial\mathfrak{S}}$ and $\|\cdot\|_{\partial^*\mathfrak{K}\mathfrak{S}^*}$ for many spaces $\mathfrak{S}$. If the norm in $\mathfrak{S}$ is already weighted (in a broad sense of the word, including the integral norms in (5) and the norms in the spaces $l_A^p(w_n)$), then the norm $\|\cdot\|_{\partial\mathfrak{S}}$ is not weighted; under mild restrictions on the class $\mathfrak{S}$ it is "homogeneous" in the sense that $\|zf\|_{\partial\mathfrak{S}} = \|f\|_{\partial\mathfrak{S}}$ for $f \in \partial\mathfrak{S}$. The value of going from a space $\mathfrak{S}$ to the boundary space $\partial\mathfrak{S}$ resides in the fact that problems connected with estimates on Taylor coefficients, convergence of series, and so forth, in the space $\mathcal{E}$ are transferred to $\partial\mathcal{E}$ where they often become easy or even trivial.

One necessary detail in computing the spaces $\partial^*\mathcal{K}\mathcal{E}^*$ is the verification of "Bernšteĭn inequalities". Here we mean relations of the form

$$\|p\|_{\mathcal{E}} \geqslant \text{const}\, \|p\|_{\partial\mathcal{E}} \|z^n\|_{\mathcal{E}}. \tag{6}$$

or

$$\|p\|_{\mathcal{E}} \leqslant \text{const}\, \|p\|_{\partial\mathcal{E}} \|z^n\|_{\mathcal{E}}, \tag{7}$$

which hold for all $n \geqslant 1$ and all polynomials in $\mathcal{P}_A$ of degree less than or equal to n. Inequality (6) is called a "conjugate Bernšteĭn inequality" in §2.3, where it is proved for spaces which contain functions having no additional smoothness properties (spaces of the type (5), $l^p_A(w_n)$ with $\sup_p w_n < \infty$, and so forth). Inequality (7), which is dual to (6), is considered only indirectly since it is not used in the computation of boundary spaces. But we note that (7) apparently holds for all norms $\|\cdot\|_{\mathcal{E}}$ which preserve some "smoothness" of functions in the closed disk $\overline{\mathbf{D}}$ and have "regularity" properties. From the calculations in §2.3 it follows that this "Bernšteĭn inequality" holds, in particular, for $\|p\|_{\mathcal{E}} = \|p'\|_E$, where E is a space of functions with an isometric shift operator ($\|zg\|_E = \|g\|_E$, $g \in E$), in which the powers $\{z^n\}_{n\geqslant 0}$ form a Cesàro basis. Here $\|p\|_{\partial\mathcal{E}} = \|p\|_E$ and inequality (7) for $E = H^p$ or $E = C_A$ coincides(3) with the classical Bernšteĭn inequality. Other special cases are contained in the works of Szegö [175], Bernšteĭn [6], and Civin [125], where they were established for all trigonometric polynomials (and not just for the polynomials in $\mathcal{P}_A$). We emphasize the fact that the requirements in §2.3 are completely satisfied by (6) and indirectly by (7) with some fixed constant, so that the important question in the constructive theory of functions concerning the magnitude of the best such constant is not considered in the present work. We note also that estimates on derivatives of polynomials in a complex domain in terms of their values on the real axis or on the sides of some angle are contained in the work [32] of Džrbašjan.

We use the same method of reduction to the boundary space $\partial\mathcal{E}$ for solving the power basis problem in the weighted class $\mathcal{E}_f$ for a special choice of $\mathcal{E}$ with $\partial\mathcal{E} = H^2$. Here we show that the powers $\{z^n\}_{n\geqslant 0}$ form a basis of E if and only if f, $1/f \in H^2$, and for the function $|f|^2$ the Helson-Szegö condition is satisfied (see [142], [144] or §2.3).

5°. *Metric criteria for weighted completeness. Spectral analysis.* Following the weak invertibility methods mentioned in 2°, in §§2.4–2.7 we consider normed classes of functions X and subclasses $X_1 \subset X$ which have some "smoothness" with respect to X. We seek the largest possible subclass X_1 consisting of functions f for which the condition $f(\zeta) \neq 0$, $\zeta \in \mathbf{D}$, implies weak invertibility in X. Under some "regularity" assumptions on the norms $\|\cdot\|_X$ and $\|\cdot\|_{X_1}$ a sufficient condition for the above situa-

---

(3) After calculating the constant, which in this case equals 1. The space C_A consists of all functions which are analytic in $\mathbf{D}$ and continuous in $\overline{\mathbf{D}}$.

tion is divergence of the series

$$\sum_{n\geqslant 1} n^{-3/2}\log\frac{\|z^n\|_{X_1}}{\|z^n\|_X}. \tag{8}$$

The proof of this criterion depends on a theorem of Bernšteĭn and Džrbašjan [7], [31] (see also [60] and [71]) on weighted polynomial approximation on a half-line. Further efforts are directed towards translating this condition into a more "natural" language for frequently encountered spaces. In §2.5 we analyze the classes of functions $A(\lambda)$, $L^p_{1/\lambda}(E)$ and $l^p_A(w_n)$. Here $A(\lambda)$ is the set of all analytic functions in $\mathbf{D}$ such that $|f(\zeta)| = o(\lambda(|\zeta|))$ as $|\zeta| \to 1-0$, equipped with a weighted sup-norm (λ is a given function which grows to $+\infty$ on $[0, 1)$); $L^p_{1/\lambda}(E)$ consists of all analytic functions with a finite norm

$$\left(\int_0^1 \|f_r\|_E^p \lambda(r)^{-p}\, dr\right)^{1/p},$$

where E is an arbitrary "Hardy-type space" H^q. If two of these spaces are determined by majorants Λ and λ respectively, where $\Lambda \geqslant \lambda$, $\log\Lambda = \log\lambda + h(\log\lambda)$, and h is an increasing function on $[0, +\infty)$, then divergence of the series (8) implies divergence of the integral

$$\int^{\infty} \frac{h(t)}{t\varkappa(t)^{1/2}}\, dt, \tag{9}$$

where κ is a convex function such that $\kappa(\mu) \geqslant \mu'$ and $\mu = \log\lambda$. If the exponential order of the majorant λ equals α, $1 \leqslant \alpha \leqslant \infty$, then the best power order of the function h which gives us divergence of the integral (9) is $1/2 + 1/2\alpha$.

For weak invertibility in the spaces $A(\lambda)$ and $L^p_{1/\Lambda}(E)$ of all nonvanishing functions f in $\mathbf{D}$ which "grow slowly" in the sense that

$$\int^1 \left(\frac{\log \max_{|\zeta|=r} |f(\zeta)|}{1-r}\right)^{1/2} dr < +\infty,$$

it suffices to have

$$\int^1 \left(\frac{\log\Lambda(r)}{1-r}\right)^{1/2} dr = +\infty.$$

This condition is necessary for weak invertibility of the inner function

$$\exp\frac{z+1}{z-1},$$

which was first considered in weighted approximation problems by Keldyš [45] and was then used by a series of authors (see [29], [149] and [171]).

On the basis of calculations in §2.7 we construct classes of functions which are unions or intersections of the spaces $A(\lambda)$, $L^p_{1/\lambda}(E)$ and $l^p_A(w_n)$, and which admit "individual" spectral analysis. We recall that this phrase means the possibility of weak inversion of an arbitrary function f in the space such that $f(\zeta) \neq 0$, $\zeta \in \mathbf{D}$. We show that the corollaries of our fundamental theorems are substantially sharper than the Beurling criterion which we mentioned in §2 for the space $\bigcup_{k=0}^{\infty} l^{\infty}_A(d_n^{-k})$. They are best possible with respect to the order of growth of the functions involved (see §2.8).

All of the results in §§2.4–2.7 are obtained under assumptions on the "regularity" of growth of the majorants λ, Λ, $\{w_n\}_0^\infty$, and so forth, as in Chapter 1.

6°. *Obstacles to weak invertibility*. As we have already noted, in 1945 Keldyš showed that the function $\exp((z + 1)/(z - 1))$ is not weakly invertible in $H^2(l_2)$. His proof was based on the fact(4) that this function decreases rapidly on $[0, 1)$ as $r \to 1 - 0$ and is large on the surrounding circle $\{\zeta: |\zeta - ½| = ½\}$. This fact underlies a criterion for the impossibility of weak invertibility in a general space X of analytic functions (see §2.8). Here we introduce a subclass of the class of "X-inner" functions, distinguished by special asymptotic behavior near a selected point on the circumference.(5) From results in §2.8 it follows that the power order $1/2 + 1/2\alpha$ of the function h, which guarantees weak invertibility of functions in $A(\lambda)$ in the space $A(\lambda e^{h(\log \lambda)})$, cannot be improved. We note also that the faster λ increases, the more difficult it becomes to construct functions in $A(\lambda)$ with the required asymptotic properties (see §2.1.1° and §2.8).

7°. *The "analysis-synthesis" problem for completely continuous operators*. Finite-dimensional operators, which admit spectral analysis and synthesis, are studied by essentially algebraic methods. When we consider completely continuous operators with a complete set of eigenvectors or root vectors we encounter difficult problems of analysis and infinite-dimensional geometry, many of which are far from being solved; some classes of operators are studied from this point of view in works on completeness of eigenvectors or root vectors. In the work of Markus [62] it is shown that the well-known completeness criteria of Keldyš [46], Lidskiĭ [55], Gohberg and Kreĭn [24], Macaev [66], Markus [61], and others are hereditary for invariant subspaces and therefore lead to the possibility of spectral synthesis. Another collection of theorems on spectral synthesis is contained in the works of Brodskiĭ [10] and Sz-Nagy and Foiaş [96], which involve dissipative or contraction operators which differ by a nuclear operator from selfadjoint or unitary operators respectively. On the other hand, Hamburger [134] showed by means of examples that an operator without eigenvectors (Volterra operator) can be part of a complete, completely continuous operator.

Chapter 3 opens with a theorem on extensions which shows that the circumstance described by Hamburger is not unusual; each completely continuous operator is a part of some complete, completely continuous operator defined in a larger space (see §3.1). This theorem, which is useful in subsequent sections, forces us to return to the study of special subclasses of the class of completely continuous operators which are closely related to other problems in mathematical analysis and the theory of operators. One of these problems is the description of unicellular operators, which are an infinite-dimensional analogue of the linear-algebraic "Jordan cells" which were defined by Brodskiĭ [9], [25] as operators with a linearly ordered family of invariant subspaces.

(4)See also Mergeljan [70], §6.

(5)After the appearance of [87], in which this approach was used, the work [149] of Kelleher and Taylor introduced the concept of an "X-inner" function for classes of the form $X = \bigcup_{n=1}^\infty A(\lambda^n)$ with a slowly increasing majorant λ.

§3.2 is devoted to one such type of operator. We consider weighted shift operators $T = S\Lambda$, $Te_n = \lambda_n e_{n+1}$, $n \geqslant 0$, where $\{e_n\}_{n \geqslant 0}$ is a natural basis of the sequence space l^p and the λ_n, $n \geqslant 0$, are preassigned factors. The operator T^* is complete if the root vectors e_n, $n \geqslant 0$, form a complete set in the corresponding space. The possibility of spectral synthesis in this situation is equivalent to unicellularity of T^* (and therefore of T), which in turn means that the discrete chain $L(e_k\colon 0 \leqslant k < n)$, $n = 0, 1, \ldots$, exhausts all T^*-invariant subspaces. The results in §3.2 show that the unicellularity property (synthesis) for weighted shift operators depends on relatively subtle "regularity" properties of the moduli $|\lambda_n|$ (it suffices, for example, to have monotone decrease on arithmetic subprogressions) and it is lost if this "regularity" becomes too weak (for instance, it is insufficient to have monotonicity of $|\lambda_n|$ on subsets of the natural numbers which are more sparse than arithmetic progressions). In a series of cases the theorems in §3.2 can be used as criteria for completeness of analytic functions.(6) For example, if J is the integral operator $(Jf)(\zeta) = \int_0^\zeta f(t)\,dt$, $|\zeta| < 1$, in the Hardy space H^2, then for completeness of the system $\{J^n f\}_{n \geqslant 0}$ it is necessary and sufficient that $f(0) \neq 0$. The methods in §3.2 enable us to give an affirmative answer to the question of Šilov on existence of nonstandard ideals in primary algebras. More precisely, in this section we construct examples of Banach algebras of formal power series in which there exists a continuum of ideals which do not coincide with the standard ideals $\{f\colon f^{(k)}(0) = 0, 0 \leqslant k < n\}$, $n = 0, 1, \ldots$.

8°. *Spectral analysis on the unitary spectrum.* In the second part of Chapter 3 the problem of spectral analysis is solved for operators with a unitary point spectrum which are in some sense close to isometries. In general the results enable us to assert that under severe restrictions on these operators, we can preserve a large part of the "Hilbert" theory of unitary operators, while at the same time we get pathological results by departing only a little from the unitary situation. We let $E_\lambda = E_\lambda^T$ denote the space of eigenvectors of T corresponding to a number $\lambda \in \mathbf{C}$, and we say that an operator T is $\mathbf{\Gamma}$-complete in X if $\mathfrak{L}(E_\lambda\colon |\lambda| = 1) = X$. For $\mathbf{\Gamma}$-complete strongly ergodic operators (that is, operators such that $\sup_{n \geqslant 0} |T^n| < +\infty$; $|\cdot|$ is the operator norm in X) it is shown in §3.3 that there exist "almost-orthogonal" (in the sense of Banach) spectral projectors onto each subgroup γ of the circle $\mathbf{\Gamma} = \partial\mathbf{D} = \{\lambda\colon |\lambda| = 1\}$; that is, projectors onto the subspace $E_\gamma = \mathfrak{L}(E_\lambda\colon \lambda \in \gamma)$, which commute with all operators that commute with T. The proof of this fact contains within itself the possibility of spectral synthesis for these operators, which was previously established in [63]. By splitting off the part of the operator which corresponds to the point spectrum (we set $\gamma = \mathbf{\Gamma}$), we can describe the invariant subspaces of a class of direct products $T_1 \times T_2$.

The existence of projectors P_λ onto the eigenspaces E_λ, $\lambda \in \mathbf{\Gamma}$, follows from the weaker requirement of A-ergodicity(7)

$$\sup_{|\lambda|>1} \| R(\lambda, T) \| \, (|\lambda| - 1) < +\infty,$$

(6)We note that a description of just the cyclic vectors of T leads to criteria for completeness of systems of the form $\{T^n x\}_{n \geqslant 0}$, $x \in X$.

(7)We use the terminology of Hille and Phillips [108].

where $R(\lambda, T) = (\lambda I - T)^{-1}$ is the resolvent of T. Thus to each vector $x \in X$ we can associate a formal Fourier series $x \sim \Sigma_\lambda P_\lambda x$ which contains no more than countably many nonzero "harmonics" $P_\lambda x$. If T is a rotation (through an angle which is not commensurable with π) in some space of (generalized) functions which contains the trigonometric monomials z^n, $n = 0, \pm 1, \ldots$ (here z is the identity mapping of $\mathbf{\Gamma}$ onto itself), then this series becomes the usual Fourier series of a function x. Keeping this analogy in mind, in §3.3 we consider questions of convergence and summability by linear methods for the series (10) in the general case of an isometric $\mathbf{\Gamma}$-complete operator.

In a "negative" direction a very strong result in §3.3 can be stated as follows: each completely continuous operator T with a spectrum contained in $\mathbf{D}$ can be extended to a $\mathbf{\Gamma}$-complete C-ergodic operator; that is, an operator such that

$$\sup_{|\lambda|=1,\, n\geqslant 0} \left\| \frac{1}{n+1} \sum_{k=0}^{n} T^k \lambda^k \right\| < +\infty .$$

Thus $\mathbf{\Gamma}$-complete C-ergodic operators do not admit spectral synthesis. The proof of this fact is based on an extension theorem in §3.1 and on estimates for norms of finite-dimensional rotation operators.

In §3.3 we also investigate the effect of the unitary point spectrum on other characteristics of the operator. We prove a theorem on localization of the spectrum; namely, the spectrum of a non-quasi-analytic ($\Sigma_{n=-\infty}^{\infty} |\log\|T^n\|| / (n^2 + 1) < \infty$) $\mathbf{\Gamma}$-complete operator coincides with the closure of the set of its eigenvalues; if the assumptions are not satisfied, the result does not follow. We obtain a generalization of the Wiener-Pitt theorem [179] on localization of the spectrum for functions of a $\mathbf{\Gamma}$-complete isometric operator. At the end of §3.3 we clarify the question of the "degree of divergence" from being an isometry for an operator T in a Hilbert space X depending on the massiveness of the set $\sigma_p(T) \cap \mathbf{\Gamma}$; here $\sigma_p(T) = \{\lambda\colon E_\lambda \neq \{0\}\}$ is the point spectrum of T. If $\gamma = \Sigma \gamma_{|n|} e^{int}$ is an integrable convex function on $(0, 2\pi)$, then positivity of the γ-capacity of the set $\sigma_p(T) \cap \mathbf{\Gamma}$ implies that $\Sigma_{n\geqslant 0} \gamma_{n+N} \|T^n\|^{-2} < +\infty$ for some $N \geqslant 0$; if the Lebesgue measure of $\sigma_p(T) \cap \mathbf{\Gamma}$ is positive, then $\Sigma_{n\geqslant 0} \|T^n\|^{-2} < \infty$. We establish that the last assertion is sharp.

9°. *Multiplicity of the spectrum and simultaneous approximation.* The question treated in the last section of Chapter 3 is only indirectly connected with the general problem of analysis-synthesis. We indicate which parameters of the original space influence the multiplicity of the spectrum of the operator which is dual to the shift operator. More precisely, we consider a space X of generalized functions on $\mathbf{\Gamma}$ which is invariant with respect to the shift S (here $Sf = zf$) and the inverse S^{-1}, and we consider the direct product $S \times S$. We show that the multiplicity of the spectrum of this operator depends on the "quality" of the space X, and in particular on the possibility of a natural realization of X by measurable functions on $\mathbf{\Gamma}$. For example, if X consists of generalized functions on $\mathbf{\Gamma}$ whose Fourier coefficients are pth power summable ($1 \leqslant p \leqslant \infty$), then for $1 \leqslant p \leqslant 2$ the multiplicity of the spectrum of the operator $S \times S$ equals 2, and for $p > 2$ it falls to 1 (the smallest possible value). The connection with simultaneous approximation is based on the definition of a cyclic vector: if the multiplicity equals 1, then there exist (generalized) functions f and g such that for arbitrary functions $h, k \in X$ we have

$$\inf_p (\| pf - h \|_X + \| pg - k \|_X) = 0,$$

where the infimum is taken over all polynomials p or only over the polynomials in $\mathcal{P}_A$ depending on the choice of the definition of multiplicity, which is discussed in detail in §3.4.

10°. *Concluding remarks.* We make use of the basic results in the theory of functions and functional analysis without special mention. More specialized results are formulated in detail. Material of an auxiliary nature is presented in Appendices I–III. The number of each section consists of two parts: §a.b means section b of chapter a. Theorems, lemmas, and formulas are numbered independently within each section, so that we will speak of "Theorem c in §a.b". The symbol • will denote the end of a proof or discussion. At the end of this work there is a list of basic symbols.

CHAPTER 1

ONE-SIDED AND MODULUS ESTIMATES OF HARMONIC FUNCTIONS IN A DISK AND IN A STRIP

§1.0. Problems and results

1°. In this chapter we will consider the following two problems.

I. Let μ be a positive function defined on the interval $[0, 1)$; let $h(\mu)$ be the set of all functions u which are harmonic in the unit disk $\mathbf{D}$ of the complex plane $\mathbf{C}$, $\mathbf{D} = \{\zeta \in \mathbf{C}\colon |\zeta| < 1\}$, and satisfy the inequality

$$|u(\zeta)| \leqslant \mu(|\zeta|), \quad |\zeta| < 1.$$

Let $v = \Gamma u$ be the harmonic conjugate function of u with $v(0) = 0$. We seek the best estimate for the supremum $M = M_\mu$,

$$M(r) = \sup_{u \in h(\mu)} |(\Gamma u)(r)|, \quad 0 \leqslant r < 1,$$

as $r \longrightarrow 1$.

II. Let $a(\mathbf{D}, \lambda)$ be the set of all functions f which are analytic in $\mathbf{D}$ and satisfy the conditions $|f(\zeta)| \leqslant \lambda(|\zeta|)$, $f(0) = 1$ and $f(\zeta) \neq 0$ for $|\zeta| < 1$, where $\lambda = \exp \mu$. We seek the best estimate for the infimum $m_{\mathbf{D}}$,

$$m_{\mathbf{D}}(r) = \inf_{f \in a(\mathbf{D}, \lambda)} |f(r)|, \quad 0 \leqslant r < 1,$$

as $r \longrightarrow 1$.

In Chapter 1 we present solutions of problems I and II which are definitive in a certain sense under the assumption that the majorant μ (or $\lambda = \exp \mu$) satisfies some conditions on "regularity" of growth. The speed of the growth of μ (or λ) remains arbitrary.

Problem I can be reformulated as a problem of describing weight functions λ_1 for which the (linear) operator of harmonic conjugation $u \longrightarrow \Gamma u$ is continuous from a linear space $H(\mu)$ with a unit ball $h(\mu)$ into a space $H(\mu_1)$ with a ball $h(\mu_1)$. Estimates to be introduced later will enable us to describe all such weight functions μ_1 and to establish bounds for the norm $\|\Gamma\|_{H(\mu) \to H(\mu_1)}$.

Problem II is equivalent to finding the best minorant in the class of all real functions u which are harmonic in $\mathbf{D}$ and such that $u(\zeta) \leqslant \log \lambda(|\zeta|)$, $u(0) = 0$. From a technical point of view, in problem II it is convenient to begin by considering functions with one singularity (for example, at the point $\zeta = 1$) and even (after a conformal transformation into the strip $\Pi = \{\zeta \in \mathbf{C}: |\mathrm{Im}\,\zeta| < \pi/2\}$) functions f which are regular in the strip Π for which $|f(\zeta)| \leqslant \lambda_\Pi(\mathrm{Re}\,\zeta)$ and $f(0) = 1$, while $f(\zeta) \neq 0$ for $\zeta \in \Pi$; here $\lambda_\Pi(t) = \lambda(1 - e^{-t})$, $t \geqslant 0$. The class of all such functions will be denoted by $a(\Pi, \lambda_\Pi)$. A function m_Π is determined by a function $m_\mathbf{D}$ in an analogous fashion.

The problem of finding estimates of types I and II for harmonic and analytic functions has attracted the attention of many analysts, beginning with the classical works of Nevanlinna on functions of a specific form, and the works of Golubev [21], Hardy and Littlewood [136], [137], and Cartwright [124] on analytic functions with a given majorant, and continuing with the recent works [67]–[69], [153] and [154]. Undoubtedly asymptotic uniqueness theorems (essentially estimating the admissible order of smallness of $|f(\zeta)|$ on or near the boundary) are closely connected with the problems considered here. A discussion of these uniqueness theorems and a vast amount of literature on this theme is outside the scope of this work. More detailed references to the literature on problems I and II will be given in §§1.1–1.4.

Before we begin a brief summary of the results in this chapter, we mention once again that the estimates which we obtain are fundamental to the investigation of the problem of "spectral analysis" and weighted approximation in spaces of functions which are analytic in **D** and grow near the boundary (see Chapter 2).

2°. *Description of results.* §1.1 contains the solution of problem I. We show that the supremum of a function which is the harmonic conjugate of a function in the class $h(\mu)$ does not exceed: 1) const $\cdot\, \mu$ if $l = \lim_{t\to\infty} \psi'(t) > 0$ ($\psi = \log\log \lambda_\Pi$); 2) const $\cdot\, \mu/\psi'(\log(1/(1-r)))$, if $l = 0$ but $k = \lim \psi'(t)t > 0$; 3) const $\mu \cdot \log(1-r)^{-1}$, if $k = 0$. The constants indicated by const depend on l and k. By constructing special harmonic functions we show that the estimates 1)–3) are best possible in the class $h(\mu)$, and in cases 2) and 3) the constants which we find are also best possible. These results are obtained under certain additional assumptions on the majorant λ, a typical assumption being convexity of the function $1/\psi'$. This requirement is stronger than the usual restrictions imposed on the pointwise order (see [53]). The conditions in the theorems in Chapter 1 are discussed briefly in 3°, below, and again in §1.5.

§§1.2 and 1.3 are devoted to the solution of problem II. Here the outcome also depends on the magnitude of the limit $l = \lim_{t\to\infty} \psi'$, whose existence is always assumed. For $l = \infty$ (that is, in the case of a rapidly growing majorant), we show that

$$f \in a(\mathbf{D}, \lambda) \Rightarrow \log|f(\zeta)| \geqslant -B \log \lambda(|\zeta|), \quad |\zeta| < 1,$$

where B is an absolute constant (calculated explicitly in §§1.2. and 1.3). First (in §1.2) we establish the sharper but "sequential" estimate

$$\varliminf_{r\to 1} \frac{1}{\varepsilon_\mathbf{D}\left(\log\frac{1}{1-r}\right)} \left\{ \frac{\log|f(re^{i\theta})|}{\log\lambda(r)} + 1 \right\} \geqslant -\frac{7\pi^2}{4},$$

if $f \in a(\mathbf{D}, \lambda)$; $\epsilon_\mathbf{D}$ is some decreasing function on $[0, +\infty)$ which as a rule is of the order $1/\psi$.

In the interval $1 < l < +\infty$ the minorant $m_\mathbf{D}$ can be estimated from below in terms of $1/\lambda^{\mathrm{const}}$, where const depends only on l and has growth of the order $(l-1)^{-2}$

as $l \to 1$. In the case $l = 1$ the function $m_{\mathbf{D}}$ decreases faster than any power of $1/\lambda$ and satisfies the estimate

$$\log m_{\mathbf{D}}(r) \geqslant -\mathrm{const}\,\frac{1}{1-r}\left(\int_0^r \sqrt{\frac{\log \lambda(t)}{1-t}}\,dt\right)^2, \quad r_0 < r < 1,$$

where const is an absolute constant. The case $l < 1$ is essentially contained in the work [124] of Cartwright and gives a minorant $m_{\mathbf{D}}$ of the type $\exp(-\mathrm{const}(1-r)^{-1})$.

In §1.4 we establish that our estimates for the minimum of the modulus of functions in $a(\mathbf{D}, \lambda)$ are sharp; we construct functions in the given class for which $\min_{|\zeta|=r} \log|f(\zeta)|$ coincides to within a multiplicative constant with the right side of the corresponding estimate for $\log m_{\mathbf{D}}$.

All of the theorems in §§1.2–1.4 are proved in two versions, namely, for the strip Π and for the disk $\mathbf{D}$; as a rule the first result is a preliminary assertion needed for the second.

The conditions imposed on the weight λ in §§1.2 and 1.3 are of the same nature as those in §1.1. But the inverse theorem for the case $l = \infty$ is impossible without very special assumptions on λ. In fact, the existence of a function $f \in a(\mathbf{D}, \lambda)$ for which $|f(\zeta)| \geqslant 1/\lambda^{\mathrm{const}}(|\zeta|)$ means that $\log \lambda$ acts, as $r \to 1-0$, like the maximum on $|\zeta| = r$ of some harmonic function in $\mathbf{D}$. The conditions on the majorant in the inverse theorem for the case $1 \leqslant l < +\infty$ are the same as in §§1.2 and 1.3.

§1.5 is an analysis of the conditions imposed on the weight λ in the theorems of the preceding sections. We use the concept of a Hardy field, introduced by Bourbaki in [12]. A Hardy field is a field of (germs of) functions defined in a neighborhood of $+\infty$ which is closed with respect to differentiation. An arbitrary function λ in an arbitrary Hardy field satisfies all of the conditions on "regularity" of growth which we encounter in this work. Also in §1.5 we show that for each weight λ we can select envelopes λ_+ and λ_-, $\lambda_- \leqslant \lambda \leqslant \lambda_+$, which satisfy the necessary conditions of "regularity" of growth.

The methods of proof are discussed in the introductory paragraphs of the corresponding sections. Numerical illustrations of the theorems to be proved are also contained there.

3°. *Notation and terminology.* By convexity of a function f we always mean monotonicity of f' in some neighborhood of the point being considered (as a rule this is either $+\infty$ or 1). All "smoothness" conditions (convexity, monotonicity, sign-definiteness, and so forth) are assumed to be satisfied only in some neighborhood of this point unless we give specific information to the contrary. The functions which we consider have, in general, only as many derivatives as is indicated in the statement of the theorem, lemma, definition, or whatever.

The symbols which we will need in this chapter are listed here:

λ is an increasing positive function on $[0, 1)$;

$\lambda_\Pi(t) = \lambda(1 - e^{-t})$, $t \geqslant 0$;

$\varphi = \log \lambda_\Pi$, $\psi = \log \varphi$, $\mu = \log \lambda$;

$a(\mathbf{D}, \lambda) = \{f\colon f$ is analytic in $\mathbf{D}$, $f(\zeta) \neq 0$, $|\zeta| < 1$, $f(0) = 1$; $|f(\zeta)| \leqslant \lambda(|\zeta|)$, $\zeta \in \mathbf{D}\}$;

$\Pi = \{\zeta: \zeta \in \mathbf{C}, |\mathrm{Im}\, \zeta| < \pi/2\}$;

$a(\Pi, \lambda_\Pi)$ (respectively, $a(\overline{\Pi}, \lambda_\Pi)$) is the set of all functions f which are analytic in the strip Π (the closed strip $\overline{\Pi}$) such that $f(\zeta) \neq 0$, $f(0) = 1$, $|f(\zeta)| \leqslant \lambda_\Pi(\mathrm{Re}\, \zeta)$; $\mathrm{Re}\, \zeta \geqslant 0$, $\zeta \in \Pi$, $h(\mu) = \{f: f \text{ is harmonic in } \mathbf{D}; |f(\zeta)| \leqslant \mu(|\zeta|), \zeta \in \mathbf{D}\}$;

Γ is the Herglotz operator of harmonic conjugation;

$I = [-\pi/2, \pi/2]$ and $I_\epsilon = [-\pi/2 + \epsilon, \pi/2 - \epsilon]$;

Z is a conformal transformation of a curvilinear strip Ω onto the strip Π which is normed in a suitable fashion.

§1.1. An estimate for the modulus of a harmonic conjugate function

1°. The simplest estimate of the type discussed above is essentially contained in the classical Schwarz formula which relates the harmonically conjugate functions u and $v = \Gamma u$:

$$u(\zeta) + iv(\zeta) = \frac{1}{2\pi} \int_{-\pi}^{\pi} u(Re^{it}) \frac{Re^{it} + \zeta}{Re^{it} - \zeta}\, dt, \quad |\zeta| < R < 1.$$

In fact, if we assume that u is bounded in the disk $\mathbf{D}$, we immediately obtain

$$|v(\zeta)| \leqslant \frac{1}{2\pi} \int_{-\pi}^{+\pi} \max |u| \frac{2|\zeta||\sin t|}{1 - 2|\zeta| \cos t + |\zeta|^2}\, dt \leqslant \max |u| \left(\frac{2}{\pi} \log \frac{1}{1 - |\zeta|} + \mathrm{const}\right).$$

This inequality can be regarded as a simplified form of the Smirnov-Zygmund theorem [39] concerning integrability on $\partial\mathbf{D}$ of a function $\exp |v|$ with an arbitrary power $p < \pi \max |u|/2$. In general the operator of harmonic conjugation Γ is comparatively well known in spaces with integral norms on the circle $\partial\mathbf{D}$ (see, for example, [5] and [39]). Concerning uniform estimates for the growth of harmonic conjugate functions in $\mathbf{D}$, we first mention the Hardy-Littlewood-Cartwright (HLC) theorem (see [124] and [136]):

$$|u(\zeta)| \leqslant \frac{1}{(1 - |\zeta|)^\alpha}, \quad |\zeta| < 1 \Rightarrow |v(\zeta)| \leqslant \frac{\mathrm{const}}{(1 - |\zeta|)^\alpha}, \quad |\zeta| < 1,$$

where $\alpha > 0$. In other words, the operator Γ is continuous in the space $H(\mu_\alpha)$, $\mu_\alpha(r) = (1 - r)^{-\alpha}$, $\alpha > 0$. Slight refinements of the HLC theorem are presented in [153] and [154].

In this section we establish (in Theorem 1) suitable estimates for $\max_{|\zeta| \leqslant r} |v(\zeta)|$ in terms of $\max_{|\zeta| \leqslant r} |u(\zeta)|$ for functions u in the class $h(\mu)$, where the weight μ satisfies some restrictions on the "regularity" of growth (but not on the speed of growth; see §1.5). The estimates are obtained by elementary methods which refine some arguments of Cartwright. The only surprise is the fact[1] that these estimates are precise in the class $h(\mu)$, and even the constants (unfortunately not calculated in [124] and [136]) in the upper estimates and in the examples of functions which achieve the supremum in the class $h(\mu)$ differ only slightly (Theorem 2).

(1) A fact which to some degree justifies the restrictions on the "regularity" of the majorant.

A convenient and traditional form of writing the majorant μ is the following: $\mu(r) = \varphi(\log(1/(1-r)))$, $0 \leqslant r < 1$, where φ is a monotone increasing function on $[0, \infty)$. All of the conditions on "regularity" of growth (monotonicity, convexity) are assumed to be satisfied in a neighborhood of $+\infty$. An analysis of these conditions is given in §1.5.

THEOREM 1. *Let* $\mu(r) = \varphi(\log(1/(1-r)))$, $0 \leqslant r < 1$, *where* φ *is an increasing positive function on* $[0, +\infty)$, *and let* $\psi = \log \varphi$. *Let* $h(\mu)$ *denote the set of all harmonic functions* u *in the disk* $\mathbf{D}$ *for which* $|u(\zeta)| \leqslant \mu(|\zeta|)$, *and*

$$M_\mu(r) = \sup\{|(\Gamma u)(r)| : u \in h(\mu)\}.$$

Then the following assertions are true:

1) *If* $1/\psi'$ *is convex and* $l = \lim_{t\to\infty} \psi' > 0$, *then for an arbitrary* $\epsilon > 0$ *there exists a number* r_ϵ, $0 \leqslant r_\epsilon < 1$, *such that*

$$M_\mu(r) \leqslant \frac{2}{\pi}\left(\left(1+\frac{1}{l}\right)^{l+1} + \varepsilon\right)\mu(r), \quad r \geqslant r_\varepsilon.$$

2) *If* $1/\psi'$ *is convex and* $\lim_{t\to 0} \psi' = 0$, *but*(2) $\lim_{t\to\infty} \psi'(t)t = k > 0$, *then for an arbitrary* $\epsilon > 0$ *there exists a number* r_ϵ, $0 \leqslant r_\epsilon < 1$, *such that*

$$M_\mu(r) \leqslant \frac{2}{\pi}\cdot\frac{k(1+\varepsilon)}{1+k}\mu(r)\frac{1}{\psi'\left(\log\frac{1}{1-r}\right)}, \quad r \geqslant r_\varepsilon.$$

3) *If the limit* $\lim_{t\to\infty} \psi'(t)t = 0$ *exists, then for an arbitrary* $\epsilon > 0$ *there exists a number* r_ϵ, $0 \leqslant r_\epsilon < 1$, *such that*

$$M_\mu(r) \leqslant \frac{2+\varepsilon}{\pi}\mu(r)\log\frac{1}{1-r}, \quad r \geqslant r_\varepsilon.$$

REMARK. Convexity of the function $1/\psi'$ in the hypothesis of Theorem 1 can be replaced by convexity of ψ and boundedness of ψ''/ψ'^2 (or another version of assertion 2): $\psi''/\psi'^2 = o(1)$). These conditions do not impose additional restrictions on the growth of φ (besides those explicitly stated in the theorem), and they allow for arbitrarily rapid or arbitrarily slow growth of the weight μ (also see §1.5). For example, all of these conditions are satisfied if μ (or φ) belongs to some Hardy field (in particular, if $\mu \in \mathbf{H}$; see §1.5).

Concerning the speed of growth of the majorant μ, the assertions 1)–3) in Theorem 1 correspond to the following list of conditions:

1′) $$\lim_{r\to 1-0} \frac{\log\mu(r)}{\log\frac{1}{1-r}} > 0;$$

2′) $$\lim_{r\to 1-0} \frac{\log\mu(r)}{\log\frac{1}{1-r}} = 0 \text{ and } \lim_{r\to 1-0} \frac{\log\mu(r)}{\log\log\frac{1}{1-r}} > 0;$$

---

(2) The existence of the limit k follows from the convexity of $1/\psi'$. For $k = +\infty$ we let $k/(k+1) = 1$.

3') $\lim\limits_{r\to 1-0} \dfrac{\log \mu(r)}{\log\log \frac{1}{1-r}} = 0.$

Thus a result of the type in the HLC theorem (that $v = \Gamma u$, as well as the function u, grows no faster than μ) cannot be established for a majorant μ which grows more slowly than all of the powers $1/(1-r)^\alpha$, $\alpha > 0$.

EXAMPLE 1. $\log \mu(r) = A(1-r)^{-\alpha} \log^\beta(1/(1-r))$, where $\alpha \geqslant 0$, and $\beta \geqslant 1$ if $\alpha = 0$; $A > 0$. If $\alpha = 0$ and $\beta = 1$, then $\mu(r) = (1-r)^{-A}$ and([3])

$$M_\mu(r) \leqslant \frac{2}{\pi}\left\{\left(1+\frac{1}{A}\right)^{A+1} + \varepsilon\right\}\mu(r) \equiv \frac{2}{\pi}\left\{\left(1+\frac{1}{A}\right)^{A+1} + \varepsilon\right\}\frac{1}{(1-r)^A};$$

in the remaining cases we have

$$M_\mu(r) \leqslant \frac{2e+\varepsilon}{\pi}\mu(r), \quad r \geqslant r_\varepsilon.$$

2. $\log \mu(r) = A \log^\beta (1-r)^{-1} (\log\log(1-r)^{-1})^\gamma$, where $0 \leqslant \beta \leqslant 1$ and $\gamma < 0$ if $\beta = 1$; $\gamma \geqslant 1$ if $\beta = 0$; $A > 0$. If $\beta = 0$ and $\gamma = 1$, then $\mu(r) = \log^A (1-r)^{-1}$ and

$$M_\mu(r) \leqslant \frac{2}{\pi}\cdot\frac{1+\varepsilon}{A+1}\mu(r)\log\frac{1}{1-r} \equiv \frac{2}{\pi}\cdot\frac{1+\varepsilon}{A+1}\log^{A+1}\frac{1}{1-r}, \quad r \geqslant r_\varepsilon;$$

if $\beta = 0$ and $\gamma > 1$, then

$$M_\mu(r) \leqslant (A\gamma)^{-1}\left(\frac{2}{\pi}+\varepsilon\right)\frac{\mu(r)\log\frac{1}{1-r}}{\left(\log\log\frac{1}{1-r}\right)^{\gamma-1}}, \quad r \geqslant r_\varepsilon;$$

if $0 < \beta \leqslant 1$, then

$$M_\mu(r) \leqslant (A\beta)^{-1}\left(\frac{2}{\pi}+\varepsilon\right)\frac{\mu(r)\log\frac{1}{1-r}}{\left(\log\frac{1}{1-r}\right)^\beta\left(\log\log\frac{1}{1-r}\right)^\gamma}, \quad r \geqslant r_\varepsilon.$$

3. $\log \mu(r) = A(\log\log(1-r)^{-1})^\gamma(\log\log\log(1-r)^{-1})^\delta$, where $0 \leqslant \gamma \leqslant 1$ and $\delta < 0$ if $\gamma = 1$; $\delta \geqslant 0$ if $\gamma = 0$. Here

$$M_\mu(r) \leqslant \left(\frac{2}{\pi}+\varepsilon\right)\mu(r)\log\frac{1}{1-r}, \quad r \geqslant r_\varepsilon;$$

if $\gamma = 0$ and $\delta = 1$, then

$$\mu(r) = \left(\log\log\frac{1}{1-r}\right)^A, \quad M_\mu(r) \leqslant \left(\frac{2}{\pi}+\varepsilon\right)\left(\log\log\frac{1}{1-r}\right)^A \log\frac{1}{1-r}, \quad r \leqslant r_\varepsilon.$$

2°. *An inverse theorem.* As we have already mentioned, the estimates in Theorem 1 are best possible. For an arbitrary sufficiently "regular" weight μ we will construct a harmonic function u which attains to within a constant the supremum $M_\mu(r)$ on the whole interval $[0, 1)$. First we consider the precision of the estimate for a rapidly growing majorant μ (assertion 1) in Theorem 1). It is clear that the *order* of the

(3) It follows that the constant in the HLC inequality (the "asymptotic norm" $\|\Gamma\|_{H((1-r)^{-A})}$ does not exceed $(2/\pi)\{(1 + 1/A)^{A+1} + \epsilon\}$.

estimate is best possible, for if there exists a weight $\mu_1 = o(\mu)$ such that $u \in H(\mu) \Rightarrow \Gamma u \in H(\mu_1)$, then it (together with μ) can be regarded as satisfying the conditions in part 1) of Theorem 1, and then by applying Theorem 1 to μ_1 we obtain $u \in H(\mu) \Rightarrow u \in H(\mu_1)$. This implication means that $\|u\|_{H(\mu_1)} \leqslant \text{const} \cdot \|u\|_{H(\mu)}$, $u \in H(\mu)$. If we set $u(\zeta) = \zeta^n$, $n = 1, 2, \ldots$, $|\zeta| < 1$, we arrive at a contradiction.

If we want to construct a simple function $u \in h(\mu)$ which attains the supremum M_μ on the whole interval $[0, 1)$, then we must significantly strengthen the requirements on μ. In fact, the existence of the function u described above means that $\Gamma u \in H(\mu)$ and $\max_{|\zeta| \leqslant r} |\Gamma u(\zeta)| > \text{const}\, \mu(r)$, $0 \leqslant r < 1$; that is, the weight μ is equivalent (in the sense $\asymp$) to a function of the form

$$M(r) = \max_{|\zeta| \leqslant r} |v(\zeta)|, \quad 0 \leqslant r < 1,$$

where v is a harmonic function in $\mathbf{D}$. Not every majorant μ in part 1) of Theorem 1 has this property, but this property is possessed, for example, by absolutely monotone majorants

$$\mu(r) = \sum_{n=0}^{\infty} a_n r^n \quad (a_n \geqslant 0).$$

On the other hand, if we use results of Videnskiĭ [16] (see also [126]) to be introduced later, we can obtain a weaker inverse theorem under the same assumptions on μ as in Theorem 1.

LEMMA 0 (see [16] and [126]). *Let $\psi = \log \varphi$, where the function $t\psi'(t)$ is increasing for each $t \in [0, \infty)$ and $\lim t\psi'(t)/\log t = +\infty$. Then there exists an entire function P of the form $P(t) = \sum_0^\infty a_n t^n$, $a_n \geqslant 0$, such that*

$$t^{-1}\varphi(t) \leqslant P(t) \leqslant \varphi(t)$$

for all sufficiently large t.

Since the conditions in Lemma 0 follow from the conditions in part 1) of Theorem 1, we can set

$$v(\zeta) = \operatorname{Re} \sum_{n=0}^{\infty} a_n \left(\log \frac{1}{1-\zeta}\right)^n, \quad |\zeta| < 1.$$

Then for $r_0 < r < 1$ we have

$$\frac{\mu(r)}{\log \dfrac{1}{1-r}} \leqslant \max_{|\zeta|=r} |v(\zeta)| = P\left(\log \frac{1}{1-r}\right) \leqslant \mu(r).$$

From Theorem 1 we obtain

$$|\Gamma v(\zeta)| \leqslant \frac{2}{\pi}\left(\left(1 + \frac{1}{l}\right)^{l+1} + \varepsilon\right) \mu(|\zeta|)$$

and if we set

$$u = \Gamma v \Big/ \left\{ \frac{2}{\pi}\left(1+\frac{1}{l}\right)^{l+1} + \varepsilon \right\} - \text{const},$$

we arrive at the following "inversion" of assertion 1) in Theorem 1: there exists a function $u \in h(\mu)$ such that

$$(\Gamma u)(r) \geqslant \mu(r) \Big/ \left\{ \frac{2}{\pi}\left(1+\frac{1}{l}\right)^{l+1} + \varepsilon \right\} \log(1-r)^{-1}, \quad r_1 < r < 1.$$

The unimprovability of the estimate for a weakly growing weight μ is contained in the following theorem.

THEOREM 2. *Let μ be an increasing function on* $[0, 1)$, $\mu(r) = \varphi(\log(1-r)^{-1})$, *where $\psi = \log\varphi$ is convex in a neighborhood of $+\infty$. Suppose that* $\lim_{t\to\infty}\psi'(t) = 0$. *Then the following assertions are true*:

1) *If we have a finite limit*(4) $\lim_{t\to\infty}\psi''/\psi'^2 \overset{\text{def}}{=} L$, *then for an arbitrary $\epsilon > 0$ we can find a function $u = u_\epsilon$, $u \in h(\mu)$, such that*

$$|(\Gamma u)(r)| > \frac{2-\varepsilon}{\pi}(1-L)^{-1}\frac{\mu(r)}{\psi'\left(\log\frac{1}{1-r}\right)}, \quad r > r_\varepsilon.$$

2) *If $\psi'(t)t = o(1)$, then for an arbitrary $\epsilon > 0$ we can find a function $u = u_\epsilon$, $u \in h(\mu)$, such that*

$$|(\Gamma u)(r)| > \frac{2-\varepsilon}{\pi}\mu(r)\log\frac{1}{1-r}, \quad r > r_\varepsilon.$$

The proof of Theorem 2 makes essential use of the asymptotic formula of Warschawski for conformal transformations of curvilinear strips. It will be convenient to introduce this formula later, in §1.4.4°, in order to avoid duplication of the proofs in §1.4 (that the lower estimates are sharp).

3°. *Proof of Theorem* 1. Let $v = \Gamma u$, $u \in h(\mu)$ and $r < R < 1$. Then, if we write v as a Poisson integral, we obtain

$$\left|\frac{\partial v}{\partial r}\right| \leqslant \frac{2(R^2-r^2)}{\pi}\mu(R)\int_0^\pi \frac{\sin t\,dt}{(R^2+r^2-2Rr\cos t)^2}$$
$$= \frac{R^2-r^2}{\pi R r}\mu(R)\left\{\frac{1}{(R-r)^2} - \frac{1}{(R+r)^2}\right\} = \frac{4\mu(R)}{\pi(R^2-r^2)} \leqslant \frac{(2+\delta)\mu(R)}{\pi(R-r)}$$

for an arbitrary $\delta > 0$ and $r > r_\delta$, $0 \leqslant r_\delta < 1$. Consequently

$$|v(r)| \leqslant \frac{2+\delta}{\pi}\int_0^r \left(\min_{R\geqslant t}\frac{\mu(R)}{R-t}\right)dt + \text{const},$$

(4) We note that the conditions in Theorem 1 imply the existence of the limit L, and (in the notation of the theorem) $k = 1/|L|$ and $(1-L)^{-1} = (1+|L|)^{-1} = k/(1+k)$.

where $r \geqslant r_\delta$ and the const depends only on μ and δ, or after a change of variable

$$|v(r)| \leqslant \frac{2+\delta}{\pi} \int_0^{\log(1-r)^{-1}} \left[\min_{\varepsilon>0} e^{\psi(x+\varepsilon)}(1-e^{-\varepsilon})^{-1}\right] dx \overset{\text{def}}{=} J(r). \tag{1}$$

The following elementary assertion is a variation of a theorem of Hardy [135] (see also [12], Chapter 5, §3).

LEMMA 1. *Let ψ be an increasing function on $[0, \infty)$ with $\psi \in C^{(2)}$ and $\psi' > 0$. Let $\mathcal{J}(x) = \int_0^x e^{\psi} dt$, $x > 0$.*

1) *If $\psi'(t) = o(1/t)$, $t \to \infty$, then $\mathcal{J}(x) \sim xe^{\psi(x)}$, $x \to +\infty$; $\mathcal{J} \leqslant xe^{\psi(x)}$.*

2) *If we have a finite limit $\lim_{t\to\infty} \psi''/\psi'^2 \overset{\text{def}}{=} L$, then*

$$\mathcal{J}(x) \sim (1-L)^{-1} \frac{1}{\psi'(x)} e^{\psi(x)}, \quad x \to +\infty.$$

PROOF. 1) From the monotonicity of ψ it follows that $\mathcal{J} \leqslant xe^{\psi(x)}$. On the other hand,

$$\mathcal{J} = \int_0^x e^{\psi(t)} dt = xe^{\psi(x)} - \int_0^x t\psi'(t) e^{\psi(t)} dt = xe^{\psi(x)} + o(\mathcal{J}),$$

and the assertion is proved.

2) We have

$$\mathcal{J} = \int_0^x \frac{de^{\psi(t)}}{\psi'(t)} = \frac{1}{\psi'(x)} e^{\psi(x)} - \text{const} + \int_0^x \frac{\psi''}{\psi'^2} e^{\psi} dt,$$

and it remains to show that $L \leqslant 0$. In the contrary case we would have $\psi''/\psi'^2 > L/2 > 0$ for large $t > t_0$; that is,

$$-1/\psi'(t) + 1/\psi'(t_0) > \frac{L}{2}(t - t_0),$$

which is impossible. ●

Returning to the estimate of the integral (1), we first consider the case $0 < l = \lim_{t\to\infty} \psi' < +\infty$. The integrand in (1) equals

$$e^{\psi(x+\varepsilon_x)}(1-e^{-\varepsilon_x})^{-1} = e^{\psi(x+\varepsilon_x)}\{1+\psi'(x+\varepsilon_x)\}, \tag{2}$$

since the extremum point ϵ_x satisfies the equation

$$\psi'(x+\varepsilon_x) - \frac{e^{-\varepsilon_x}}{1-e^{-\varepsilon_x}} = 0,$$

or

$$\varepsilon_x = \log(1 + 1/\psi'(x+\varepsilon_x)).$$

From the equation

$$\exp\{\psi(x+\varepsilon_x)\} = \exp\{\psi(x) + \varepsilon_x \psi'(\xi)\}, \quad x \leqslant \xi \leqslant x + \varepsilon,$$

it follows that

$$\exp\psi(x+\varepsilon_x)\sim\left(1+\frac{1}{l}\right)^l\exp\psi(x)$$

as $x\to+\infty$. If we apply Lemma 1 and note that in this case we have $\lim\psi''/\psi'^2=0$, we obtain

$$J(r)\sim\frac{2+\delta}{\pi}(l+1)\left(1+\frac{1}{l}\right)^l\int\limits_0^{\log\frac{1}{1-r}}e^{\psi(x)}\,dx$$

$$\sim\frac{2+\delta}{\pi}(l+1)\left(1+\frac{1}{l}\right)^l\frac{1}{\psi'\left(\log\frac{1}{1-r}\right)}\mu(r)\sim\frac{2+\delta}{\pi}\left(\frac{l+1}{l}\right)^{l+1}\mu(r).$$

If $l=+\infty$, then $\lim_{x\to\infty}\epsilon_x=0$ and

$$\log\frac{\psi'(x+\varepsilon_x)}{\psi'(x)}=\varepsilon_x\frac{\psi''}{\psi'}(\xi)\leqslant\frac{1}{\psi'(x+\varepsilon_x)}\frac{\psi''}{\psi'}(\xi)\leqslant\frac{\psi''(\xi)}{\psi'^2(\xi)}=o(1),$$

since $x\leqslant\xi\leqslant x+\epsilon_x$ and the function ψ' is increasing in this case. Consequently

$$J(r)=\frac{2+\delta}{\pi}\int\limits_0^{\log\frac{1}{1-r}}\{1+\psi'(x+\varepsilon_x)\}\,e^{\psi(x)}e^{\varepsilon_x\psi'(\xi)}dx$$

$$\sim\frac{2+\delta}{\pi}\int\limits_0^{\log\frac{1}{1-r}}\psi'(x)\,e^{\psi(x)}e\,dx=\frac{(2+\delta)e}{\pi}\mu(r).$$

This proves assertion 1) in Theorem 1.

Assertion 2) is obtained from (2) if we note that $\lim_{x\to\infty}\epsilon_x=+\infty$ and

$$\min_{\varepsilon>0}e^{\psi(x+\varepsilon)}(1-e^{-\varepsilon})^{-1}=e^{\psi(x+\varepsilon_x)}(1-e^{-\varepsilon_x})^{-1}\geqslant e^{\psi(x)}(1-e^{-\varepsilon_x})^{-1}\sim e^{\psi(x)},$$

$$\min_{\varepsilon>0}e^{\psi(x+\varepsilon)}(1-e^{-\varepsilon})^{-1}\leqslant\left(1-e^{-\sqrt{\frac{1}{\psi'(x)}}}\right)^{-1}e^{\psi\left(x+\sqrt{\frac{1}{\psi'(x)}}\right)}\tag{3}$$

$$\sim e^{\psi(x)}e^{\sqrt{\frac{1}{\psi'(x)}}\psi'(\xi)}\sim e^{\psi(x)},$$

since $\xi\geqslant x$ and the function ψ', which is decreasing, converges to zero (ψ''/ψ'^2 cannot maintain a positive sign since $\psi'>0$ and $\lim\psi'=0$). These facts give us the asymptotic relation

$$J(r)\sim\frac{2+\delta}{\pi}\int\limits_0^{\log\frac{1}{1-r}}e^{\psi(x)}dx,\quad r\to1-0.$$

We will show that the limit $L=\lim_{u\to\infty}\psi''/\psi'^2$ (the existence of which follows from convexity of $1/\psi'$) is finite: if $L=-\infty$, then if we integrate the relation $1=o(-\psi''/\psi'^2)$, we obtain $x=o(1/\psi'(x))$. This contradicts the assumptions in assertion 2) of Theorem 1. Now we can use Lemma 1:

$$J(r) \sim \frac{2+\delta}{\pi}(1-L)^{-1}\frac{1}{\psi'\left(\log\frac{1}{1-r}\right)}\mu(r), \quad r \to 1-0.$$

Assertion 2) is proved.

The last inequality in Theorem 1 follows from the same kind of arguments, but in proving the relation which is analogous to (3) we replace $\widetilde{\epsilon}_x = \sqrt{1/\psi'(x)}$ by an $\widetilde{\epsilon}_x$ such that

$$\lim_{x\to\infty} \varepsilon_x = +\infty, \quad \lim_{x\to\infty} \varepsilon_x \inf_{t>x} \psi'(t) = 0.$$

§1.2. A sequential lower estimate

1°. A bounded analytic function f which does not vanish in the disk **D** admits the following elementary lower estimate:

$$\log|f(r)| = \lim_{R\to 1-0} \frac{1}{2\pi}\int_{-\pi}^{\pi} \frac{(R^2-r^2)\log|f(Re^{i\theta})|}{R^2-2Rr\cos\theta+r^2}\,d\theta$$

$$\geqslant -\frac{1}{1-r}\varlimsup_{R\to 1-0}\frac{1}{\pi}\int_{-\pi}^{\pi}|\log|f(Re^{i\theta})||\,d\theta.$$

Similar estimates for a function f with a given growth majorant λ become more complicated the faster the weight λ increases (see §1.3.1°). An intermediate step in obtaining such inequalities is finding sequential estimates of the type

$$\varlimsup_{r\to 1-0}(1-r)\log|f(r)| > -\infty, \quad f \in a(\mathbf{D}, \text{ const}).$$

If we abandon the requirement $f(\zeta) \neq 0$, $|\zeta| < 1$, these estimates can be regarded naturally as uniqueness theorems:

$$\varlimsup_{r\to 1-0}(1-r)\log|f(r)| = -\infty \Rightarrow f \equiv 0 \ (f \in A\ (\mathbf{D}, \text{const})).$$

This is a significantly more complicated assertion than the ones we will deal with; for various classes of analytic functions which are characterized by restrictions on growth such uniqueness theorems are contained in the works [3], [33], [36], [38], [105], [106] and [111], where there are additional references. We note that in all of these works the behavior of $|f(\zeta)|$ *on the boundary of the domain* (for instance, on the boundary of the disk **D**) is investigated when a given point (for instance, $\zeta = 1$) is approached, with the exception of the works [38] and [111]. In the latter works different types of classes of functions are considered (namely, classes of functions which are integrable with some weight $\log^+|f|$).

In this section we prove theorems on estimates for the classes $a(\mathbf{D}, \lambda)$ and $a(\Pi, \lambda_\Pi)$ under conditions of sufficient regularity and speed of the growth of the majorant λ. These theorems give us lower estimates (for some sequence of values of the argument) which are more precise than the "global" inequalities which we proved in the preceding section. For slowly increasing majorants λ the corresponding inequalities will be obtained in §1.3 for all sufficiently large values of the argument. We now state the theorems to be proved and briefly discuss the estimates which they contain.

THEOREM 1. *Suppose that* $\psi = \log\log \lambda_\Pi$, $\psi' \uparrow$, $\lim \psi' > 2$, $1/\sqrt{\psi'}$ *is a convex function, and* $f \in a(\Pi, \lambda_\Pi)$. *Then the following assertions are true*:

i) *For each* θ, $|\theta| < \pi/2$, *we have*

$$T(f) \overset{\text{def}}{=} \varlimsup_{\sigma\to+\infty} \frac{\log|f(\sigma+i\theta)|}{\log \lambda_\Pi(\sigma)} \geqslant -1.$$

ii) *If* $\lim \psi' = +\infty$, *and if* ψ'/ψ'' *and* ψ''/ψ' *are convex, then for each* θ, $|\theta| < \pi/2$, *we have*

$$\varlimsup_{\sigma\to+\infty} \frac{1}{\varepsilon_\Pi(\sigma)}\left(\frac{\log|f(\sigma+i\theta)|}{\log\lambda_\Pi(\sigma)}+1\right) \geqslant -\frac{5}{4}\pi^2,$$

where

$$\varepsilon_\Pi(\sigma) = \int_\sigma^\infty \frac{\psi''^2}{\psi'^3}\,dt + \frac{32}{5}\frac{1}{\psi(\sigma)}, \quad \sigma > 0.$$

THEOREM 2. *Suppose that the hypothesis of Theorem* 1 *is satisfied and* $f \in a(\mathbf{D}, \lambda)$. *Then the following assertions are true*:

i) *For each* θ, $0 \leqslant \theta < 2\pi$, *we have*

$$\varlimsup_{r\to1-0} \frac{\log|f(re^{i\theta})|}{\log\lambda(r)} \geqslant -1.$$

ii) *If* $\lim \psi' = +\infty$ *and if* ψ''/ψ' *and* ψ'/ψ'' *are convex, then for each* θ, $0 \leqslant \theta < 2\pi$, *we have*

$$\varlimsup_{r\to1-0} \frac{1}{\varepsilon_{\mathbf{D}}\left(\log\frac{1}{1-r}\right)}\left\{\frac{\log|f(re^{i\theta})|}{\log\lambda(r)}+1\right\} \geqslant -\frac{7\pi^2}{4},$$

where

$$\varepsilon_{\mathbf{D}}(\sigma) = \max\left(\varepsilon_\Pi(\sigma), \frac{1}{\psi'(\sigma)}\right), \quad \sigma > 0.$$

REMARK. The conditions on λ in Theorems 1 and 2 imply that λ grows no slower than $\exp(1-r)^{-2-\epsilon}$ for some $\epsilon > 0$, and in assertion ii) of Theorems 1 and 2 we must have

$$\varlimsup_{r\to1} \frac{\log\log\lambda(r)}{\log\frac{1}{1-r}} = +\infty.$$

The requirement of "regularity" of growth in assertion i) can be weakened to convexity of $1/\psi'$ and boundedness of $(\psi''/\psi'^2)\log^2\psi'$ without changing the proofs. The nature of these "regularity" assumptions is discussed in §1.5.

2°. *Discussion of the results.* Theorems 1 and 2 can be stated in a simpler and more "explicit" form. We will do this for part ii) of Theorem 2.

THEOREM 2'. *Suppose that the majorant λ satisfies the hypotheses of Theorems 1 and 2, and let f be a function which is analytic in $\mathbf{D}$ and such that $|f(\zeta)| \leqslant \lambda(|\zeta|)$ and $f(\zeta) \neq 0$, $|\zeta| < 1$. Then there exists a sequence of numbers $\{r_n\}_1^\infty$, $0 < r_n < 1$, $\lim r_n = 1$, such that for an arbitrary $\delta > 0$ we have*

$$|f(r_n)| > \frac{1}{\lambda(r_n)} \exp\left\{-\left(\frac{7\pi^2}{4}+\delta\right)\varepsilon_{\mathbf{D}}\left(\log\frac{1}{1-r_n}\right)\log\lambda(r_n)\right\}$$

for all $n > n(\delta)$.

THEOREM 2''. *Suppose that the majorant λ satisfies the hypotheses of Theorems 1 and 2, and let u be a real function which is harmonic in $\mathbf{D}$ and such that $u(\zeta) \leqslant \log\lambda(|\zeta|)$, $|\zeta| < 1$. If for some $\delta > 0$ and some r_δ with $0 < r_\delta < 1$ we have*

$$u(r) < -\log\lambda(r) - \left(\frac{7\pi^2}{4}+\delta\right)\varepsilon_{\mathbf{D}}\left(\log\frac{1}{1-r}\right)\log\lambda(r), \quad r > r_\delta,$$

then $u \equiv -\infty$.

On the other hand, for highly "regular" weights λ we can find functions in $a(\lambda)$ which decrease like $1/\lambda$.

Suppose, for example, that the majorant λ is of the form

$$\log\lambda(r) = \sum_{n=0}^{\infty} a_n r^n, \quad 0 \leqslant r < 1 \quad (a_n \geqslant 0).$$

If we set $f(\zeta) = \exp(-\Sigma_0^\infty a_n \zeta^n)$, $\zeta \in \mathbf{D}$, we have $f \in a(\mathbf{D}, \lambda)$ and $f(r) = 1/\lambda(r)$, $0 \leqslant r < 1$.

If we omit the conditions in Theorems 1 and 2, then from Lemma 0 in §1.1 it only follows that there exists a function(5) $f \in a(\mathbf{D}, \lambda)$ for which

$$|f(r)| \leqslant \exp\left(-\frac{\log\lambda(r)}{\log(1-r)^{-1}}\right), \quad r_1 < r < 1.$$

Thus the first assertions in Theorems 1 and 2 are sharp in the sense that $\inf_{f\in a(\lambda)} T(f) = -1$. Later we will introduce some considerations on the order of smallness of the functions ϵ_Π and $\epsilon_{\mathbf{D}}$. The behavior of ϵ_Π and $\epsilon_{\mathbf{D}}$ depends on peculiarities of the growth of λ (perhaps even on the speed of growth). We consider two simple cases.

1) The function $\log\psi$ ($\equiv \log\log\log\lambda_\Pi$) is convex downward. Then ψ'/ψ is decreasing, and $\psi''\psi \leqslant \psi'^2$; together with $\psi'' > 0$ this gives us $\psi''^2/\psi'^3 \leqslant \psi'/\psi^2$; that is,

$$\int_\sigma^\infty \frac{\psi''^2}{\psi'^3}\,dt \leqslant \frac{1}{\psi(\sigma)}.$$

2) The function $\log\psi$ is convex upward. Then ψ'/ψ is increasing, $\psi''\psi \geqslant (\psi')^2$, and

(5) In the notation of Lemma 0 in §1.1 this is the function $f(\zeta) = \exp(-P(\log(1/(1-\zeta))))$, $|\zeta| < 1$.

$$\int_\sigma^\infty \frac{\psi''^2}{\psi'^3}\,dt \geqslant \frac{1}{\psi(\sigma)}.$$

If we impose different conditions on ψ, then we can use a theorem of Hardy-Dubois-Raymond (see [12]) on asymptotics of derivatives and obtain a sharper result. Namely, under the conditions in Theorems 1 and 2 we obtain:

1) if ψ'/ψ'' is convex and comparable to x as $x \to \infty$, and if

$$\lim_{x\to+\infty} \frac{\log \psi'}{\log x} = +\infty,$$

then

$$\int_\sigma^\infty \frac{\psi''^2}{\psi'^3}\,dt \sim \frac{1}{\psi(\sigma)}, \quad \sigma \to +\infty;$$

2) if ψ'''/ψ'' is comparable to $1/x$, where(6) the order p of ψ'' with respect to x is finite, then

$$\int_\sigma^\infty \frac{\psi''^2}{\psi'^3}\,dt \sim \left(\frac{p+1}{p+2}\right)^2 \frac{1}{\psi(\sigma)}, \quad \sigma \to +\infty.$$

Thus for a sufficiently "regular" majorant λ (for instance, if λ belongs to some Hardy field $\mathfrak{R}$ (see [12])) the remainder term ϵ_Π is of the order $1/\psi$.

This last remark bears upon the method of proof of Theorems 1 and 2. The method is fairly standard and consists in going to a subdomain $\Omega \subset \Pi$ (respectively, $\Omega \subset \mathbf{D}$) whose boundary contains the line $\operatorname{Im} w = \theta$ (respectively, $\arg \zeta = \theta$) and then using a boundary uniqueness theorem for Ω. The width of the strip Ω is chosen according to the speed of growth of λ.

3°. *Auxiliary assertions.* In this section we present an alternate version of a formula of Carleman and a refinement of the asymptotic estimates of Warschawski which we will need for the proofs of Theorems 1 and 2. The estimate for the remainder in the Warschawski formula might be of interest in itself.

LEMMA 1 (T. Carleman). *If a function F is analytic in a closed strip $\overline{\Pi}$ and satisfies the condition that $F(w) \neq 0$, $w \in \Pi$, $\operatorname{Re} w \geqslant \sigma_0$, then*

$$e^{-\sigma} \int_{-\frac{\pi}{2}}^{\frac{\pi}{2}} \log|F(\sigma+i\tau)| \cos\tau\,d\tau + \frac{1}{2}\int_{\sigma_0}^{\sigma} \log\left|F\left(t+i\frac{\pi}{2}\right)F\left(t-i\frac{\pi}{2}\right)\right|$$
$$\times (e^{-2t} - e^{-2\sigma})\,e^t\,dt = O(1), \quad \sigma \to +\infty.$$

This relation is obtained by means of a conformal "transformation" of the classical Carleman formula for a half-plane (see, for example, [53], Chapter 5, §1, Theorem 1, and Appendix I below). We can show that the quantity $O(1)$ in Lemma 1 has a limit as $\sigma \to +\infty$ equal to

(6) The existence of the limit follows from comparability of ψ'''/ψ'' and $1/x$.

$$\operatorname{Im} \frac{1}{2} \int_{\operatorname{Re} w=\sigma_0} \log F(w) \frac{dw}{e^w}.$$

For $\sigma_0 = 0$ the modulus of this number is called the Carleman constant and is denoted by Car F.

THEOREM 3. *Let s be a function defined on the whole real line* $\mathbf{R}$ *and such that*: a) $\lim_{t\to\infty} s = 0$, b) *the functions s and s/s' are convex in a neighborhood of* $+\infty$, c) $\int^\infty s^{-1}(s')^2\, dt < +\infty$, *and* d) $\overline{\lim}_{t\to+\infty} 2ss'' < 1$. *Let Z be a conformal mapping of the curvilinear strip* $\Omega = \{\sigma + i\tau\colon |\tau| < s(\sigma)\}$ *onto the strip* Π, *and let* $Z(+\infty) = +\infty$. *Then*

$$\operatorname{Re} Z(\sigma + i\tau) = \frac{\pi}{2} \int_0^\sigma \frac{dt}{s(t)} + r(\sigma, \tau),$$

$$|r(\sigma, \tau)| \leqslant \frac{7}{4}(1+\varepsilon)\pi \int_\sigma^\infty \frac{s'^2}{s}\, dt + \frac{5\pi^2}{2}(1+\varepsilon)\frac{1}{\psi(\sigma)}, \quad \sigma \geqslant \sigma_0,$$

where $\psi(\sigma) = (\pi/2)\int_0^\sigma (1/s(t))\, dt$, $\sigma > 0$.

The proof of this theorem consists in refining some steps in the arguments of Warschawski in [15]. It will be presented in Appendix II.

4°. *Proof of Theorem* 1. We consider the strip(7) $\Omega_\theta = \{\sigma + i\tau\colon \theta < \tau < \theta + s(\sigma)\}$, where the function s satisfies $\varphi(\sigma) = \exp(\pi \int_0^\sigma (1/s(t))\, dt)$, $\sigma > \sigma_0$; that is, $s = \pi\varphi/\varphi'$. Later we will show that for the function s all of the conditions of the theorem(8) of Warschawski [15] are satisfied, and thus for the conformal mapping Z of Ω onto Π we have

$$\operatorname{Re} Z(\sigma + i\tau) = \pi \int_0^\sigma \frac{dt}{s(t)} + o(1), \quad \sigma \to +\infty.$$

We set $F(w) = f(Z^{-1}(w))$, $w \in \Pi$. At a point $w = x + iy = Z(\sigma + i\tau)$ we obtain

$$|F(x+iy)| = |f(\sigma + i\tau)| \leqslant \exp \varphi(\sigma),$$

$$\log |F(x+iy)| \leqslant \varphi(\sigma) = \exp\left(\pi \int_0^\sigma \frac{dt}{s(t)}\right) = e^{x + \alpha(x, y)}, \tag{1}$$

and $\alpha(x, y) = o(1)$ as $x \to \infty$ (uniformly with respect to y). Consequently

$$e^{-x} \int_{-\pi/2}^{\pi/2} \log |F(x+iy)| \cos y\, dy = O(1), \quad x \to +\infty,$$

and by Lemma 1 we have

$$\int_0^x \left\{\log\left|F\left(t + i\frac{\pi}{2}\right)\right| + \log\left|F\left(t - i\frac{\pi}{2}\right)\right|\right\}(e^{-2t} - e^{-2x})\, e^t\, dt = O(1).$$

---

(7) Without loss of generality we assume that $-\pi/2 < \theta \leqslant 0$. If $0 < \theta < \pi/2$, then we consider the strip $\Omega_\theta = \theta - \Omega_0$, so that we always have $\Omega_\theta \subset \Pi$.

(8) A statement of this theorem is given in Appendix I.

If we now assume that there exists a number $x_0 > 0$ and also a function δ, $\delta \downarrow 0$, $\int_0^\infty \delta(t)\,dt = +\infty$, such that

$$\frac{\log\left|F\left(x-i\frac{\pi}{2}\right)\right|}{e^{x+\alpha(x)}\,(1+\delta(x))} \leqslant -1, \quad x > x_0,$$

with

$$\alpha(x) = \max\left\{\alpha\left(x,\ i\frac{\pi}{2}\right),\ \alpha\left(x,\ -i\frac{\pi}{2}\right)\right\},$$

then we arrive at a contradiction to Carleman's formula:

$$\int_{x_0}^{x}\left\{\log\left|F\left(t+i\frac{\pi}{2}\right)\right| + \log\left|F\left(t-i\frac{\pi}{2}\right)\right|\right\}(e^{-2t}-e^{-2x})\,e^t\,dt$$
$$\leqslant \int_{x_0}^{x}\{e^{t+\alpha(t)} - e^{t+\alpha(t)} - \delta(t)\,e^{t+\alpha(t)}\}(e^{-2t}-e^{-2x})\,e^t\,dt,$$

$$\lim_{x\to+\infty}\int_0^x \delta(t)\,e^{2t}(e^{-2t}-e^{-2x})\,dt \geqslant \lim_{x\to+\infty}\int_0^{\frac{x}{2}}\delta(t)(1-e^{-x})\,dt = +\infty.$$

Thus there exist arbitrarily large[9] values of x and σ for which

$$\frac{\log\left|F\left(x-i\frac{\pi}{2}\right)\right|}{e^{x+\alpha(x)}\,(1+\delta^{*}(x))} = \frac{\log|f(\sigma+i\theta)|}{e^{\alpha(x)-\alpha\left(x,\,-i\frac{\pi}{2}\right)}\,\varphi(\sigma)\,(1+\delta(\operatorname{Re} Z(\sigma+i\theta)))} > -1, \tag{2}$$

and therefore

$$\varlimsup_{\sigma\to+\infty}\frac{\log|f(\sigma+i\theta)|}{\varphi(\sigma)} \geqslant -1.$$

In order to verify the second assertion in Theorem 1, for the same values of σ we estimate the sum

$$\frac{\log|f(\sigma+i\theta)|}{\varphi(\sigma)} + 1,$$

using (2) and Theorem 3:

$$\frac{\log|f(\sigma+i\theta)|}{\varphi(\sigma)} + 1 > 1 - e^{\alpha(x)-\alpha\left(x,\,-i\frac{\pi}{2}\right)}(1+\delta(\log\varphi(\sigma)+o(1)))$$
$$> 1-(1+2|\alpha(x)|+O(\alpha(x)^2))(1+\delta(\log\varphi(\sigma)+o(1)))$$
$$= -2|\alpha(x)| - \delta(\log\varphi(\sigma)+o(1)) + O(|\alpha|\delta+\alpha^2).$$

---

(9) From our arguments it is clear that the distance from a given number σ_* to the closest values σ here mentioned is determined only by the values $f(\sigma+i\tau)$, $\sigma \leqslant \sigma_*$, $\theta \leqslant \tau \leqslant \theta + s(\sigma_*)$, and the quantity $O(1)$ in the Carleman formula. In the final analysis the distance depends only on f, σ_*, and $|\theta - \pi/2|$ and $|\theta + \pi/2|$.

From Theorem 3([10]) we have

$$|\alpha(x)| \leqslant \frac{5}{8}(\pi+\varepsilon)\int_{\sigma}^{\infty}\frac{s'^2}{s}\,dt + 4\pi^2(1+\varepsilon)\frac{1}{\psi(\sigma)}, \quad \sigma > \sigma_0, \tag{3}$$

and for $\delta(t) = 1/t \log t$ we have the relation

$$\delta(t+o(1)) \sim \delta(t), \quad t \to +\infty, \quad \delta(\operatorname{Re} Z(\sigma+i\theta)) = o\left(\frac{1}{\psi(\sigma)}\right),$$

$$\frac{\log|f(\sigma+i\theta)|}{\varphi(\sigma)} + 1 > -2|\alpha(x)| + o\left(\frac{1}{\psi(\sigma)}\right)$$

$$> -\frac{5}{4}\pi^2(1+\varepsilon)\,\varepsilon_{\Pi}(\sigma) \overset{\text{def}}{=} -\frac{5}{4}\pi^2(1+\varepsilon)\left\{\frac{1}{\pi}\int_{\sigma}^{\infty}\frac{s'^2}{s}\,dt + \frac{32}{5}\cdot\frac{1}{\log\varphi(\sigma)}\right\}.$$

Consequently

$$\varliminf_{\sigma}\frac{1}{\varepsilon_{\Pi}(\sigma)}\left\{\frac{\log|f(\sigma+i\theta)|}{\varphi(\sigma)} + 1\right\} \geqslant -\frac{5}{4}\pi^2(1+\varepsilon)$$

for each $\epsilon > 0$, and this gives us the required inequality.

In order to finish the proof of Theorem 1 we have to verify the conditions of the Warschawski theorem and Theorem 3.

LEMMA 2. 1) *If* $s = \pi/\psi'$ *and for the function* ψ *the conditions in part* 1) *of Theorem* 1 *are satisfied, then*

$$\lim_{t\to\infty} s' = 0, \qquad \int^{\infty}\frac{s'^2}{s} < \infty.$$

2) *If the conditions in part* 2) *of Theorem* 1 *are satisfied, then the function* $s = \pi/\psi'$ *satisfies the requirements in Theorem* 3.

PROOF OF THE LEMMA. 1) By the definition of s we have $s = \pi/\psi'$ and $s' = -\pi\psi''/\psi'^2$. Since $\psi''/\psi'^{3/2}$ is monotone, we have

$$\lim_{t\to\infty}\psi''/\psi'^{3/2} = 0$$

and

$$\lim_{t\to\infty} s' = 0, \quad \int^{\infty}\frac{s'^2}{s} = \int^{\infty}\frac{\psi''^2}{\psi'^3} \leqslant \text{const}\int^{\infty}\frac{\psi''}{\psi'^{3/2}} < +\infty.$$

([10]) In this (nonsymmetric) case of the strip $\Omega_\theta = \{\sigma + i\tau \colon \theta < \tau < \theta + s\}$, Theorem 3 remains valid. We just note that in the statement of the theorem s denotes half of the width of Ω and in the nonsymmetric case of Ω_θ the original formula of Warschawski is of the form

$$\operatorname{Re} Z = \pi\int_0^{\sigma}\frac{dt}{s(t)} + \frac{\pi}{4}\int_0^{\sigma}\frac{s'^2}{s}\,dt + \cdots,$$

and this leads to (3).

2) Since s' is monotone, it follows that the sign of s'' does not change and $\int^{\infty}|s''| < +\infty$. We still have to show that $\overline{\lim}_{t\to\infty} 2ss'' < 1$. Since $s'/s = -\psi''/\psi'$ is convex, there are only two possibilities: a) s'/s is decreasing, and then $0 \leqslant s''s < (s')^2$; that is, $s''s = o(1)$; b) s'/s is increasing, and then $|s'|/s$ is decreasing and convex downward, which implies downward convexity of the function $|s'|s = (|s'|/s)(s)^2$. From monotonicity of the derivative $(s's)' = ss'' + (s')^2$ it follows that $ss'' = o(1)$ (since the contrary assertion immediately leads to a contradiction: $(s's)' \geqslant c > 0 \Rightarrow s's > 0$ for large values of the argument). ●

5°. The proof of Theorem 2 depends on the following elementary lemma.

LEMMA 3. *Let O_r be the image of the circle $\{\zeta\colon |\zeta| = r\}$, $0 \leqslant r < 1$, under the transformation $\zeta \longrightarrow \log(1/(1-\zeta))$. If $\tau = \tau(\sigma)$ is the equation of the upper half of O_r, and if $t = \log(1/(1-r))$, and $d = t - \sigma$, then*

$$\tau = \arccos \frac{1}{2}\left[e^{\sigma-t}(2-e^{-t}) + e^{-\sigma}\right],$$

$$\tau = (2d)^{1/2}(1 + O(d)) \text{ as } d \to 0,$$

where $O(d)$ is uniform with respect to $t \geqslant 0$.

PROOF OF THE LEMMA. The image of the circle $|\zeta| = r$ under the linear-fractional transformation $\zeta \longrightarrow 1/(1-\zeta)$ is the circle with center $C = (1-r^2)^{-1}$ and the radius $R = r(1-r^2)^{-1}$, having the equation

$$|w - (1-r^2)^{-1}|^2 = r^2(1-r^2)^{-2}.$$

If we set $w = \rho e^{i\tau}$, $-\pi/2 < \tau < \pi/2$, we have

$$\rho \cos \tau = \frac{\rho^2 + (1-r^2)^{-1}}{2(1-r^2)^{-1}},$$

and for the image of $w \longrightarrow \log w = \log \rho + i\tau = \sigma + i\tau$, we obtain

$$\tau = \arccos \frac{1}{2}\left[e^{\sigma-t}(2-e^{-t}) + e^{-\sigma}\right]$$

$$= \arccos\left\{1 - \left[1 - e^{-d} - \frac{1}{2}e^{-t}(e^{d} - e^{-d})\right]\right\} = (2d)^{1/2}(1 + O(d)),$$

since $\arccos(1-\epsilon) = (2\epsilon)^{1/2} + O(\epsilon^{3/2})$. ●

Now let $f \in a(\mathbf{D}, \lambda)$. Without loss of generality we can assume that $\theta = 0$. We consider the function $f_1 = f \circ p^{-1}$, where $p(\zeta) = \log(1/(1-\zeta))$, $|\zeta| < 1$, is a conformal transformation of the disk $\mathbf{D}$ onto a curvilinear substrip Π_1 of the strip Π (see Figure 1). Then

$$\max_{w \in O_r} |f_1(w)| \leqslant \exp \varphi(t), \quad t = \log \frac{1}{1-r}.$$

We argue as in the proof of Theorem 1, considering the strip Ω_0, the transformation Z, and the function F, $F(w) = f_1(Z^{-1}(w))$, $w \in \Pi$, $\operatorname{Re} w > w_0$. We note that the composition $f_1 \circ Z^{-1}$ makes sense for $\operatorname{Re} w > w_0$ (for $w_0 > 0$ sufficiently large), since

$$\lim_{t\to\infty} s = \lim_{t\to\infty} \frac{\pi}{\psi'} < \frac{\pi}{2} \quad \text{and} \quad \Omega \cap \{w : \operatorname{Re} w > w_0\} \subset \Pi_1 \cap \{w : \operatorname{Re} w > w_0\}.$$

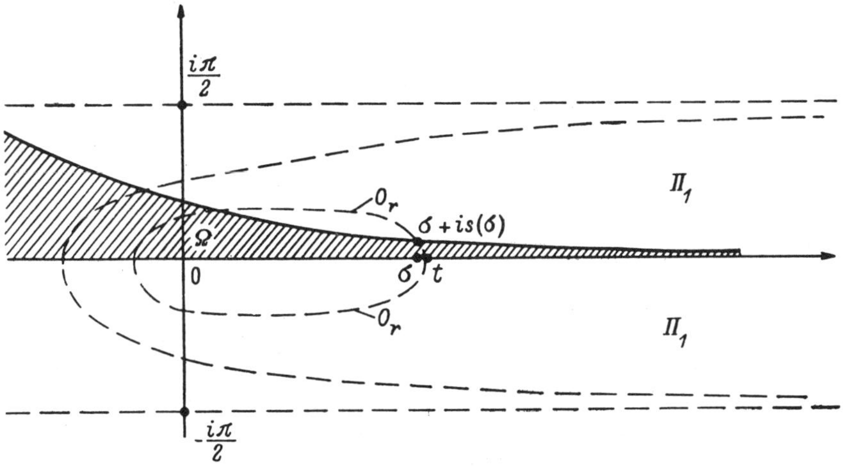

FIGURE 1

LEMMA 4. *Suppose that* $f \in a(\mathbf{D}, \lambda)$, $p(\zeta) \equiv \log(1/(1-\zeta))$ $(|\zeta| < 1)$, $f_1 = f \circ p^{-1}$ *and the function* λ *satisfies the conditions in Theorem* 2. *Then*

$$\log |f_1(\sigma + i\tau)| \leqslant \varphi(\sigma)\, e^{\beta(\sigma)}, \quad 0 \leqslant \tau \leqslant \frac{\pi}{\psi'(\sigma)},$$

where $\beta(\sigma) \leqslant \pi^2/\psi'(\sigma)$.

PROOF OF THE LEMMA. Using the notation just introduced, at the point $w = x + iy = Z(\sigma + i\tau)$, $0 \leqslant \tau \leqslant s(\sigma)$, we have

$$|F(x+iy)| = |f_1(\sigma + i\tau)| \leqslant \exp \varphi(t),$$

where $t = \sigma + d$ is a number such that $\sigma + i\tau \in O_r$, $t = \log(1/(1-r))$. We estimate the increment d, using Lemma 3:

$$\tau = (2d)^{1/2}(1 + O(d)) \leqslant s(\sigma),$$

$$d < \left(\frac{1}{2} + \varepsilon\right) s(\sigma)^2, \quad \sigma > \sigma_0.$$

Consequently

$$\log |F(x+iy)| \leqslant \varphi\left(\sigma + \left(\frac{1}{2} + \varepsilon\right) s(\sigma)^2\right). \tag{4}$$

Furthermore,

$$\psi(\sigma + \bar{d}) - \psi(\sigma) = \bar{d}\psi'(\sigma + \xi) = \bar{d}\psi'(\sigma)\frac{\psi'(\sigma+\xi)}{\psi'(\sigma)}.$$

If we set $A = 1/\psi'$, we can rewrite this equation in the form

$$\psi(\sigma + \bar{d}) - \psi(\sigma) = \bar{d}\psi'(\sigma)\frac{A(\sigma)}{A(\sigma+\xi)} = \frac{\bar{d}\psi'(\sigma)}{\left\{1 + \xi\frac{A'(\sigma+\bar{\xi})}{A(\sigma)}\right\}},$$

where $0 \leqslant \bar{\xi} \leqslant \xi \leqslant \bar{d}$. By hypothesis

$$|A'(t)| = \frac{\psi''}{(\psi')^2}(t) = \frac{\psi''}{(\psi')^{3/2}} \cdot \frac{1}{(\psi')^{1/2}}.$$

is a decreasing function which is of order $o(1/\psi'^{1/2})$, and $|A'(\sigma + \bar{\xi})| \leqslant |A'(\sigma)|$, so that

$$\psi(\sigma + \bar{d}) - \psi(\sigma) \leqslant \frac{\bar{d}\psi'(\sigma)}{\{1 - \bar{d}\psi'(\sigma)\,|A'(\sigma)|\}} \leqslant \frac{\bar{d}\psi'(\sigma)}{\left\{1 - \bar{d}\psi'(\sigma)\, o\left(\frac{1}{\psi'(\sigma)^{1/2}}\right)\right\}}.$$

If we continue inequality (4), we obtain

$$\log|F(x+iy)| \leqslant \varphi(\sigma)\, e^{\beta(\sigma)} = e^{x+\alpha(x,y)+\beta(\sigma)}, \tag{5}$$

where

$$\beta(\sigma) = \bar{d}\psi'(\sigma)/\{1 - \bar{d}\psi'(\sigma) \cdot o(1/\psi'(\sigma)^{1/2})\}, \quad \bar{d} = \left(\frac{1}{2} + \varepsilon\right) s(\sigma)^2, \quad \sigma = \sigma(x, y). \quad \bullet$$

Now we can use the argument following the estimate (1), replacing this estimate by the inequality (5).[11]

Thus there exist arbitrarily large values of σ for which

$$\frac{\log|f_1(\sigma)|}{e^{\alpha(x)-\alpha\left(x,\, -i\frac{\pi}{2}\right)+\beta(\sigma)}\varphi(\sigma)\{1 + \delta(\operatorname{Re} Z(\sigma))\}} > -1, \tag{6}$$

and therefore

$$\overline{\lim_{r\to 1-0}}\, \frac{\log|f(r)|}{\varphi\left(\log\frac{1}{1-r}\right)} \equiv \overline{\lim_{r\to 1-0}}\, \frac{\log|f(r)|}{\log\lambda(r)} \geqslant -1.$$

In order to verify the second assertion in Theorem 2 we estimate the sum

$$\frac{\log|f_1(\sigma)|}{\varphi(\sigma)} + 1$$

as we did in the proof of Theorem 1. We obtain (for the same values of σ as in (6))

$$\begin{aligned}\frac{\log|f_1(\sigma)|}{\varphi(\sigma)} + 1 &> 1 - e^{2|\alpha(x)|+\beta(\sigma)}(1 + \delta(\log\varphi(\sigma) + o(1)))\\ &> 1 - \{1 + 2|\alpha(x)| + \beta(\sigma) + O(|\alpha|^2 + \beta^2)\}(1 + \delta(\log\varphi(\sigma) + o(1)))\\ &= -2|\alpha(x)| - \beta(\sigma) + O(|\alpha|^2 + \beta^2) + O(\delta|\alpha| + \delta\beta).\end{aligned}$$

We define the function $\epsilon_{\mathbf{D}}$ by $\epsilon_{\mathbf{D}} = \max(\epsilon_{\Pi}, 1/\psi')$. Then from (3), the preceding equation, and the definition of $\beta(\sigma)$ it follows that

$$\frac{\log|f_1(\sigma)|}{\varphi(\sigma)} + 1 > -\left(\frac{5\pi^2}{4} + \frac{\pi^2}{2}\right)\varepsilon_{\mathbf{D}}(\sigma)(1+\epsilon). \quad \bullet$$

---

(11) Here we should note that $\beta = o(1)$, since $s = \pi/\psi'$, and $\bar{d}\psi' \sim \pi^2/2\psi'$ and $\beta(\sigma) \sim \pi^2/2\psi'(\sigma)$ as $\sigma \longrightarrow +\infty$.

REMARK. If $\epsilon_\Pi = o(1/\psi')$, then the right side of the preceding inequality can be replaced by $\pi^2(1+\epsilon)\epsilon_D/2$, and if $1/\psi' = o(\epsilon_\Pi)$, then it can be replaced by $5\pi^2(1+\epsilon)\epsilon_D/4$.

§1.3. A lower estimate for a harmonic function

1°. This section is devoted to the solution of problem II which was stated in §1.0. As we already mentioned, we are concerned with an asymptotic estimate for $\min_{|\zeta|\le r} u(\zeta)$ in terms of $\max_{|\zeta|\le r} u(\zeta)$ as $r \to 1-0$, where u is a real function which is defined and harmonic in the disk **D**. The results in this section are based on the fundamental inequality of Carathéodory (see [53]): if f is an analytic function in **D** and $0 \le r < R < 1$, then

$$|f(\zeta) - f(0)| \le \frac{2r}{R-r}\left(\max_{|\zeta|\le R} \operatorname{Re} f(\zeta) - \operatorname{Re} f(0)\right)$$

for all ζ, $|\zeta| \le r$. Another approach to obtaining the necessary estimates is by means of the asymptotic formula of Warschawski [15] for conformal transformations of curvilinear strips and the refinement of this formula (with an estimate of the remainder term) introduced in §1.2.

The first result in this area,(12) which is now classical, is due to Cartwright. In [124] she showed that for an arbitrary $\alpha > 1$ there exists a constant $c = c(\alpha)$ such that

$$f \in a\left(\mathbf{D}, \exp\frac{1}{(1-r)^\alpha}\right) \Rightarrow |f(re^{i\theta})| \ge \exp\left\{-\frac{c(\alpha)}{(1-r)^\alpha}\right\}, \quad r > r_0.$$

For $\alpha < 1$ we know that the estimate

$$|f(re^{i\theta})| \ge \exp\left(-\frac{c}{1-r}\right), \quad r > r_0,$$

is best possible in these classes, as we see from the example $f\colon \zeta \to \exp(\zeta-1)^{-1}$, $|\zeta| < 1$. The case $\alpha = 1$ is also analyzed in [124]; here

$$\log m_{\mathbf{D}}(r) \asymp -(1-r)^{-1}\log^2(1-r)^{-1}$$

as $r \to 1-0$.

Later the works of Linden [153], [154] and Macaev [67], [68] appeared. Linden obtained lower estimates for cases where the "order α" is replaced by the "generalized order α"; that is,

$$\lambda_\rho(r) = \exp((1-r)^{-\rho(r)}), \qquad \lim_{r\to 1} \rho(r) = \alpha < +\infty.$$

(12) We are concerned with lower estimates in classes of functions with a given growth majorant. The elementary estimate

$$\log|f(re^{i\theta})| \ge -\text{const}\,(1-r)^{-1}, \; f \in a(\mathbf{D}, \text{const}),$$

introduced in §1.2.1°, is due to Nevanlinna.

Under certain restrictions on the "regularity" of growth of the function ρ (consistent with subsequent uses) he showed (without calculating c_ρ) that

$$1<\alpha<\infty,\ f\in a(\mathrm{D},\lambda_\rho)\Rightarrow m_{\mathrm{D}}(r)\geqslant \exp(-c_\rho(1-r)^{-\rho(r)}).$$

In the hardest case $\alpha = 1$, the work [154] only dealt with weights of the form

$$\exp\left(\frac{1}{1-r}\prod_{k=1}^{n}\log_k^{\alpha_k}\frac{1}{1-r}\right),$$

where $\log_1 = \log$, $\log_k = \log\circ\log_{k-1}$, $k\geqslant 2$, and the α_k are real numbers, or with weights of the form

$$\exp\frac{1}{1-r}\chi\left(\frac{1}{1-r}\right)^2,\ \chi\downarrow 0,\ \int^{\infty}\frac{\chi(t)}{t}\,dt<+\infty.$$

For the corresponding classes of functions lower estimates were found; that they are sharp was not verified, but many specific examples were discussed. In [67] and [68] Macaev found an estimate of the type $m_{\mathrm{D}}(r)\geqslant\lambda^{-c}(1-\gamma+\gamma r)$, where $c=c(\lambda)>0$ and $1>\gamma=\gamma(\lambda)>0$, and the class of majorants under consideration was distinguished by the condition $1<\alpha\leqslant+\infty$. We should note that a proportional "shift of the argument" ($r\longrightarrow 1-\gamma+\gamma r$) in the majorant λ manifests itself more strongly than does raising to an arbitrary power, at least for

$$\alpha=\lim_{r\to 1-0}\frac{\log\log\lambda(r)}{\log(1-r)^{-1}}=+\infty.$$

This effect becomes more noticeable the faster λ grows; as examples we can use the weights $\lambda=\exp\exp\cdots\exp(1/(1-r))$. We will return to this fact in §2.7 with a discussion of a theorem of Beurling.

Later we will establish lower estimates for functions in $a(\mathbf{D},\lambda)$ and $a(\Pi,\lambda_\Pi)$ without any restrictions on the speed of growth of the majorants, but under essential conditions on "regularity of growth". The nature of these conditions is discussed in §1.5. That the estimates obtained in this section are sharp will be verified in §1.4.

Hereafter we will always use the notation $\lambda_\Pi(t)=\lambda(1-e^{-t})$, $t\geqslant 0$; $\varphi=\log\lambda_\Pi$, $\psi=\log\varphi$; that is,

$$\lambda(r)=\exp\varphi\left(\log\frac{1}{1-r}\right),\ 0\leqslant r<1;\ I=\left(-\frac{\pi}{2},\frac{\pi}{2}\right),\ I_\varepsilon=\left[-\frac{\pi}{2}+\varepsilon,\frac{\pi}{2}-\varepsilon\right].$$

The nature of a lower estimate for functions in $a(\mathbf{D},\lambda)$ and $a(\Pi,\lambda_\Pi)$ depends on the magnitude of the limit $\psi'_\infty=\lim_{t\to\infty}\psi'(t)$, whose existence is assumed in the hypotheses of all theorems in this section. Here the case $\psi'_\infty<1$ is covered by the aforementioned theorem of Cartwright since $\psi(t)\leqslant\text{const}+ct$, $c<1$, and the best minorant in the disk is $\exp(-k(1-r)^{-1})$, and in a substrip of the strip Π the best minorant is $\exp(-ke^{\sigma})$ for some $k>0$.

We note that for functions in $a(\Pi,\lambda_\Pi)$ we may be concerned with a lower estimate only in some substrip (in general, curvilinear), since the conditions $|f(\sigma+i\tau)|\leqslant\exp\varphi(\sigma)$

and $f(\sigma + i\tau) \neq 0$, $\sigma + i\tau \in \Pi$, do not forbid the "vanishing" of a function f on the boundary $\partial\Pi$. It is true that this remark is not relevant to the sequential lower estimate which we proved in the preceding section, and it is consistent with the fact that such "boundary zeros" cannot be very numerous ($\int_0^x \log|f(\sigma \pm i\pi/2)|\,d\sigma > -\infty$ for each $x > 0$).

THEOREM 1. *Let* $\psi = \log\log\lambda_\Pi$ *and* $f \in a(\Pi, \lambda_\Pi)$.

I. *If* $\lim_{t\to\infty}\psi' = +\infty$ *and* $1/\sqrt{\psi'}$ *is a convex function, then for each* $\epsilon > 0$ *there exists a number* σ_ϵ *such that*

$$|f(\sigma + i\theta)| \geqslant \frac{1}{\lambda_\Pi(\sigma)^B}, \quad \sigma > \sigma_\epsilon, \quad \theta \in I_\epsilon, \quad B = 1 + 2\pi(e^{\pi+1} + 1).$$

II. *If* $f \in a(\overline{\Pi}, \lambda_\Pi)$, $1 < \lim_{t\to\infty}\psi' \overset{\text{def}}{=} \psi'_\infty < \infty$ *and* ψ' *is a slowly oscillating function (that is, the limit* $\lim_{t\to\infty}\psi''$ *exists), then for each* $\delta > 0$ *there exist numbers* $\sigma_{\epsilon,\delta}$ *and* $\alpha(\delta)$ *such that for* $0 < \epsilon < \alpha(\delta)$ *we have*

$$|f(\sigma + i\theta)| \geqslant \frac{1}{\lambda_\Pi(\sigma)^B}, \quad \sigma > \sigma_{\epsilon,\delta}, \quad \theta \in I_\epsilon,$$

where

$$B = B_{\epsilon,\delta} = e^{\frac{2\pi}{\epsilon}}\left\{\frac{e^{-4}}{\psi'^2_\infty - 1} + e^{-4+\psi'_\infty\epsilon} + \delta\right\}.$$

THEOREM 2. *Let* $f \in a(D, \lambda)$ *and* $\psi = \log\log\lambda_\Pi$. *Under the conditions in Theorem* 1 *the following assertions hold*:

I. *If* $\lim_{t\to\infty}\psi' = +\infty$, *then*

$$|f(re^{i\theta})| \geqslant \frac{1}{\lambda(r)^B}, \quad r > r_0,$$

where $B = 1 + 2\pi(e^{\pi+1} + 1)$.

II. *If* $1 < \lim_{t\to\infty}\psi' \equiv \psi'_\infty < +\infty$, *then*

$$|f(re^{i\theta})| \geqslant \frac{1}{\lambda(r)^B}, \quad r > r_0,$$

where $r_0 = r_0(\psi'_\infty)$ *and* $B = (4\pi + e^{\sqrt{2}} + 1)e^{5/11}$ *for* $\psi'_\infty > 2$, *and*

$$B = \frac{3 + 2\sqrt{2}}{\psi'_\infty - 1}\left(\frac{2\pi + 1}{\psi'_\infty - 1} + e^{\sqrt{2}} + 2\pi\right) \quad \text{for} \quad \psi'_\infty \leqslant 2.$$

We will make several remarks about Theorems 1 and 2. The proof of part I of the first theorem uses the theorem in §1.2 on the sequential lower estimate, and part II uses the Carleman formula directly. This enables us to conclude (in part I) that

$$\log|f(\sigma_n + i\theta)| \geqslant -\varphi(\sigma_n)(1 + o(1)), \quad \theta \in I,$$

for some σ_n, $\sigma_n \to +\infty$ ($\sigma_n = \sigma_n(\theta)$). Then in the strip $\Omega_\theta = \{\sigma + i\tau\colon 0 < \tau - \theta < \pi/\psi'(\sigma)\}$ we construct a "supporting" function $\mathcal{S}$ which is small on $\partial\Omega_\theta$ and large on the real axis, so that $|\mathcal{S}f| \leqslant 1$ on $\partial\Omega_\theta$, and $|f(\sigma_n)\mathcal{S}(\sigma_n)| > 1$. By the maximum principle, on each segment $\{\sigma + i\tau \colon \theta < \tau < \theta + \pi/\psi'(\sigma)\}$ there exists a point $\zeta = \sigma + i\tau(\sigma)$ such that $|f(\zeta)\mathcal{S}(\zeta)| > 1$. If we apply the Carathéodory inequality to the function $\log f$ with center at the point ζ, we obtain the needed lower estimate for $\log|f(\sigma + i\theta)|$.

After a conformal transformation this same method of proof works in part I of Theorem 2 if we note that here it is necessary to consider the distortion of the majorant in the strip Ω_θ caused by the transformation.

In part II the Carleman formula leads at once to an estimate of the form

$$\log|f(\sigma + i\theta_\sigma)| \geqslant -\varphi(\sigma)\,\mathrm{const},$$

which we then extend to all θ in an interval I_ϵ, $\epsilon > 0$, by means of repeated applications of the Carathéodory inequality ($\sim$const/ϵ times) with a small R (see (7) below). This process, which is similar to the procedure for defining the number e, is employed in Lemma 2.

The requirements imposed on the growth of the functions λ and λ_Π in Theorems 1 and 2 consist, roughly speaking, of the following:

$$\text{I)}\qquad \lim_{\sigma\to\infty} \frac{\log\varphi(\sigma)}{\sigma} = +\infty, \qquad \text{or} \qquad \lim_{r\to1-0} \frac{\log\log\lambda(r)}{\log\dfrac{1}{1-r}} = +\infty;$$

$$\text{II)}\qquad 1 < \lim_{\sigma\to\infty} \frac{\log\varphi(\sigma)}{\sigma} < \infty, \qquad \text{or} \qquad 1 < \lim_{r\to1-0} \frac{\log\log\lambda(r)}{\log\dfrac{1}{1-r}} < +\infty;$$

that is,

$$\exp\left(\frac{1}{(1-r)^{c_1}}\right) \leqslant \lambda(r) \leqslant \exp\left(\frac{1}{(1-r)^{c_2}}\right),$$

for some $c_1, c_2 > 1$. The other conditions in Theorems 1 and 2 are connected not to the speed of growth of λ but to its "regularity" (see §1.5). As in §1.2, the requirements in assertion I can be weakened to convexity of $1/\psi'$ and boundedness of $(\psi''/\psi'^2)\log^2\psi'$.

Finally, we consider the magnitude of the constants B in Theorems 1 and 2. The statements of the theorems do not always contain the best constant B for the method of proof which we use; sometimes this constant is overstated for the convenience of description. The method which we use does not enable us to obtain in assertion I the inequality $B \leqslant 1 + \epsilon$ with a small $\epsilon > 0$. A constant $B < 1$ is clearly impossible in view of the reflexivity of the argument for f and $1/f$. Concerning the dependence of B on ψ'_∞ in assertion II of Theorems 1 and 2, in the first case for ψ'_∞ near 1 it is proportional to $(\psi'_\infty - 1)^{-1}$, and in the second case it is proportional to $(\psi'_\infty - 1)^{-2}$. The constant B in part II of Theorem 1 contains the term $e^{\psi'_\infty\epsilon}$, the appearance of which depends upon the immediate application of the Carleman formula, which is disadvantageous for large ψ'_∞. If for $\psi'_\infty > 2$ we use the method of proof of assertion I,[13] we can obtain

[13] For $1 < \psi'_\infty < 2$ an application of this method is not possible, and to keep the treatment uniform we use the Carleman formula throughout the whole range $1 < \psi'_\infty < \infty$.

an estimate with $B = \text{const} \cdot e^{2\pi/\epsilon}$, where const is an absolute constant. We note, finally, that for $\log|f| \leqslant \epsilon\varphi$, where $\lim_{t\to\infty}\epsilon = 0$, Theorems 1 and 2 give us a lower estimate of the same type: $\log|f| = o(\varphi)$.

We turn to majorants λ which are of exponential order 1 (with respect to $1/(1-r)$).

THEOREM 3. *Suppose that $f \in a(\overline{\Pi}, \lambda_\Pi)$, $\varphi = \log\lambda_\Pi$, $\psi = \log\log\lambda_\Pi$ and $\lim_{t\to\infty}\psi'(t) = 1$. For each ϵ with $0 < \epsilon < 4/9$ there exists a number $\sigma(\epsilon)$ such that*

$$|f(\sigma + i\tau)| \geqslant \exp\left(-Be^\sigma \int_0^\sigma \varphi(t)e^{-t}dt\right)$$

for $\sigma > \sigma(\epsilon)$ and $\tau \in I_\epsilon$, where $B = B_\epsilon$ is determined by the equations

$$B_\epsilon = e^{\frac{2\pi}{\epsilon}-4}, \quad \textit{if} \quad \int_0^\infty \varphi e^{-t}dt = \infty,$$

$$B_\epsilon = e^{\frac{2\pi}{\epsilon}-4}\left(1 + \operatorname{Car} f \Big/ \int_0^\infty \varphi(t)e^{-t}dt\right), \quad \textit{if} \quad \int_0^\infty \varphi e^{-t}dt < \infty,$$

where Car f *is the Carleman constant*([14]) *of f.*

THEOREM 4. *Suppose that $f \in a(\mathbf{D}, \lambda)$, $\varphi = \log\lambda_\Pi$, $\psi = \log\log\lambda_\Pi$ and $\lim_{t\to\infty}\psi' = 1$. Set $a(t) = \psi(t) - t$, $t \geqslant 0$, and assume that a is convex in a neighborhood of $+\infty$. Then*

$$\log|f(re^{i\theta})| \geqslant -\frac{2}{\pi}B_{2/5}\frac{1}{1-r}\left(\int_0^{\log\frac{1}{1-r}} \sqrt{\varphi(t)e^{-t}}\,dt\right)^2$$

$$= -\frac{2}{\pi}B_{2/5}\frac{1}{1-r}\left(\int_0^r \sqrt{\frac{\log\lambda(x)}{1-x}}\,dx\right)^2, \quad r > r_0,$$

where $B_{2/5}$ is the constant in Theorem 3.([15])

The method of proof for Theorem 3 is the same as for part II of Theorem 1. Theorem 4 follows from Theorem 3 in the same way that Theorem 2 follows from Theorem 1, except that the distortion from the conformal transformation in this case is comparable in size with the contribution in the lower estimate from the "rectilinear" case of Theorem 3.

---

(14) The definition is given after Lemma 1 in §1.2.

(15) The Carleman constant in $B_{2/5}$ for $\int_0^\infty \varphi e^{-t}dt < +\infty$ is calculated for the function $f \circ q^{-1} \circ Z^{-1}$ from the proof of Theorem 4. As an upper estimate for this constant we can use

$$A\left|\int_{-1/2}^{1/2} \log f\left(re^{i\left(\theta+\frac{\pi}{2}\right)}\right)dr\right| \quad \text{or} \quad A\max_\theta\left|\int_{-1/2}^{1/2}\log f(re^{i\theta})\,dr\right|,$$

where A is an absolute constant.

In §1.4 we will show that the estimates in Theorems 3 and 4 are best possible([16]) at least to within the magnitudes of the constants B_ϵ and $2B_{2/5}/\pi$.

The requirements imposed on the growth of λ and λ_Π in Theorems 3 and 4 consist, roughly speaking, of the following:

$$\varphi(t) = \log \lambda_\Pi(t) = e^{(1+o(1))t} \quad \text{or} \quad \log \lambda(r) = \frac{1}{(1-r)^{1+o(1)}};$$

in the notation of these theorems

$$\lambda(r) = \exp\left(\frac{1}{1-r} \exp a\left(\log \frac{1}{1-r}\right)\right) \quad \text{for} \quad \lim_{t\to\infty} a'(t) = 0.$$

In the rest of this section we will consider some typical classes of majorants λ and λ_Π for which the minorants in Theorems 3 and 4 can be calculated explicitly. We will use the notation

$$\Delta_1(\sigma) = e^\sigma \int_0^\sigma \varphi(t) e^{-t} dt = e^\sigma \int_0^\sigma e^{a(t)} dt, \quad \sigma > 0,$$

$$\Delta_2(\sigma) = e^\sigma \left(\int_0^\sigma \sqrt{\varphi(t) e^{-t}} dt \right)^2 = e^\sigma \left(\int_0^\sigma e^{\frac{1}{2} a(t)} dt \right)^2, \quad \sigma > 0.$$

The first question here is the following: for which λ (λ_Π) will a lower estimate be of the order $\exp(-k(1-r)^{-1})$ (respectively, $\exp(-ke^\sigma)$); that is, the same as for $\lambda \equiv \text{const}$ or $\lambda \equiv \exp(1-r)^{-\alpha}$, $\alpha < 1$? The answer was obtained by Linden in [154] (without asserting that the estimates are sharp; see §1.4); the answer follows immediately from Theorems 3 and 4.

COROLLARY 1.

$$\Pi)\ \Delta_1 \asymp e^\sigma \Leftrightarrow \int_0^\infty \varphi(t) e^{-t} dt < +\infty.$$

$$\text{D})\ \Delta_2 \asymp e^\sigma \Leftrightarrow \int_0^\infty (\varphi(t) e^{-t})^{1/2} dt < +\infty.$$

EXAMPLE.

$\Pi)$ $\log|f(\sigma+i\tau)| \leqslant e^\sigma(\sigma \log \sigma \ldots \log_n^{1+\delta}\sigma)^{-1}$, $\delta > 0$, $f \neq 0$ $\Rightarrow \log|f(\sigma+i\tau)| \geqslant -\text{const}\, e^\sigma$, $\tau \in I_\varepsilon$, $\text{const} = \tilde{c}(\delta, \varepsilon, f)$.

D) $\log|f(re^{i\theta})| \leqslant \frac{1}{1-r}\left(\log\frac{1}{1-r} \cdots \log_n^{1+\delta}\frac{1}{1-r}\right)^{-2}$, $f \neq 0 \Rightarrow \log|f(re^{i\theta})| \geqslant -\frac{\text{const}}{1-r}$, $\text{const} = c(\delta, f)$.

COROLLARY 2. *The minorants m_Π and m_D in the hypotheses of Theorems 3 and 4 respectively have a higher order of smallness than $1/\lambda$ and $1/\lambda_\Pi$. More precisely*:

([16]) With regard to Theorems 1 and 2 see the corresponding remark after Theorem 2″ in §1.2.

$$\Pi)\ \varphi = o(\Delta_1),\ but \int_1^\infty \varphi/\Delta_1 d\sigma = +\infty; \qquad D)\ \varphi = o(\Delta_2),\ but \int_1^\infty (\varphi/\Delta_2)^{1/2} d\sigma = +\infty.$$

Both assertions in Corollary 2 follow from Lemma 4, which will be stated and proved in subsection 4° of this section.

In the remaining corollaries to Theorems 3 and 4 we calculate the magnitude of the additional increasing factor $\Delta_{1,2}/\varphi$. In order to clarify the asymptotic behavior of the integrals $\int_0^\sigma e^a\,dt$ and $\int_0^\sigma e^{a/2}\,dt$, we need new assumptions on the majorants λ and λ_Π. We will assume the existence of the limits(17)

$$\alpha = \lim_{t\to\infty} a'(t)\,t, \qquad L = \lim_{t\to\infty} a''/a'^2 \quad (\text{in the case } a' > 0). \tag{1}$$

Corollary 3. *Suppose that for the function* $a(\sigma) = \psi(\sigma) - \sigma$, $\sigma \geqslant 0$, *the conditions in Theorem* 4 *are satisfied and the limits in* (1) *exist. Then*

1) $\alpha = +\infty \Rightarrow \Delta_1(\sigma) \sim \dfrac{\varphi(\sigma)}{\psi'(\sigma)-1}$, $\Delta_2(\sigma) \sim \dfrac{4\varphi(\sigma)}{(\psi'(\sigma)-1)^2}$;

2) $-1 < \alpha < +\infty \Rightarrow \Delta_1(\sigma) \sim \varphi(\sigma)\dfrac{\sigma}{1+\alpha}$, $\quad -2 < \alpha < +\infty \Rightarrow \Delta_2(\sigma) \sim \varphi(\sigma)\left(\dfrac{2\sigma}{2+\alpha}\right)^2$;

3) $\alpha < -1 \Rightarrow \Delta_1(\sigma) \asymp e^\sigma$, $\alpha < -2 \Rightarrow \Delta_2(\sigma) \asymp e^\sigma$.

The proof in the case $a' > 0$ (for large σ) follows immediately from Lemma 1 in §1.1. If $\alpha < 0$, we can use the formula

$$\int_0^\sigma e^{a(t)}dt = \sigma e^{a(t)} - \int_0^\sigma a'(t)\,te^{a(t)}dt.$$

Assertion 3) follows from Corollary 1. ●

We note, finally, that the case $\alpha = -1$ or -2 requires special treatment (see below) and that the behavior of the function φ for $|\alpha| < \infty$ is described by the equation $\varphi(\sigma) = e^\sigma \sigma^{\alpha + o(1)}$.

Example. Suppose that

$$\varphi(\sigma) = e^{\sigma + \sum_{k=1}^n c_k\sigma^{\gamma_k}} \prod_{k=0}^m \log_k^{\alpha_k}\sigma, \qquad \log_0\sigma \overset{\text{def}}{=} \sigma, \qquad c_k \neq 0,$$

$$1 > \gamma_1 > \ldots > \gamma_n,\ f \neq 0.$$

Then

$\Pi)\ \log|f(\sigma+i\tau)| \leqslant \varphi(\sigma) \Rightarrow$

$$\log|f(\sigma+i\tau)| \geqslant -e^{\frac{2\pi}{\varepsilon}-4}\varphi(\sigma)\frac{\sigma^{1-\gamma_1}}{c_1\gamma_1},\quad \tau\in I_\varepsilon,\ \gamma_1>0,\ c_1>0;$$

$$\log|f(\sigma+i\tau)| \geqslant -e^{\frac{2\pi}{\varepsilon}-4}\varphi(\sigma)\frac{\sigma}{1+\alpha_0},\quad \tau\in I_\varepsilon,\ \gamma_1=0,\ \alpha_0>-1;$$

$$\log|f(\sigma+i\tau)| \geqslant -B_\varepsilon e^\sigma,\quad \tau\in I_\varepsilon,\ \gamma_1>0,\ c_1<0 \text{ or } \alpha_0<-1.$$

(17) In fact, the existence of the limit L implies the existence of α (see the remark after Theorem 2 in §1.1).

Here B_ϵ is the constant from Theorem 3.

D) $\log|f(re^{i\theta})| \leqslant \varphi\left(\log\frac{1}{1-r}\right) \Rightarrow$

$$\log|f(re^{i\theta})| \geqslant -\frac{2}{\pi} B_{2/5}\varphi\left(\log\frac{1}{1-r}\right)\frac{4\log^{2(1-\gamma_1)}(1-r)^{-1}}{c_1^2\gamma_1^2}, \quad \gamma_1>0,\ c_1>0;$$

$$\log|f(re^{i\theta})| \geqslant -\frac{2}{\pi} B_{2/5}\varphi\left(\log\frac{1}{1-r}\right)\frac{4\log^2\frac{1}{1-r}}{(2+\alpha_0)^2}, \quad \gamma_1=0,\ \alpha_0>-2;$$

$$\log|f(re^{i\theta})| \geqslant -\frac{2}{\pi} B_{2/5}\frac{1}{1-r}, \quad \gamma_1>0,\ c_1<0 \ \text{ or } \ \alpha_0<-2.$$

Returning to the cases $\alpha = -1$ (-2), which we skipped, we note that here we cannot hope to find an asymptotic formula for Δ_1 (Δ_2) which contains only φ and ψ'. There are two possible analyses of the integrals Δ_1 and Δ_2: we can require regularity of the behavior of the highest derivatives of the function $a(\sigma) = \psi(\sigma) - \sigma$ and thereby find asymptotics for Δ_1 and Δ_2 (for example, we can consider the limit

$$\beta = \lim_{\sigma\to\infty}(a''(\sigma) - a'^2(\sigma))\sigma^2)$$

or we can assume the existence of an asymptotic expansion of the function a with respect to some congruence scale $\mathfrak{S}$ (see [12]) and then investigate the analogous integrals of the terms in the expansion. These two methods are roughly equivalent. We will use the second, choosing for $\mathfrak{S}$ the logarithmic scale $\mathrm{Log} = \{\log_k : k = 1, 2, \dots\}$. We will use the following elementary lemma.

LEMMA 0. *If b is a function such that*

$$0 < \underline{l} = \varliminf_{t\to\infty}(a'-b')e^{-b} \leqslant \varlimsup_{t\to\infty}(a'-b')e^{-b} = \overline{l} < +\infty,$$

then

$$\int_0^\sigma e^{a(t)}dt \asymp e^{a(\sigma)-b(\sigma)}, \quad \sigma\to+\infty,$$

and for $\int_0^\infty e^a = +\infty$ and $\underline{l} = \overline{l} = l$ we have

$$\int_0^\sigma e^a dt \sim \frac{1}{l}e^{a(\sigma)-b(\sigma)}, \quad \sigma\to\infty.$$

The proof follows easily from the obvious equation

$$\int_0^\sigma e^a(a'-b')e^{-b}dt = e^{a-b} + \text{const.} \quad \bullet$$

We will say (see [12]) that a function a has an asymptotic expansion of order 1 to within $\log_n$, $n \geqslant 1$ (with respect to the scale Log) if in a neighborhood of $+\infty$ we have

$$a' = \sum_{k=1}^{n} c_k \log_k' + o(\log_n').$$

In this language the requirements in Corollary 3 mean that a has an asymptotic expansion of order 1 to withing log; that is, $a' \sim \alpha/\sigma$ as $\sigma \longrightarrow +\infty$, and $\alpha \neq -1$ (in the case Π) and $\alpha \neq -2$ (in the case **D**). We consider the case $\alpha = -1$ (-2).

COROLLARY 4. *Suppose that the function $\alpha(\sigma) = \psi(\sigma) - \sigma$, $\sigma \geqslant 0$, has an asymptotic expansion of order* 1 *to within*

$$\log_n, \quad n \geqslant 1, \quad a = \sum_{k=1}^{n} c_k \log_k + o(\log_n).$$

1) *If $(c_1, \ldots, c_n) \prec (-1, \ldots, -1)$ lexicographically, then $\Delta_1(\sigma) \asymp e^\sigma$.*
2) *If $(c_1, \ldots, c_n) \prec (-2, \ldots, -2)$ then $\Delta_2(\sigma) \asymp e^\sigma$.*
3) *If $(c_1, \ldots, c_n) \succ (-1, \ldots, -1)$, where $c_1 = \cdots = c_{m-1} = -1$, $1 \leqslant m \leqslant n$, and $c_m > -1$, then*

$$\Delta_1(\sigma) \sim \varphi(\sigma) \frac{\sigma \log \sigma \ldots \log_{m-1} \sigma}{c_m + 1}.$$

4) *If $(c_1, \ldots, c_n) \succ (-2, \ldots, -2)$, where $c_1 = \cdots = c_{m-1} = -2$, $1 \leqslant m \leqslant n$, and $c_m > -2$, then*

$$\Delta_2(\sigma) \sim \varphi(\sigma) \left(\frac{\sigma \log \sigma \ldots \log_{m-1} \sigma}{c_m/2 + 1}\right)^2.$$

Assertions 1) and 2) follow from Corollary 1. We verify 3) and 4) by means of Lemma 0. To do this we set

$$b(t) = -\sum_{k=1}^{m} \log_k t,$$

and then

$$(a' - b') e^{-b} = \left\{(c_m + 1) \frac{1 + o(1)}{t \log t \ldots \log_{m-1} t}\right\} t \log t \ldots \log_{m-1} t;$$

the rest follows from Lemma 0. ●

As an example we consider the function $\varphi(\sigma) = e^\sigma \Pi_{k=0}^n \log_k^{\alpha_k} \sigma$ and use an expansion for $a = \log \varphi - \sigma$ to within $\log_{n+2}$ (with a coefficient $c_{n+2} = 0$).

EXAMPLE. Suppose that $f \neq 0$. Then

$$\Pi) \ \log|f(\sigma + i\tau)| \leqslant e^\sigma \prod_{k=1}^{m} \log_k^{-1} \sigma \cdot \prod_{k=m+1}^{n} \log_k^{\alpha_k} \sigma, \quad 0 \leqslant m \leqslant n \text{ and } \alpha_{m+1}$$

$$\neq -1 \Rightarrow \log|f(\sigma + i\tau)| \geqslant -e^{\frac{2\pi}{\varepsilon} - 4} \frac{e^\sigma}{\alpha_{m+1} + 1} \prod_{k=m+1}^{n} \log_k^{\alpha_k} \sigma, \quad \text{if } \tau \in I_\varepsilon, \ \alpha_{m+1} > -1,$$

and $\log|f(\sigma + i\tau)| \geqslant -B_\varepsilon e^\sigma$, if $\tau \in I_\varepsilon$, $\alpha_{m+1} < -1$.

D) $\log|f(re^{i\theta})| \leqslant \frac{1}{1-r}\prod_{k=0}^{m}\log_{k+1}^{-2}\frac{1}{1-r}\prod_{k=m+1}^{n}\log_{k+1}^{\alpha_k}\frac{1}{1-r}$, $0 \leqslant m \leqslant n$ and $\alpha_{m+1} \neq -2 \Rightarrow \log|f(re^{i\theta})| \geqslant -\frac{2}{\pi}B_{2/5}\frac{1}{\left(\frac{1}{2}\alpha_{m+1}+1\right)^2}\cdot\frac{1}{1-r}\prod_{k=m+1}^{n}\log_{k+1}^{\alpha_k}\frac{1}{1-r}$, if $\alpha_{m+1} > -2$, and $\log|f(re^{i\theta})| \geqslant -\frac{2}{\pi}B_{2/5}\frac{1}{1-r}$, if $\alpha_{m+1} < -2$.

By analogy we can also treat the more general case $\varphi(\sigma) = e^{\sigma}\Pi_{k=0}^{n}\exp(\alpha_k\log_{k+1}^{\beta_k}\sigma)$. We end our discussion of Theorems 3 and 4 with one more assertion.

COROLLARY 5. *Suppose that the weight λ belongs to the Hardy field* **H** *(see §1.5) and in the expansion of the function $a(\sigma) = \log\log\lambda_{\Pi}(\sigma) - \sigma$ in an asymptotic series with respect to the scale* Log *not all of the coefficients equal* -1 *(or* -2*). Let c_m be the first of these. If $c_m < -1$ ($c_m < -2$), then for a function f in $a(\Pi, \lambda_{\Pi})$ ($a(\mathbf{D}, \lambda)$) we have a lower estimate of type* 1) *in Corollary* 4 *(type* 2)*); if $c_m > -1$ ($c_m > -2$), then we have an estimate of type* 3) *(type* 4)*).*

2°. The proof of Theorem 1 splits into two parts: I) $\lim\psi' = +\infty$; II) $\lim\psi' < +\infty$.

I. We consider the strip $\Omega_\theta = \{\sigma + i\tau\colon |\tau - \theta| < s(\sigma)\}$, where s is a function such that $\varphi(\sigma) = \exp(\pi\int_0^\sigma(1/s(t))\,dt)$, $\sigma \geqslant 1$; that is, $s = \pi\varphi/\varphi' = \pi/\psi'$. Since for s the conditions of Warschawski's theorem (Lemma 2 in §1.2) are satisfied, the conformal mapping Z of the strip Ω_θ onto the strip Π has asymptotics

$$\operatorname{Re} Z(\sigma + i\tau) = \frac{\pi}{2}\int_0^\sigma\frac{dt}{s(t)} + r(\sigma, \tau),$$

where $r(\sigma, \tau) = o(1)$ as $\sigma \to +\infty$. We define functions G and F_1 by the equations

$$G = \exp(p\exp 2Z),\quad F_1 = fG \text{ in } \Omega_\theta \cap \Pi.$$

It is clear that there exists a number σ_θ such that $x > \sigma_\theta$ and $x + iy \in \Omega_\theta$ imply that $x + iy \in \Pi$. Thus for $w = x + iy \in \partial\Omega_\theta$, $x > \sigma_\theta$, we have

$$\begin{aligned}|F_1(x+iy)| &= |f(x+i(\theta \pm s(x)))|\,|G(x+i(\theta \pm s(x)))|\\ &\leqslant \lambda(x)\exp\{p[\exp 2\operatorname{Re} Z(x+i(\theta \pm s(x)))]\cos 2\operatorname{Im} Z(x+i(\theta \pm s(x)))\}\\ &= \lambda(x)\exp\{-p\varphi(x)e^{r_\pm(x)}\},\end{aligned}$$

where

$$|r_\pm(x)| \leqslant \max\{|r(x, \tau)| : |\tau - \theta| \leqslant s(x)\} \overset{\text{def}}{=} r(x).$$

Now suppose that $\sigma_1 > \sigma_\theta$ and g is an outer function (see [23]) in the strip $\Omega_\theta \cap \{\sigma + i\tau\colon \sigma > \sigma_1\} = \Omega_{\theta,\sigma_1}$, the modulus of which on the vertical segment $\sigma = \sigma_1$ equals

$$1/\max_{|\theta - \tau| \leqslant s(\sigma_1)}|F_1(\sigma + i\tau)|$$

and equals 1 on the sides $\tau = \theta \pm s(\sigma)$, $\sigma > \sigma_1$. We set $F = gF_1$ in Ω_{θ,σ_1}, and $p = \max_{x \geqslant \sigma_1} e^{r(x)}$. Then

$$|F(x+i(\theta \pm s(x)))| \leqslant \lambda(x) \exp\{-p\varphi(x) e^{r\pm(x)}\} \leqslant 1, \quad x \geqslant \sigma_1,$$
$$|F(\sigma_1 + i\tau)| \leqslant 1, \quad |\theta - \tau| \leqslant s(\sigma_1).$$

On the other hand, by Theorem 1 in §1.2 we know that there exists a sequence of numbers $\{\sigma_n\}_2^\infty$ such that

$$|F(\sigma_n + i\theta)| = |f(\sigma_n + i\theta)|\,|G(\sigma_n + i\theta)|\,g(\sigma_n + i\theta)|$$
$$> \exp\{-\varphi(\sigma_n)(1 + o_n(1)) + p\varphi(\sigma_n)(1 + o_n(1)) + o_n(1)\},$$

where $\lim_{n\to\infty} o_n(1) = 0$. Since $p > 1$, we have (for sufficiently large N) $|F(\sigma_n + i\theta)| > 1$. Consequently for $\sigma > \sigma_N$ there exists a point $\sigma + i\tau(\sigma)$, $|\theta - \tau| < s(\sigma)$, such that $|F(\sigma + i\tau)| > 1$; that is

$$|f(\sigma + i\tau(\sigma))| > \exp\{-p\varphi(\sigma)(1 + o_\sigma(1)) + o_\sigma(1)\}.$$

From the fact that $p \equiv p(\sigma_1) = 1 + o(1)$ as $\sigma_1 \longrightarrow \infty$, we conclude that for some $\bar{\sigma}_N$ we have

$$|f(\sigma + i\tau(\sigma))| > \exp\{-\varphi(\sigma)(1 + o(1))\}, \quad \sigma > \bar{\sigma}_N. \tag{2}$$

Now assertion I of Theorem 1 follows from the next lemma.

LEMMA 1. *Suppose that* $\log|f(\sigma + i\tau)| \leqslant \varphi(\sigma)$, $|\tau - \theta| < 3\pi/\psi'$, *and for large values of* σ *there exists* $\tau(\sigma)$, $|\theta - \tau(\sigma)| \leqslant \pi/\psi'(\sigma)$, *such that*

$$\log|f(\sigma + i\tau(\sigma))| \geqslant -\varphi(\sigma)(1 + o(1)).$$

Then for the same σ *we have*

$$\log|f(\sigma + i\theta)| \geqslant -\varphi(\sigma)(A + o(1)),$$

where

$$A = 1 + \min_{u>0} 2\left\{\frac{1}{u} e^{\pi(u+1)} + \frac{1}{u}\right\}.$$

PROOF OF THE LEMMA. The Carathéodory inequality, applied to the function $h = \log f$ in the disk $|t - \zeta| \leqslant r < R$, $\zeta = \sigma + i\tau(\sigma)$, gives us

$$\begin{aligned} |\operatorname{Re} h(t)| &\leqslant |\operatorname{Re} h(\zeta)| + \frac{2r}{R-r}\Big\{\max_{|w-\zeta|\leqslant R} \operatorname{Re} h(w) - \operatorname{Re} h(\zeta)\Big\} \\ &\leqslant \frac{R+r}{R-r}|\operatorname{Re} h(\zeta)| + \frac{2r}{R-r}\varphi(\sigma + R) \leqslant \frac{R+r}{R-r}\varphi(\sigma)(1 + o(1)) + \frac{2r}{R-r}\varphi(\sigma + R). \end{aligned} \tag{3}$$

We set $R = lr = c/\psi'(\sigma + R)$, $l > 1$ (here, as always, $\psi = \log\varphi$ and $\psi' = \pi/s$), and we note that

$$\varphi(\sigma + R) = \varphi(\sigma)\exp\left\{\psi'(\xi)\frac{c}{\psi'(\sigma + R)}\right\} \leqslant \varphi(\sigma)e^c.$$

Thus

$$|\operatorname{Re} h(t)| \leqslant \frac{R+r+2re^c}{R-r}(1+o(1))\,\varphi(\sigma) = \frac{2e^c+l+1}{l-1}(1+o(1))\,\varphi(\sigma). \quad (4)$$

If $c/l > \pi$, then

$$|\sigma + i\theta - \zeta| = |\theta - \tau(\sigma)| \leqslant s(\sigma) = \frac{\pi}{\psi'(\sigma)} < r = \frac{c}{l\psi'(\sigma+R)}.$$

In fact,

$$\frac{\psi'(\sigma+R)}{\psi'(\sigma)} = \exp\{\log\psi'(\sigma+R) - \log\psi'(\sigma)\}$$
$$= \exp R\frac{\psi''(\xi)}{\psi'(\xi)} \leqslant \exp c\,\frac{\psi''(\xi)}{\psi'(\xi)^2},\quad \sigma \leqslant \xi \leqslant \sigma + R,$$

and by the hypothesis of the theorem

$$\lim_{\sigma\to\infty} \psi'(\sigma+R)/\psi'(\sigma) = 1.$$

Therefore inequality (4) is valid, in particular, for $t = \sigma + i\theta$, and so

$$|\log|f(\sigma+i\theta)|| \leqslant \frac{2e^{(\pi+o(1))l}+l+1}{l-1}\varphi(\sigma).$$

If we set

$$A = 1 + \min_{u>0} 2\left\{\frac{1}{u}e^{\pi(u+1)} + \frac{1}{u}\right\},$$

we arrive at the required result:

$$|f(\sigma+i\theta)| \geqslant \exp(-(A+o(1))\,\varphi(\sigma)) = \lambda_\Pi(\sigma)^{-A+o(1)},\quad \sigma > \bar\sigma_N. \quad \bullet$$

We note, finally, that the constant $\bar\sigma_N = \bar\sigma_N(\theta)$ is bounded with respect to θ on each segment $[-\pi/2+\epsilon, \pi/2-\epsilon]$, since it depends only on the numbers σ_θ and σ_N which are chosen in the proof. The required property for $\sigma_N = \sigma_N(\theta)$ was mentioned in the proof of Theorem 1 in §1.2 (see footnote 9 there).

II. In the case $1 < \lim_{t\to\infty}\psi'(t) \overset{\text{def}}{=} \psi'_\infty < \infty$, in order to obtain an inequality analogous to (2) it is convenient to apply the Carleman formula at once, and then to use an estimate of the type (3) repeatedly. This is done in Lemmas 2 and 3.

LEMMA 2. *Suppose that* $f \in a(\Pi, \lambda_\Pi)$, $\psi = \log\log\lambda_\Pi$, *and the limit* $\lim_{t\to\infty}\psi'(t) = \psi'_\infty < +\infty$ *exists. Suppose also that* $0 < \epsilon < \delta < \pi/2$, Δ *is a positive function on* $[0,\infty)$, $\Delta \geqslant \varphi \equiv \log\lambda_\Pi$, *and that for all sufficiently large* σ *there exists* $\tau = \tau(\sigma)$, $\tau \in I_\delta$, *such that*

$$\log|f(\sigma + i\tau(\sigma))| \geqslant -\Delta(\sigma).$$

Then for an arbitrarily small $\bar\epsilon > 0$ *and for large* σ *we have*

$$\log|f(\sigma+i\tau)| \geqslant -e^{\frac{2\pi}{\varepsilon}-4}\Delta(\sigma) - e^{\frac{2\pi}{\varepsilon}+(\psi'_\infty+\bar\varepsilon)\varepsilon-4}\varphi(\sigma),\quad \tau \in I_\varepsilon.$$

PROOF OF THE LEMMA. We apply the Carathéodory inequality to the function $\log f$ in the disk $\{t\colon |t-\zeta| \leqslant r < R\}$, where $\zeta = \sigma + i\tau(\sigma)$ and $R \leqslant \delta$:

$$
\begin{aligned}
|\log|f(t)|| &\leqslant \frac{R+r}{R-r}|\log|f(\zeta)|| + \frac{2r}{R-r}\max_{|w-\zeta|\leqslant R}\log|f(w)| \\
&\leqslant \frac{R+r}{R-r}\Delta(\sigma) + \frac{2r}{R-r}\varphi(\sigma+R).
\end{aligned} \tag{5}
$$

We estimate $\varphi(\sigma+R)$:

$$\varphi(\sigma+R) = \varphi(\sigma)\exp\{\psi'(\xi)R\} \leqslant \varphi(\sigma)\exp\{(\psi'_\infty+\bar{\varepsilon})R\}, \tag{6}$$

where $\bar{\epsilon}$ is an arbitrary positive constant and σ is sufficiently large. We set $R = lr$, $l > 1$, and $R - r = \epsilon$. Then $r = \epsilon/(l-1)$. If we apply (5) and (6) successively with centers at the points $\zeta = \sigma + i\tau(\sigma)$, $\sigma + i(\tau \pm r)$, $\sigma + i(\tau \pm 2r)$, . . . , then after n steps we find that for $\tau \in [\tau(\sigma) - nr, \tau(\sigma) + nr]$ we have

$$
\begin{aligned}
&|\log|f(\sigma+i\tau)|| \\
&\leqslant \left(\frac{l+1}{l-1}\right)^n \Delta(\sigma) + \frac{2}{l-1}e^{(\psi'_\infty+\bar{\varepsilon})R}\left\{\left(\frac{l+1}{l-1}\right)^{n-1} + \dots + 1\right\}\varphi(\sigma) \\
&= \left(\frac{l+1}{l-1}\right)^n \Delta(\sigma) + e^{(\psi'_\infty+\bar{\varepsilon})R}\left\{\left(\frac{l+1}{l-1}\right)^n - 1\right\}\varphi(\sigma).
\end{aligned} \tag{7}
$$

This process terminates when the disk of radius $R = l\epsilon/(l-1)$ with center at $\sigma + i[\tau(\sigma) \pm nr]$ no longer fits in the strip Π; that is, when the distance from $\sigma + i[\tau(\sigma) \pm nr]$ to the boundary of Π becomes less than R. Therefore after no more than

$$n = [(\pi - 2\varepsilon)/r] + 1 = [(l-1)(\pi - 2\varepsilon)/\varepsilon] + 1$$

steps the inequality (7) is established for all τ in $[-\pi/2 + R, \pi/2 - R]$. If we take the limit as $l \longrightarrow \infty$, we obtain

$$|\log|f(\sigma+i\tau)|| \leqslant e^{\frac{2\pi}{\varepsilon}-4}\Delta(\sigma) + e^{(\psi'_\infty+\bar{\varepsilon})\varepsilon}\left(e^{\frac{2\pi}{\varepsilon}-4} - 1\right)\varphi(\sigma),$$

where $\tau \in I_\epsilon$. ●

LEMMA 3. *Suppose that $f \in a(\overline{\Pi}, \lambda_\Pi)$, $\varphi = \log \lambda_\Pi$, $c(f) > \operatorname{Car} f$ and*

$$\Delta(\sigma) = e^\sigma\left(c(f) + \int_0^\sigma \varphi(t)(e^{-2t} - e^{-2\sigma})e^t\,dt\right), \quad \sigma > 0.$$

Define a set $E_{\sigma,K}$ $(K > 1/2)$ by the formula

$$E_{\sigma,K} = \left\{\tau : \tau \in \left(-\frac{\pi}{2}, \frac{\pi}{2}\right), \ \log|f(\sigma+i\tau)| \geqslant -K\Delta(\sigma)\right\}.$$

Then

$$\sin\frac{1}{2}|E_{\sigma,K}| \geqslant \frac{\left(K - \frac{1}{2}\right)\Delta(\sigma)}{K\Delta(\sigma) + \varphi(\sigma)}.$$

PROOF OF THE LEMMA. By the Carleman formula (Lemma 1 in §1.2) we have

$$\int_{-\pi/2}^{\pi/2} \log|f(\sigma+i\tau)|\cos\tau d\tau$$

$$> -e^{\sigma}\left\{c(f)+\frac{1}{2}\int_0^{\sigma}\left(\log\left|f\left(t+i\frac{\pi}{2}\right)f\left(t-\frac{i\pi}{2}\right)\right|\right)(e^{-2t}-e^{-2\sigma})e^t dt\right\}$$

$$\geqslant -e^{\sigma}\left\{c(f)+\int_0^{\sigma}\varphi(t)(e^{-2t}-e^{-2\sigma})e^t dt\right\} = -\Delta(\sigma).$$

We find an upper estimate for the left side of the inequality, using the definition of the set $E_{\sigma,K}$ and the membership statement $f\in a(\Pi,\lambda_\Pi)$:

$$\int_{-\pi/2}^{\pi/2}\log|f|\cos\tau d\tau = \int_{E_{\sigma,K}} + \int_{CE_{\sigma,K}} \leqslant \varphi(\sigma)\int_{E_{\sigma,K}}\cos\tau d\tau - K\Delta(\sigma)\int_{CE_{\sigma,K}}\cos\tau d\tau$$

$$= \varphi(\sigma)\int_{E_{\sigma,K}}\cos\tau d\tau - 2K\Delta(\sigma) + K\Delta(\sigma)\int_{E_{\sigma,K}}\cos\tau d\tau$$

$$= \{\varphi(\sigma)+K\Delta(\sigma)\}\int_{E_{\sigma,K}}\cos\tau d\tau - 2K\Delta(\sigma).$$

Therefore

$$2\sin(|E_{\sigma,K}|/2) = 2\int_0^{|E|/2}\cos\tau d\tau \geqslant \int_{E_{\sigma,K}}\cos\tau d\tau \geqslant \frac{(2K-1)\Delta(\sigma)}{K\Delta(\sigma)+\varphi(\sigma)}. \bullet$$

Returning to the proof of the theorem, we use Lemma 1 in §1.1 to calculate $\Delta(\sigma)$:

$$\Delta(\sigma) = c(f)e^{\sigma} + e^{\sigma}\int_0^{\sigma}\varphi(t)e^{-t}dt - e^{-\sigma}\int_0^{\sigma}\varphi(t)e^t dt$$

$$= c(f)e^{\sigma} + (1+o(1))\frac{1}{\psi'_\infty - 1}\varphi(\sigma) - (1+o(1))\frac{1}{\psi'_\infty+1}\varphi(\sigma)$$

$$\sim \frac{2}{\psi'^2_\infty - 1}\varphi(\sigma), \quad \sigma\to+\infty.$$

We set $\mathfrak{S} = 2/(\psi'^2_\infty - 1)$, $c = \mathfrak{S}K$, and choose $\alpha > 0$. Then

$$\frac{|E_{\sigma,K}|}{2} \geqslant \sin\frac{|E_{\sigma,K}|}{2} \geqslant \left(K-\frac{1}{2}\right)(1-\alpha)\frac{\mathcal{E}}{\mathcal{E}K+1}, \quad \sigma > \sigma(\alpha),$$

and for $\tau\in E_{\sigma,K}$ we have

$$\log|f(\sigma+i\tau)| \geqslant -c(1+\alpha)\varphi(\sigma), \quad \sigma>\sigma(\alpha).$$

If we apply Lemma 2 with $\delta < (K-1/2)\mathfrak{S}/(\mathfrak{S}K+1)$, we obtain

$$\log|f(\sigma+i\tau)|>-\varphi(\sigma)e^{\frac{2\pi}{\varepsilon}-4}\left(c(1+\alpha)+e^{(\psi'_\infty+\bar{\varepsilon})\varepsilon}\right)$$

for $0<\epsilon<\delta$, $\sigma>\sigma(\epsilon,\alpha)$ and $\tau\in[-\pi/2+\epsilon,\pi/2-\epsilon]$, $\bar{\epsilon}=\bar{\epsilon}(\sigma)=o(1)$. Thus for an arbitrary $\beta>0$ there exist numbers $\sigma(\beta,\epsilon)>0$ and $\delta(\beta)>0$ such that

$$\log|f(\sigma+i\tau)|\geqslant-\varphi(\sigma)e^{\frac{2\pi}{\varepsilon}-4}\left\{\frac{1}{\psi'^2_\infty-1}+e^{\psi'_\infty\varepsilon}+\beta\right\}$$

for

$$\tau\in\left[-\frac{\pi}{2}+\varepsilon,\ \frac{\pi}{2}-\varepsilon\right],\ 0<\varepsilon<\delta(\beta),\ \sigma>\sigma(\beta,\varepsilon). \bullet$$

3°. The proof of Theorem 2 uses the same methods as the proof of Theorem 1.

I. In the case $\lim\psi'=+\infty$ we again consider the strip $\Omega_0=\{\sigma+i\tau\colon|\tau|<s(\sigma)\}$, where $s=\pi/\psi'$, and we define functions G, F_1, g and F by the same formulas (replacing f by $f_1=f\circ p^{-1}$, where $p(\zeta)=\log(1/(1-\zeta))$) as in the proof of Theorem 1. In order to estimate $|f(\sigma+i\tau)|$ for $\sigma+i\tau\in\Omega'_0\cap\Pi_1$ ($\Pi_1=p(\mathrm{D})$, $\Omega'_0=\{\sigma+3i\tau\colon\sigma+i\tau\in\Omega\}$), we use the estimates in Lemma 4 of §1.2. We obtain

$$\begin{gathered}\log|f_1(\sigma+i\tau)|\leqslant\varphi\left(\sigma+\left(\frac{1}{2}+\varepsilon\right)s(\sigma)^2\right)\leqslant\varphi(\sigma)\exp\beta(\sigma),\\ \sigma+i\tau\in\Omega'_0\cap\Pi_1,\end{gathered}\tag{8}$$

where $\beta(\sigma)=o(1)$ as $\sigma\longrightarrow+\infty$. Consequently

$$\log|F_1(x+iy)|\leqslant\varphi(x)e^{\beta(x)}-p\varphi(x)e^{r\pm(x)}$$

for $x+iy\in\partial\Omega_0$, $x>\sigma_0$ (here we use the notation in the proof of Theorem 1). If we choose $\sigma_1>\sigma_0$ and $p=\max_{x\geqslant\sigma_1}e^{r(x)+\beta(x)}$ and use the same arguments as in the proof of Theorem 1, we arrive at inequality (2) (with f replaced by f_1). Now the required result is obtained by applying Lemma 1.

II. In the case $1<\lim_{t\to\infty}\psi'\equiv\psi'_\infty<+\infty$ we consider the function f_2, $f_2(w)=f_1(\gamma w)$, $w\in\Pi\cap\{w\colon\operatorname{Re}w>w_0\}$, where, as before, $f_1=f\circ p^{-1}$ and γ is a number such that $1>\gamma>1/\psi'_\infty$. We use Lemma 3 in §1.2 in order to estimate $\max_{\operatorname{Re}w=\sigma}|f_2(w)|$, $\sigma>w_0$. Clearly we have

$$|f_2(\sigma+i\tau)|=|f_1(\gamma\sigma+i\gamma\tau)|\leqslant\exp\varphi(t),\quad\tau\in I,$$

where $t=\gamma\sigma+d$ is a number such that $\gamma\sigma+i\gamma\tau\in\operatorname{Int}O_r$; $t=\log(1/(1-r))$. We estimate the increment d by using Lemma 3 in §1.2. To do this, we note that the smallest admissible value of t is determined by the equation $\gamma\sigma+i\gamma\pi/2\in O_r$; that is,

$$\gamma\frac{\pi}{2}=\arccos\frac{1}{2}[e^{-d}(2-e^{-\gamma\sigma-d})+e^{-\gamma\sigma}].$$

Thus it is clear that $d=d(\sigma)$ is bounded as $\sigma\longrightarrow+\infty$, and

$$\cos\gamma\frac{\pi}{2}=e^{-d}(1+o(1)),\quad d=\log\frac{1}{\cos\frac{\gamma\pi}{2}}+o(1),\quad\sigma\to+\infty.$$

Consequently

$$|f_2(\sigma+i\tau)|\leqslant \exp\varphi(\gamma\sigma+d),\quad \tau\in I,$$
$$\log|f_2(\sigma+i\tau)|\leqslant\varphi(\gamma\sigma+d)=\varphi(\gamma\sigma)\exp d\psi'(\xi)$$
$$\leqslant\varphi(\gamma\sigma)\exp\left\{\psi'_\infty\log\frac{1}{\cos\frac{\gamma\pi}{2}}(1+o(1))\right\}=\varphi(\gamma\sigma)\left(\frac{1}{\cos\frac{\gamma\pi}{2}}\right)^{\psi'_\infty}(1+o(1)),$$

where $\tau\in I$ and $\sigma\longrightarrow\infty$. Since

$$\{\psi(\gamma\sigma)\}'=\gamma\psi'(\gamma\sigma)\to\gamma\psi'_\infty>1$$

as $\sigma\longrightarrow\infty$, we can apply Theorem 1. This gives us

$$\log|f_2(\sigma+i\tau)|\geqslant-B\varphi(\gamma\sigma)\left(\frac{1}{\cos\frac{\gamma\pi}{2}}\right)^{\psi'_\infty},\quad \tau\in I_\varepsilon,\quad \sigma>\sigma(\varepsilon,\beta),$$

where

$$B=B_{\varepsilon,\beta}=e^{\frac{2\pi}{\varepsilon}}\left\{\frac{e^{-4}}{\gamma^2\psi'^2_\infty-1}+e^{-4+\gamma\psi'_\infty\varepsilon}+\beta\right\},\quad 0<\varepsilon<\delta(\beta).$$

If we choose β, ϵ and γ in a suitable fashion (see the formulas at the end of the proof of Theorem 1) we obtain

$$\log|f_1(\gamma\sigma)|=\log|f_2(\sigma)|\geqslant-B_0\varphi(\gamma\sigma),$$
$$\log|f_1(\sigma)|\geqslant-B_0\varphi(\sigma),\quad \sigma>\sigma_0,$$

where $B_0=(4\pi+e^{\sqrt{2}}+1)e^{5/11}$ if $\psi'_\infty>2$, and

$$B_0=\frac{3+2\sqrt{2}}{\psi'_\infty-1}\left(\frac{2\pi+1}{\psi'_\infty-1}+e^{\sqrt{2}}+2\pi\right),$$

if $\psi'_\infty\leqslant 2$. Then we return to the function $f=f_1\circ p$. ●

4°. *Proof of Theorem* 3. We apply Lemma 3 to the function f and use the notation of that lemma. We arrive at the inequality

$$\log|f(\sigma+i\tau)|\geqslant-K\Delta(\sigma),\quad \tau\in E_{\sigma,K},$$

where $K>1/2$ and

$$\sin\frac{1}{2}|E_{\sigma,K}|\geqslant\frac{\left(K-\frac{1}{2}\right)\Delta(\sigma)}{K\Delta(\sigma)+\varphi(\sigma)}\overset{\text{def}}{=}\delta(\sigma).$$

LEMMA 4. *Let a be a differentiable function on* $[0,\infty)$ *with* $a'(t)=o(1)$ *as* $t\longrightarrow+\infty$.

1) $$\exp a=o\left(\int_0^\sigma\exp a\,dt\right),\quad \sigma\to+\infty;$$

$$2)\qquad \int_0^\sigma \exp(a+2t)\,dt = o\left(\int_0^\sigma \exp a\,dt \exp 2\sigma\right), \quad \sigma \to +\infty.$$

For if $c > 0$ is arbitrary, we have

$$\int_{\sigma-c}^{c} e^a dt = ce^{a(\xi)} = ce^{a(\sigma)+a'(\xi)(\xi-\sigma)}$$
$$\geqslant c \exp\{a(\sigma) - c \max_{u \geqslant \sigma - c} |a'(u)|\} \sim c \exp a(\sigma), \quad \sigma \to +\infty.$$

Therefore

$$\overline{\lim_{\sigma\to\infty}}\, e^{a(\sigma)} \Big/ \int_0^\sigma e^a dt \leqslant \overline{\lim_{\sigma\to\infty}}\, e^{a(\sigma)} \Big/ \int_{\sigma-c}^{\sigma} e^a dt \leqslant 1/c,$$

and assertion 1) is proved. Analogously we can prove 2):

$$e^{-2\sigma} \int_0^\sigma e^{a+2t} dt = e^{-2\sigma} \int_0^{\sigma-c} e^{a+2t} dt + e^{-2\sigma} \int_{\sigma-c}^{\sigma} e^{a+2t} dt$$
$$\leqslant e^{-2\sigma+2(\sigma-c)} \int_0^\sigma e^a dt + (\max_{\sigma-c\leqslant t\leqslant\sigma} e^{a(t)})\, e^{-2\sigma} \frac{1}{2} (e^{2\sigma} - e^{2\sigma-2c})$$
$$\leqslant e^{-2c} \int_0^\sigma e^a dt + \frac{1}{2}(1 - e^{-2c}) \exp\{a(\sigma) + c \max_{\sigma-c\leqslant u\leqslant\sigma} |a'(u)|\},$$

and thus

$$\overline{\lim_{\sigma\to\infty}}\, e^{-2\sigma} \int_0^\sigma e^{a+2t} dt \Big/ \int_0^\sigma e^a dt \leqslant e^{-2c}. \quad \bullet$$

If we set $\varphi(t)e^{-t} = e^{a(t)}$ (that is, $a(t) = \psi(t) - t$), we see that

$$\int_0^\infty \varphi(t)(e^{-2t} - e^{-2\sigma}) e^t dt \sim \int_0^\sigma \varphi(t) e^{-t} dt, \quad \sigma \to +\infty,$$

and

$$\Delta(\sigma) \sim e^\sigma \int_0^\sigma \varphi(t) e^{-t} dt, \quad \text{if} \quad \int_0^\infty \varphi e^{-t} dt = +\infty,$$

or

$$\Delta(\sigma) \sim c_1(f) e^\sigma, \quad \text{if} \quad c_1(f) = c(f) + \int_0^\infty \varphi e^{-t} dt < +\infty.$$

By Lemma 4 we see that in both cases we have $\pm\varphi(\sigma) = o(\Delta(\sigma))$ as $\sigma \to +\infty$. Consequently

$$\delta(\sigma) = 1 - \frac{1}{2K} + o(1), \quad \sigma \to +\infty.$$

Now we can use Lemma 1 with an arbitrary δ, $0 < \delta < 1 - 1/2K$ (and with Δ replaced by $K\Delta$). We obtain

$$\log|f(\sigma + i\tau)| \geqslant -e^{\frac{2\pi}{\varepsilon}-4} K\Delta(\sigma) - e^{\frac{2\pi}{\varepsilon}+2\varepsilon-4}\varphi(\sigma)$$

for $\sigma > \sigma_\epsilon$, $\tau \in I_\epsilon$ and $0 < \epsilon < \delta$. Let $K = 0.9$. Since $\varphi = o(\Delta)$ as $\sigma \to +\infty$, we have

$$\log|f(\sigma + i\tau)| \geqslant -0.99 e^{\frac{2\pi}{\varepsilon}-4} \Delta(\sigma) \tag{9}$$

for $\sigma > \sigma(\epsilon)$, $\tau \in I_\epsilon$ and $0 < \epsilon < 4/9$. If we use the asymptotic equation

$$\Delta(\sigma) \sim \bar{B} e^{\sigma} \int_0^{\sigma} \varphi(t) e^{-t} dt, \quad \sigma \to \infty,$$

(where $\bar{B} = 1$ for $\int_0^\infty \varphi e^{-t}\, dt = \infty$ and $\bar{B} = 1 + c(f)/\int_0^\infty \varphi e^{-t}\, dt$ otherwise) and in (9) enlarge 0.99 to 1 when we switch to the asymptotics,(18) we arrive at the required assertion. ●

REMARK. The condition $f \in a(\overline{\Pi}, \lambda_\Pi)$ is only needed in order to enable us to apply the Carleman formula. Therefore this condition can be replaced by $f \in a(\Pi, \lambda_\Pi)$ and $\inf_{\sigma \leqslant \Sigma, \tau \in I} |f(\sigma + i\tau)| > 0$ for an arbitrary $\Sigma > 0$.

5°. The proof of Theorem 4 is based on using a conformal mapping which is different from the mapping p: $\zeta \to \log(1 - \zeta)^{-1}$, $|\zeta| < 1$, which we applied in the preceding theorems. This is because $p(\mathbf{D}) \neq \Pi$, and in Theorem 4 we have to consider a function which is "transplanted" out of the disk near the boundary of its domain of definition. On the other hand, the choice of the mapping p in the preceding theorems did not lead to additional technical difficulties and can be justified by the traditional "naturalness" of the representation $\lambda = \exp(\varphi \circ p)$ (see also [126], [153] and [154]). Later we will use the mapping q defined by

$$q(\zeta) = \log \frac{1}{2} \cdot \frac{1+\zeta}{1-\zeta}, \quad |\zeta| < 1,$$

and the representation $\lambda = \exp(\varphi_* \circ q)$; that is,

$$\varphi_*(t) = \log \lambda\left(\left(e^t - \frac{1}{2}\right)\Big/\left(e^t + \frac{1}{2}\right)\right), \qquad t \geqslant \log \frac{1}{2}.$$

For this mapping we have the following obvious analogue of Lemma 3 in §1.2.

LEMMA 5. *Let O_r^* be the image of the circle $\{\zeta\colon |\zeta| = r\}$, $0 \leqslant r < 1$, under the mapping q. If $\tau = \tau_t(\sigma)$ is the equation of the upper half of O_r^* and*

$$t = \log \frac{1}{2} \cdot \frac{1+r}{1-r},$$

(18) And replace $c(f)$ by Car f in the case $\int_0^\infty \varphi e^{-t}\, dt < +\infty$; it is easy to see that $c(f)$ and Car f are close to each other, since in Lemma 3 we have $c(f)$ equal to an arbitrary number larger than Car f.

then

$$\tau(\sigma) = \tau_t(\sigma) = \arccos \frac{e^{\sigma+\log 2} + e^{-\sigma-\log 2}}{e^{t+\log 2} + e^{-t-\log 2}} = \frac{\pi}{2} - \arcsin e^{\sigma-t}\left(\frac{1+e^{-2(\sigma+\log 2)}}{1+e^{-2(t+\log 2)}}\right),$$

where $-t - 2\log 2 \leqslant \sigma \leqslant t$.

Let $f \in a(\mathbf{D}, \lambda)$ and $f_1 = f \circ q^{-1}$. We define a function s by means of the equations

$$\frac{1}{s(x)} = \frac{2}{\pi}\left(1 + e^{\frac{1}{2}a(x)} \bigg/ \int_0^x e^{\frac{1}{2}a(t)}\,dt\right), \quad x \geqslant 1,$$

$$s(x) = s(0), \quad x < 0,$$

and on the interval $[0, 1]$ we choose s such that $s \in C^2(\mathbf{R})$, $0 < s < \pi/2$, $s(0) = \log 3/2$ and

$$\frac{\pi}{2}\int_0^1 \frac{dt}{s(t)} = 1 + \log \int_0^1 e^{\frac{1}{2}a(t)}\,dt,$$

where $a(t) = \psi(t) - t$. Let d be the positive function on $[1, \infty)$ defined by the equation $s(\sigma) = \tau_{\sigma+d(\sigma)}(\sigma)$, and let $\tau_{\sigma+d(\sigma)}(\,\cdot\,)$ be the function in Lemma 5, constructed for the value $t = \sigma + d(\sigma)$; let $\Phi(\sigma) = \varphi_*(\sigma + d(\sigma))$ and let $\Omega = \{\sigma + i\tau\colon |\tau| < s(\sigma)\}$ be a curvilinear strip, $\Omega \subset \Pi$, and let $f_1 = f \circ q^{-1}$. Then (see Figure 2) we have $\log|f_1(\sigma + i\tau)| \leqslant \Phi(\sigma)$, $\sigma + i\tau \in \Omega$. Let Z denote a conformal mapping of Ω onto Π, and let $F = f_1 \circ Z^{-1}$.

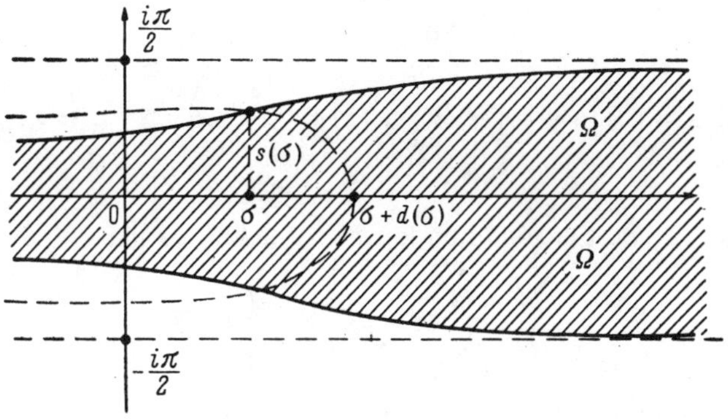

FIGURE 2

Later we will show that

(i) $\lim_{\sigma\to\infty} d(\sigma) = +\infty$,

(ii) $\lim_{t\to\infty} (\Phi_0'/\Phi_0)(t) = 1$, where $\Phi_0 = \Phi \circ k$, $k = j^{-1}$, $j(\sigma) = (\pi/2)\int_0^\sigma (1/s(t))dt$;

(iii) $\Phi_0(x) = (1 + o(1))\max_{y\in I} \Phi(\operatorname{Re} Z^{-1}(x + iy))$ as $x \to +\infty$;

(iv) for the strip Ω (that is, for the function s) the conditions in Warschawski's theorem are satisfied, and therefore

$$\operatorname{Re} Z(\sigma+i\tau)=\frac{\pi}{2}\int_0^\sigma \frac{dt}{s(t)}+r(\sigma,\tau),$$

$r = o(1)$ as $\sigma \longrightarrow +\infty$.

It now follows that

$$\log|F(x+iy)|=\log|f_1(Z^{-1}(x+iy))| \\ \leqslant \Phi(\operatorname{Re} Z^{-1}(x+iy)) \leqslant (1+o(1))\,\Phi_0(x), \quad x\to+\infty,$$

and by Theorem 3[19] (with φ replaced by Φ_0) we have

$$\log|F(x+iy)| \geqslant -(B_{2/5}-\varepsilon)\,e^x\int_0^x \Phi_0(t)\,e^{-t}dt, \quad x>x_{2/5}, \quad y\in I_{2/5},$$

where the constant $B_{2/5}$ is defined as in Theorem 3, and ϵ is sufficiently small. Returning to the function f_1, we obtain

$$\log|f_1(\sigma)|=\log|F(j(\sigma)+o(1))| \\ \geqslant -(B_{2/5}-\varepsilon)\,e^{j(\sigma)+o(1)}\int_0^{j(\sigma)+o(1)} \Phi_0(t)\,e^{-t}dt, \quad \sigma>\sigma_0.$$

We extend this inequality, noting that by Lemma 4 we have

$$\int_{j(\sigma)}^{j(\sigma)+o(1)} \Phi_0(t)\,e^{-t}dt=o(1)\,\Phi_0(j(\sigma))\,e^{-j(\sigma)}\exp\left\{o(1)\left(\frac{\Phi_0'(\xi)}{\Phi_0(\xi)}-1\right)\right\} \\ =o\left(\int_0^{j(\sigma)} \Phi_0(t)\,e^{-t}dt\right), \quad \sigma\to+\infty,$$

and therefore

$$\log|f_1(\sigma)| \geqslant -\left(B_{2/5}-\frac{\varepsilon}{2}\right)e^{j(\sigma)}\int_0^{j(\sigma)} \Phi_0(t)\,e^{-t}dt \\ =-\left(B_{2/5}-\frac{\varepsilon}{2}\right)e^{j(\sigma)}\int_0^{\sigma} \Phi(x)\,e^{-j(x)}j'(x)\,dx, \quad \sigma>\sigma_1.$$

From Lemma 4 and the condition $\lim a' = 0$ it follows that

$$s(x)\sim\frac{\pi}{2}, \quad j'(x)=\frac{\pi}{2}\cdot\frac{1}{s(x)}\sim 1$$

as $x \longrightarrow +\infty$. This gives us

$$\log|f_1(\sigma)| \geqslant -\left(B_{2/5}-\frac{\varepsilon}{3}\right)e^{j(\sigma)}\int_0^{\sigma} \Phi(t)\,e^{-j(t)}dt, \quad \sigma>\sigma_2.$$

By the definition of the functions j and s, for $\sigma \geqslant 1$ we have

---

(19) Theorem 3 is applicable to the function F even though this function might not be analytic on $\overline{\Pi}$ (see the remark after the proof of Theorem 3).

$$j(\sigma)=\frac{\pi}{2}\int_0^{\sigma}\frac{2}{\pi}\left(1+e^{\frac{1}{2}a}\Big/\int_0^t e^{\frac{1}{2}a(x)}\,dx\right)dt=\sigma+\log\int_0^{\sigma}e^{\frac{1}{2}a}\,dt$$

$$+\frac{\pi}{2}\int_0^1\frac{dt}{s(t)}-1-\log\int_0^1 e^{\frac{1}{2}a}\,dt=\sigma+\log\int_0^{\sigma}e^{\frac{1}{2}a}\,dt.$$

Thus for $\sigma>\sigma_2$, $\sigma>1$, we have

$$\log|f_1(\sigma)|\geqslant-\left(B_{2/5}-\frac{\varepsilon}{3}\right)\left(\int_0^{\sigma}e^{\frac{1}{2}a}\,dt\right)e^{\sigma}\int_0^{\sigma}e^{\psi_*(t+d(t))-t}\left(\int_0^t e^{\frac{1}{2}a}\,dx\right)^{-1}dt, \quad (10)$$

where $\psi_*=\log\varphi_*$. By the definition of the functions φ and φ_* we have $\psi_*(t)=\psi(t+O(e^{-t}))=\psi(t)+o(1)$ as $t\longrightarrow+\infty$; this enables us to replace asymptotically in (10) the function ψ_* by ψ, increasing the constant $B_{2/5}-\epsilon/3$. Then we use the definition of d: since by Lemma 5 we have

$$s(\sigma)=\frac{\pi}{2}-\arcsin e^{-d(\sigma)}(1+o(1)),$$

it follows that

$$\frac{1}{s(\sigma)}=\frac{2}{\pi}\left(1-\frac{2}{\pi}e^{-d(\sigma)}(1+o(1))\right)^{-1}=\frac{2}{\pi}\left(1+\frac{2}{\pi}e^{-d(\sigma)}(1+o(1))\right)$$

and therefore

$$e^{\frac{1}{2}a(\sigma)}\Big/\int_0^{\sigma}e^{\frac{1}{2}a}\,dt=\frac{2}{\pi}e^{-d(\sigma)}(1+o(1)).$$

We estimate the right side of (10) from below:

$$\log|f_1(\sigma)|\geqslant-\left(B_{2/5}-\frac{\varepsilon}{4}\right)\left(\int_0^{\sigma}e^{\frac{1}{2}a}\,dt\right)e^{\sigma}\int_0^{\sigma}e^{a(t+d(t))}\frac{2}{\pi}e^{-\frac{1}{2}a(t)}\,dt, \quad (11)$$

$$\sigma>\sigma_3.$$

We now show that the conditions $a'>0$, $a'\downarrow 0$, give us the asymptotic equation

$$a(t+d(t))=a(t)+o(1),\quad t\to+\infty.$$

In fact, from the condition $a'\downarrow 0$ we have

$$|a(t+d(t))-a(t)|=|a'(\xi)|\,d(t)\leqslant|a'(t)|\,d(t),$$

and it suffices to show that $\exp d=\exp(o(1/a'(\sigma)))$ or

$$e^{-\frac{1}{2}a(\sigma)}\int_0^{\sigma}e^{\frac{1}{2}a}\,dt=\exp\left(o\left(1/a'(\sigma)\right)\right),\quad \sigma\to+\infty. \quad (12)$$

Since $a' \downarrow 0$, we have $a'' < 0$ and

$$\int_0^\sigma e^{\frac{1}{2}a}dt = \frac{2}{a'(\sigma)} e^{\frac{1}{2}a(\sigma)} + 2\int_0^\sigma e^{\frac{1}{2}a}\frac{a''}{a'^2}\,dt < \frac{2}{a'(\sigma)} e^{\frac{1}{2}a(\sigma)} + \text{const},$$

which implies (12).

Under the contrary condition $a' \leqslant 0$ we have $a(t + d(t)) \leqslant a(t)$. In both cases, if we extend (11) we obtain

$$\log|f_1(\sigma)| \geqslant -\frac{2}{\pi}\left(B_{2/5} - \frac{\varepsilon}{5}\right)e^\sigma\left(\int_0^\sigma e^{\frac{1}{2}a}dt\right)^2, \quad \sigma > \sigma_4.$$

In switching to the function f (with the help of q) we use the following elementary lemma.

LEMMA 6. *If* $\lim_{t\to\infty} a' = a'_\infty$ *and* $0 \leqslant a'_\infty < \infty$, *then*

$$\int_0^\sigma e^a dt \sim \int_0^{\sigma+o(1)} e^a dt, \quad \sigma \to +\infty.$$

Thus

$$\log|f(r)| \geqslant -\frac{2}{\pi}\left(B_{2/5} - \frac{\varepsilon}{5}\right)e^{\log\frac{1}{1-r}+o(1)}\left(\int_0^{\log\frac{1}{1-r}+o(1)} e^{\frac{1}{2}a}dt\right)^2$$

$$\geqslant -\frac{2}{\pi}B_{2/5}\frac{1}{1-r}\left(\int_0^{\log\frac{1}{1-r}} e^{\frac{1}{2}a(t)}dt\right)^2, \quad r > r_0.$$

In order to complete the proof, we still have to verify properties (i)–(iv) of the functions s and d by means of the assumptions in the theorem.

(i) By Lemma 4 we have $\lim_{x\to+\infty} s(x) = \pi/2$, and by Lemma 5 we have

$$s(\sigma) = \frac{\pi}{2} - \arcsin e^{-d(\sigma)}(1 + o(1)),$$

so that $\lim_{\sigma\to\infty} d(\sigma) = +\infty$.

(ii) We set $A(t) = e^{t+\log 2} + e^{-t-\log 2}$, $t \geqslant 0$. Lemma 5 and the definition of the function d give us

$$\cos s(\sigma) = \frac{A(\sigma)}{A(\sigma + d(\sigma))}, \quad \sigma \geqslant 1.$$

We first show that $\lim_{\sigma\to\infty} d'(\sigma) = 0$, and we use this fact in proving (ii). We have

$$-s'\tan s = \frac{A'}{A} - \frac{A'(\sigma + d(\sigma))(1 + d'(\sigma))}{A(\sigma + d(\sigma))},$$

and therefore (since $A'/A = 1 + o(1)$)

$$-s' \tan s = 1 + o(1) - (1 + o(1))(1 + d'). \tag{13}$$

Since $\tan s \sim (\pi/2 - s)^{-1}$ as $\sigma \longrightarrow \infty$, we have $-s' \tan s \sim [\log(\pi/2 - s)]'$, where

$$\frac{\pi}{2} - s(\sigma) = \frac{\pi}{2} e^{\frac{1}{2}a(\sigma)} \Bigg/ \left(e^{\frac{1}{2}a(\sigma)} + \int_0^\sigma e^{\frac{1}{2}a} dt \right).$$

Consequently

$$-s' \tan s \sim \left\{ \frac{1}{2} a' - \frac{\frac{1}{2} a' e^{\frac{1}{2}a} + e^{\frac{1}{2}a}}{e^{\frac{1}{2}a} + \int_0^\sigma e^{\frac{1}{2}a} dt} \right\} = o(1)$$

by virtue of Lemma 4. Equation (13) now gives us $d' = o(1)$.

We set $\Psi = \log \Phi$ and $\Psi_0 = \log \Phi_0$, and we verify (ii): $\Psi_0' = 1 + o(1)$ as $\sigma \longrightarrow +\infty$. We have

$$\Psi_0 = \Psi \circ k, \quad \Psi_0' = (\Psi' \circ k) k', \quad \Psi_0' \circ j = \Psi' \frac{1}{j'}.$$

Since $j' = (\pi/2) \cdot (1/s) = 1 + o(1)$ and

$$\Psi'(\sigma) = \psi_*'(\sigma + d(\sigma))(1 + d'(\sigma))$$

where $\psi_* = \log \varphi_*$, it remains to show that $\psi_*' = 1 + o(1)$. This equation is clear:

$$\psi_*(t) = \psi\left(t + \log\left(1 + \frac{1}{2} e^{-t}\right)\right),$$

and

$$\psi_*'(t) = \psi'\left(t + \log\left(1 + \frac{1}{2} e^{-t}\right)\right) \left\{1 - \frac{1}{2} e^{-t}\left(1 + \frac{1}{2} e^{-t}\right)^{-1}\right\} = 1 + o(1)$$

by hypothesis.

(iii) Let

$$\max_{y \in I} \Phi(\operatorname{Re} Z^{-1}(x + iy)) = \Phi(\operatorname{Re} Z^{-1}(x + i\bar{y})), \quad Z^{-1}(x + i\bar{y}) = \sigma + i\tau, \quad \sigma = \sigma(x).$$

Then $x = \operatorname{Re} Z(\sigma + i\tau) = f(\sigma) + o(1)$ as $\sigma \longrightarrow +\infty$; that is, as $x \longrightarrow \infty$ (here we use property (iv) of the strip Ω; the proof of this property, to be carried out later, does not use (iii)). Now the asymptotic equation to be proved can be written in the form

$$\Phi(\sigma) \sim \Phi_0(j(\sigma) + o(1)), \quad \sigma \to \infty.$$

Since $k(j(\sigma) + o(1)) = \sigma + k'(\xi)o(1) = \sigma + o(1)$ and $\Phi_0 = \Phi \circ k$, we have

$$\begin{aligned}\Phi_0(j(\sigma) + o(1)) &= \Phi(\sigma + o(1)) = \varphi_*(\sigma + o(1) + d(\sigma + o(1)))\\ &= \varphi_*(\sigma + o(1) + d(\sigma) + o(1)) = \varphi_*(\sigma + d(\sigma))(1 + o(1))\\ &= \Phi(\sigma)(1 + o(1))\end{aligned}$$

as $\sigma \longrightarrow \infty$, where the third inequality is a consequence of $d' = o(1)$ and the fourth involves $\psi_*' = 1 + o(1)$. Assertion (iii) is proved.

(iv) We first note that in the case we are considering ($\lim_{x\to\infty} s = \pi/2$) both conditions in the Warschawski theorem (that is, $\int^\infty s'^2/s < \infty$ and $\lim s' = 0$) follow from the sign-definiteness of s' for large values of the argument and the equation $\lim_{t\to\infty} s' = 0$. By the definition of s we have

$$s' = \frac{\pi}{2}\cdot\frac{e^a - {}^1\!/_2 a' e^{\frac{1}{2}a}\int\limits_0^x e^{\frac{1}{2}a}\,dt}{\left(e^{\frac{1}{2}a} + \int\limits_0^x e^{\frac{1}{2}a}\,dt\right)^2}, \quad x \geqslant 1.$$

From the obvious equation(20)

$$\int\limits_0^x e^{\frac{1}{2}a}\,dt = \frac{2}{a'}e^{\frac{1}{2}a} + 2\int\limits_0^x e^{\frac{1}{2}a}\cdot\frac{a''}{a'^2}\,dt \tag{14}$$

we obtain

$$1 - \frac{1}{2}a' e^{-\frac{1}{2}a}\int\limits_0^x e^{\frac{1}{2}a}\,dt = -a' e^{-\frac{1}{2}a}\int\limits_0^x e^{\frac{1}{2}a}\frac{a''}{a'^2}\,dt.$$

Consequently the sign of s' does not change (for large values of the argument). The second condition ($\lim_{t\to\infty} s' = 0$) follows from Lemma 4. ●

§1.4 Inverse theorems

1°. The estimates obtained in the preceding section are of the form $\log|f| \geqslant -C_\lambda\Delta_\lambda$, where $f \in a(\Pi, \lambda_\Pi)$ or $f \in a(\mathbf{D}, \lambda)$, C_λ is a constant and Δ_λ is a function (minorant) defined in terms of the weight λ. The assertion that such an estimate is precise means that for arbitrary $\overline{C}_\lambda < C_\lambda$ or $\overline{\Delta}_\lambda = o(\Delta_\lambda)$ the corresponding inequality is not true for a suitable choice of f and a suitable choice of (arbitrarily large) values of the argument. A similar converse of the classical Nevanlinna theorem is well known: If $K > 0$ and

$$f_K = \exp\left(-\frac{2K\zeta}{1-\zeta}\right), \quad |\zeta| < 1,$$

then

$$|f_K| \leqslant e^K, \quad f_K(0) = 1 \text{ and } \log|f_K(r)| \leqslant -\frac{2K-\varepsilon}{1-r}, \quad r > r_\varepsilon. \tag{1}$$

This example demonstrates the fact that the estimate of Cartwright is best possible for the class $a(\mathbf{D}, \exp(1/(1-r)^\alpha))$, $0 < \alpha < 1$. There is another example in [124]; there it is shown that the extremal decrease for the class $a(\mathbf{D}, \exp(1/(1-r)))$ is possessed by the function

$$\exp\left(-\frac{\log^2(1-\zeta)^{-1}}{1-\zeta}\right), \quad |\zeta| < 1.$$

(20) We always assume that $a', a'' \not\equiv 0$. Otherwise property (iv) is obtained by a direct calculation.

This simple example shows that growth of the majorant λ increases the difficulty involved in constructing the corresponding examples. Thus in the work [153] of Linden, where certain special weights(21) are considered with $\psi'_\infty = 1$, an extremal example is indicated only for $\log \lambda(r) = (1-r)^{-1}\log_m(1-r)^{-1}$, and a comment is made on the precision of the estimates in the other cases (see [153], p. 60).

In this section, for $1 \leqslant \psi'_\infty < \infty$ we will construct functions in the classes $a(\Pi, \lambda_\Pi)$ and $a(\mathbf{D}, \lambda)$ which realize the infima m_Π and $m_\mathbf{D}$ (to within a constant C_λ) for all sufficiently large values of the argument.(22) For a rapidly growing majorant ($\psi'_\infty = +\infty$) a discussion of an analogous inverse theorem was presented after Theorem 2″ in §1.2. We recall that if $\log \lambda$ is absolutely monotone, we can find a function f in $a(\mathbf{D}, \lambda)$ (or $a(\Pi, \lambda_\Pi)$) which has a decrease on the radius of the order $1/\lambda(r)$, $r \longrightarrow 1-0$, and for weights λ with $1/\psi'$ convex we can only find a function f in the same class for which

$$|f(r)| \leqslant \exp\left(-\frac{\log \lambda(r)}{\log(1-r)^{-1}}\right).$$

Before we state our inverse theorems precisely, we exclude one simple case from the discussion. We are concerned with majorants λ and λ_Π for which $\int^\infty e^{a(t)/2}\,dt < \infty$ and $\int^\infty e^{a(t)}\,dt < \infty$, respectively, where $a(t) = \psi(t) - t$, $t \geqslant 0$. In this case Theorems 3 and 4 in §1.3 give us minorants $\exp(-\text{const}/(1-r))$ and $\exp(-\text{const}\cdot e^\sigma)$ respectively. The example (1) shows that these minorants are best possible (to within const(23)). Therefore the assumption of divergence of the integral $\int^\infty e^{a/2}\,dt$ in Theorem 2 below does not reduce the generality of the result, but it imposes a restriction from below on the growth of the weight λ. By the same principle, we always consider weights with $\psi'_\infty \geqslant 1$, since $\psi'_\infty < 1$ clearly implies that $\int^\infty e^{a(t)}\,dt = \int^\infty e^{\psi(t)-t}\,dt < \infty$.

THEOREM 1. *Suppose that* $\psi = \log\log\lambda_\Pi$, $a(\sigma) = \psi(\sigma) - \sigma$ *and the limit* $\lim_{t\to\infty}\psi' = \psi'_\infty$ *exists, where* $1 < \psi'_\infty < +\infty$. *Suppose that the function* a *is monotone in a neighborhood of* $+\infty$, *and that in the case* $a' > 0$ *it is convex in the same neighborhood. Then for each* $\epsilon > 0$ *there exists a function* $f_\epsilon \in a(\Pi, \lambda_\Pi)$ *such that*

$$|f_\varepsilon(\sigma)| \leqslant \exp\left(-\frac{2(1-\varepsilon)}{\pi} e^\sigma \int_0^\sigma \varphi(t) e^{-t} dt\right), \quad \sigma > \sigma_\varepsilon,$$

in the case $\psi'_\infty = 1$,

$$|f_\varepsilon(\sigma)| \leqslant \exp\left(-\frac{2(1-\varepsilon)}{\pi(\psi'_\infty - 1)}\varphi(\sigma)\right), \quad \sigma > \sigma_\varepsilon,$$

in the case $1 < \psi'_\infty < 1 + 2/\pi$, *and*

$$|f_\varepsilon(\sigma)| \leqslant \exp(-(1-\varepsilon)\varphi(\sigma)), \quad \sigma > \sigma_\varepsilon,$$

in the case $\psi'_\infty \geqslant 1 + 2/\pi$.

---

(21) $\psi = \log\log\lambda_\Pi$, $\psi'_\infty = \lim_{t\to\infty}\psi'(t)$.

(22) And also we will present the proof of Theorem 2 in §1.1.

(23) But the difference between the constants $\int_0^\infty e^{a/2}\,dt$ and $2K$ in (1) can apparently be reduced by adjusting the functions in (1).

THEOREM 2. *Suppose that* $\psi = \log\log\lambda_\Pi$, $a(\sigma) = \psi(\sigma) - \sigma$ *and the limit* $\lim_{t\to\infty}\psi' = \psi'_\infty$ *exists, where* $1 \leqslant \psi'_\infty < +\infty$. *Suppose that* $\int^\infty e^{a/2}\,dt = +\infty$ *and the function* a *is convex in a neighborhood of* $+\infty$, *and in the case* $a' < 0$ *the function* $|a(\sigma)|$ *is comparable to* σ^α *for some* α, $0 < \alpha \leqslant 1/3$. *Then for each* $\epsilon > 0$ *there exists a function* $F_\epsilon \in a(\mathbf{D}, \lambda)$ *such that*

$$\log|F_\varepsilon(r)| \leqslant -\left(\frac{4}{\pi^2} - \varepsilon\right)\frac{1}{1-r}\left(\int_0^r \sqrt{\frac{\log\lambda(x)}{1-x}}\,dx\right)^2, \quad r > r_\varepsilon,$$

in the case $\psi'_\infty = 1$,

$$\log|F_\varepsilon(r)| \leqslant -\frac{8}{\pi^3}\cdot\frac{1}{(\psi'_\infty - 1)^2}\log\lambda(r), \quad r > r_\varepsilon,$$

in the case $1 < \psi'_\infty < 1 + 3/\pi$, *and*

$$\log|F_\varepsilon(r)| \leqslant -(1-\varepsilon)\log\lambda(r), \quad r > r_\varepsilon,$$

in the case $\psi'_\infty \geqslant 1 + 3/\pi$.

REMARK. The conditions imposed on the weight λ in Theorems 1 and 2 are of the same type as those in the preceding sections on the regularity of growth (see also §1.5). The restriction on the speed of growth in Theorem 1 says, roughly speaking, that $\lambda_\Pi(\sigma) \leqslant \exp\exp K\sigma$, $K \geqslant 1$, and in Theorem 2 that

$$\lambda(r) \leqslant \exp\frac{1}{(1-r)^K}, \quad K \geqslant 1, \quad \text{and} \quad \int^1 \left(\frac{\log\lambda(r)}{1-r}\right)^{1/2} dr = +\infty.$$

Comparability of the functions $|a(\sigma)|$ and σ^α $(0 < \alpha \leqslant 1/3)$ for $a' < 0$ in Theorem 2 can be replaced, for example, by the requirement of convexity of the function a^3 (or existence of $\lim_{t\to\infty} a^2(t)a'(t)$).

The proofs of Theorems 1 and 2 are carried out by considering a suitable curvilinear strip $\Omega = \{\sigma + i\tau : |\tau| < s(\sigma)\}$ and a conformal mapping Z which takes Ω onto $\Pi = \{x + iy : |y| < \pi/2\}$. The function $\exp(-Z^{-1})$ (or $\exp\{-kw - Z^{-1}(w)\}$, $w \in \Pi$, depending on the case) furnishes an example of the requirements in Theorems 1 and 2.

This method breaks down in the case $\psi'_\infty = +\infty$, since a conformal mapping of the type Z^{-1} does not have a sufficiently high order of growth (see [20]).

2°. *Proof of Theorem* 1. We define functions k, j, and s by means of the equations

$$\exp k(\sigma) = \frac{2}{\pi} e^\sigma \int_0^\sigma \varphi(t) e^{-t}\,dt, \quad \sigma \geqslant 1, \quad j = k^{-1},$$

$$s(x) = \frac{\pi}{2}\cdot\frac{1}{j'(x)}, \quad x \geqslant k(1),$$

and on $[0, k(1)]$ we choose s so that

$$j(k(1)) = \frac{\pi}{2}\int_0^{k(1)} \frac{dt}{s(t)},$$

and finally we let $s(x) = s(0)$ for $x < 0$. Then

$$j(x) = \frac{\pi}{2} \int_0^x \frac{dt}{s(t)}, \quad x \geqslant k(1).$$

Later we will show that s satisfies the conditions $\lim s' = 0$ and $\int_0^\infty (s'^2/s)\, dt < \infty$ which occur in Warschawski's theorem [15],(24) and thus for the conformal mapping Z of the strip $\Omega = \{\sigma + i\tau \colon |\tau| < s(\sigma)\}$ onto the strip Π we have

$$\operatorname{Re} Z(\sigma + i\tau) = \frac{\pi}{2} \int_0^\sigma \frac{dt}{s(t)} + r(\sigma, \tau),$$

where $r(\sigma, \tau) = o(1)$ as $\sigma \longrightarrow +\infty$. We set $\log f_0(w) = \exp Z^{-1}(w)$, $w \in \Pi$, and we show that the function $f = 1/f_0$ leads to the construction of a function f_ϵ with the properties described in Theorem 1.

Let $w = x + iy = Z(\sigma + i\tau)$ and $x = j(\sigma) + o(1)$. Then

$$\log |f_0(x + iy)| = e^\sigma \cos \tau = e^{k(x + o(1))} \cos \tau,$$

where the convergence of $o(1)$ to zero as $x \longrightarrow +\infty$ is uniform with respect to $y \in I$, while $\sigma = \sigma(x, y)$ and $\tau = \tau(x, y)$. Therefore $\log|f_0(x)| = e^{k(x + o(1))}$, since the symmetry of the strip Ω makes the values $Z(\sigma)$ real for all real σ.

From the definition of k it follows that(25)

$$k'(\sigma) = 1 + \frac{\varphi(\sigma) e^{-\sigma}}{\int_0^\sigma \varphi(t) e^{-t}\, dt} = \psi'_\infty + o(1),$$
$$j'(\sigma) = \frac{1}{\psi'_\infty} + o(1), \quad s = \frac{\pi}{2} \psi'_\infty + o(1). \tag{2}$$

Therefore $\sigma = k(x + o(1)) = k(x) + o(1)$ and $\log|f_0(x)| = e^{k(x)}(1 + o(1))$ as $x \longrightarrow +\infty$. On the other hand

$$\min_{y \in I} \log |f_0(x + iy)| \geqslant -e^\sigma$$

for $\psi'_\infty \geqslant 1$ and

$$\min_{y \in I} \log |f_0(x + iy)| \geqslant e^\sigma \cos s(\sigma)$$

for $\psi'_\infty < 2$, since $s(\sigma) \geqslant \tau(x, y) \geqslant -s(\sigma)$ and $\pi/2 < s(\sigma) < \pi$. Consequently

$$\min_{y \in I} \log |f_0(x + iy)| \geqslant -e^{k(x)} (1 + o(1)) \quad \text{for } \psi'_\infty \geqslant 1,$$
$$\min_{y \in I} \log |f_0(x + iy)| \geqslant -e^{k(x) + o(1)} \left[s(k(x + o(1))) - \frac{\pi}{2} \right] \quad \text{for } \psi'_\infty < 2. \tag{3}$$

(24) This theorem is also stated in Appendix I.

(25) This asymptotic equation follows from Lemma 4 in §1.3 if $\psi'_\infty = 1$. If $\psi'_\infty > 1$, we use the following elementary lemma: if a and g are functions on $[0, \infty)$ such that $\underline{\lim}\, a' > 0$ and $g > 0$, $\int_0^\infty g\, dt < \infty$, then $\int_0^x e^a g\, dt = o(e^{a(x)})$ as $x \longrightarrow +\infty$.

Now we note that

$$s=\frac{\pi}{2}k'\circ j,\quad k(x+o(1))=k(x)+o(1),\quad j(t+o(1))=j(t)+o(1)$$

and $k'(x+o(1))-1\sim k'(x)-1$. The last asymptotic equation follows from (2) and the boundedness of the logarithmic derivatives $\varphi'/\varphi=\psi'$ and $\varphi(x)e^{-x}/\int_0^x\varphi e^{-t}\,dt$ (which clearly does not exceed some constant for $\psi'_\infty>1$; and we use Lemma 4 in §1.3 to see this for $\psi'_\infty=1$ also). We extend inequality (3), using the definition of k:

$$\min_{y\in I}\log|f_0(x+iy)|\geqslant -e^{k(x)+o(1)}\left[\frac{\pi}{2}k'\circ j(k(x+o(1)))-\frac{\pi}{2}\right]$$

$$=-e^{k(x)+o(1)}\frac{\pi}{2}[k'(x)-1]=-\varphi(x)(1+o(1)).$$

It now follows that for $\psi'_\infty<2$

$$\log|f(x+iy)|\leqslant\varphi(x)(1+o(1)),$$

$$\log|f(x)|=-\frac{2(1+o(1))}{\pi}e^{x}\int_0^x\varphi(t)e^{-t}\,dt$$

and for $\psi'_\infty\geqslant 1$

$$\log|f(x+iy)|\leqslant\frac{2(1+o(1))}{\pi}e^{x}\int_0^x\varphi e^{-t}\,dt,$$

$$\log|f(x)|=-\frac{2(1+o(1))}{\pi}e^{x}\int_0^x\varphi e^{-t}\,dt.$$

The assertion in the theorem follows from this fact and the equation

$$e^{x}\int_0^x\varphi e^{-t}\,dt=\varphi(x)\frac{1+o(1)}{\psi'_\infty-1},$$

which is true for $\psi'_\infty>1$.

We still have to verify the conditions of Warschawski's theorem for the function $s(x)=(\pi/2)k'\circ j$. We note first that $\int_0^\infty(s'^2/s)\,dt<\infty$ if the derivative $s'(t)$ is bounded and of one sign for large t and $\lim_{t\to\infty}s>0$. We have

$$s'=\frac{\pi}{2}(k''\circ j)j'=\frac{\pi}{2}\frac{(k''\circ j)}{k'\circ j},$$

$$\frac{k''}{k'}(x)=\frac{e^{a(x)}(a'(x)+1)}{\int_0^x e^{a}dt+e^{a(x)}}-\frac{e^{a(x)}}{\int_0^x e^{a}dt},\quad\text{where } a(t)=\psi(t)-t.$$

As we mentioned in footnote 25, we have

$$T\overset{\text{def}}{=}\frac{e^{a(x)}}{\int_0^x e^{a}dt}=a'(x)+o(1),$$

and therefore

$$\frac{k''}{k'}=T\frac{a'+1}{1+T}-T=o(1),$$

and also $s'=o(1)$. We verify the sign-definiteness of $s'\circ k=(\pi/2)k''/k'$:

$$\frac{k''}{k'}=T\frac{a'-T}{1+T},$$

and from (14) in §1.3 we have

$$\frac{1}{T(x)}=\frac{1}{a'(x)}+e^{-a(x)}\int_0^x e^a\frac{a''}{a'^2}\,dt.$$

If $a'\leqslant 0$ for large x, then automatically $s'\circ k=k''/k'<0$. If $a'>0$, the difference $a'-T$ is of one sign since $\int_0^x e^a(a''/a'^2)\,dt$ is of one sign for large values of x. ●

3°. We first present the proof of Theorem 2 for the case $\psi'_\infty=1$. We define functions a, k, j and s by means of the equations

$$a(t)=\psi(t)-t,$$

$$\exp k(\sigma)=\frac{2}{\pi}\int_0^\sigma e^{\frac{1}{2}a(t)}\,dt,\quad \sigma\geqslant 1,$$

$$j=k^{-1},\quad s(x)=\frac{\pi}{2}\cdot\frac{1}{j'(x)},\quad x\geqslant k(1),$$

and on $[0, k(1)]$ we choose s so that

$$j(k(1))=\frac{\pi}{2}\int_0^{k(1)}\frac{dt}{s(t)},$$

and we let $s(x)=s(0)$ for $x<0$. Then

$$j(x)=\frac{\pi}{2}\int_0^x\frac{dt}{s(t)},\quad x\geqslant k(1).$$

Later we will show that s satisfies the conditions $\lim s'=0$ and $\int^\infty(s'^2/s)\,dt<+\infty$ which occur in Warschawski's theorem [15],(26) and thus for the conformal mapping Z of the strip $\Omega=\{\sigma+i\tau\colon |\tau|<s(\sigma)\}$ onto the strip Π we have

$$\operatorname{Re} Z(\sigma+i\tau)=\frac{\pi}{2}\int_0^\sigma\frac{dt}{s(t)}+o(1),\quad \sigma\geqslant k(1). \tag{4}$$

We set $\log f(w)=-\exp\{w+2Z^{-1}(w)\}$, $w\in\Pi$, and we show that the function

$$F=f\circ q,\quad q(\zeta)=\log\frac{1}{2}\cdot\frac{1+\zeta}{1-\zeta},\quad |\zeta|<1,$$

has the properties described in Theorem 2.

We estimate

(26) This theorem is also stated in Appendix I.

$$\max_{w\in O_t^*}(-\operatorname{Re}\exp\{w+2Z^{-1}(w)\})\overset{\text{def}}{=}M(t)$$

from above; here O_t^* is the image of the circle $\{\zeta: |\zeta| = r\}$ under the transformation q, and $t = \log(\tfrac{1}{2}(1+r)/(1-r))$.

LEMMA 1. 1) *There exists a function A such that $\lim_{t\to\infty}A(t) = +\infty$ and*

$$M(t)=\max_{\substack{w\in O_t^*\\ A(t)\leqslant \operatorname{Re} w\leqslant t-A(t)}}(-\operatorname{Re}\exp\{w+2Z^{-1}(w)\}).$$

2) *If $x+iy\in O_t^*$, then $y=\pi/2-e^{-(t-x)}(1+o(x;t))$, where $o(x;t)$ is a function of x and t such that $\lim_{t\to\infty}o(x;t)=0$ uniformly with respect to $x\in[A(t),t-A(t)]$.*

PROOF OF THE LEMMA. If $x\leqslant c_1=\text{const}$, then

$$|\operatorname{Re}\exp\{x+iy+2Z^{-1}(x+iy)\}|\leqslant e^{x+2\operatorname{Re}Z^{-1}(x+iy)}\leqslant e^{c_1+c_2}$$

by virtue of (4). If $x+iy\in O_t^*$ and $\sigma\geqslant t-c_1$, then by Lemma 5 in §1.3 we have $|\tau|\leqslant\pi/2-\epsilon$, $\epsilon=\epsilon(c_1)>0$. At the same time

$$|\operatorname{Im}Z^{-1}(x+iy)|\leqslant s(\operatorname{Re}Z^{-1}(x+iy)),$$

where $\operatorname{Re}Z^{-1}(x+iy)$ converges to $+\infty$ together with t (by virtue of the inequality $x\geqslant t-c_1$ and formula (4)), and by Lemma 4 in §1.3 we have $\lim_{\sigma\to\infty}s(\sigma)=0$ since

$$s(\sigma)=\frac{\pi}{2}k'\circ j(\sigma),\quad k'(x)=\frac{e^{\frac{1}{2}a(x)}}{\int_0^x e^{\frac{1}{2}a(t)}\,dt}. \tag{5}$$

Therefore

$$|\operatorname{Im}(x+iy+2Z^{-1}(x+iy))|<\frac{\pi}{2}$$

for large t and $\operatorname{Re}\exp(x+iy+2Z^{-1}(x+iy))>0$. It is clear that $M(t)\uparrow+\infty$; this proves assertion 1) in Lemma 1. Assertion 2) follows immediately from Lemma 5 in §1.3. ●

Now let $x\in[A(t),t-A(t)]$, where A is the function in Lemma 1 and $\sigma+i\tau=Z^{-1}(x+iy)$, $x+iy\in O_t^*$. Then

$$y=\frac{\pi}{2}-e^{-(t-x)}(1+o(1)),\quad |\tau|\leqslant s(\sigma),$$

$$\cos(y+2\tau)\geqslant\cos\left(\frac{\pi}{2}-e^{-(t-x)}(1+o(1))+2s(\sigma)\right),$$

since cos decreases on $[0,\pi]$. From (4) we have $x=j(\sigma)+o(1)$ and $\sigma=k(x+o(1))$, and therefore $\sigma=k(x)+o(1)$. Thus we have

$$\begin{aligned}\log|f(x+iy)|&=-e^{x+2\sigma}\cos(y+2\tau)\\&\leqslant -e^{x+2k(x+o(1))}\cos\left\{\frac{\pi}{2}-e^{-(t-x)}(1+o(1))+2s(k(x+o(1)))\right\}\\&\leqslant e^{x+2k(x+o(1))}\{2s(k(x+o(1)))-e^{-(t-x)}(1+o(1))\},\end{aligned}$$

or

$$\log|f(x+iy)| \leqslant 0,$$

depending on the sign of the difference

$$2s(k(x+o(1))) - e^{-(t-x)}(1+o(1)).$$

In this expression $o(1)$ is used to denote various functions of x and y which converge to zero (uniformly with respect to y) as $x \longrightarrow \infty$. In the concluding part of the proof of the theorem it will be shown that $s' < 0$ for large values of the argument and $\lim s' = 0$. Thus if we rewrite the aforementioned difference in the form

$$V = 2s(k(x+o_1(1))) - e^{-t+x+o_2(1)},$$

we see that

$$V \leqslant 2s(k(x+o_*(1))) - e^{-t+x+o_*(1)},$$

where $o_* = o_1(1)$ if $o_1 \leqslant o_2$, and $o_* = o_2(1)$ if $o_2 \leqslant o_1$. Since $M(t) \uparrow +\infty$ as $t \longrightarrow +\infty$, we can restrict ourselves to the case $V > 0$; thus

$$\log|f(x+iy)| \leqslant e^{x_*+2k(x_*)+o(1)}\{2s\circ k(x_*) - e^{-(t-x_*)}\}$$
$$= e^{x_*+2k(x_*)+o(1)}\left\{2\frac{\pi}{2}k'(x_*) - e^{-(t-x_*)}\right\},$$

where $x_* = x + o(1)$. We replace $A(t)$ by $A(t) - 1$ if necessary and use the above notation; we thereby obtain

$$M(t) = \max_{0_t^*} \log|f(x+iy)|$$
$$= (1+o(1)) \max_{A(t)\leqslant x\leqslant t-A(t)} e^{x+2k(x)}\{2s\circ k(x) - e^{-(t-x)}\}.$$

A stationary point $\bar{x}$ is determined by the equation

$$(1+2k')(2s\circ k - e^{-(t-\bar{x})}) + \{2(s'\circ k)k' - e^{-(t-\bar{x})}\} = 0,$$

or

$$e^{-(t-\bar{x})} = \pi k' \frac{1+2k'+\frac{2}{\pi}k's'\circ k}{2(1+k')} = \frac{\pi}{2}k'(1+o(1)). \tag{6}$$

Consequently

$$M(t) = e^{\bar{x}+2k(\bar{x})}\left\{\pi k'(\bar{x}) - \frac{\pi}{2}k'(\bar{x})(1+o(1))\right\}$$
$$= e^{\bar{x}+k(\bar{x})}\frac{\pi}{2}k'(\bar{x})e^{k(\bar{x})}(1+o(1)) = e^{\bar{x}+k(\bar{x})+\frac{1}{2}a(\bar{x})}(1+o(1)),$$

where the last equality is a consequence of the definition of k. We now show that $\bar{x} + k(\bar{x}) + \frac{1}{2}a(\bar{x}) \leqslant t + a(t) + o(1)$. Formulas (5) and (6) enable us to rewrite this inequality in the form

$$\frac{\pi}{2} k'(\bar{x}) e^{k(\bar{x})+\frac{1}{2}a(\bar{x})} \leqslant e^{a(t)+o(1)},$$

or $a(\bar{x}) \leqslant a(t) + o(1)$. This last inequality is obvious if $a' \geqslant 0$ for large values of the argument. Since a' is of one sign in a neighborhood of $+\infty$, we just have to consider the case $a' < 0$, and then by hypothesis we have $|a'| \downarrow$. Thus

$$a(\bar{x}) - a(t) = (\bar{x} - t) a'(\xi) \leqslant (t - \bar{x}) |a'(\bar{x})|,$$

$$t - \bar{x} = \log\left\{e^{-\frac{1}{2}a(\bar{x})} \int_0^{\bar{x}} e^{\frac{1}{2}a(t)} dt\right\} + O(1).$$

LEMMA 2. *Suppose that the hypothesis of Theorem 2 is satisfied and $a' < 0$. Then*

$$a'(\bar{x}) \log^2\left\{e^{-\frac{1}{2}a(\bar{x})} \int_0^{\bar{x}} e^{\frac{1}{2}a(t)} dt\right\} = o(1), \quad \bar{x} \to +\infty.$$

PROOF OF THE LEMMA. We have

$$e^{-\frac{1}{2}a(\bar{x})} \int_0^{\bar{x}} e^{\frac{1}{2}a} dt = \frac{2}{a'(\bar{x})} + 2e^{-\frac{1}{2}a(\bar{x})} \int_0^{\bar{x}} e^{\frac{1}{2}a} \frac{a''}{a'^2} dt$$

$$< O(1) e^{-\frac{1}{2}a(\bar{x})} \int_0^{\bar{x}} \frac{a''}{a'^2} dt = O(1) e^{-\frac{1}{2}a(\bar{x})} \frac{1}{|a'(\bar{x})|},$$

$$a'(\bar{x}) | \log^2\left\{e^{-\frac{1}{2}a(\bar{x})} \int_0^{\bar{x}} e^{\frac{1}{2}a} dt\right\} \leqslant |a'(\bar{x})| \left(\log\frac{1}{|a'(\bar{x})|} - \frac{1}{2}a(\bar{x}) + O(1)\right)^2$$

$$= \frac{1}{4} |a'(\bar{x})| a(\bar{x})^2 + o(1).$$

Also,

$$|a(\bar{x})| = \left|\text{const} + \int_0^{\bar{x}} a' dt\right| \geqslant -\text{const} + \bar{x}|a'(\bar{x})| \quad \text{and} \quad |a'(\bar{x})| \leqslant \frac{|a(\bar{x})| + \text{const}}{\bar{x}}.$$

Comparability of the functions $|a(t)|$ and t^α (for some α with $0 < \alpha \leqslant 1/3$) implies that $|a(t)| = o(t^{1/3})$. In fact, the reverse inequality $|a(t)| \geqslant \text{const}\, t^\alpha$ $(0 < \alpha \leqslant 1/3)$ under the condition $a' < 0$ would mean that $a(t) \leqslant -\text{const}\, t^\alpha$ and $\int^\infty e^{a/2} dt < +\infty$, which contradicts the hypothesis of the lemma. Therefore $|a(t)| = o(t^{1/3})$ and

$$|a'(\bar{x})| a(\bar{x})^2 \leqslant \frac{|a(\bar{x})| + \text{const}}{\bar{x}} a(\bar{x})^2 = o(1). \bullet$$

The lemma implies that $(t - \bar{x})|a'(\bar{x})| = o(1)$ as $t \to \infty$, since $A(t) \leqslant \bar{x}$ and $\lim_{t\to\infty} A(t) = +\infty$. Returning to the estimation of $\log f$, we obtain

$$\log|f(x+iy)| \leqslant e^{t+a(t)+o(1)} = \varphi(t)(1+o(1)) \tag{7}$$

for all $x + iy \in O_t^*$.

On the other hand,[27]

$$\log|f(x)| = -\exp\{x + 2Z^{-1}(x)\} = -\exp\{x + 2k(x + o(1))\}$$

$$= -\exp\{x + 2k(x) + o(1)\} = -\left(\frac{4}{\pi^2} + o(1)\right) e^x \left(\int_0^x e^{\frac{1}{2}a(t)} dt\right)^2. \quad (8)$$

Estimates (7) and (8) and Lemma 6 in §1.3 mean that the function $F = f \circ q$ has the properties described in the theorem (for the case $\psi'_\infty = 1$).

We now verify the conditions in Warschawski's theorem and also the inequality $s' < 0$ by means of the assumptions in Theorem 2. In order to do this, it suffices to show that (i) $\lim_{t\to\infty} s' = 0$; (ii) $s' \log^2 1/s$ is a bounded function; and (iii) $s' < 0$ for large values of the argument. In fact, convergence of the integral $\int^\infty (s'^2/s)\, dt$ follows immediately from (i), (ii), and (iii):

$$\int^\infty \frac{s'^2}{s} = \int^\infty \frac{|s'|}{s \log^2 1/s} |s'| \log^2 \frac{1}{s} \leqslant \text{const} \int^\infty \frac{|s'|}{s \log^2 1/s} = \text{const} \int_0 \frac{ds}{s \log^2 1/s} < +\infty.$$

We recall that

$$s = \frac{\pi}{2} k' \circ j, \quad k'(x) = \frac{e^{\frac{1}{2}a(x)}}{\int_0^x e^{\frac{1}{2}a(t)} dt}, \quad s' = \frac{\pi}{2} \frac{k'' \circ j}{k' \circ j}.$$

Assertion (i) is contained in Lemma 4 in §1.3. For the proof of (iii) it suffices to show that the sign of s' is stabilized; we have

$$\frac{k''}{k'}(x) = \frac{1}{2} a'(x) - \frac{e^{\frac{1}{2}a(x)}}{\int_0^x e^{\frac{1}{2}a(t)} dt} = \left(\frac{1}{2} a'(x) e^{-\frac{1}{2}a(x)} \int_0^x e^{\frac{1}{2}a(t)} dt - 1\right) \frac{e^{\frac{1}{2}a(x)}}{\int_0^x e^{\frac{1}{2}a(t)} dt}$$

$$= \left(a'(x) e^{-\frac{1}{2}a(x)} \int_0^x e^{\frac{1}{2}a(t)} \frac{a''(t)}{a'^2(t)} dt\right) \frac{e^{\frac{1}{2}a(x)}}{\int_0^x e^{\frac{1}{2}a(t)} dt}.$$

The last expression is clearly of one sign for $a' \leqslant 0$, and for $a' > 0$ this fact follows from the hypothesis of the theorem by virtue of the monotonicity of

$$\int_0^x e^{\frac{1}{2}a(t)} \frac{a''(t)}{a'^2(t)} dt$$

for large x.

---

[27] Here we use the fact that $Z^{-1}(x)$ is real for real x (symmetry of the strip Ω).

We now verify (ii); that is, we show that the function

$$\frac{k''}{k'}\log^2\frac{1}{k'}=\left(\frac{1}{2}a'(x)-k'(x)\right)\log^2\frac{1}{k'(x)}$$

is bounded for $x \geqslant 1$. From (i) and (iii) we see that it suffices to prove boundedness of $\frac{1}{2}a' \log^2 1/k'$ for $a' < 0$, and this in turn follows from Lemma 2. The case $\psi'_\infty = 1$ in Theorem 2 is completely proved.

The proof for the case $1 < \psi'_\infty < \infty$ is obtained in a similar fashion, but the functions k, j, s and f are defined somewhat differently:

$$\exp k(\sigma)=\frac{2}{\pi}\int_0^\sigma e^{a(t)}dt, \quad \sigma\geqslant 1,$$

$j = k^{-1}$ and $s = (\pi/2)\cdot(1/j')$ (with the same "compatibility condition" on the segment $[0, k(1)]$ and $\log f(w) = \exp(w + Z^{-1}(w))$, $w \in \Pi$, where Z is a conformal mapping of the strip $\{\sigma + i\tau\colon |\tau| < s(\sigma)\}$ onto the strip Π. Arguments which are completely analogous to those above show that[28]

$$\max_{w\in O_t^*}\log|f(w)|=M(t)\leqslant e^{\bar{x}+k(\bar{x}+o(1))}\{s(k(\bar{x}+o(1)))-e^{\bar{x}+o(1)-t}\}$$

where the point $\bar{x}$ in the case $\psi'_\infty < 1 + 3/\pi$ can be determined from the equation

$$e^{\bar{x}-t}=\frac{\pi}{2}\cdot\frac{\psi'_\infty(\psi'_\infty-1)}{\psi'_\infty+1}+o(1). \tag{9}$$

If we perform the same transformations on this expression as in the case $\psi'_\infty = 1$, we obtain the estimate

$$M(t)\leqslant e^{\bar{x}+k(\bar{x})+o(1)}\frac{\frac{\pi}{2}k'(\bar{x})}{\varphi'_\infty+1}=e^{t+a(t)}e^{\bar{x}-t+a(\bar{x})-a(t)}\frac{1+o(1)}{\psi'_\infty+1}.$$

Now we can once more use (9) and the asymptotic formula

$$a(\bar{x})-a(t)=(\bar{x}-t)a'(\xi)=(\bar{x}-t)(\psi'_\infty-1)+o(1),$$

which is true as $t \to \infty$:

$$M(t)\leqslant\varphi(t)\left(\frac{\pi}{2}\cdot\frac{\psi'_\infty(\psi'_\infty-1)}{\psi'_\infty+1}\right)^{\psi'_\infty+o(1)}\frac{1}{\psi'_\infty+1}\leqslant\varphi(t)\frac{\pi^2+o(1)}{4}(\psi'_\infty-1).$$

On the other hand, on the real line we have

$$\log|f(x)|=-e^{x+k(x)+o(1)}=-\frac{2}{\pi}(1+o(1))e^x\int_0^x e^{a(t)}dt=-\frac{2}{\pi}(1+o(1))\frac{\varphi(x)}{\psi'_\infty-1}.$$

Together with Lemma 6 in §1.3 this completes the proof of Theorem 2. ●

(28) In the case $\psi'_\infty < 1 + 3/\pi$. If $\psi'_\infty \geqslant 1 + 3/\pi$, then for f we simply take the function in Theorem 1; this can be done since $a(\Pi, \lambda_\Pi)\circ q \subset a(D, \lambda)$.

4°. *Proof of Theorem* 2 *in* §1.1. We define the functions k, j and s by means of the equations

$$\exp k(x) = \frac{2}{\pi}\int_0^x \varphi(t)\,dt, \quad x \geqslant 1, \quad j = k^{-1}, \quad s = \frac{\pi}{2}\cdot\frac{1}{j'},$$

and we extend s (smoothly) onto the interval $[0, k(1)]$ so that

$$j(\sigma) = \frac{\pi}{2}\int_0^\sigma \frac{dt}{s(t)}, \quad \sigma \geqslant k(1),$$

and we let $s(t) = s(0)$ for $t < 0$. Let Z be a conformal mapping of the strip $\Omega = \{\sigma + i\tau\colon |\tau| < s(\sigma)\}$ onto the strip Π. It has already been shown (see the proof of Theorem 2 in §1.4) that under the conditions of Theorem 2 in §1.1 we can apply Warschawski's theorem to Ω, and we find that

$$\operatorname{Re} Z(\sigma + i\tau) = \frac{\pi}{2}\int_0^\sigma \frac{dt}{s(t)} + o(1), \quad \sigma \to \infty.$$

We set $F = \exp Z^{-1}$. Then at a point $x + iy = Z(\sigma + i\tau)$ we have

$$|\operatorname{Im} F(x + iy)| = e^\sigma |\sin\tau| \leqslant e^\sigma |\tau| \leqslant e^\sigma s(\sigma) = \varphi(x)(1 + o(1)),$$

since the equations $x = j(\sigma) + o(1)$, $\sigma = k(x + o(1))$, $k' = o(1)$ (see Lemma 4 in §1.3) and $k''/k' = o(1)$ (see the proof of Theorem 2 in §1.4) lead to

$$e^\sigma s(\sigma) = e^{k(x+o(1))} s\circ k(x + o(1)) = \frac{\pi}{2} e^{k(x)+o(1)} k'(x + o(1))$$
$$= \frac{\pi}{2} e^{k(x)+o(1)} k'(x) = \varphi(x)(1 + o(1)).$$

On the other hand,

$$\operatorname{Re} F(x) = \exp Z^{-1}(x) = e^\sigma = e^{k(x)+o(1)} = \frac{2 + o(1)}{\pi}\int_0^x \varphi(t)\,dt.$$

Now we define harmonic functions u and v on the disk $\mathbf{D}$ by means of the equations $u = \operatorname{Im} F \circ p$ and $v = \operatorname{Re} F \circ p - F(0)$, where $p(\zeta) = \log(1 - \zeta)^{-1}$, $|\zeta| < 1$. From previous calculations it follows that for $|\zeta| \leqslant r$ we have

$$|u(\zeta)| = |\operatorname{Im} F(p(\zeta))| \leqslant \max_{w \in O_r} |\operatorname{Im} F(w)| \leqslant \varphi\left(\log\frac{1}{1-r}\right)(1 + o(1)),$$

where O_r is the image of the disk $\{\zeta\colon |\zeta| = r\}$ under the mapping p, and

$$|v(r)| = \operatorname{Re} F\left(\log\frac{1}{1-r}\right) - F(0) = \frac{2 + o(1)}{\pi}\int_0^{\log\frac{1}{1-r}} \varphi(t)\,dt.$$

In order to obtain the assertion in the theorem we now apply Lemma 1 in §1.1. ●

§1.5. On the conditions imposed on the majorant

1°. A typical condition which we impose on the regularity of growth is the requirement of convexity of the function $1/\sqrt{\psi'}$ or $1/\psi'$, where $\psi = \log\log\lambda_\Pi$. We recall that we agreed that in this chapter (see §1.0) convexity of f means monotonicity (and existence) of f' in some neighborhood of $+\infty$ (or of 1, if we are dealing with the interval $[0, 1)$). We will show that: 1) membership of λ_Π in some Hardy field guarantees that the conditions on "regularity" of growth which we use in this chapter are satisfied; and 2) none of these conditions (except, perhaps, the conditions in the second parts of Theorems 1 and 2 in §1.2 and Theorem 2 in §1.4) impose any new restrictions on the speed of growth; that is, there always exist functions λ_+ and $\lambda_-, \lambda_- \leqslant \lambda \leqslant \lambda_+$, which have the necessary "regularity" and which together with λ satisfy the remaining conditions of the corresponding theorem.

2°. *Hardy fields.* Following Bourbaki [12], we call a set $\mathfrak{R}$ of real (germs of) functions defined in a neighborhood of $+\infty$ a Hardy field if $\mathfrak{R}$ is a field with respect to the operations of pointwise addition and multiplication and $\mathfrak{R}$ is closed with respect to differentiation. These conditions mean that if f and g are in $\mathfrak{R}$, then $f \pm g$, fg, and f' are also in $\mathfrak{R}$, and if $f \not\equiv 0$ in a neighborhood of $+\infty$, then $f(t) \neq 0$ in some (other) neighborhood of $+\infty$ and $1/f \in \mathfrak{R}$. Consequently the sign of an arbitrary function in $\mathfrak{R}$ stabilizes near the point $+\infty$. Real constants are assumed to be contained in $\mathfrak{R}$.

We note that all of the conditions on regularity of growth in this chapter are requirements of monotonicity (or sign-definiteness) of some rational function of ψ, ψ', ψ'', ... (and possibly of the independent variable t). Since a given Hardy field $\mathfrak{R}$ can be extended (see [12]) to a skew field $\mathfrak{R}_1$ which contains along with λ_Π the functions ψ and t, assertion 1) stated above is thereby proved.

We most often use the Hardy field **H**, the smallest Hardy field which contains all rational functions and is closed with respect to the operations log and exp.

3°. *"Regularity" and speed of growth.* We consider the conditions in Theorem 1 in §1.1 (typical of the conditions in this chapter). Here $\psi' > 0$ and the "growth requirement" consists of the existence of the limits $l = \lim_{t\to\infty}\psi'(t)$ and $k = \lim_{t\to\infty}\psi'(t)t$, and the "regularity requirement" reduces to convexity of $1/\psi'$ if $k > 0$. We set $g = 1/\psi'$ and construct convex functions g_+ and g_- with $g_- \leqslant g \leqslant g_+$, having the same values l and k for an arbitrary pair of numbers (l, k), $l \geqslant 0$, $k > 0$.

Suppose that $0 < l < +\infty$. The existence of the required functions follows from the next lemma.

LEMMA 1. *Let G be a positive continuous function on $[0, +\infty)$ with $G \downarrow 0$. Then there exist piecewise-linear convex functions G_+ and G_- such that* $\lim G_+ = 0$ *and* $0 < G_- < G < G_+$.

The proof is similar to the construction of a Hadamard polygonal line [14], but it differs in several details. We choose arbitrary numbers $G_+(0)$ and $G_-(0)$ such that $0 < G_-(0) < G(0) < G_+(0)$. Let L_1^- be a line which is tangent to the graph $\Gamma(G)$ of the function G and passes through the point $(0, G_-(0))$, and let L_1^+ be a line through $(0, G_+(0))$ which has a negative slope ϵ_1 which is small enough so that the point of intersection of L_1^+ with $\Gamma(G)$ (the point A_1 in Figure 3) has an abscissa x_1 which is bigger than 1.

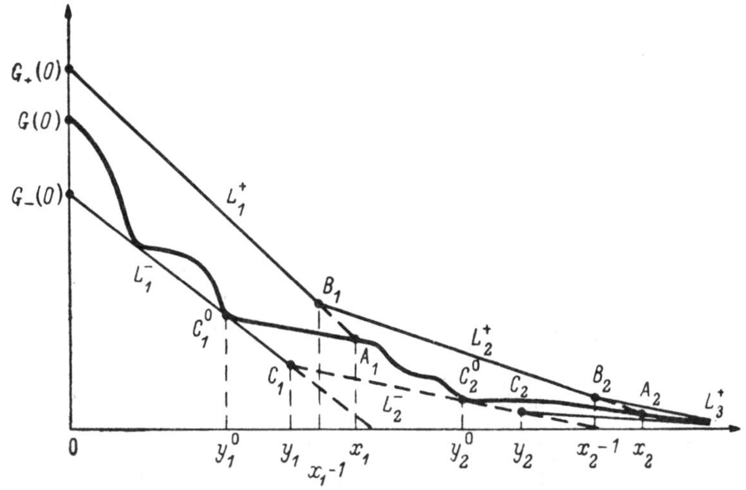

FIGURE 3

In constructing the envelope G_- we let C_1^0 denote the point in $\Gamma(G) \cap L_1^-$ with the largest abscissa (above y_1^0 in Figure 3). On the line L_1^- we choose a point C_1, with abscissa $y_1 > y_1^0$, which is close enough to the real axis $\mathbf{R}$ so that the line through it which is tangent to $\Gamma(G) \cap \{(x, y): x \geqslant y_1\}$ hits the graph at a point C_2^0 with an abscissa $y_2^0 > y_1^0 + 1$. By continuing this process we complete the polygonal line G_-.

To construct the next segment of G_+ we choose on L_1^+ the point B_1 with abscissa $x_1 - 1$ and put through it a line L_2^+ with a negative slope ϵ_2, with $\epsilon_1 < \epsilon_2 < 0$, which is small enough so that the point of intersection of L_2^+ and $\Gamma(G)$ (the point A_2 in Figure 3) has an abscissa x_2 which is bigger than 2. By continuing this process we complete G_+.

We return to the construction of the envelopes g_+ and g_-, setting

$$G^1(x) = \frac{1}{l} + \max\left\{0, \sup_{t \geqslant x}\left(g(t) - \frac{1}{l}\right)\right\},$$

$$G^2(x) = \frac{1}{l} - \max\left\{0, \sup_{t \geqslant x}\left(\frac{1}{l} - g(t)\right)\right\}.$$

Then $G^2 \leqslant g \leqslant G^1$ and $G^2 \leqslant 1/l \leqslant G^1$, where $G^1 \downarrow$ and $G^2 \uparrow$. If we apply Lemma 1 to the functions $G^1 - 1/l$ and $1/l - G^2$, we obtain the required envelopes: $g_+ = (G^1 - 1/l)_+ + 1/l$ and $g_- = 1/l - (1/l - G^2)_+$.

Suppose that $l = +\infty$. This means that $\lim_{t\to\infty} g = 0$; then $g_+ = G_+^1$ and $g_- = G_-^2$, where $G^1(x) = \sup_{t \geqslant x} g(t)$ and $G^2(x) = \inf_{t \leqslant x} g(t)$.

Suppose that $l = 0$ and $0 < k < +\infty$. Then

$$\lim_{t\to\infty} g(t) = +\infty, \quad \lim_{t\to\infty} \frac{g(t)}{t} = \frac{1}{k}.$$

In this case we can let

$$g_+(x) = \int_1^x \left\{\sup_{u \geqslant t} \frac{g(u)}{u}\right\} dt + \text{const}, \quad g_-(x) = \int_1^x \left\{\inf_{u \geqslant t} \frac{g(u)}{u}\right\} dt + \text{const}. \tag{1}$$

In fact,

$$g_+(x) \geqslant (x-1) \sup_{u \geqslant x} \frac{g(u)}{u} + \text{const} \geqslant g(x) - \frac{g(x)}{x} + \text{const} \geqslant g(x),$$

$$g_-(x) \leqslant (x-1) \inf_{u \geqslant x} \frac{g(u)}{u} + \text{const} \leqslant g(x) - \frac{g(x)}{x} + \text{const} \leqslant g(x),$$

and clearly $g_\pm$ are convex functions. It is also clear that $\lim_{x\to\infty} g_\pm(x)/x = 1/k$.

If $l = 0$ and $k = +\infty$, then $\lim_{t\to\infty} g = +\infty$, but $\lim_{t\to\infty} g(t)/t = 0$. For the upper envelope g_+ we can again take the function in (1). The lower envelope g_- is constructed from the function $\inf_{x \geqslant t} g(x)$ in the same way that G_+ in Lemma 1 was constructed from G. ●

Thus the "regularity conditions" in §1.1 have been fully analyzed. In the same way we consider the conditions in Theorems 1 and 2 in §1.2 (part 1)), all of the theorems in §1.3, and Theorem 1 in §1.4. For Theorem 2 in §1.4, if we choose a lower envelope $a_- \leqslant a$, $a(t) = \psi(t) - t$, it can turn out that $\int^\infty e^{a_-/2} < \infty$ but $\int^\infty e^{a/2} = +\infty$.

CHAPTER 2

POLYNOMIAL APPROXIMATION IN SPACES WITH AN "ANALYTIC" WEIGHT

§2.0. Problems and results

1°. Suppose that X is a space of analytic functions in the disk $\mathbf{D} = \{\zeta\colon |\zeta| < 1\}$ of the complex plane, z is the identity mapping of $\mathbf{D}$ onto itself ($z(\zeta) \equiv \zeta$, $|\zeta| < 1$), and $\mathcal{P}_A$ is the set of all polynomials in z. For which functions $f \in X$ is the set $\mathcal{P}_A$ dense in the weighted class $X_f \overset{\text{def}}{=\!=} \{g\colon g \text{ is analytic in } \mathbf{D} \text{ and } gf \in X\}$? The problem of "spectral analysis" of the operator of multiplication by the "independent variable" z leads to the above statement of the problem of density of polynomials in the weighted case. We recall (see Chapter 0) that the "individual" version of this problem consists of describing all functions f which are cyclic for the operator $f \longrightarrow zf$; that is, those functions f for which the system of functions $\{z^n f\}_0^\infty$ is dense in X. For the sake of definiteness we assume that $zX \subset X$ and $\mathcal{P}_A$ is dense in X. If X is a normed space, then the weighted norm in the class X_f is defined by the formula $\|g\|_f = \|gf\|_X$, $g \in X_f$. In this situation density of the polynomials in the space X_f is equivalent to the relation

$$\inf_{p \in \mathcal{P}_A} \| pf - \mathbf{1} \|_X = 0,$$

where $\mathbf{1}(\zeta) \equiv 1$, $|\zeta| < 1$. It is convenient to call such functions f "weakly invertible functions" (see [8], [113] and [170]). It is clear that a similar concept makes sense in nonnormed spaces of analytic functions.

The goal of this chapter is to find a metric criterion for weak invertibility of functions f in the space X; that is, to show what additional "smoothness" with respect to the norm (metric, topology) $\|\cdot\|_X$ the function f must have in order to be weakly invertible in X. The results in this chapter admit the following qualitative interpretation: if f belongs to some subclass X_1 which is infinitely differentiable and quasi-analytic with respect to X and $f(\zeta) \neq 0$, $|\zeta| < 1$, then f is weakly invertible in X; conversely, absence of quasi-analyticity implies existence of a function f, $f(\zeta) \neq 0$, $|\zeta| < 1$, $f \in X_1$, which is not weakly invertible in X. The present chapter is devoted to a precise treatment of these results; see, in particular, §2.4.

We now give a short description of the various sections in Chapter 2. References will be kept to a minimum in order to avoid repetition of the information in Chapter 0 and in §§2.1–2.7.

2°. *Description of results.* §2.1 is devoted to the "trivialization" of the method of weak invertibility, which was referred to in Chapter 0 as pushing by means of the powers f^t, $0 \leqslant t \leqslant 1$. The theorems in this section are applicable to all spaces of analytic functions which satisfy several simple axioms. A typical result can be described as follows: if $f(\zeta) \neq 0$, $|\zeta| < 1$, and $fg \in X$ for an arbitrary analytic function g with $|g(\zeta)| \leqslant (\min_{|\xi|=|\zeta|}|f(\xi)|)^{-\epsilon}$ for some previously chosen $\epsilon > 0$, then f is weakly invertible in X. In spite of the simplicity of the tests in §2.1, they (together with the estimates in Chapter 1) include all of the known sufficient conditions for weak invertibility in normed spaces X. We will consider weak invertibility of inner and outer functions in the sense of Beurling; in particular, an outer function is weakly invertible in all spaces which satisfy the axioms in this section. Indirectly we consider the connection between weak invertibility and the problem of simultaneous approximation in the disk **D** and on the circle ∂**D**.

In §2.2 we give a complete description of the functions f in L^∞ which are weakly invertible with respect to bounded convergence almost everywhere ((*)-convergence). More precisely, we find the set $E(f)$ of all functions $g \in L^\infty$ such that $g(\zeta) = \lim_n p_n(\zeta)f(\zeta)$ for almost all $\zeta \in \mathbf{\Gamma}$, where $p_n \in \mathfrak{P}_A$ and $\sup_n\|p_nf\|_\infty < +\infty$. In the description of $E(f)$ an essential role is played by the Smirnov class $\mathfrak{D}$ (see [95]) which is related to approximation problems in H^p-theory. It also turns out that "smoothness" of f is essential (for example, for (*)-weak invertibility of a function $f \in H^\infty$ it is necessary and sufficient that f be an outer function and $s_{|f|}|f|^{-1} \in L^\infty$, where $s_{|f|}$ is the best upper semicontinuous majorant of $|f|$). Sets which are similar to $E(f)$ are also described for the cases where instead of polynomials p_n we choose functions $p_n \in H^\infty$ or $p_n \in \mathfrak{P}$.

In §2.3 we present estimates for Taylor coefficients of functions in the weighted space $X = \mathfrak{S}_f$ and briefly discuss the question of whether the sequence of powers $\{z_n\}_0^\infty$ is a basis in this space. The results are stated in terms of the "boundary norm" (also introduced in this section)

$$\|f\|_{\partial\mathcal{E}} \overset{\text{def}}{=} \overline{\lim_n} \frac{\|z^nf\|_{\mathcal{E}}}{\|z^n\|_{\mathcal{E}}},$$

if this limit is finite. The boundary space $\partial\mathfrak{S}$ is also used for solving the problem of a power basis in some spaces with an "analytic" weight. The space $\partial\mathfrak{S}$ and the "conjugately" defined space $\partial^*\mathfrak{K}\mathfrak{S}^*$ are calculated explicitly for several classes $\mathfrak{S}$. In terms of the norm $\|\cdot\|_{\partial\mathfrak{S}}$ it is possible to write "inequalities of Bernšteĭn type" which are used in calculating boundary spaces:

$\|p\|_{\mathfrak{S}} \leqslant \text{const}\,\|z^n\|_{\mathfrak{S}}\,\|p\|_{\partial\mathfrak{S}}$ (where $\mathfrak{S}$ is the space of "smooth" functions in $\overline{\mathbf{D}}$);

$\|p\|_{\mathfrak{S}} \geqslant \text{const}\,\|z^n\|_{\mathfrak{S}}\,\|p\|_{\partial\mathfrak{S}}$ (where $\mathfrak{S}$ is the space of "functions which grow near ∂**D**");

here p is a polynomial in z of degree less than or equal to n. We obtain the classical Bernšteĭn inequality if we set $\|p\|_{\mathfrak{S}} = \max_{|\zeta|\leqslant 1}|p'(\zeta)|$; then $\|p\|_{\partial\mathfrak{S}} = \max_{|\zeta|\leqslant 1}|p(\zeta)|$ and const $= 1$. We make an application to the estimation of $\mathfrak{S}_f$-orthogonal polynomials.

§2.4 contains preliminaries for the rest of Chapter 2. We split the problem of weak invertibility into two parts: approximation of the functions D^nf, $n \geqslant 0$ ($Df = zf'$), by weighted polynomials $\Sigma_0^N a_kz^kf$, and approximation of the function **1** by linear combinations of "derivatives" $D^n f$, $n \geqslant 0$. This division into two parts was used by Beurling in [118] for a class of topological algebras; it is an essential fact that in this section and later sections we are dealing with approximation with respect to a fixed norm. We will discuss the connection with quasi-analyticity and the approximation problem of Bernšteĭn on a half-line. The results in §2.4 reduce the question of weak invertibility of a function f, $f(\zeta) \neq 0$, $|\zeta| < 1$, in a subclass $X_1 \subset X$ to investigation of the series

$$\sum_{1}^{\infty} n^{-3/2} \log \frac{\| z^n \|_{X_1}}{\| z^n \|_X}.$$

Technical lemmas contain estimates of Wiman-Valiron type for the "central term" of nonanalytic functions

$$\sup_{n \geqslant 0} n^k d_n r^n \leqslant \lambda(r) (\log \lambda(r))^{k\alpha},$$

where $d_n = \inf_{0 \leqslant r < 1} r^{-n} \lambda(r)$, $n \geqslant 0$, λ is a "properly" increasing function on $[0, 1)$, and $\alpha \geqslant 1$ is a number which depends on the degree of this "properness".

Here we will make some general remarks concerning §§2.4–2.8. On all "weight" functions λ, Λ, $\{w_n\}_0^\infty$, h, κ, and so forth, which we encounter in the theorems in this chapter, we impose certain requirements of "regularity" of growth (or of decrease). The most severe of these requirements are like those in Chapter 1; all of them are discussed as they appear in the succeeding sections.

§2.5 contains a fundamental metric criterion for weak invertibility in the spaces

$$A(\lambda) = \{f : f \text{ is analytic in } \mathbf{D} \text{ and } |f(\zeta)| = o(\lambda(|\zeta|)),\ |\zeta| \to 1\};$$

$$L^p_{1/\lambda}(E) = \left\{ f : f \text{ is analytic in } \mathbf{D} \text{ and } \int_0^1 \frac{\| f_r \|_E^p}{\lambda(r)^p} dr < +\infty \right\},$$

where E is an arbitrary space of the Hardy space type H^r, $1 \leqslant r \leqslant \infty$; and also the spaces

$$l^p_A(w_n) = \{f : f \text{ is analytic in } \mathbf{D} \text{ and } \{\hat{f}(n) w_n\}_0^\infty \in l^p\},$$

where $\{w_n\}_0^\infty$ is a decreasing sequence of positive numbers.

One possible version of the fundamental theorem consists of the fact that the conditions $\log \Lambda = \log \lambda + h(\log \lambda)$, $\mu = \log \lambda$, $\kappa(\mu) \geqslant \mu'$ and

$$\int^{\infty} \frac{h(t)}{t \kappa(t)^{1/2}} dt = +\infty \tag{1}$$

imply weak invertibility in $A(\Lambda)$ (or in $L^p_{1/\Lambda}(E)$) of an arbitrary function f, $f(\zeta) \neq 0$, $|\zeta| < 1$, $f \in A(\lambda)$ (or $f \in L^p_{1/\lambda}(E)$). If the exponential order of the majorant λ equals α, $1 \leqslant \alpha \leqslant +\infty$, then the best power order of the function h for divergence of the above integral is $\frac{1}{2} + 1/2\alpha$.

In §2.6 we investigate weak invertibility of "slowly increasing" functions f; that is, functions such that

$$\int_0^1 \left(\frac{\log \max_{|\zeta|=r} |f(\zeta)|}{1-r} \right)^{1/2} dr < +\infty.$$

It is shown that for weak convergence of all such functions in the space $A(\Lambda)$ (or in $L^p_{1/\Lambda}(E)$) it is necessary and sufficient that

$$\int_0^1 \left(\frac{\log \Lambda(r)}{1-r} \right)^{1/2} dr = +\infty.$$

§2.7 contains examples which illustrate the theorems in §§2.5 and 2.6, equivalent formulations of the fundamental theorem, a comparison of this theorem with the results in §2.6 and with a theorem of Beurling in [118], and applications to "spectral analysis" in inductive and projective limits. Among the results of the latter type we have the following: if $X_\lambda = A(\lambda)$ or $X_\lambda = L^p_{1/\lambda}(E)$ and the function h satisfies (1), then in the spaces

$$X_{\mathrm{int}} = \bigcup_{\varepsilon>0} X_{\lambda \exp(-\varepsilon h(\log \lambda))} \text{ and } X_{\mathrm{ext}} = \bigcap_{\varepsilon>0} X_{\lambda \exp(\varepsilon h(\log \lambda))}$$

individual spectral analysis is possible; that is, each function of this space which is not an annihilator in the disk **D** is weakly invertible. Concerning the concept of convergence in these spaces, we just note that under the conditions $\lambda, h \uparrow +\infty$ (which are fulfilled in this chapter) the spaces X_{int} and X_{ext} and spaces similar to them are, respectively, inductive and projective limits with completely continuous imbeddings (see [43]). Therefore convergence of a sequence of functions in these spaces reduces (see [43]) to convergence in one (respectively, in each) of the "limiting" spaces $X_{\lambda \exp(\epsilon h(\log \lambda))}$.

In §2.8 we establish "inverse" theorems. They are deduced from a general sufficient condition for noninvertibility of a function in the space X; this condition, which develops an idea of Keldyš, determines a subclass of the class of "X-inner" functions. The condition implies that the power order $1/2 + 1/2\alpha$ of the function h which guarantees weak invertibility of functions in $A(\lambda)$ in the space $A(\lambda \exp h(\log \lambda))$ cannot be improved.

Appendix III contains an elementary example which demonstrates the insufficiency of lower estimates "with a shift in the argument" for obtaining precise criteria for weak invertibility.

3°. *Notation and terminology.* We preserve all of the notation and terminology in §1.0. A list of fundamental symbols to be used throughout this chapter now follows:

$\mathcal{P}_A$ is the set of all polynomials in the function z, $z(\zeta) \equiv \zeta$, $|\zeta| \leq 1$;

$\mathcal{P}$ is the set of all polynomials in z and z^{-1};

$\mathfrak{L}_X(V)$ is the closed linear hull of a set V in a space X;

$E_X(f) = \mathfrak{L}_X(z^n f\colon n = 0, 1, \ldots)$; $E_X(f) = E(f)$, if this is not ambiguous;

$A(\mathbf{D})$ is the space of all analytic functions in **D** with the topology of compact convergence;

$A(\lambda)$ is the space of all functions in $A(\mathbf{D})$ such that $|f(\zeta)| = o(\lambda(|\zeta|))$ as $|\zeta| \to 1 - 0$ and $\|f\|_{A(\lambda)} = \sup_{\mathbf{D}} |f(\zeta)|/\lambda(|\zeta|)$;

$H^p(\mathfrak{M})$ is the space of all functions in $A(\mathbf{D})$ with a finite norm $\|f\|_p = (\int_{\mathbf{D}} |f(\zeta)|^p \, d\mathfrak{M}(\zeta))^{1/p}$, $1 \leqslant p \leqslant \infty$; $\mathfrak{M}$ is a finite measure on $\mathbf{D}$;

l_2 is normalized Lebesgue measure on $\mathbf{D}$;

l_1 is the same for the circle $\partial\mathbf{D}$; $l_1(e) = |e|$;

$l^p_A(w_n)$ is the space of all functions $f \in A(\mathbf{D})$ with a finite norm

$$\|f\|_{p,w} = \left(\sum_{n=0}^{\infty} |\hat{f}(n)|^p w_n^p\right)^{1/p}, \quad 1 \leqslant p \leqslant \infty;$$

$\{w_n\}_0^\infty$ is a given sequence of numbers; $l^p_A(1) \equiv l^p_A$;

$c_0(w_n)$ is the space of all sequences $x = \{x_n\}_0^\infty$ of complex numbers such that $x_n w_n = o(1)$ as $n \to \infty$; $\|x\|_{c_0(w_n)} = \sup_{n \geqslant 0} |x_n w_n|$;

C_A is the space of all functions in $A(\mathbf{D})$ which have continuous extensions to the closure $\overline{\mathbf{D}}$;

H^p is the Hardy space of all functions $f \in A(\mathbf{D})$ with

$$\|f\|_p = \sup_{0 \leqslant r < 1} \left(\int_{\partial\mathbf{D}} |f(r\zeta)|^p \, dl_1(\zeta)\right)^{1/p} < +\infty, \quad 1 \leqslant p \leqslant \infty;$$

$|T|_X$ is the norm of a linear operator T in a space X.

§2.1. Elementary tests for weak invertibility

1°. The impossibility of weak inversion of a function f is connected, roughly speaking, with the existence of deep and closed "hollows"(1) on the surface of the modulus of f going to the unit circle only at one point. This fact was noticed in an example of Keldyš, and it is verified by the contents of this chapter. To be more precise, we consider a function f in some space X of analytic functions in $\mathbf{D}$ with $f(\zeta) \neq 0$, $|\zeta| < 1$. If points ζ where the quantity $|f(\zeta)|$ is small are located sufficiently close to points ξ where $|f(\xi)|$ is large, then a sequence of polynomials $\{p_n\}_1^\infty$ which inverts f (that is, $p_n f \xrightarrow{X} 1$) will be large at the points ζ and by virtue of the small "hilliness" of the surface of the modulus $|p_n|$ it will also be large at the points ξ. This makes it impossible for the sequence $\{p_n(\xi)f(\xi)\}_{n=1}^\infty$ to converge to unity. On the contrary, if the function $|f|$ is not "too big" or "too small" (with respect to X), it can turn out to be weakly invertible. Arguments of a similar type underlie the criteria of invertibility in this section. They are fairly coarse criteria since they involve global characteristics of the growth ($\max_{|\zeta|=r} |f(\zeta)|$ and $\min_{|\zeta|=r} |f(\zeta)|$) of f. They are fairly efficient since they are very simple and are of a general nature. More refined criteria will be established in §§2.5–2.7.

We mention several unsolved problems. The results in subsections 2° and 4° of this section are developments of work of Smirnov and Lebedev (Chapter III, §§2.1–2.2 in [100]) and Shapiro [113], [170].

---

(1) Or "cavities" if we use the geological terminology.

2°. *General criteria for weak invertibility with applications to the spaces $H^p(\mathfrak{M})$ and* $A(\lambda)$. The principle which we just discussed will be realized for a reasonably large class of spaces X. To fix ideas, we will always have in mind one of the following spaces:

$H^p(\mathfrak{M})$, the space of all functions f which are analytic in $\mathbf{D}$ and such that

$$\|f\|_p^p \overset{\text{def}}{=} \int_{\mathbf{D}} |f(\zeta)|^p \, d\mathcal{M}(\zeta) < +\infty,$$

where $\mathbf{D} = \{\zeta: |\zeta| < 1\}$, $\mathfrak{M}$ is a finite (nonnegative) measure(2) in $\mathbf{D}$, and $1 \leqslant p < +\infty$;

$A(\lambda)$, the space of all functions f which are analytic in $\mathbf{D}$ and such that

$$\|f\|_\lambda \overset{\text{def}}{=} \sup_{\zeta \in \mathbf{D}} |f(\zeta)| / \lambda(|\zeta|) < +\infty,$$
$$|f(\zeta)| = o(\lambda(|\zeta|)), \quad |\zeta| \to 1, \tag{1}$$

where λ is a function on $[0, 1)$ which grows without bound.

Countable unions or intersections of spaces of this type will also be used (for details see §2.7).

We consider a family $\mathfrak{A}$ of locally convex spaces X with the following properties:

(A) Each space X consists of functions which are analytic in $\mathbf{D}$, and the functional $f \longrightarrow f(\zeta)$, $f \in X$, is continuous in X for each ζ, $|\zeta| < 1$.

(B) For each pair of spaces X, $Y \in \mathfrak{A}$ the closed graph theorem (see [43]) holds.

The elements of $\mathfrak{A}$ will be called spaces of analytic functions in $\mathbf{D}$. For such a space X we will prove the following facts:

(C) X is determined by the behavior of the modulus if for arbitrary analytic functions f, f_1, f_2 in $\mathbf{D}$ the conditions $f_{1,2} \in X$ and $|f(\zeta)| \leqslant |f_1(\zeta)| + |f_2(\zeta)|$, $\zeta \in \mathbf{D}$, imply that $f \in X$, and the set of all such functions f is bounded in X.

(D) Convergence in X does not depend on the boundary $\partial\mathbf{D}$ if an arbitrary bounded sequence $\{f_n\}_1^\infty$ in X for which(3) $\lim_n f_n(\zeta) = 0$, $|\zeta| < 1$, is weakly convergent to zero.

By density ((*)-density) of polynomials in a space X we mean that

(E) for each $f \in X$ there exists a bounded sequence of polynomials $\{p_n\}_1^\infty$, $p_n = \Sigma_0^n c_{k,n} z^k \in X$, converging (perhaps weakly) to f (respectively, converging pointwise in $\mathbf{D}$ to f).

As examples of such families we use the following: let $\mathfrak{A}$ be the set of all Banach (or Fréchet) spaces with property (A); a reflexive space in $\mathfrak{A}$ has property (D).(4) Density of polynomials in X is easily obtained from invariance of X with respect to inversion; that is, from the property

(2) The measure $\mathfrak{M}$ will always be chosen in this section so that $H^p(\mathfrak{M})$ is a Banach (that is, complete) space.

(3) Such a sequence will be said to be (*)-convergent in X, and simply (*)-convergent if $X = H^\infty$, the space of all regular and bounded functions in D with the usual sup-norm.

(4) Since the family of functionals φ_ζ: $f \longrightarrow f(\zeta)$, $|\zeta| < 1$, is weakly continuous with respect to ζ, we have $\sup_{|\zeta|<1-\delta} \|\varphi_\zeta\| < +\infty$, $\delta > 0$; under the conditions in (D) the sequence $\{f_n\}$ is weakly compact in X and therefore has a weak limit point, and by hypothesis the only such point is 0.

(F) the mapping $\alpha \longrightarrow f_\alpha$, $|\alpha| = 1$ ($f_\alpha(\zeta) \stackrel{\text{def}}{=} f(\alpha\zeta)$, $|\zeta| < 1$), is continuous in X for each $f \in X$; or from the property

(F*) the mapping $\alpha \longrightarrow f_\alpha$ is (*)-continuous and X is (*)-sequentially complete (that is, it contains limits of (*)-convergent sequences).

The arithmetic mean $\sigma_n(f)$ of the Taylor series of a function $f \in X$ is obtained by a convolution of f with the Fejér kernel K_n:

$$\sigma_n(f) = \frac{1}{2\pi i} \int_\Gamma f_{\bar{\alpha}} K_n(\alpha) \frac{d\alpha}{\alpha}.$$

The integral can be understood as a limit in X (respectively, (*)-limit in X) of Riemann sums. Therefore $\sigma_n(f) \in X$, the sequence $\{\sigma_n(f)\}_1^\infty$ is bounded, and $(\sigma_n f)(\zeta) \longrightarrow f(\zeta)$, $|\zeta| < 1$. In the case (F) we still have to show that for each continuous functional Φ the mapping $\alpha \longrightarrow \langle f_\alpha, \Phi\rangle$ is continuous and therefore $\langle \sigma_n f, \Phi\rangle \longrightarrow \langle f, \Phi\rangle$; that is, $\sigma_n f \xrightarrow{\text{w.}X} f$.

In particular, the spaces $H^p(\mathfrak{M})$ and $A(\lambda)$ which were described at the beginning of this section have all of the properties (A)–(D). We can also include the space $A_\infty(\lambda)$, which consists of all analytic functions in $\mathbf{D}$ with a finite norm (1) if by weak convergence in property (D) we mean (*)-convergence. The space $A(\lambda)$ has property (E) and $A_\infty(\lambda)$ has property (E*); the connection between (E) and the problems in this chapter (in particular, for the spaces $H^p(\mathfrak{M})$) was clarified in §2.0.

We note that in an arbitrary space X of analytic functions which has property (C) we have a continuous operator of multiplication by z ($z(\zeta) \equiv \zeta$, $|\zeta| < 1$): $f \longrightarrow zf$, $f \in X$.

In later theorems we will use the following strengthened version of property (C):

(C′) If the sequence $\{f_n\}_2^\infty$ is bounded in X, and $f_1 \in X$, and g_n ($n = 2, 3, \ldots$) are analytic functions in $\mathbf{D}$ for which $|g_n(\zeta)| = |f_1(\zeta)| + |f_n(\zeta)|$, $n = 2, 3, \ldots$, then $g_n \in X$, $n \geqslant 2$, and the sequence $\{g_n\}_2^\infty$ is bounded in X.

Property (C′) holds for each space X with a monotone norm (monotone system of seminorms); that is, in each space X such that

$$g \in A(\mathbf{D}),\ f \in X \text{ and } |g| \leqslant |f| \Rightarrow g \in X \text{ and } \|g\| \leqslant \|f\|.$$

For instance, this property holds for the spaces $H^p(\mathfrak{M})$, $A(\lambda)$ and $A_\infty(\lambda)$.

THEOREM 1. *Let X and $\mathfrak{M}$ be spaces of analytic functions, where X has properties* (C) *and* (D), *and $\mathfrak{M}$ contains* **1** *and has properties* (C) *and* (E). *Suppose that f is a function in* **D** *such that*

1) $f \in X$, $f(\zeta) \neq 0$, $|\zeta| < 1$;
2) $\mathfrak{M} \subset X$, $f\mathfrak{M} \subset X$;
3) $\exists \varepsilon > 0$: $f^{-\varepsilon} \in \mathfrak{M}$.

Then f is weakly invertible in X.

THEOREM 1′. *Let X and $\mathfrak{M}$ be spaces of analytic functions, where X has properties* (C) *and* (D), *and $\mathfrak{M}$ contains* **1** *and has properties* (C′) *and* (F) (*or* (F*)).

Suppose that f is a function in **D** *such that*

$$1)\ f\in X,\ f(\zeta)\neq 0,\ |\zeta|<1;\ 2)\ f\,\mathfrak{M}\subset X;\ 3)\ \exists\varepsilon>0: f^{-\varepsilon}\in\mathfrak{M}.$$

Then f is weakly invertible in X.

The proofs of Theorems 1 and 1′ differ only in one detail, and therefore they can be carried out together. Condition 2) defines a linear operator of multiplication from $\mathfrak{M}$ into X, namely, $g\longrightarrow fg$, which clearly is closed and therefore continuous. Since the space $\mathfrak{M}$ contains **1** and has property (C), the inequalities

$$|f^{-\delta}(\zeta)|\leqslant 1+|f^{-\varepsilon}(\zeta)|,\ |\zeta|<1,\ 0<\delta\leqslant\varepsilon,$$

imply that $f^{-\delta}\in\mathfrak{M}$ for $0<\delta\leqslant\epsilon$. We choose δ in the interval $0<\delta<\min(\epsilon, 1)$ such that $1/\delta$ is a large integer. By virtue of the density of polynomials in $\mathfrak{M}$ and possibly property (A) we can find in $\mathfrak{M}$ a bounded sequence $\{p_n\}_1^\infty$, $p_n=\Sigma_0^n c_{k,n}z^k$, $p_n \xrightarrow{(*)} f^{-\delta}$. Then the sequence $\{p_nf\}_1^\infty$ is bounded in X and $p_n(\zeta)f(\zeta)\longrightarrow f^{1-\delta}(\zeta)$, $|\zeta|<1$. Consequently $f^{-\delta}\in E(f)=\mathfrak{L}(z^nf)_{n\geqslant 0}$.

Now we note that by the hypothesis of Theorem 1 we have $f^{1-\delta}\mathfrak{M}\subset X$, $0\leqslant\delta\leqslant 1$; in fact, if $g\in\mathfrak{M}$, then

$$|f^{1-\delta}g|\leqslant(1+|f|)|g|=|g|+|fg|,$$

and then we just have to quote property (C) of X. Since $f^{-\epsilon(1-\delta)}\in\mathfrak{M}$, to the function $f^{1-\delta}$ we can apply the arguments in the preceding paragraph, and if we repeat them $1/\delta$ times we find that $\mathbf{1}\in E(f)$. Theorem 1 is proved.

Under the conditions in Theorem 1′ the proof is somewhat different. For the approximating sequence $\{p_n\}_1^\infty$ we use the Abel means of $f^{-\delta}$:

$$p_n=\frac{1}{2\pi i}\int\limits_{|\alpha|=1}(f^{-\delta})_{\bar\alpha}\,P_{1-\frac{1}{n}}(\alpha)\frac{d\alpha}{\alpha},$$

where P_r, $0\leqslant r<1$, is the Poisson kernel. By property (F) we have $p_n\xrightarrow{(*)} f^{-\delta}$. Also

$$p_n(\zeta)=f^{-\delta}\left(\left(1-\frac{1}{n}\right)\zeta\right)\neq 0,\ |\zeta|<1 \text{ and } |p_n^{1-\delta}|\leqslant 1+|p_n|.$$

By virtue of (C′) the sequence $\{p_n^{1-\delta}\}_1^\infty$ is bounded in $\mathfrak{M}$, and therefore $p_n^{1-\delta}f^{(1-\delta)} \xrightarrow{(*)} f^{(1-\delta)^2}$. Consequently $f^{(1-\delta)^2}\in E(f)$. By induction we have

$$(*)\text{-}\lim_{k\to\infty} p_k^{(1-\delta)^{n-1}}f^{(1-\delta)^{n-1}}=f^{(1-\delta)^n}\in E(f),\ n\geqslant 2, \text{ and } |f^{(1-\delta)^n}|\leqslant 1+|f|.$$

Therefore from properties (C) and (D) of the space X it follows that $f^{(1-\delta)^n}\xrightarrow{\text{w.}X}\mathbf{1}$, and thus $\mathbf{1}\in E(f)$. ●

COROLLARY 1. *Let X be a space of analytic functions having properties* (C) *and* (D). *Suppose that $f\in X$, $f(\zeta)\neq 0$, $|\zeta|<1$, and*

$$m_f(r)\overset{\text{def}}{=}\min_{|\zeta|=r}|f(\zeta)|,\quad 0\leqslant r<1.$$

If for each f *we have* $A(1/m_f^\delta) \subset X$, *where* $\delta > 0$, *then* f *is weakly invertible in* X.

For if $i = \inf_{0 \leqslant r < 1} m_f(r) > 0$, then by (*)-approximating the (bounded) function $1/f$ by polynomials p_n, $n = 1, 2, \ldots$, we find from properties (C) and (D) that $p_n f \xrightarrow{w.X} 1$; that is, f is weakly invertible in X. If $i = 0$, then $m_f^\delta = o(m_f^{\delta/2})$, and consequently $m_f^\delta(|\zeta|) = o(|f(\zeta)|^{\delta/2})$, or $1/f^{\delta/2} \in A(1/m_f^\delta)$. If we apply Theorem 1′ we now obtain Corollary 1. ●

Another special case of Theorem 1′ is the following refinement of a theorem of Hedberg [139].

COROLLARY 2. *Let* $\mathfrak{M}$ *be a finite measure in* **D** *which "depends only on the radius"; that is,* $d\mathfrak{M} = h\,dl_2$, $h(\zeta) = h(|\zeta|)$, $|\zeta| < 1$. *If* $1 \leqslant p < +\infty$ *and for some* $\delta > 0$ *we have* $f \in H^{p+\delta}(\mathfrak{M})$ *and* $f^{-\delta} \in H^1(\mathfrak{M})$, *then* f *is weakly invertible in* $H^p(\mathfrak{M})$. *This means* (*see* §2.0) *that polynomials are dense in the space* $H^p(|f|^p\mathfrak{M})$.

In the work of Hedberg $\mathfrak{M} = l_2$, but the role of the disk **D** is played by an arbitrary Carathéodory domain. Corollary 2 is obtained from Theorem 1′ with $X = H^p(\mathfrak{M})$ and $\mathfrak{N} = H^q(\mathfrak{M})$, where $q = p(1 + p/\delta)$ and $\epsilon = \delta/q$. Here we should also mention the work of Lisin [56] where the Hedberg theorem is generalized in another direction.

EXAMPLE 1. If f is an analytic function in **D** with $f(\zeta) \neq 0$, $|\zeta| < 1$, and

$$\int_{\mathbf{D}} \frac{|f|^p}{m_f^\delta}\, d\mathcal{M} < +\infty \tag{2}$$

for some $\delta > 0$, then f is weakly invertible in $H^p(\mathfrak{M})$.

REMARK. In the case where $p = 2$, $\delta = 2$, and $\mathfrak{M} = l_2$ is planar Lebesgue measure in **D**, this assertion is Theorem 3 in [100], Chapter III, §2, and for $p = 2$, $\delta > 0$, and $\mathfrak{M} = l_2$ it is Theorem 1 in [171], where it is also assumed that for $\mathfrak{M} = l_2$ we have

$$\int_{\mathbf{D}} m_f^{-\delta}\, d\mathcal{M} < +\infty. \tag{3}$$

It is clear that for measures $\mathfrak{M}$ which "depend only on the radius" (that is, $d\mathfrak{M} = h(r)dl_2$), from (2) we obtain (3):

$$\int_{\mathbf{D}} |f(re^{i\theta})|^p\, m_f^{-\delta}(r)\, r h(r)\, dr\, d\theta \geqslant \text{const} \int_0^1 m_f^{-\delta}(r)\, h(r)\, dr = \text{const} \int_{\mathbf{D}} m_f^{-\delta}\, d\mathcal{M}.$$

Therefore the Shapiro theorem also follows from Theorem 1.

EXAMPLE 2. If $\lim_{r \to 1-0} \lambda(r) = +\infty$, $f \in A(\lambda)$, $M_f(r) \stackrel{\text{def}}{=} \max_{|\zeta|=r} |f(\zeta)|$, $0 \leqslant r < 1$, and[5] $M_f m_f^{-\epsilon} = O(\lambda)$ for some $\epsilon > 0$, then f is weakly invertible in $A(\lambda)$. In other words, each function f which is analytic and does not vanish in **D** is weakly invertible in $A(M_f m_f^{-\epsilon})$, $\epsilon > 0$, if it belongs to this space.

---

(5) This relation corresponds to the inclusions $A(m_f^{-\epsilon}) \subset X$ and $fA(m_f^{-\epsilon}) \subset X$, where $X = A(\lambda)$, and it means that f is not "too small" and not "too big" (see subsection 1°) with respect to X.

REMARK. Example 2 expresses the content of Theorem 4 in [87]. Other applications of Theorems 1 and 1′ (to nonnormable unions and intersections of spaces of the type $H^p(\mathfrak{M})$ and $A(\lambda)$) are contained in §2.7. We note also that from the results in Chapter 1 it follows automatically that we have the necessary lower estimates for f in Corollary 2 and Examples 1 and 2 if the measure $\mathfrak{M}$ and the weight $1/\lambda$ decrease with sufficient speed and "regularity" upon approach to the circumference. This fact is discussed in detail in §2.4.

If we use known lower estimates for inner functions (see [159] and [170]), we can find explicit sufficient conditions for their weak invertibility; see also [113].

EXAMPLE 3. Suppose that f is an inner function which is determined by a singular measure $\rho \geqslant 0$; set

$$f(\zeta) = \exp\left(-\int_{\partial D} (t+\zeta)(t-\zeta)^{-1} d\rho\right), \quad |\zeta| < 1;$$

and let $\omega(h) = \sup \rho(\Delta)$ be the modulus of continuity of ρ (the supremum is taken over all arcs Δ, $l_1(\Delta) \leqslant h$, $h > 0$). In [159] and [170] it is shown that

$$\log 1/m_f(r) = O((1-r)^{-1}\tilde{\omega}(1-r)), \quad r \to 1 \Leftrightarrow \tilde{\omega}(h) = O(\omega(h)), \quad h \to 0.$$

Therefore if

$$\int_D |f|^p \exp\left(\delta \frac{\omega(1-|\zeta|)}{1-|\zeta|}\right) d\mathcal{M} < +\infty$$

for some $\delta > 0$, then f is weakly invertible in $H^p(\mathfrak{M})$, $1 \leqslant p < \infty$; if $d\mathfrak{M} = h(r) dl_2$, then this condition can be written in the form

$$\int_0^1 \exp\left(\delta \frac{\omega(1-r)}{1-r}\right) h(r)\, dr < +\infty.$$

For "smooth" singular measures with $\omega(h) = Ch \log(1/\omega_1(h))$, where $\omega_1(0) = \omega(+0) = 0$ (for instance, $\omega_1(h) \equiv h$), we obtain the condition

$$\int_0^1 \frac{h(r)}{\omega_1(1-r)^\delta} dr < +\infty.$$

In particular, singular functions f with $\omega_1(h) \equiv h$ are weakly invertible in $H^2((1-r)^\alpha dl_2)$, $\alpha > -1$; the latter assertion is contained in [113].

For the space $A(\lambda)$, $\lim_{r \to 1-0} \lambda = +\infty$, a sufficient condition for weak invertibility is

$$\frac{\omega(1-r)}{1-r} = O(\log \lambda(r)), \quad r \to 1-0.$$

In particular, functions f with $\omega_1(h) \equiv h$ are weakly invertible in $A((1-r)^{-\alpha})$ for each $\alpha > 0$. In the space $A_\infty(1)$ this is not so.

3°. *Weak invertibility in the nonseparable space $A_\infty(\lambda)$, and the connection with the problem of simultaneous approximation in* **D** *and* ∂**D**. We return to the investigation of the exceptional case $f \notin A(M_f m_f^{-\epsilon})$, which was not covered by Example 2. The assumption which we made is equivalent to the condition $\inf_{\mathbf{D}} |f| > 0$. In this case weak invertibility of f is possible in an arbitrary space $A(\lambda)$ with $M_f = o(\lambda)$. It is also clear that such a function is $(*)$-invertible in $A_\infty(M_f)$:

$$\sup_n \|\sigma_n f\|_{A_\infty(M_f)} < +\infty \quad \text{and } \sigma_n(\zeta) f(\zeta) \to 1, \quad |\zeta| < 1,$$

where σ_n is the arithmetic mean of the Taylor series for $1/f$.

The problem of weak invertibility of f in the "small" space $A_\infty(M_f)$ is of a special type since this space is not separable. This last fact follows from the next lemma.

LEMMA. *If λ is a positive function on* $[0, 1)$ *and* $\inf \lambda > 0$, *then the space $A_\infty(\lambda)$ is not separable.*

PROOF. We consider a "sparse" sequence of powers $\{z^{n_k}\}_{k=1}^\infty$, $z(\zeta) \equiv \zeta$, $|\zeta| < 1$. We have

$$\overline{\lim_n} \|z^n\|_{A_\infty(\lambda)}^{1/n} = 1, \tag{4}$$

since otherwise $r^n/\lambda(r) \leqslant \text{const } R^n$ for some R with $0 < R < 1$ and for all $0 \leqslant r < 1$ and $n = 1, 2, \ldots$. Then we would have $\lambda(r) = +\infty$ if $r > R$. Without loss of generality we can assume that $\lim_{r\to 1-0} \lambda(r) = +\infty$, since for $\underline{\lim}_{r \to 1-0} \lambda(r) < +\infty$ we have $A_\infty(\lambda) = A_\infty(1)$, and this space was shown to be nonseparable in [23]. If $n_1 < \cdots < n_k$ and $R_1 < \cdots < R_{k-1} < 1$ have already been chosen, then we define the number R_k with $R_{k-1} < R_k < 1$ by the condition

$$\frac{r^{n_k}}{\lambda(r) \|z^{n_k}\|_{A_\infty(\lambda)}} < 2^{-k} \quad \text{for} \quad r \geqslant R_k, \tag{5}$$

and then from (4) we know that there exists a number n_{k+1} such that

$$\frac{r^{n_{k+1}}}{\lambda(r) \|z^{n_{k+1}}\|_{A_\infty(\lambda)}} < \frac{R_k^{n_{k+1}}}{c \|z^{n_{k+1}}\|_{A_\infty(\lambda)}} < 2^{-k} \quad \text{for} \quad r < R_k, \tag{6}$$

where $c = \inf_{0 \leqslant r < 1} \lambda(r)$. Let $\{c_k\}_1^\infty$ be a sequence of real numbers with $0 \leqslant c_k \leqslant 1$. Then for each $k \geqslant 1$ we have

$$c_k = c_k \sup_r \frac{r^{n_k}}{\lambda(r) \|z^{n_k}\|} \leqslant \sup_{|\zeta|<1} \left| \sum_{j=1}^\infty c_j \frac{\zeta^{n_j}}{\|z^{n_j}\|} \right| \frac{1}{\lambda(|\zeta|)} = \left\| \sum_{j=1}^\infty c_j \frac{z^{n_j}}{\|z^{n_j}\|} \right\|.$$

On the other hand, if $R_{k-1} \leqslant r < R_k$, then from (5) and (6) we obtain the inequalities

$$\begin{aligned} \sum_{j=1}^\infty c_j \frac{r^{n_j}}{\lambda(r) \|z^{n_j}\|} &\leqslant c_k + \sum_{j=1}^{k-1} c_j \frac{r^{n_j}}{\lambda(r) \|z^{n_j}\|} + \sum_{j=k+1}^\infty c_j \frac{r^{n_j}}{\lambda(r) \|z^{n_j}\|} \\ &\leqslant c_k + \sum_{j=1}^{k-1} c_j 2^{-j} + \sum_{j=k+1}^\infty c_j 2^{-j+1} \leqslant 2 \sup_{k \geqslant 1} c_k. \end{aligned}$$

Consequently

$$\sup_{k\geqslant 1} c_k \leqslant \left\| \sum_{j=1}^{\infty} c_j \frac{z^{n_j}}{\|z^{n_j}\|} \right\| \leqslant \text{const} \sup_{k\geqslant 1} c_k,$$

and nonseparability of $A_\infty(\lambda)$ follows from the analogous property for the space l^∞ of all bounded sequences. ●

Nonseparability of the space $A_\infty(M_f)$ leads to a series of pathological features of the problem which we are considering; we will describe a few of them together with those few cases where the required approximation is automatically possible.

THEOREM 2. *Let* $f \in A(\mathbf{D})$, $f(\zeta) \neq 0$, $|\zeta| < 1$, $M_f(r) = \max_{|\zeta|=r} |f(\zeta)|$, $\sup_{\mathbf{D}} |f| = +\infty$ *and* $m = \inf_{\mathbf{D}} |f| > 0$. *The following assertions are true for approximations in* $A_\infty(M_f)$:

1) $E(f) \neq A_\infty(M_f)$.

2) *If the function* $1/f$ *is uniformly continuous in* $\mathbf{D}$, *then* $E(f)$ *contains the set* C_A *of all functions with this property (and* f *is weakly invertible in* $A_\infty(M_f)$).

3) *Let*

$$e = e(f) \equiv \left\{ \zeta : |\zeta| = 1,\ \varlimsup_{\substack{\xi\to\zeta \\ |\xi|<1}} |f(\xi)| / M_f(|\xi|) > 0 \right\}.$$

If $|e| = 0$, *then* $C_A \subset E(f)$; *in particular, if the canonical factorization of* $1/f$ *contains a singular factor of the form*

$$\exp\left(-\alpha \frac{\zeta_0 + z}{\zeta_0 - z}\right), \quad |\zeta_0| = 1,$$

then $C_A \subset E(f)$.

4) *There exists a function* f *which satisfies the conditions in the theorem and is such that*

$$\sup_{|\zeta|<1} \frac{|pf - g|}{M_f} \geqslant \max_{|\zeta|\leqslant 1} |p|$$

for an arbitrary polynomial p *and for an arbitrary function* g *in* $A(M_f)$; *in particular,* $A(M_f) \cap E(f) = \{0\}$.

PROOF. 1) This follows from the nonseparability of $A_\infty(M_f)$.

2) If $g \in C_A$ and $\|p_n - g/f\|_{C_A} \to 0$, then

$$\sup_{\mathbf{D}} \frac{|p_n f - g|}{M_f} \leqslant \sup_{\mathbf{D}} |p_n - g/f| = \|p_n - g/f\|_{C_A} \to 0$$

as $n \to \infty$. Therefore $C_A \subset E(f)$.

3) If we let

$$e_n = \left\{ \zeta : |\zeta| = 1,\ \varlimsup_{\substack{\xi\to\zeta \\ |\xi|<1}} |f(\xi)| / M_f(|\xi|) \geqslant \frac{1}{n} \right\},$$

then $e = \bigcup_1^\infty e_n$, each of the sets e_n is closed, and $|e_n| = 0$ for $n = 1, 2, \ldots$. Let $e_{n,\epsilon}$ be an ϵ-neighborhood (in $\overline{\mathbf{D}}$) of the set e_n (here $\epsilon > 0$). If we use the Fatou theorem in [95], we can construct a function $\varphi_n \in C_A$ such that $\varphi_n | e_n = 0$ and $1 \geqslant |\varphi_n(\zeta)| > 0$ if $\zeta \in \overline{\mathbf{D}} \backslash e_n$. We now show that

$$\operatorname{dist}(\varphi_n^\alpha, E(f)) \leqslant \frac{1}{n} \cdot \frac{1}{m} \tag{7}$$

for an arbitrary $\alpha > 0$, where $m = \inf |f|$. If we choose $\alpha_n \longrightarrow 0$ such that $\varphi_n^{\alpha_n} \overset{(*)}{\longrightarrow} \mathbf{1}$, we find that(6) $\|\varphi_n^{\alpha_n} - \mathbf{1}\|_{A_\infty(M_f)} \longrightarrow 0$ and $\mathbf{1} \in E(f)$, and then $C_A \subset E(f)$. In order to prove (7) we first note that

$$\lim_{\varepsilon \to 0} \max_{e_{n,\varepsilon}} |\varphi_n^\alpha| = 0.$$

On the other hand, for each $\epsilon > 0$ there exists a number r_ϵ with $0 < r_\epsilon < 1$ such that $1/M_f(r_\epsilon) < \epsilon$ and $|f(\xi)|/M_f(|\xi|) < 1/n$ for $\xi \in \mathbf{D} \backslash e_{n,\epsilon}$, $|\xi| > r_\epsilon$. Let $\{p_n\}_1^\infty$ be a sequence of polynomials which is (*)-convergent to $1/f$ (to fix ideas, let $p_n = \sigma_n(f)$, the arithmetic mean of the Taylor series of $1/f$), and let $n(\epsilon)$ be so large that

$$\max_{|\zeta| \leqslant r_\varepsilon} |p_n f - 1| < \varepsilon.$$

Then

$$\frac{|\varphi_n^\alpha - p_n \varphi_n^\alpha f|}{M_f} \leqslant \frac{\varepsilon}{M_f(0)},$$

if $|\zeta| \leqslant r_\epsilon$,

$$\frac{|\varphi_n^\alpha - p_n \varphi_n^\alpha f|}{M_f} \leqslant \frac{|1 - p_n f|}{M_f} \leqslant \frac{1}{M_f} + \frac{|p_n||f|}{M_f} < \varepsilon + \frac{1}{n} \cdot \frac{1}{m},$$

if $|\zeta| > r_\epsilon$ and $\zeta \in \mathbf{D} \backslash e_{n,\epsilon}$, and

$$\frac{|\varphi_n^\alpha - p_n \varphi_n^\alpha f|}{M_f} \leqslant |\varphi_n^\alpha| \left(\frac{1}{M_f} + \frac{|p_n f|}{M_f} \right) \leqslant \max_{e_{n,\varepsilon}} |\varphi_n^\alpha| \left(\frac{1}{M_f(r_\varepsilon)} + \frac{1}{m} \right),$$

if $|\zeta| > r_\epsilon$ and $\zeta \in e_{n,\epsilon}$. Consequently

$$\|\varphi_n^\alpha - \varphi_n^\alpha p_n f\|_{A_\infty(M_f)} \leqslant \frac{\varepsilon}{M_f(0)} + \varepsilon + \frac{1}{nm} + \max_{e_{n,\varepsilon}} |\varphi_n^\alpha| \left(\frac{1}{M_f(r_\varepsilon)} + \frac{1}{m} \right).$$

Since $\epsilon > 0$ was arbitrary and since $\varphi_n^\alpha p_n f \in E(f)$, inequality (7) now follows. Assertion 3) is proved.

4) We construct an inner (see [23]) function f, defining it by the formula

$$f(\zeta) = \exp\left(\int_{\partial \mathbf{D}} \frac{t + \zeta}{t - \zeta} q(t)\, dl_1(t) \right), \quad |\zeta| < 1,$$

where q is an integrable nonnegative function on $\partial\mathbf{D}$, and l_1 is normalized Lebesgue

(6) We use the fact that $\lim_{r \to 1-0} M_f(r) = +\infty$.

measure. If we use the fact that for all ζ, $|\zeta| = 1$, we have

$$\lim_{\substack{\xi\to\zeta\\ |\xi|<1}} \frac{|f(\xi)|}{M_f(|\xi|)} = 1, \tag{8}$$

then it is easy to finish the proof of the theorem. In fact, for arbitrary functions $g \in A(M_f)$ and $p \in \mathcal{P}_A$ we have

$$\max_{|\xi|\leqslant 1} |p(\xi)| = \max_{|\zeta|=1} \lim_{\xi\to\zeta} |p(\xi)| = \max_{|\zeta|=1}\left(\overline{\lim_{\xi\to\zeta}} \frac{|p(\xi)f(\xi)|}{M_f(|\xi|)} - \overline{\lim_{\xi\to\zeta}} \frac{|g(\xi)|}{M_f(|\xi|)}\right)$$
$$\leqslant \max_{|\zeta|=1} \lim_{\xi\to\zeta} \frac{|pf-g|}{M_f} \leqslant \sup_{|\xi|<1} \frac{|pf-g|}{M_f} = \|pf-g\|_{A_\infty(M_f)}.$$

Therefore the equation $\lim_n \|p_n f - g\|_{A_\infty(M_f)} = 0$ implies that $\lim_n \|p_n\|_{C_A} = 0$, and since $\lim_n p_n = g/f$ in $\mathbf{D}$, we have $g \equiv 0$.

We now construct an outer function f with the property (8) by setting $q_0 = 0$ and $q = \Sigma_0^\infty q_n$, where we define the functions q_n, $n \geqslant 1$, by induction. The basis of our argument will be the elementary fact that the magnitude of the Poisson integral

$$\int_{\partial\mathbf{D}} \frac{1-r^2}{1-2r\operatorname{Re}\left(\frac{\zeta}{t}\right)+r^2}\, d\mathcal{M}(t) \quad (r\zeta \in \mathbf{D},\ |\zeta|=1)$$

for small r depends only on the total mass of the measure $\mathfrak{M}$, and for r close to 1 it depends on the magnitude of the derivative $|(d\mathfrak{M}/dl_1)(\zeta)|$.

We will use the following notation: k and p are natural numbers;

$$s_n = \left\{\zeta\colon\ |\zeta|=1,\ 0 \leqslant \arg\zeta < 2\pi\frac{1}{n}\right\};$$

$$s(k,\ p,\ j) = s_{2^{k+p}} \exp(2\pi i j 2^{-k}),\quad j = 0,\ 1,\ \ldots,\ 2^k - 1;$$

$$\mathcal{E}_{k,\,p} = \bigcup_{j=0}^{2^k-1} s(k,\ p,\ j);$$

$$P_{r\zeta}(t) = (1-r^2)(1-2r\operatorname{Re}(\zeta/t)+r^2)^{-1}$$

—is the Poisson kernel ($|\zeta| = |t| = 1$; $0 \leqslant r < 1$); $\chi_{\mathcal{E}}$ is the characteristic function of the set $\mathcal{E}$; and the d_k are positive numbers such that $d_k \uparrow +\infty$.

Suppose that we have already chosen numbers $p_1 \leqslant \cdots \leqslant p_n$ and $r_1 < \cdots < r_n$, and functions $q_1, \ldots, q_n$ of the form

$$q_k = \sum_{j=0}^{2^k-1} c_j \chi_{s(k,\,p_k,\,j)} \ (c_j \geqslant 0) \quad\text{and}\quad Q_k = \sum_{j=0}^{k} q_j$$

such that

a) $Q_k | \mathfrak{S}_{k,p_k} \equiv d_k$;

b) $(1-r_{k-1})^{-1} \int_{\partial D} q_k dl_1 \leqslant 2^{-k}$, $k \geqslant 1$ ($r_0 \overset{\text{def}}{=} 0$);

c) $\int_{\partial D} P_{r_k \zeta_j} Q_k dl_1 \geqslant \max_{|\zeta|=1} \int_{\partial D} P_{r_k \zeta} Q_k dl_1 - 1/2k > d_k - 1/k$, where

$$\zeta_j = \exp\{2\pi i\,(j2^{-k} + 2^{-k-p_k-1})\},\ j = 0, \ldots, 2^k - 1,$$

are the centers of the arcs $s(k, p_k, j)$.

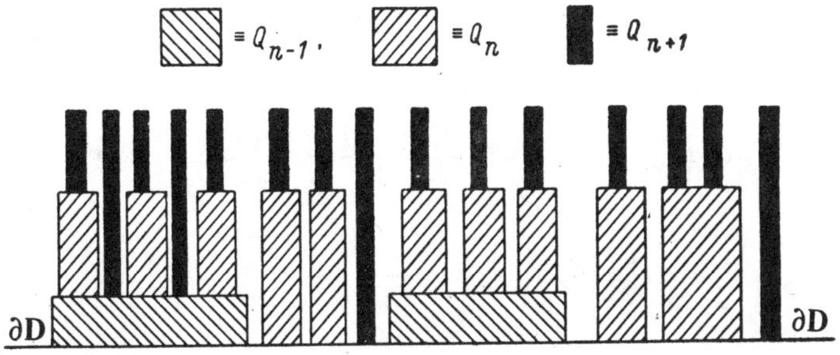

FIGURE 4

We set

$$q_{n+1} = \sum_{j=0}^{2^{n+1}-1} c_j \chi_{s(n+1,\ p_{n+1},\ j)},$$

where the coefficients $c_j \geqslant 0$ are chosen in accord with the condition(7) (see Figure 4)

$$(Q_n + q_{n+1}) \,|\, \mathcal{E}_{n+1,\ p_{n+1}} \equiv d_{n+1},$$

and the numbers p_{n+1} with $p_{n+1} \geqslant p_n$ are chosen so that

$$(1 - r_n)^{-1} \int_{\partial D} q_{n+1} dl_1 < 2^{-n-1}.$$

This is possible since

$$\int_{\partial D} q_{n+1} dl_1 = \sum_{j=0}^{2^{n+1}-1} c_j 2^{-(n+1)-p_{n+1}},$$

and for the chosen c_j this number becomes arbitrarily small for large p_{n+1}.

In order to find r_{n+1} with $r_n < r_{n+1} < 1$, we note that

$$\max_{|\zeta| \leqslant 1} \int_{\partial D} P_\zeta Q_{n+1} dl_1 = \max_{|\zeta|=1} Q_{n+1} = d_{n+1},$$

and this maximum is attained at points in the set $\mathfrak{S}_{n+1,p_{n+1}}$. Therefore it is clear that the harmonic function $\int_{\partial D} P_\zeta Q_{n+1} dl_1$ tends along the radii

(7) It is easy to see that: 1) each arc $s(n+1, p_{n+1}, j)$ either is completely contained in $s(k, p_k, j')$, $k \leqslant n$, or these sets do not intersect (this makes the choice of c_j possible); 2) the numbers c_j depend only on j, n and d_n, and not on p_{n+1} (see Figure 4).

$$\arg\zeta = 2\pi(j2^{-n-1} + 2^{-n-1-p_{n+1}-1})\ (j = 0, \ldots, 2^{n+1}-1)$$

to a limit d_{n+1}, and there exists a number r_{n+1} for which condition c) is satisfied with $k = n + 1$. This completes the construction of the sequence $\{q_n\}_0^\infty$.

The set of limit points for the sequence $r_n\zeta_{j,n}$, $n \geqslant 1, j = 0, 1, \ldots, 2^n - 1$ ($\zeta_{j,n}$ is the center of the arc $s(n, 2^n, j)$), is the circle $\partial\mathbf{D}$. Therefore (8) follows from the estimates

$$\max_{|\zeta|=1} |f(r_n\zeta)| = \exp\left(\max \int_{\partial\mathbf{D}} P_{r_n\zeta}\left(\sum_{j=0}^{\infty} q_j\right) dl_1\right)$$

$$\leqslant \exp\left(\max \int P_{r_n\zeta} Q_n dl_1 + \frac{1}{1-r_n} \sum_{j=n+1}^{\infty} \int q_j dl_1\right)$$

$$\leqslant \exp\left(\int P_{r_n\zeta_{j,n}} Q_n dl_1 + \frac{1}{2n} + \sum_{j=n+1}^{\infty} 2^{-j}\right)$$

$$\leqslant \exp\left(\int P_{r_n\zeta_{j,n}}\left(\sum_{j=0}^{\infty} q_j\right) dl_1 + \frac{1}{2n} + 2^{-n}\right) = |f(r_n\zeta_{j,n})| \exp\left(\frac{1}{2n} + 2^{-n}\right). \ \bullet$$

We now make a few comments on the theorem which we just proved. The outer function f which we constructed is pathological since it is the case that near each point of the circumference the value $|f(\zeta)|$ is as large as $\max_{|t|=|\zeta|}|f(t)|$. Moreover, the set $e(f)$ in the third assertion in the theorem coincides in this case with $\partial\mathbf{D}$. If in the definition of $e(f)$ we restrict ourselves to angular limits (that is, if we set

$$e_*(f) = \{\zeta: |\zeta| = 1, \varlimsup_{\xi\to\zeta,\ \xi\in\Delta_\zeta} |f(\xi)|/M_f(|\xi|) > 0\},$$

where Δ_ζ is some angle with vertex ζ and bisector $\{r\zeta: 0 \leqslant r < 1\}$ and with an opening less than π), then $|e_*(f)| = 0$ for each unbounded function f in the class of Nevanlinna (see [95]) and therefore also for the function f in Theorem 2. It is not clear whether we can have the result of assertion 4) in Theorem 2 (that is, $e(f) = \partial\mathbf{D}$) for singular functions $1/f$; if the singular part of $1/f$ contains a factor of the form

$$\exp\left(-\alpha\frac{\zeta_0 + z}{\zeta_0 - z}\right), \quad |\zeta_0| = 1, \quad \alpha > 0,$$

then the set $e(f)$ is finite. It is also not clear whether the property $e(f) = \partial\mathbf{D}$ (or $|e(f)| > 0$) depends on the speed of growth of $\max_{|\zeta|=r}|f(\zeta)|$ as $r \longrightarrow 1 - 0$.

Finally we comment on the connection between Theorem 2 and the problem of simultaneous approximation. The inequality $\|pf - 1\|_{A_\infty(\lambda)} < \epsilon$ means that

$$\left|p - \frac{1}{f}\right| < \varepsilon\frac{\lambda}{|f|} \tag{9}$$

and on the set

$$e_m(f) = \left\{\zeta : |\zeta| = 1, \varlimsup_{\xi\to\zeta} \frac{\lambda(|\xi|)}{|f(\xi)|} \leqslant m\right\}$$

we have $|p| < \epsilon m$. Thus under the conditions in Theorem 2 weak invertibility of a function f in $A_\infty(\lambda)$, $\lambda \equiv M_f$, implies the possibility of simultaneous approximation(8)

$$\inf_n \left(\left\| p - \frac{1}{f} \right\|_{A_\infty(\lambda)} + \max_{e_m} |p| \right) = 0, \tag{10}$$

where $e_m = e_m(f)$ and the infimum is taken over all polynomials p. From (10) it follows that the pair $(1/f, \mathbf{0})$ belongs to the closure of the diagonal $D = \{(p, p)\}$ in the space $A_\infty(\lambda) \times C(e_m)$, and therefore $A(\lambda) \times \{\mathbf{0}\}$ also belongs to this closure, so that for each pair of functions (g, φ), $g \in A(\lambda)$, $\varphi \in C(e_m)$, we can achieve the simultaneous approximation

$$\inf_p (\| p - g \|_{A(\lambda)} + \| p - \varphi \|_{C(e_m)}) = 0.$$

For example, such an approximation is possible if the condition in assertion 3) of Theorem 2 is satisfied; that is, if $|e(f)| = 0$ (and consequently $|e_m| = 0$). This fact follows from a general theorem of Havin [104] on completeness of the diagonal D in spaces $X \times C(e)$, where e is a closed subset of $\partial\mathbf{D}$ with $|e| = 0$, and X consists of analytic functions in $\mathbf{D}$ and contains the polynomials as a dense subset and satisfies the condition

$$\sup \| p_n \|_{C_A} < +\infty, \quad \lim_n p_n(\zeta) = 0 \, (\zeta \in \overline{\mathbf{D}} \setminus e) \Rightarrow \lim_n \| p_n \|_X = 0.$$

On the other hand, the reasoning in part 3) of Theorem 2 also proves the Havin theorem as follows: since the uniform norm $\| \cdot \|_{C_A}$ is stronger than the norm in X, we have $(\varphi^\alpha, \mathbf{0}) \in \overline{D}$ (the closure is taken in $X \times C(e)$; $\varphi = \varphi_e$ is the function in the proof of Theorem 2) and $\|\varphi^\alpha - \mathbf{1}\|_X \longrightarrow 0$ as $\alpha \longrightarrow 0$. Consequently $(\mathbf{1}, \mathbf{0}) \in \overline{D}$, and $(X, \mathbf{0}) \subset \overline{D}$, and $\overline{D} = X \times C(e)$. As examples of such spaces X we can use (see [104]) $H^p(\mathfrak{M})$ with $\mathfrak{M}(e) = 0$, and $A(\lambda)$ with $\lim \lambda = +\infty$ and $\inf \lambda > 0$. This problem of simultaneous approximation was considered earlier by Kegejan [44] and then by Hruščev [109]. In [44] it was shown that there exist sets $e \subset \partial\mathbf{D}$ of measure arbitrarily close to 1 for which the diagonal D is dense in $H^2(l_2) \times C(e)$. We also mention the connection between problems of this sort and the theory of multiplicity of the spectrum in spaces of generalized functions (see [84] and §3.4 in the present work) and uniqueness theorems and approximations "with additional restrictions" (for example, restrictions on the magnitude of the Taylor coefficients of the approximating polynomials; see [107], [172], [173]).

4°. *Properties of weakly invertible functions. Functions in the class* $\mathfrak{D}$. In this subsection and the next we introduce certain general properties of weakly invertible functions and present examples of such functions. The results which we obtain show that in an arbitrary space X these functions are a natural analogue of the outer functions

---

(8) This transition is a significant weakening of the fact of weak invertibility of f, since the approximation (9) takes into account the smallness of $|p - 1/f|$ also in a neighborhood of the set $e_m(f)$, but for (10) this smallness is not needed.

of Beurling. The following theorem generalizes some thorems in [100], where $X = H^2(l_2)$.

THEOREM 3. *Let X be a space of analytic functions which has properties* (C) *and* (D). *Then the following assertions are true*:

1) *If* $f_1 \in X$, $|f(\zeta)| \leqslant \text{const}|f_1(\zeta)|$, $|\zeta| < 1$, *and f is weakly invertible in X, then f_1 has the same property.*

2) *If* f_n, $n = 1, 2, \ldots$, *are weakly invertible functions in X and* $|f(\zeta)| \leqslant \text{const}|f_n(\zeta)|$, $|\zeta| < 1$, $n \geqslant 1$, *and* $\lim_n f_n(\zeta) = f(\zeta) \neq 0$ *for* $|\zeta| < 1$, *then f is also weakly invertible in X.*

3) *If $f \in X$ and $1/f$ is a function in the Smirnov*(9) *class* $\mathfrak{D}$, *then f is weakly invertible in X. In particular, an outer function f is weakly invertible in an arbitrary space of the type which we are considering.*

PROOF. 1) Let σ_n be the arithmetic mean of the partial sums of the Taylor series of f/f_1; then $\sigma_n \xrightarrow{(*)} f/f_1$. From properties (C) and (D) we have $\sigma_n f_1 \xrightarrow{X} f$, and therefore $f \in E(f_1)$. Weak invertibility of f implies that $\mathbf{1} \in E(f_1)$.

2) Since f/f_n is bounded in $\mathbf{D}$, properties (A)–(C) give us continuity of the multiplication operator $g \longrightarrow gf/f_n$, $g \in X$. By hypothesis we have $\mathbf{1} \in E(f_n)$, and for an arbitrary polynomial p we have

$$fp = f/f_n + (pf_n - \mathbf{1})f/f_n.$$

Therefore $f/f_n \in E(f)$. Since $f/f_n \xrightarrow{(*)} \mathbf{1}$, properties (C) and (D) imply that $\mathbf{1} \in E(f)$.

3) Let $f = f_1/f_2$, where the f_i $(i = 1, 2)$ are bounded functions in $\mathbf{D}$ and f_1 is an outer function (see [95]). From the same arguments on $(*)$-convergence as in parts 1) and 2) we see that $qf \in E(f)$ for each bounded (analytic) function q in $\mathbf{D}$. Let $q_n = f_2 g_n$, where g_n is an outer function such that $|g_n| = \min(n, |f_1|^{-1})$. Then $|q_n f| = |g_n f_1| = \min(|f_1|n, 1) \leqslant 1$ and $g_n f_1 \xrightarrow{(*)} \mathbf{1}$, as is clear from the integral representation

$$(g_n f_1)(\zeta) = \exp\left(\int_{\partial\mathbf{D}} \frac{t+\zeta}{t-\zeta} \log|g_n f_1(t)|\, dl_1(t)\right)$$

and Levy's theorem on taking limits under the integral. Consequently $q_n f \in E(f)$, $q_n f \xrightarrow{X} \mathbf{1}$, and $\mathbf{1} \in E(f)$. ●

REMARK. From the proof it is clear that in case 3) there exist sequences of polynomials $\{p_n\}_1^\infty$ and $\{Q_n\}_1^\infty$ such that $\sup_n \|p_n\|_\infty < +\infty$, $p_n f(\zeta) \longrightarrow f_1(\zeta)$ for $|\zeta| < 1$ and $\sup_n \|Q_n f_1\|_\infty < +\infty$, $Q_n f_1(\zeta) \longrightarrow 1$, $|\zeta| < 1$.

COROLLARY. *Let $\mathfrak{M}$ be a finite real measure in* $\mathbf{C}\backslash\mathbf{D}$, *and set*

$$f(\zeta) = \exp\left\{\int_{\mathbf{C}\backslash\mathbf{D}} \log\left(1 - \frac{\zeta}{t}\right) d\mathcal{M}(t)\right\}, \quad |\zeta| < 1.$$

(9) That is, $1/f$ is a function of a certain form (or else is in the Nevanlinna class) for which the canonical factorization has no factors of the form $1/I$, where I is an inner function (see [23] and [95]).

Then the function f is weakly invertible in any space X in Theorem 3 in which it is contained.

For the proof it suffices to note that f is an outer function.(10) We will show the following: $f = f^+/f^-$, where $f^\pm$ are bounded functions in $\mathbf{D}$ corresponding to the decomposition of the measure $\mathfrak{M} = \mathfrak{M}_+ - \mathfrak{M}_-$ into positive and negative components. Also, from geometric considerations it is clear that

$$|r\zeta - t| \geqslant \min\left(1, \frac{1}{\sqrt{2}}|\zeta - t|\right) \quad \text{for } |t| \geqslant 1,\ |\zeta| = 1,\ 0 \leqslant r < 1,$$

and therefore

$$\log|\zeta - t| - \log 2 \leqslant \log\left|1 - \frac{r\zeta}{t}\right| \leqslant \log 2$$

for the same ζ and r with $1 \leqslant |t| \leqslant 2$. Since

$$\int\limits_{\mathbf{C}\setminus\mathbf{D}} \left(\int\limits_{\partial\mathbf{D}} \left|\log\left(1 - \frac{\zeta}{t}\right)\right| dl_1(\zeta)\right) d|\mathcal{M}(t)| < +\infty,$$

the family of functions

$$\left\{\log\left|1 - \frac{rz}{t}\right| : 0 \leqslant r < 1\right\}$$

has for l_1-almost all ζ a majorant which is symmetric with respect to $d|\mathfrak{M}|$, and by the Lebesgue limit theorem

$$\lim_{r\to 1-0} |f^\pm(r\zeta)| = \exp\left(\int\limits_{\mathbf{C}\setminus\mathbf{D}} \log\left|1 - \frac{\zeta}{t}\right| d\mathcal{M}_\pm(t)\right)$$

for almost all $\zeta \in \partial\mathbf{D}$. Since the function $\zeta \longrightarrow 1 - \zeta/t$, $|\zeta| < 1$, is outer for each t with $|t| \geqslant 1$, we have

$$\int\limits_{\partial\mathbf{D}} \log|f^\pm(\zeta)|\, dl_1 = \int\limits_{\mathbf{C}\setminus\mathbf{D}} \left(\int\limits_{\partial\mathbf{D}} \log\left|1 - \frac{\zeta}{t}\right| dl_1(\zeta)\right) d\mathcal{M}_\pm(t) = 0.$$

Since $\log|f^\pm(0)| = 0$, by virtue of a criterion in [23] we conclude that f is an outer function.

In the case

$$X = H^2(l_2),\ f(\zeta) = \prod_{k=1}^{\infty}\left(1 - \frac{\zeta}{t_k}\right)^{\alpha_k},\ |t_k| \geqslant 1,\ \sum|\alpha_k| < +\infty,\ \alpha_k > -1$$

and $\alpha_k \geqslant 0$ for all but finitely many k, this corollary coincides with Theorem 6 in §2 of Chapter 3 in [100], where it is proved by a different method.

5°. *Multiplicative properties of weakly invertible functions and inner functions.* To conclude this section we consider multiplicative properties of weakly invertible

(10) Theorem 3 says almost nothing about weak invertibility of inner functions. This gap is partly filled by Example 3, Theorem 4, and Example 4 (below); see also [113].

("outer") functions in spaces X which satisfy the conditions in Theorem 3.

THEOREM 4. *Let X be a space of analytic functions having properties* (C) *and* (D). *Let $f_1, f_2 \in X$ and $f = f_1 f_2 \in X$. Then the following assertions are true*:

1) *If one of the functions f_i, $i = 1, 2$, is bounded or one of the functions $1/f_i$, $i = 1, 2$, is in the class $\mathfrak{D}$, then weak invertibility of f_1 and f_2 implies weak invertibility of f.*

2) *If $f_1 \in \mathfrak{D}$ and $1/f_1$ is bounded, then weak invertibility of f implies weak invertibility of f_2.*

PROOF. 1) If, for example, f_1 is bounded, then we choose polynomials p_n such that $\lim_n p_n f_2 = \mathbf{1}$ and use the properties of X, and thereby obtain $\lim_n p_n f = f_1$. This gives us $f_1 \in E(f)$, and from weak invertibility of f_1 we conclude that $\mathbf{1} \in E(f)$.

If $1/f_1 \in \mathfrak{D}$, then $f_1 = F_1/F_2$, where F_1 is an outer function and F_2 is a bounded function. By the remark after Theorem 3 there exist sequences of functions $\{p_n\}_1^\infty$ and $\{q_n\}_1^\infty$ such that

$$\begin{gathered} \sup_n \|p_n\|_\infty < \infty, \; p_n(\zeta) f_1(\zeta) \to F_1(\zeta), \; |\zeta| < 1, \\ \sup_n \|q_n F_1\|_\infty < \infty, \; q_n(\zeta) F_1(\zeta) \to 1, \; |\zeta| < 1. \end{gathered} \tag{11}$$

Consequently

$$|p_n f| \leqslant \text{const}\, |f|, \quad p_n(\zeta) f(\zeta) \to F_1(\zeta) f_2(\zeta),$$

and by the hypothesis of the theorem $p_n f \to F_1 f_2$. Therefore $F_1 f_2 \in E(f)$. Analogously $|q_n F_1 f_2| \leqslant \text{const}\, |f_2|$ and $q_n F_1 f_2 \to f_2$ in $\mathbf{D}$, and $f_2 \in E(f)$. Since f_2 is weakly invertible, we have $\mathbf{1} \in E(f)$, and assertion 1) is proved.

2) If $1/f_1$ is bounded and $p_n f \xrightarrow{X} f_1$, then $p_n f_2 \xrightarrow{X} \mathbf{1}$, and f_2 is weakly invertible. If $f_1 \in \mathfrak{D}$, then with the notation of assertion 1) we construct for the function $1/f_1 = F_1/F_2$ the sequences (11) with f_1 replaced by $1/f_1$. Then $|p_n f_2| \leqslant \text{const}\, |f_2|$ and $p_n f_2 \to F_1 f$ in $\mathbf{D}$; that is, $F_1 f \in E(f_2)$. Also $|q_n F_1 f| \leqslant \text{const}\, |f|$ and $q_n F_1 f \to f$ in $\mathbf{D}$, and therefore $f \in E(f_2)$ and $\mathbf{1} \in E(f_2)$. •

EXAMPLE 4. Let f be an inner function and let X be a space as in Theorem 4.

1) If $1/f \in X$, then f is weakly invertible in X. For $X = H^p(\mathfrak{M})$ or $X = A(\lambda)$ the condition is of the form

$$\int_{\mathbf{D}} \left|\frac{1}{f}\right|^p d\mathcal{M} < +\infty, \quad \frac{1}{|f(\zeta)|} = o(\lambda(|\zeta|)), \; |\zeta| \to 1, \tag{12}$$

respectively.

2) If $f = f_1 f_2$, where f_i, $i = 1, 2$, are inner and weakly invertible (for example, as in 1)), then f is weakly invertible. In particular, if $1/f^{1/n} \in X$ we again obtain weak invertibility of the function $f = f^{1/n} \cdots f^{1/n}$ (the inductive argument in part 1 of Theorem 4). For $X = H^p(\mathfrak{M})$ this gives us a stronger assertion than in Example 1;[11]

[11] Sometimes the assertions coincide; for example, this is the case when

$$\mathcal{M} = l_2, \; f(\zeta) = \exp\left(-a\frac{1+\zeta}{1-\zeta}\right), \; |\zeta| < 1, \; a > 0.$$

for $X = A(\lambda)$ we get the assertion in Example 2 (see also Example 3). To verify (2) and (12) (with f replaced by $f^{1/n}$) we can use a special case of the Gol'dberg inequality (see [22] and [106])

$$\int_{\partial\mathbf{D}} |f(r\zeta)|^{-\delta}\, dl_1(\zeta) \leqslant \int_{\partial\mathbf{D}} \exp\left(\frac{\delta(1-r^2)\log|f(0)|^{-1}}{1-2r\operatorname{Re}\zeta + r^2}\right) dl_1(\zeta),$$

where $\delta > 0$ and $0 \leqslant r < 1$.

The most typical case which is not covered in Theorem 4 is the case where f_1 and f_2 are relatively prime inner functions. Examples 3 and 4 contain sufficient conditions for weak invertibility of the numerator of such a quotient. It is not clear in which spaces X the denominator (also an inner function) can compensate for noninvertibility of the numerator.

The full result is easy only for the Hardy spaces H^p, $1 \leqslant p < \infty$ (by virtue of factorization theorems and the theorem of Beurling): $f = f_1 f_2$ is weakly invertible if and only if f_1 and f_2 are weakly invertible (here, as in Theorem 4, we assume that f_1, $f_2 \in X$; $X = H^p$).

In the case where we impose no additional restrictions on the factors f_1 and f_2, a similar factorization theorem is impossible. In §2.8 we will construct majorants λ and functions f in $A(\lambda)$ which are not weakly invertible but such that the factor $\sqrt{f}$ is weakly invertible ($f = \sqrt{f}\sqrt{f}$).

A very interesting special case of the problem which we are considering is the question of weak invertibility of a function $f \in X$ if it is known that $1/f \in X$. For spaces X which contain rapidly increasing functions an affirmative answer would give us a very strong theorem on weighted approximations, but even for $X = H^2(l_2)$ the question is not completely answered by the results in this section (or other known results). From Theorem 4 we only know that the answer is affirmative if f is bounded or $1/f \in \mathfrak{D}$. In the general case we can establish an elementary criterion([12]) which is similar to Theorem 1: if X and $\mathfrak{M}$ are spaces of analytic functions, the polynomials are dense in $\mathfrak{M}$, and $1/f \in \mathfrak{M}$ and $f\mathfrak{M} \subset X$, then f is weakly invertible in X. In fact, $p_n \xrightarrow{\mathfrak{M}} 1/f$ implies that $p_n f \xrightarrow{X} \mathbf{1}$.

We mention a similar question for the spaces $H^p(l_2)$: do there exist functions $f \in H^{p_1}(l_2)$ which are weakly invertible in $H^{p_2}(l_2)$, $p_2 < p_1$, but not in $H^{p_1}(l_2)$?

§2.2. Bounded approximation almost everywhere and invariant subspaces of a shift operator

1°. The questions treated in this section involve applications of theorems on invariant subspaces to approximation problems (see Chapter 0). Our basic goal is to find a complete description of weakly invertible functions in L^∞ and H^∞ with respect to $*$-convergence. The treatment is in accord with the author's work [77].

We let L^p denote the set of all (classes of) functions([13]) which are pth power

---

([12]) It applies, for example, to spaces of the type $l^p_A(w_n)$, which we will consider later.

([13]) We will say "functions" instead of "classes of functions which differ on a set of Lebesgue measure zero" whenever this will not lead to confusion.

integrable ($1 \leqslant p \leqslant \infty$) on the circle $\mathbf{\Gamma} = \partial \mathbf{D} = \{\zeta \in \mathbf{C}\colon |\zeta| = 1\}$ (essentially bounded functions if $p = \infty$). The Hardy space H^p will be identified (see [23]) with the subspace of L^p which consists of all functions f whose Fourier coefficients

$$\hat{f}(n) = (2\pi)^{-1} \int_{\mathbf{\Gamma}} f(\zeta)\bar{\zeta}^n dl_1(\zeta), \quad n = 0, \pm 1, \ldots,$$

equal zero for negative n. In connection with this, the identity mappings of the disk $\mathbf{D}$ onto itself and the circle $\mathbf{\Gamma}$ onto itself will both be denoted in this section by the symbol z. Invariant subspaces of the shift operator S with $Sf = zf$ in L^p and H^p, $1 \leqslant p < \infty$, have been investigated by many authors (see [23] and [140]). First we will discuss some of our results in connection with weighted approximation problems.

As a simple corollary of the Helson-Lowdenslager theorem [141] on invariant subspaces of S in L^2 we have the following special case of the Szegö-Kolmogorov-Kreĭn approximation theorem (see Chapter IV in [23]). Let $f \in L^2$. The system of functions

$$\{S^n f\}_{n \geqslant 0} \tag{1}$$

is complete in L^2 if and only if $\log |f| \notin L^1$ and $f(t) \neq 0$ for almost all $t \in \mathbf{\Gamma}$. It is easy to see that for density of the system (1) it is sufficient (and, of course, necessary) that there exist a sequence of polynomials $\{p_n\}_{n \geqslant 1}$, $p_n(\zeta) = \Sigma_0^n a_{m,n}\zeta^m$, $\zeta \in \mathbf{\Gamma}$, such that $\|p_n f - S^{-1}f\|_{L^2} \longrightarrow 0$ as $n \longrightarrow \infty$. If $f \in L^\infty$, the following natural questions arise: 1) Can we choose this sequence so that $p_n f \longrightarrow S^{-1}f$ in the sense of convergence almost everywhere (a precise definition will follow)? 2) Which elements are approximated in this sense by sequences of the form $\{p_n f\}_{n \geqslant 1}$? 3) Does the function $\mathbf{1}$, $\mathbf{1}(\zeta) = 1$, $\zeta \in \mathbf{\Gamma}$, belong to this set?

The answers to these questions will be discussed in this section.

It turns out that the answer to question 1) is affirmative if and only if it is possible to approximate $S^{-1}f$ uniformly; that is, $\lim_n \|\widetilde{p}_n f - S^{-1}f\|_{L^\infty} = 0$ for some sequence of polynomials $\{\widetilde{p}_n\}_{n \geqslant 1}$.

Now we discuss the connection between the above questions and a description of invariant subspaces of the shift operator.

We regard L^∞ as a space of operators acting in L^2. Namely, an element $v \in L^\infty$ generates an operator (which we will denote by the same letter) $(vf)(\zeta) = v(\zeta)f(\zeta)$, $\zeta \in \mathbf{\Gamma}$, $f \in L^2$; and here $\left| v \right| = \|v\|_{L^\infty}$. Such operators are called multipliers (see [121]). The connection between multipliers and a description of irreducible invariant subspaces of S in L^2 is given by the Helson-Lowdenslager theorem; namely, each such subspace is of the form vH^2 for some isometric multiplier v. Here we are interested in subspaces $E \subset L^\infty$ which are invariant with respect to S and are closed in the following two senses:

1) in the strong operator topology([14]) ((s)-topology),
2) with respect to bounded convergence almost everywhere ($*$-convergence).

([14]) It is easy to see that on L^∞ this topology is defined by a family of weighted seminorms of the form $\|v\|_w^2 = \int_{\mathbf{\Gamma}} |v|^2 w\, dl_1$, where $w \geqslant 0$, $w \in L^1$.

Here a sequence $\{v_n\}_{n\geqslant 1}$, $v_n \in L^\infty$, is said to be $*$-convergent to an element $v \in L^\infty$ if

a) there exists a constant K such that $\|v_n\|_\infty \equiv \|v_n\|_{L^\infty} \leqslant K$, $n \geqslant 1$; and

b) $\lim_n v_n(\zeta) = v(\zeta)$ for almost all $\zeta \in \Gamma$.

It turns out (see Theorem 1 below) that such subspaces are in one-to-one correspondence with the S-invariant subspaces of L^2, and each such irreducible subspace is of the form $E = vH^\infty$, where v is an isometric multiplier. It is clear that the approximation problems for $*$-convergence which were stated at the beginning of this section are significantly more subtle than this description of $*$-closed invariant subspaces. The difference from the case L^2 (where these problems are equivalent) arises from the fact that the (s)-topology in L^∞ cannot be described by (countable) convergent sequences.

The interest in $*$-convergence in L^∞ is also explained by the following fact: if $*\text{-}\lim_n v_n = v$, $v_n, v \in L^\infty$, then for each unitary operator U with an absolutely continuous spectrum (see [2]) we have $\lim_n v_n(U) = v(U)$ in the sense of strong convergence in the space of operators.

It is remarkable that the converse is also true: if $\lim_n v_n(U) = v(U)$ in the sense of strong convergence in the space of operators, then we can choose a sequence $\{v_{n_k}\}_{k\geqslant 1}$ such that $*\text{-}\lim_k v_{n_k}\chi_e = v\chi_e$, where χ_e is the characteristic function of some "spectral" set $e = e_U$.

2°. *Invariant subspaces.* We now make precise our notion of a $*$-closed subspace of L^∞. We mean a subspace E such that if $v_n \in E$, $n \geqslant 1$, and $*\text{-}\lim_n v_n = v$, where $v \in L^\infty$, then $v \in E$.

THEOREM 1. *Let E be a $*$-closed or (s)-closed subspace of L^∞ which is invariant with respect to the operator S (that is, $SE \subset E$). Then one of the following two assertions is true*:

1) *$S^{-1}E \subset E$, and then there exists a measurable (unique to within a set of Lebesgue measure zero) set $e \subset \Gamma$ such that $E = \chi_e L^\infty$, where χ_e is the characteristic function of e; or*

2) *$S^{-1}E \not\subset E$, and then $E = vH^\infty$, where v is an isometric multiplier*(15) *(that is, $v \in L^\infty$ and $|v(\zeta)| = 1$ almost everywhere on Γ); such a multiplier is unique to within a constant factor.*

For the proof it suffices to work with $*$-closed subspaces E (since (s)-closedness implies $*$-closedness).

1) In this case $vE \subset E$ for each v in L^∞, and therefore $E = \chi_e L^\infty$ for some $e \subset \Gamma$.

2) We will show that E is closed in the weak topology on L^∞ which is induced by functionals in L^1. By a theorem of Banach (see [11], Chapter IV, §2, Theorem 5) it suffices to show that if v is a point of adherence (in the weak topology) of the set $E \cap W$, where W is a ball in L^∞, then $v \in E$. But under these assumptions v is a point of adherence of $E \cap W$ in the weak topology of L^2. Since $E \cap W$ is convex, v is also a

---

(15) If $v \in H^\infty$, such a function is said to be inner, following Beurling [117].

point of adherence of this set in the norm topology of L^2; that is, there exists a sequence $\{v_n\}_{n\geq 1} \subset E \cap W$ such that $\lim_n \|v_n - v\|_{L^2} = 0$. It is clear that $\{v_n\}_{n\geq 1}$ is bounded in L^∞, and (if necessary we pass to a subsequence) we obtain $*\text{-}\lim_k v_{n_k} = v$. Thus $v \in E$.

We consider the polar

$$E^\perp = \left\{f : f \in L^1, \int_\Gamma f\bar{g}\,dl_1 = 0, \quad \text{if} \quad g \in E\right\}.$$

It is clear that $S^{-1}E^\perp \subset E^\perp$, but $SE^\perp \not\subset E^\perp$. By a theorem of Forelli [130] we have

$$E^\perp = \Phi\overline{H^1} = \{f : f \in L^1,\ f = \Phi\bar{h},\ h \in H^1\},$$

where $\Phi \in L^\infty$, $|\Phi(\zeta)| = 1$ for almost all $\zeta \in \Gamma$, and H^1 is the Hardy subspace of L^1 (the bar denotes complex conjugation). Since E is weakly closed in L^∞ (in the sense explained above), we have

$$E = E^{\perp\perp} = \left\{g : g \in L^\infty \text{ and } \int_\Gamma g\bar{f}\,dl_1 = 0, \quad \text{if} \quad f \in E^\perp\right\}$$

$$= \left\{g : g \in L^\infty, \int_\Gamma g\bar{\Phi}h\,dl_1 = 0, \quad \text{if} \quad h \in H^1\right\}.$$

Thus if $g \in E$, then by a theorem of Riesz (see [23]) we have $S^{-1}(g\bar{\Phi}) \in H^1$. Since $g\bar{\Phi} \in L^\infty$, we have $S^{-1}(g\bar{\Phi}) \in H^\infty$. Conversely, each element $g \in L^\infty$ such that $S^{-1}(g\bar{\Phi}) \in H^\infty$ belongs to E. Therefore $E = FH^\infty$, where $F = S\Phi$. ●

3°. *Weighted approximation by polynomials.* We introduce some notation. Let C be the set of all continuous functions on Γ. Let $C_A = C \cap H^\infty$ and $\mathfrak{P}_A = \mathfrak{P} \cap H^\infty$, where $\mathfrak{P}$ is the set of all trigonometric polynomials (that is, functions p of the form $p(\zeta) = \Sigma_{-n}^n c_k\zeta^k$, $\zeta \in \Gamma$). As in §2.1, we will say that an element $v \in L^\infty$ is weakly $*$-invertible if there exists a sequence $\{p_n\}_{n\geq 1}$, $p_n \in \mathfrak{P}_A$, such that $*\text{-}\lim_n p_n v = 1$.

We let $\mathfrak{D}$ denote the class of functions, introduced by Smirnov, consisting of all measurable functions v on Γ which are boundary values of functions h_v of the form $h_v = gF$, where g is an outer analytic function (not necessarily in H^1) and F is an inner function in H^∞ (see [95] or [23]). Let f be a measurable real function on Γ. We set

$$s_f(\zeta) = \lim_{\delta\to+0} \operatorname*{ess\,sup}_{\arg\zeta-\delta<\arg t<\arg\zeta+\delta} f(t), \quad \zeta \in \Gamma.$$

Now we give precise formulations of the questions on $*$-approximations which we mentioned in subsection 1°. Let $v \in L^\infty$. Which functions $h \in L^\infty$ are $*$-limits of sequences of the form $\{p_n v\}_{n\geq 1}$, $p_n \in \mathfrak{P}_A$? What are the weakly $*$-invertible elements of L^∞? The answer to the first question gives an effective description of the set $E(v) = \{h \in L^\infty$: there exist $p_n \in \mathfrak{P}_A$, $n \geq 1$, such that $*\text{-}\lim_n p_n v = h\}$.

We will obtain other interesting results as corollaries of this description.

THEOREM 2. *Let $v \in L^\infty$.*

1) *If $s_{\log|v|} \in L^1$, then $E(v) = \{hv$: $h \in \mathfrak{D}$ and $hs_{|v|} \in L^\infty\}$.*

2) *If $s_{\log|v|} \notin L^1$, then $E(v) = \{hv$: h is a measurable function on Γ and $hs_{|v|} \in L^\infty\}$.*

REMARK. The Smirnov class $\mathfrak{D}$ arises naturally in various problems of approximation by analytic functions (see [95]), and Theorem 2 can apparently be related to investigations of Tumarkin on convergence of sequences of boundary values of analytic functions (see [102]). We should also note that the description of the set $\widetilde{E}(v) = \{g$: there exists $\{p_n\}_{n\geq 1} \subset \mathcal{P}$ such that $*\text{-lim}_n p_n v = g\}$ can be simplified. Namely, we have $\widetilde{E}(v) = \{Hv$: H is measurable and $Hs_{|v|} \in L^\infty\}$.

THEOREM 3. *Let $v \in L^\infty$. The following assertions are equivalent:*

1) *There exists a sequence $\{p_n\}_{n\geq 1}$, $p_n \in \mathcal{P}_A$, such that $\lim_n p_n v = S^{-1}v$ with respect to the norm in L^∞.*

2) *There exists a sequence $\{q_n\}_{n\geq 1}$, $q_n \in \mathcal{P}_A$, such that $*\text{-lim}_n q_n v = S^{-1}v$.*

3) $s_{\log|v|} \notin L^1$.

4) $\log|v| \notin L^1$ *and there exists an upper semicontinuous function s such that $s(\zeta) \geq \log|v(\zeta)|$ for almost all $\zeta \in \Gamma$ and $\int_\Gamma s\,dl_1 = -\infty$.*

If v is continuous, then 1)–4) *are equivalent to the condition*:

5) $\log|v| \notin L^1$.

REMARK. Condition 3) is satisfied if $\log|v| \notin L^1$ and

$$s_{\log|v|} \leqslant \log|v| + K, \tag{2}$$

where K is a constant; this is the case, for instance, if $\log|v|$ is an upper semicontinuous function (see Lemma 1 below). This inequality can be interpreted to mean that the function $\log|v|$ does not have "very large jumps". Theorem 3 develops the aforementioned theorem of Szegö, Kolmogorov and Kreĭn [23].

THEOREM 4. *Let $v \in L^\infty$. Then v is weakly $*$-invertible if and only if one of the following is true*:

1) $\log|v| \in L^1$, *$v = g/u$, where g is an outer and u is an inner function in H^∞, and for the function* $\log|v|$ *condition* (2) *is satisfied* (*see also Lemma* 1, *which contains other equivalent conditions*); *or*

2) $\log|v| \notin L^1$, *for the function* $\log|v|$ *condition* (2) *is satisfied, and $v(\zeta) \neq 0$ for almost all $\zeta \in \Gamma$.*

Now we proceed to the proof of Theorems 2–4, and then we formulate several corollaries to these theorems which will describe the set $E(v)$ in more detail. First we prove some lemmas.

LEMMA 1. *Let f be a measurable real function on Γ with $f(\zeta) \leqslant K$ for almost all $\zeta \in \Gamma$ and for some K. The following conditions are equivalent*:

1) *There exist a constant K_1 and an upper semicontinuous function s such that $s(\zeta) \leqslant f(\zeta) \leqslant s(\zeta) + K_1$ almost everywhere.*

2) *There exists a constant K_1 such that $s_f(\zeta) \leqslant f(\zeta) + K_1$ almost everywhere.*

3) *There exist a constant K_1 and a sequence of real polynomials $\{p_n\}_{n\geq 1}$, $p_n \in \mathcal{P}$, such that $\lim_n p_n(\zeta) = f(\zeta)$ for almost all $\zeta \in \Gamma$, $f(\zeta) \leqslant p_n(\zeta) + K_1$ for almost all ζ and all $n \geqslant 1$, and $p_n(\zeta) \leqslant 2K$, $\zeta \in \Gamma$.*

The proof of this lemma is elementary and will therefore be omitted. We note that condition 1) in Lemma 1 is obviously satisfied if the function f is upper semicontinuous.

LEMMA 2.(16) *Let $v \in L^\infty$ and $f \in L^\infty$; set $\|f\|_v = \|fv\|_\infty$. Define a functional Φ on H^∞ by means of the formula $\Phi(u) = u(0)$. Then the following assertions are true*:

1) *The functional Φ is continuous in the norm $\|\cdot\|_v$ on the set C_A (or $\mathcal{P}_A$) if and only if $s_{\log|v|} \in L^1$.*

2) *The functional Φ is continuous in the norm $\|\cdot\|_v$ on the set H^∞ if and only if $\log|v| \in L^1$.*

PROOF. 1) *Sufficiency.* Let $s_{\log|v|} \in L^1$. We first note that $\|uv\|_\infty = \|us_{|v|}\|_\infty$ if $u \in C_A$. In fact, the inequality $|u(\zeta)v(\zeta)| \leqslant |u(\zeta)s_{|v|}(\zeta)|$ holds for almost all $\zeta \in \Gamma$ for an arbitrary function u. Therefore $\|uv\|_\infty \leqslant \|us_{|v|}\|_\infty$. If $u \in C$ (and therefore if $u \in C_A$) the inequality $|u(\zeta)v(\zeta)| \leqslant \|uv\|_\infty$ for almost all $\zeta \in \Gamma$ implies(17) that

$$|s_{|v|}(\zeta)u(\zeta)| \leqslant \|uv\|_\infty, \quad \|us_{|v|}\|_\infty \leqslant \|uv\|_\infty.$$

Thus we have $\|uv\|_\infty = \|us_{|v|}\|_\infty$ if $u \in C$. Let h_μ be an outer function in H^∞ with boundary values μ such that $|\mu| = s_{|v|}$ almost everywhere (such a function exists since $\log s_{|v|} = s_{\log|v|} \in L^1$; see [23] or [95]). Now for a function $u \in C_A$ we have

$$|\Phi(u)| = \frac{1}{|h_\mu(0)|}|h_\mu(0)\Phi(u)| \leqslant \frac{1}{|h_\mu(0)|}\|h_\mu u\|_\infty = \frac{1}{|h_\mu(0)|}\|us_{|v|}\|_\infty = \frac{\|u\|_v}{|h_\mu(0)|}.$$

Thus the functional Φ is continuous on C_A in the norm $\|\cdot\|_v$, and $\|\Phi\| \leqslant 1/|h_\mu(0)|$.(18)

Necessity. Suppose that Φ is continuous on C_A in the norm $\|\cdot\|_v$ and $\log s_{|v|} = s_{\log|v|} \notin L^1$. Since $s_{|v|}$ is upper semicontinuous, $1/s_{|v|}$ is lower semicontinuous (and bounded from below). Therefore there exists a sequence of continuously differentiable functions $\{g_n\}_{n\geqslant 1}$ such that $g_n \leqslant g_{n+1}$, $n \geqslant 1$, and $\lim_n g_n(\zeta) = 1/s_{|v|}$ for almost all $\zeta \in \Gamma$. Since $1/s_{|v|} \geqslant 1/\|v\|_\infty$, we can assume that $\int_\Gamma \log|g_n|\,dl_1 > -\infty$ for all n. For $n \geqslant 1$ let G_n be an outer function in H^∞ such that $|G_n| = |g_n|$ almost everywhere. Then $G_n \in C_A$, $n \geqslant 1$ (see [23]), and

$$\|G_n\|_v = \|G_n s_{|v|}\|_\infty = \|g_n s_{|v|}\|_\infty \leqslant 1$$

for $n \geqslant 1$. On the other hand,

$$\Phi(G_n) = \exp\left(\int_\Gamma \log|g_n|\,dl_1\right),$$

(16) This lemma can be compared with the Szegö-Kolmogorov-Kreĭn theorem which asserts that Φ is continuous in the metric of $L^2(\nu)$ if and only if $\log(d\nu/dl_1) \in L^1$ (here ν is a finite measure on Γ).

(17) If $u(\zeta) \neq 0$, $\zeta \in \Gamma$, we can see this by virtue of the inequality $|v(\zeta)| \leqslant \|uv\|_\infty/|u(\zeta)|$ and the relation $s_{1/|u|} = 1/|u|$. At points where $u(\zeta) = 0$, the required inequality is clear.

(18) We can see that in fact the equality sign holds.

and by the theorem on limits with monotone convergence we have $\lim_n|\Phi(G_n)| = \infty$, which contradicts the continuity of Φ. Thus $s_{\log|v|} \in L^1$.

2) Sufficiency is proved in this case just as it was in 1), but with the condition $|\mu| = s_{|v|}$ replaced by $|\mu| = |v|$.

The proof of necessity also uses the methods in 1), but for the functions g_n, $n \geqslant 1$, we use

$$g_n(\zeta) = \begin{cases} \dfrac{1}{v(\zeta)}, & \text{if } |v(\zeta)| \geqslant \dfrac{1}{n}, \\ n, & \text{if } |v(\zeta)| < \dfrac{1}{n}, \end{cases} \quad \zeta \in \Gamma. \; \bullet$$

LEMMA 3. *Let $v \in H^\infty$ be an outer function. Then v is weakly $*$-invertible (that is, $\mathbf{1} \in E(v)$) if and only if the function $f = \log|v|$ satisfies the equivalent conditions* 1)–3) *in Lemma* 1.

PROOF. *Sufficiency*. Suppose that $\log|v|$ satisfies the conditions 1)–3) in Lemma 1. Then there exists a sequence $\{q_n\}_{n\geqslant 1}$, $q_n \in \mathcal{P}_A$, such that

$$\lim_n \operatorname{Re} q_n(\zeta) = \log|v(\zeta)|$$

for almost all $\zeta \in \Gamma$ and

$$\log|v(\zeta)| \leqslant \operatorname{Re} q_n(\zeta) + K_1 \leqslant 2\log\|v\|_\infty + K_1$$

for almost all $\zeta \in \Gamma$, where K_1 is some constant. Since the functions $g_n = v\exp(-q_n)$ are outer, $n \geqslant 1$, for all complex numbers $t \in \mathbf{D}$ we have

$$g_n(t) = \exp\left(\int_\Gamma \frac{\zeta + t}{\zeta - t} \log|g_n(\zeta)|\, dl_1(\zeta)\right).$$

By virtue of the choice of the sequence $\{q_n\}_{n\geqslant 1}$ we have

$$-2\log\|v\|_\infty \leqslant -\operatorname{Re} q_n \leqslant K_1 - \log|v|;$$

therefore the Lebesgue theorem on taking limits under the integral, applied to

$$-\int_\Gamma \frac{\zeta + t}{\zeta - t} \operatorname{Re} q_n(\zeta)\, dl_1(\zeta),$$

gives us $\lim_n g_n(t) = 1$, $|t| < 1$. On the other hand, by the same theorem applied to

$$\int_\Gamma \exp\{2(\log|v(\zeta)| - \operatorname{Re} q_n(\zeta))\}\, dl_1(\zeta), \quad n \geqslant 1,$$

we have $\lim_n \|g_n\|_2 = 1$. Now we can conclude that the sequence $\{g_n\}_{n\geqslant 1}$ converges to the function $\mathbf{1}$ in the weak topology of H^2, and therefore also with respect to the norm in this space (see [2]). Therefore (if necessary we pass to a subsequence) we have $\lim_n g_n(\zeta) = 1$ for almost all $\zeta \in \Gamma$. Together with this we have $\|g_n\|_\infty \leqslant \exp K_1$, $n \geqslant 1$, and now we only have to choose a sequence $\{p_n\}_{n\geqslant 1}$, $p_n \in \mathcal{P}_A$, such that $\|\exp(-q_n) - p_n\|_\infty < 1/n$, $n \geqslant 1$. Then $*\text{-}\lim_n p_n v = \mathbf{1}$, which is what we wanted to prove.

Necessity. Suppose that $p_n \in \mathcal{P}_A$, $n \geqslant 1$, and $*\text{-}\lim_n p_n v = 1$. Without loss of generality we can assume that $p_n(\zeta) \neq 0$ for all $\zeta \in \Gamma$ and for all $n \geqslant 1$, since otherwise we could select polynomials ("hardly shifting" the zeros on the unit circle) $\widetilde{p}_n \in \mathcal{P}_A$, $n \geqslant 1$, with the necessary property, so that $\|\widetilde{p}_n - p_n\|_\infty < 1/n$. Consequently the functions $\log|p_n|$, $n \geqslant 1$, are continuous and there exist real polynomials $q_n \in \mathcal{P}$ such that $\|\log|p_n| + q_n\|_\infty < 1/n$, $n \geqslant 1$. Now we have $\lim_n q_n(\zeta) = \log|v(\zeta)|$ for almost all $\zeta \in \Gamma$, and $\log|v(\zeta)| \leqslant q_n(\zeta) + \log K + 1$, where $K = \sup_n \|vp_n\|_\infty$. Thus the function $f = \log|v|$ satisfies condition 3) of Lemma 1. ●

PROOF OF THEOREM 2. 1) Let $v \in L^\infty$ and $s_{\log|v|} \in L^1$. We have to show that

$$E(v) = \{hv : h \in \mathcal{D} \text{ and } hs_{|v|} \in L^\infty\} \overset{\text{def}}{=} E'(v).$$

First we show that $E(v) \subset E'(v)$. Without loss of generality we can assume that $\|v\|_\infty \leqslant 1$.

Let $g \in E(v)$; that is, $g = *\text{-}\lim_n p_n v$, where $p_n \in \mathcal{P}_A$, $n \geqslant 1$. Then there exists a constant K such that $|p_n(\zeta)v(\zeta)| \leqslant K$ for almost all $\zeta \in \Gamma$ and for all $n \geqslant 1$. We introduce the temporary notation $s(f) = -s_{-f}$. Since $|p_n(\zeta)| \leqslant K/|v(\zeta)|$ for almost all $\zeta \in \Gamma$, we have([19])

$$\log^+|p_n| \leqslant \log^+ K + \log^+ 1/|v|$$

and thus

$$s(\log^+|p_n|) \leqslant s\left(\log^+ K + \log^+ \frac{1}{|v|}\right) = \log^+ K + s\left(\log^+ \frac{1}{|v|}\right)$$
$$= \log^+ K + s\left(\log \frac{1}{|v|}\right) = \log^+ K + |s_{\log|v|}|,$$

where we used the inequality $\|v\|_\infty \leqslant 1$. It follows that the family of functions $\{\nu_n\}_{n\geqslant 1}$, where $\nu_n e = \int_e \log^+|p_n|\,dl_1$, $e \subset \Gamma$, is absolutely equicontinuous (for the definition see, for example, [30]). Also $\lim_n p_n = g/v$ for almost all ζ, $v(\zeta) \neq 0$.

Now we apply the theorem of Tumarkin ([95], Chapter II, §8.2) which implies that there exists a function $H \in \mathfrak{D}$ such that $H(\zeta) = g(\zeta)/v(\zeta)$ for almost all ζ, $v(\zeta) \neq 0$. Consequently $g = Hv$. As we noted in the proof of Lemma 2, $\|p_n v\|_\infty = \|p_n s_{|v|}\|_\infty$, and if $\Phi_{s_{|v|}}$ is an outer function with modulus $s_{|v|}$ (such a function exists since $\log s_{|v|} \in L^1$), then $\|p_n \Phi_{s_{|v|}}\|_\infty \leqslant K$, $n \geqslant 1$. Therefore $|p_n(t)\Phi(t)| \leqslant K$ for all $t \in \mathbf{D}$. By the Hinčin-Ostrowski theorem (see [95], Chapter II, Theorem 7.1) we have $\lim_n p_n(t) = H(t)$, $t \in \mathbf{D}$, and $|H(t)\Phi(t)| \leqslant K$ for all $t \in \mathbf{D}$. Consequently $|H\Phi_{s_{|v|}}| = |Hs_{|v|}| \leqslant K$. Thus $g \in E'(v)$.

We now prove the reverse inclusion $E'(v) \subset E(v)$. Since the function $s_{\log|v|}$ is upper semicontinuous, by Lemma 3 the function $\Phi = \Phi_{s_{|v|}}$ is weakly $*$-invertible. In other words, there exists a sequence $\{p_n\}_{n\geqslant 1}$, $p_n \in \mathcal{P}_A$, such that $*\text{-}\lim_n p_n \Phi_{s_{|v|}} = 1$. If $H \in \mathfrak{D}$ and $Hs_{|v|} \in L^\infty$, then $H\Phi \in L^\infty$, and since Φ is outer, we have $H\Phi \in H^\infty$ (see [23]). Therefore there exists a sequence $\{q_n\}_{n\geqslant 1}$, $q_n \in \mathcal{P}_A$, with $*\text{-}\lim_n q_n = \Phi H$. If

(19) Recall that $\log^+ a = \log a$ if $a \geqslant 1$, and $\log^+ a = 0$ if $0 \leqslant a \leqslant 1$.

we set $\pi_n = p_n q_n$, $n \geqslant 1$, then $\pi_n \in \mathcal{P}_A$, and $*\text{-lim}_n \pi_n v = vH$ (convergence almost everywhere is clear; also $\|\pi_n v\|_\infty \leqslant \|q_n\|_\infty \|p_n \Phi\|_\infty$). Thus $Hv \in E(v)$, and consequently $E'(v) \subset E(v)$. The first part of the theorem is proved.

2) Let $v \in L^\infty$ and $s_{\log|v|} \notin L^1$. We have to show that $E(v) = \{Hv$: H is a measurable function on Γ and $Hs_{|v|} \in L^\infty\} \overset{\text{def}}{=} E''(v)$. First we note that $E(v) \subset E''(v)$. For if $g \in E(v)$, then $g = *\text{-lim}_n p_n v$, where $p_n \in \mathcal{P}_A$, $n \geqslant 1$, and, as we have already noted, $\sup_n \|p_n s_{|v|}\|_\infty < \infty$. Now we set

$$H(\zeta) = \begin{cases} \dfrac{g(\zeta)}{v(\zeta)}, & \text{if } v(\zeta) \neq 0, \\ 0, & \text{if } v(\zeta) = 0. \end{cases}$$

Clearly H is measurable, $Hs_{|v|} \in L^\infty$, and $g = Hv$; that is, $g \in E''(v)$. We will prove the reverse inclusion $E''(v) \subset E(v)$. Let $g = Hv$, $g \in E''(v)$. We set

$$h(\zeta) = \begin{cases} \dfrac{v(\zeta)}{s_{|v|}(\zeta)}, & \text{if } s_{|v|}(\zeta) \neq 0, \\ 0, & \text{if } s_{|v|}(\zeta) = 0. \end{cases}$$

Then $g = Hv = Hs_{|v|}h$ (since $s_{|v|}(\zeta) = 0$ implies that $v(\zeta) = 0$). There exists a sequence $\{g_n\}_{n \geqslant 1}$, $g_n \in C$, such that $*\text{-lim}_n g_n = Hs_{|v|}$ (since $Hs_{|v|} \in L^\infty$). Also, since $1/s_{|v|}$ is lower semicontinuous (and since $1/s_{|v|}(\zeta) \neq -\infty$, $\zeta \in \Gamma$), there exists a sequence $\{f_n\}_{n \geqslant 1}$, $f_n \in C$, such that $f_{n+1}(\zeta) \geqslant f_n(\zeta)$, $\zeta \in \Gamma$, $n \geqslant 1$, and $\lim_n f_n(\zeta) = 1/s_{|v|}(\zeta)$, $\zeta \in \Gamma$. Consequently

$$h = *\text{-}\lim_n f_n v$$

(boundedness: $|f_n(\zeta)v(\zeta)| \leqslant |v(\zeta)/s_{|v|}(\zeta)| \leqslant 1$). Therefore

$$g = Hs_{|v|}h = *\text{-}\lim_n g_n f_n v.$$

Now we apply Lemma 2 to the function v. Since $s_{\log|v|} \notin L^1$, by the lemma the functional $u \longrightarrow u(0)$, $u \in \mathcal{P}_A$, is not continuous in the norm $\|\cdot\|_v$. Therefore the function $\mathbf{1}$ belongs to the closure (in the norm $\|\cdot\|_v$) of the hyperplane of zeros of this functional. Therefore $S^{-1}\mathbf{1} \in \overline{\mathcal{P}}_A^{\|\cdot\|_v}$, and (by induction) $S^{-n}\mathbf{1} \in \overline{\mathcal{P}}_A^{\|\cdot\|_v}$, $n \geqslant 0$ (here n is an integer); that is, $\mathcal{P} \subset \overline{\mathcal{P}}_A^{\|\cdot\|_v}$. If we use the fact that $\|f\|_v \leqslant \|f\|_\infty \|v\|_\infty$ for each function $f \in L^\infty$ and the set of all trigonometric polynomials $\mathcal{P}$ is dense in C, then we obtain $C \subset \overline{\mathcal{P}}_A^{\|\cdot\|_v}$. This inclusion shows that there exists a sequence $\{p_n\}_{n \geqslant 1}$, $p_n \in \mathcal{P}_A$, such that

$$\|p_n - f_n g_n\|_v = \|p_n v - f_n g_n v\|_\infty < 1/n.$$

Together with the equation $g = *\text{-lim}_n f_n g_n v$ this gives us $g = *\text{-lim}_n p_n v$; that is, $g \in E(v)$. Thus $E''(v) \subset E(v)$. ●

REMARK. In case 1) of Theorem 2 we have

$$E(v) = v\Phi_{1/s_{|v|}} H^\infty. \tag{3}$$

In case 2) of Theorem 2 we have

$$E(v) = \frac{v}{s_{|v|}} L^\infty. \tag{4}$$

In this equation, as in the proof of Theorem 2, we let $(v/s_{|v|})(\zeta) = 0$ if $s_{|v|}(\zeta) = 0$, $\zeta \in \Gamma$. We let $\Phi_{1/s_{|v|}}$ denote an outer function (not necessarily in H^1) with modulus $1/s_{|v|}$.

In order to verify the first equation we note that the conditions $hs_{|v|} \in L^\infty$ and $h\Phi_{s_{|v|}} \in L^\infty$ are equivalent, and together with $h \in \mathfrak{D}$ they imply that $g = h\Phi_{s_{|v|}} \in H^\infty$. Then we note that $\Phi_{1/s_{|v|}} = 1/\Phi_{s_{|v|}}$, and therefore $hv = vg/\Phi_{s_{|v|}} = vg\Phi_{1/s_{|v|}}$. Thus we have $E(v) \subset v\Phi_{1/s_{|v|}}H^\infty$. The reverse inclusion is clear.

The second equation follows immediately from the fact that $s_{|v|}(v/s_{|v|}) = v$.

Proof of Theorem 3. Equivalence of assertions 2) and 3) follows from Theorem 2 and the fact that $S^{-1}\mathbf{1} \notin \mathfrak{D}$.

Equivalence of assertions 3) and 4) is clear since $\log|v| \leqslant s_{\log|v|}$ and $s_f = f$ for each (real) upper-semicontinuous function f.

In order to conclude the proof of Theorem 3 we note that assertion 1) is stronger than assertion 2), and by Lemma 2 we know that assertion 3) implies assertion 1), as in the proof of Theorem 2. ●

Proof of Theorem 4. *Necessity*. Let $v \in L^\infty$ and $\mathbf{1} \in E(v)$. Since the set $E(v)$ is defined by either (3) or (4), we have $\mathbf{1} = vg/s_{|v|}$, where $g \in L^\infty$. Consequently $s_{|v|}(\zeta) \neq 0$ and $v(\zeta) \neq 0$ for almost all $\zeta \in \Gamma$, and $s_{|v|}/v \in L^\infty$; that is, there exists a constant K such that $s_{\log|v|} \leqslant \log|v| + K$, and this is condition (2) for $\log|v|$. Thus in the case $\log|v| \notin L^1$ the assertion is proved.

Also we conclude that the functions $\log|v|$ and $s_{\log|v|}$ are simultaneously integrable or not integrable (on Γ). If $\log|v| \in L^1$, then from (3) we obtain $\mathbf{1} = v\Phi_{1/s_{|v|}}f$, where $f \in H^\infty$, so that $\Phi_{s_{|v|}} = vf$. We factor (see [23]) f into outer and inner functions: $f = f_1 u$. Since

$$|f| = |f_1| = \frac{|\Phi_{s_{|v|}}|}{|v|} = \frac{s_{|v|}}{|v|} \geqslant 1,$$

we have $\Phi_{s_{|v|}}/f_1 = g \in H^\infty$. Consequently $v = g/u$, where g is an outer and u is an inner function in H^∞.

Sufficiency. First we note that if $v \in L^\infty$ and for $\log|v|$ we have condition (2), then $s_{\log|v|}$ and $\log|v|$ are simultaneously integrable or not integrable (on Γ). Thus in case 2) we have $s_{\log|v|} \notin L^1$, and we can apply (4).

In our case (for $v(\zeta) \neq 0$, $\zeta \in \Gamma$) condition (2) for the function $\log|v|$ implies that $s_{|v|}/v \in L^\infty$. Thus $\mathbf{1} \in E(v)$.

In case 1), since $|v| = |g|$, by Lemma 3 we can choose a sequence $\{p_n\}_{n\geqslant 1}$, $p_n \in \mathfrak{P}_A$, such that $*\text{-}\lim_n p_n g = \mathbf{1}$. We also choose a sequence $\{q_n\}_{n\geqslant 1}$, $q_n \in \mathfrak{P}_A$, such that $*\text{-}\lim_n q_n = u$. Then we have

$$*\text{-}\lim_n p_n q_n v = *\text{-}\lim_n p_n g \frac{q_n}{u} = \mathbf{1}. \quad ●$$

As corollaries of the theorems which we have just proved, we consider some cases in which the set $E(v)$ is very easy to describe.

COROLLARY 1. *Let $v \in L^\infty$ and $s_{\log|v|} \in L^1$. Then $E(v) = vH^\infty$ if and only if $1/s_{|v|} \in L^\infty$ (that is, if and only if the norms $\|\cdot\|_\infty$ and $\|\cdot\|_v$ are equivalent on $\mathcal{P}_A$).*

In fact, from (3) it follows that the equations $E(v) = vH^\infty$ and $\Phi_{s_{|v|}}H^\infty = H^\infty$ are equivalent, and then it follows that $1/\Phi_{s_{|v|}} \in H^\infty$; that is, $1/s_{|v|} \in L^\infty$. If for a function $v \in L^\infty$ we have $1/s_{|v|} \in L^\infty$, then

$$\left\| \frac{1}{s_{|v|}} \right\|_\infty^{-1} \| p \|_\infty \leqslant \| p s_{|v|} \|_\infty = \| pv \|_\infty = \| p \|_v \leqslant \| p \|_\infty \| v \|_\infty,$$

if $p \in \mathcal{P}_A$. This inequality means that the norms $\|\cdot\|_\infty$ and $\|\cdot\|_v$ are equivalent on $\mathcal{P}_A$ (as well as on C). Conversely, if $1/s_{|v|} \notin L^\infty$, then $1/\Phi_{s_{|v|}} \notin H^\infty$. We choose a sequence $\{p_n\}_{n\geqslant 1}$, $p_n \in \mathcal{P}_A$, such that $*\text{-}\lim_n p_n\Phi = 1$ (such a sequence exists by virtue of Lemma 3 or Theorem 4). Then

$$\sup_n \| p_n \Phi_{s_{|v|}} \|_\infty = \sup_n \| p_n \|_v < +\infty \quad \text{and} \quad \sup_n \| p_n \|_\infty = +\infty. \bullet$$

COROLLARY 2. *Let $v \in L^\infty$ and $\log|v| \in L^1$. Write v (see [23]) as a product $v = v_0 v_1$, where $v_0 \in L^\infty$ with $|v_0| = 1$, and v_1 is an outer function in H^∞. Here $E(v) = v_0 H^\infty$ if and only if $s_{|v|}/|v| \in L^\infty$; that is, if and only if one of the conditions 1)–3) in Lemma 1 is satisfied for $\log|v|$.*

In fact, from (3) we can see that the equation $E(v) = v_0 H^\infty$ is equivalent to the equation $v_1 \Phi_{1/s_{|v|}} H^\infty = H^\infty$. Therefore $s_{|v|}/|v| \in L^\infty$. $\bullet$

We note in passing that it is useful to compare Corollary 2 (in the case $v \in H^\infty$) with Theorem 1 on $*$-closed S-invariant subspaces.

COROLLARY 3. *Let $v \in L^\infty$ and $s_{\log|v|} \notin L^1$. Let $\widetilde{v}$ be a function which is a representative of the equivalence class v, and let $e = \{\zeta\colon \zeta \in \Gamma, \widetilde{v}(\zeta) \neq 0\}$.*(20) *Then:*

1) *$E(v) = vL^\infty$ if and only if $s_{|v|}(\zeta) \geqslant \delta > 0$ for all $\zeta \in e$ and for some (fixed) δ;*
2) *$E(v) = \chi_e L^\infty$ if and only if $s_{|v|}(\zeta)/|v(\zeta)| \leqslant K$ for all $\zeta \in e$ and for some (fixed) K.*

In fact, in case 1) formula (4) implies that the equation $E(v) = vL^\infty$ is equivalent to the equation $(v/s_{|v|})L^\infty = vL^\infty$, and this equation is equivalent to the condition that $f/s_{|v|}$ is bounded on the set e for an arbitrary $f \in L^\infty$; thus assertion 1) is proved.

In case 2) we argue analogously. $\bullet$

4°. *Weighted approximation by analytic functions.* We consider some questions which are connected with uniform and $*$-approximations by sequences of functions in H^∞.

Let $v \in L^\infty$. We set $\hat{E}(v) = \{g\colon g \in L^\infty$ and there exists a sequence $\{H_n\}_{n\geqslant 1}$, $H_n \in H^\infty$, such that $*\text{-}\lim_n H_n v = g\}$. Clearly $\hat{E}(v) \supset E(v)$. Since the modulus of a function

(20) The set e is determined by v uniquely to within a set of Lebesgue measure zero.

in H^∞ can be "almost any" function in L^∞, in describing the set $\hat{E}(v)$ we do not have to impose any smoothness conditions on v (as we did for $E(v)$). We have the following assertion.

THEOREM 5. *Let $v \in L^\infty$.*

1) *If $\log|v| \in L^1$, then*

$$\hat{E}(v) = (v\mathfrak{D}) \cap L^\infty = \{gv : g \in \mathfrak{D},\ \ gv \in L^\infty\}.$$

2) *If $\log|v| \notin L^1$, then*

$$\hat{E}(v) = \chi_e L^\infty = \{gv : g \text{ is measurable and } gv \in L^\infty\},$$

where $e = \{\zeta : \zeta \in \Gamma, \tilde{v}(\zeta) \neq 0\}$ and $\tilde{v}$ is a representative of the equivalence class v.

The proof of this theorem uses the methods in the proof of Theorem 2 and will be presented shortly.

If we go from $*$-convergence to convergence in the norm of L^∞, then the corresponding approximation questions turn out to be significantly more complicated (as are the questions concerning closed S-invariant subspaces of L^∞ or H^∞). In this area we will establish only a negative result, which shows that in Theorem 2 (or Theorem 5) we cannot replace $*$-convergence by uniform convergence even in the simple case $v = \chi_e$, $e \subset \Gamma$. We have the following assertion.

THEOREM 6. *If e is a set of positive Lebesgue measure, $e \subset \Gamma$, then $\overline{\chi_e H^\infty} \neq \chi_e L^\infty$, where the bar denotes the closure of a set in the metric of L^∞. But if the measure of the complement $\Gamma \backslash e$ is positive, then all functions $\chi_e f$, where $f \in C$, belong to $\overline{\chi_e H^\infty}$.*

PROOF. The second assertion follows immediately from 2) in Lemma 2. In order to prove the first assertion, it suffices to note that there exists a multiplicative functional on the Banach algebra $\overline{\chi_e H^\infty}$ which cannot be extended to a multiplicative functional on $\chi_e L^\infty \overset{\text{def}}{=\!=} L^\infty(e)$. Let $\lambda \in e$ be a density point of the set e (with respect to the Lebesgue measure l_1) and let $\zeta_k = r_k\lambda$, $0 < r_k < 1$, $k \geqslant 1$, where $\Sigma_{k\geqslant 1}(1 - r_k) < +\infty$. Let $E = \{f : f \in H^\infty$ and there exists $\lim_k f(\zeta_k) \overset{\text{def}}{=\!=} \Phi(f)\}$. It is clear that the functional Φ is continuous on E in the norm $\|\cdot\|_{\chi_e}$. Therefore there exists a multiplicative functional Ψ on $\overline{\chi_e H^\infty}$ such that $\Psi \neq 0$ and $\Psi(\chi_e f) = 0$ if $\Phi(f) = 0$. If B is a Blaschke product (see [23]) with zeros ζ_k, $k \geqslant 1$, then $\Psi(\chi_e B) = 0$. Therefore $\chi_e B$ would not be invertible in the algebra $L^\infty(e)$ if the functional Ψ had a multiplicative extension onto this algebra. ●

PROOF OF THEOREM 5. 1) Let $v \in L^\infty$ and $\log|v| \in L^1$. We have to show that $\hat{E}(v) = (v\mathfrak{D}) \cap L^\infty$. First we establish the inclusion $\hat{E}(v) \subset (v\mathfrak{D}) \cap L^\infty$. Without loss of generality we can assume that $\|v\|_\infty \leqslant 1$. Let $g \in \hat{E}(v)$; that is, $g = *\text{-}\lim_n H_n v$, where $H_n \in H^\infty$, $n \geqslant 1$. Then there exists a constant K such that $|H_n(\zeta)v(\zeta)| \leqslant K$, $\zeta \in \Gamma$, or $|H_n(\zeta)| \leqslant K/|v(\zeta)|$, $\zeta \in \Gamma$, and $n \geqslant 1$. Therefore

$$\log^+|H_n(\zeta)| \leqslant \log^+ K + \log^+\frac{1}{|v(\zeta)|} = \log^+ K + \log\frac{1}{|v(\zeta)|}.$$

It follows that the family of functions $\{\nu_n\}_{n\geqslant 1}$, where $\nu_n e = \int_e \log^+|H_n|\,dl_1$, $e \subset \Gamma$, is absolutely equicontinuous. Also, $\lim_n H_n(\zeta) = g(\zeta)/v(\zeta)$ for almost all $\zeta \in \Gamma$. By a theorem of Tumarkin([21]) ([95], Chapter II, §8.2) there exists a function $H \in \mathfrak{D}$ such that $H(\zeta) = g(\zeta)/v(\zeta)$ for almost all $\zeta \in \Gamma$; that is, $g = Hv$. Clearly $g \in L^\infty$, and consequently $g \in (v\mathfrak{D}) \cap L^\infty$.

Now we prove the reverse inclusion $(v\mathfrak{D}) \cap L^\infty \subset \hat{E}(v)$. Let $\Phi_{|v|}$ be an outer function in H^∞ whose modulus is $|v|$. If $vH \in (v\mathfrak{D}) \cap L^\infty$, where $H \in \mathfrak{D}$, then $\Phi_{|v|}H \in L^\infty$ and $\Phi_{|v|}H \in \mathfrak{D}$, which implies that $\Phi_{|v|}H \in H^\infty$. Thus for the proof of the relation $vH \in \hat{E}(v)$ it suffices to show that there exists a sequence $\{H_n\}_{n\geqslant 1}$, $H_n \in H^\infty$, such that $*\text{-}\lim_n H_n\Phi = \mathbf{1}$ (since we then have $*\text{-}\lim_n(\Phi H)H_n v = Hv$). But it is easy to see that for the H_n, $n \geqslant 1$, we can choose outer functions with moduli $|v_n|$, defined by the formulas

$$|v_n|(\zeta) = \begin{cases} |v(\zeta)|^{-1}, & \text{if} \quad |v(\zeta)| \geqslant \frac{1}{n}, \quad \zeta \in \Gamma, \\ n \ \text{otherwise}. \end{cases}$$

This completes the proof of part 1) of Theorem 5.

2) Let $v \in L^\infty$ and $\log|v| \notin L^1$. The inclusion $\hat{E}(v) \subset \chi_e L^\infty$ is clear. We will prove the reverse inclusion $\chi_e L^\infty \subset \hat{E}(v)$. Let $g \in \chi_e L^\infty$; that is, $g = Hv$, where H is some measurable function. We introduce the following functions: I is a function in L^∞ with $|I(\zeta)| = 1$ almost everywhere on Γ and $|H| = HI$;

$$g_n(\zeta) = \begin{cases} |H(\zeta)|, & \text{if} \quad \frac{1}{n} < |H(\zeta)| < n, \\ \frac{1}{n}, & \text{if} \quad |H(\zeta)| \leqslant \frac{1}{n}, \\ n, & \text{if} \quad |H(\zeta)| \geqslant n, \end{cases}$$

where $\zeta \in \Gamma$ and $n \geqslant 1$; Φ_{g_n} is an outer function in H^∞ with modulus g_n, $n \geqslant 1$; I_n is a function in L^∞ with $|I_n| = 1$ almost everywhere on Γ, $n \geqslant 1$, and $\Phi_{g_n}(\zeta)I_n(\zeta) = |H(\zeta)|$ for almost all ζ for which $1/n < |H(\zeta)| < n$, $\zeta \in \Gamma$. Then we have

$$\|\Phi_{g_n}v\|_\infty = \|g_n v\|_\infty \leqslant \sup_\Gamma \left|\left(|H(\zeta)| + \frac{1}{n}\right)v(\zeta)\right| \leqslant \|Hv\|_\infty + \|v\|_\infty, \quad n \geqslant 1.$$

Consequently

$$|H|v = *\text{-}\lim_n \Phi_{g_n} I_n v \quad \text{or} \quad Hv = *\text{-}\lim_n \Phi_{g_n}\frac{I_n}{I}v.$$

By Egorov's theorem we can choose a sequence of sets $\{e'_k\}_{k\geqslant 1}$, $e'_k \subset \Gamma$, and a sequence $\{G_k\}_{k\geqslant 1}$, $G_k = \Phi_{g_{n_k}} I_{n_k} I^{-1} v$, such that $l_1(e'_k) > 1 - 1/2k$, $k \geqslant 1$, and $|G_k(\zeta) - H(\zeta)v(\zeta)| < 1/2k$ for all $\zeta \in e'_k$. Also, we construct a sequence of functions $\{f_{p,n}\}_{p\geqslant 1}$, $n \geqslant 1$, such that

---

([21]) This theorem is applicable since for functions H_n in H^∞ absolute equicontinuity of the family $\{v_n\}_{n\geqslant 1}$ implies absolute equicontinuity of the family

$$\{\nu_n^r\}_{n\geqslant 1,\, 0\leqslant r<1}, \text{ where } \nu_n^r(e) = \int_e \log^+|H_n(r\zeta)|\, dl_1(\zeta), \ e \subset \Gamma.$$

1) $\lim f_{p,n}(\zeta) = I_n(\zeta)I^{-1}(\zeta)$ for almost all $\zeta \in \Gamma$;
2) $f_{p,n} \in C$, $n \geqslant 1$, $p \geqslant 1$;
3) $\|f_{p,n}\|_\infty \leqslant 1$, $n \geqslant 1$, $p \geqslant 1$.

(For the $\{f_{p,n}\}_{p \geqslant 1}$ we can choose, for example, the corresponding sequence of Cesàro means for the Fourier series of $I_n I^{-1}$.)

We again use Egorov's theorem, choosing a sequence of sets $\{e_k''\}_{k\geqslant 1}$, $e_k'' \subset \Gamma$, and a sequence of natural numbers $\{p_k\}_{k \geqslant 1}$ such that

$$l_1(e_k'') > 1 - \frac{1}{2k} \text{ and } |f_{p,n_k}(\zeta) - I_{n_k}I^{-1}(\zeta)| < \frac{1}{2n_k \|v\|_\infty k}$$

for all $\zeta \in e_k''$ and $p \geqslant p_k$. Then for $\zeta \in e_k' \cap e_k''$ we have

$$|\Phi_{g_{n_k}}(\zeta) f_{p_k, n_k}(\zeta) v(\zeta) - H(\zeta) v(\zeta)|$$

$$\leqslant |\Phi_{g_{n_k}}(\zeta) f_{p_k, n_k}(\zeta) v(\zeta) - G_k(\zeta)| + |G_k(\zeta) - H(\zeta) v(\zeta)|$$

$$\leqslant \|\Phi_{g_{n_k}}\|_\infty |f_{p_k, n_k}(\zeta) - I_{n_k}I^{-1}(\zeta)| \, |v(\zeta)| + \frac{1}{2k}$$

$$\leqslant n_k \frac{1}{2n_k \|v\|_\infty k} \|v\|_\infty + \frac{1}{2k} = \frac{1}{k}, \quad k \geqslant 1.$$

Also

$$\|\Phi_{g_{n_k}} f_{p_k, n_k} v\|_\infty \leqslant \|f_{p_k, n_k}\|_\infty \|\Phi_{g_{n_k}} v\|_\infty \leqslant \|Hv\|_\infty + \|v\|_\infty.$$

Since $l_1(e_k' \cap e_k'') > 1 - 1/k$, this gives us

$$Hv = *\text{-}\lim_k \Phi_{g_{n_k}} f_{p_k, n_k} v.$$

Now we apply assertion 2) of Lemma 2 to the functions $\Phi_{g_{n_k}} v$, $k \geqslant 1$. Since $\log|\Phi_{g_{n_k}} v| \notin L^1$, by virtue of the lemma there exists a function $F_k \in H^\infty$ such that

$$\|F_k - f_{p_k, n_k}\|_{\Phi_{g_{n_k}} v} = \|\Phi_{g_{n_k}} v F_k - \Phi_{g_{n_k}} v f_{p_k, n_k}\|_\infty < \frac{1}{k}$$

(since $f_{p,n} \in C$ for all n and p, and $C \subset \overline{H^\infty}^{\|\cdot\|_{\Phi v}}$, where $\Phi = \Phi_{g_{n_k}}$).

Consequently, if we set $H_k = \Phi_{g_{n_k}} F_k$, then $*\text{-}\lim_k H_k v = Hv$ and $H_k \in H^\infty$, $k \geqslant 1$. Therefore $Hv \in \hat{E}(v)$.

5°. *Concluding remarks.* 1. It is easy to see (as in the proof of Theorem 1) that Theorems 2–5 remain valid if we replace $*$-convergence by (s)-convergence or convergence in the weak topology of L^∞ which is induced by the duality between L^1 and L^∞.

2. Apparently we should not expect an effective description of all S-invariant subspaces of L^∞ which are norm-closed. In fact, even closed ideals in L^∞ (or maximal ideals) cannot be described in the simple terms that we used for the subspaces in Theorem 1. But such a description is probably possible for "smooth" invariant subspaces (that is, subspaces of the form $\overline{v\mathcal{P}_A}$, where $v \in H^\infty$ and the bar denotes closure in H^∞) or for principal ideals (that is, ideals of the form $\overline{vH^\infty}$, $v \in H^\infty$).

§2.3. An estimate for Taylor coefficients and power bases in spaces with weighted norms

1°. In order to obtain theorems on weak invertibility which are more precise than the results in §2.1, we have to compare the given space X with some standard classes of analytic functions. In addition to these classes we also consider the spaces $l^p_A(w_n)$, $w_n \geqslant 0$; $l^p_A(w_n)$ is the family of all analytic functions f in $\mathbf{D}$ with a finite norm

$$\|f\|_{p,w} = \Big(\sum_{n\geqslant 0} |\hat{f}(n)|^p w_n^p\Big)^{1/p} < +\infty,$$

if $1 \leqslant p < \infty$, and

$$\|f\|_{\infty,w} = \sup_{n\geqslant 0} |\hat{f}(n)|\, w_n < +\infty; \quad l^p_A(1) = l^p_A.$$

LEMMA 1. *If X is a Banach space of analytic functions (see properties* (A) *and* (B) *in* §2.1) *and $z^n \in X$, then*

$$l^1_A(\|z^n\|_X) \subset X.$$

PROOF. If $f \in l^1_A(\|z^n\|_X)$, then the series $\Sigma_{n\geqslant 0}\hat{f}(n)z^n$ converges in X and

$$\|f\|_X \leqslant \sum_{n\geqslant 0} |\hat{f}(n)|\, \|z^n\|_X = \|f\|_{l^1_A(\|z^n\|_X)}. \ \bullet$$

LEMMA 2. *If E is a Banach space of analytic functions which is invariant with respect to rotations (see properties* (F) *and* (F*) *in* §2.1), *then*

$$\|\hat{f}(n)\, z^n\|_{\mathcal{E}} \leqslant \text{const}\, \|f\|_{\mathcal{E}}, \quad n \geqslant 0, \ f \in \mathcal{E}, \tag{1}$$

and therefore $l^\infty_A(\|z^n\|_{\mathfrak{E}}) \supset \mathfrak{E}$.

PROOF. If $f \in \mathfrak{E}$ and $f_\alpha(\zeta) = f(\alpha\zeta)$, $|\zeta| < 1$, $|\alpha| = 1$, then

$$\hat{f}(n)\, z^n = \int_{\partial \mathbf{D}} f_\alpha \alpha^{-n} dl_1(\alpha), \quad \|\hat{f}(n)\, z^n\| \leqslant \int_{\partial \mathbf{D}} \|f_\alpha\|\, dl_1 \leqslant \text{const}\, \|f\|. \ \bullet$$

We note that for $X = \overline{\mathcal{P}}^X_A$ or $X = \overline{\mathcal{P}}^*_A$ the inclusion

$$X \subset l^\infty_A(\|z^n\|_X) \tag{2}$$

is equivalent to so-called uniform minimality of the sequence $\{z^n\}_{n\geqslant 0}$; that is, to the condition

$$d = \inf_{n\geqslant 0} \operatorname{dist}\Big(\frac{z^n}{\|z^n\|},\ \mathcal{L}(z^k : k \neq n)\Big) > 0. \tag{3}$$

In fact, if we let Φ_n denote the functional in X defined by the formula $\Phi_n(g) = \hat{g}(n)$, $n \geqslant 0$, $g \in X$, and let $\Phi_{n,0}$ denote its restriction to $\mathfrak{L}(z^k\colon k \geqslant 0)$, then by the Hahn-Banach theorem we have

$$\|\Phi_{n,0}\|^{-1} = \operatorname{dist}(z^n,\ \mathcal{L}(z^k : k \neq n)), \quad n \geqslant 0.$$

On the other hand, each functional $\Phi_{n,0}$ is continuous in the topology of compact convergence on $\mathbf{D}$ and therefore has a unique extension Φ_n to the whole space X (with preservation of the norm: $\|\Phi_n\| = \|\Phi_{n,0}\|$). Since the inclusion (2) is equivalent to the estimate

$$\sup_{n\geqslant 0} \| z^n \|_X \| \Phi_n \| < \infty,$$

our assertion is proved. Also,

$$\frac{1}{d} = \sup_{n\geqslant 0} \| z^n \|_X \| \Phi_n \|. \tag{4}$$

A large portion of this section (subsections 2°–6°) is devoted to the construction of spaces X with property (2), starting with a given collection of spaces $\mathfrak{S}$ which already have this property (for instance, those in Lemma 2). We consider the spaces $X = \mathfrak{S}_f$,

$$\mathcal{S}_f = \{g : g \in A(\mathbf{D}),\ \ gf \in \mathcal{S}\},$$

with the weighted norm

$$\| g \|_f = \| gf \|_{\mathcal{S}},\ \ g \in \mathcal{S}_f.$$

Under the natural restrictions $f \in \mathfrak{S}$ and $f(\zeta) \neq 0$, $|\zeta| < 1$, we find necessary and sufficient conditions on the "weight" f for the system of powers $\{z^n\}_{n\geqslant 0}$ to be uniformly minimal in $\mathfrak{S}_f$. These conditions are expressed in terms of the "boundary spaces" $\partial\mathfrak{S}$ and $\partial^*\mathfrak{K}\mathfrak{S}^*$, which we will define. For many spaces $\mathfrak{S}$ these boundary spaces are calculated explicitly. For example, if $\mathfrak{S}$ is a Hilbert space which is invariant with respect to rotations and satisfies some mild "norm regularity" conditions, then $\partial\mathfrak{S} = H^2$, the familiar Hardy space. An explanation of the term "boundary space" can be found in subsections 4°–6° of this section.

The method which we will present can apparently be used for estimating not only one-dimensional projectors $\hat{\Phi}_n$, $\hat{\Phi}_n g = \hat{g}(n)z^n$, $g \in \mathfrak{S}_f$, but also other linear means of Taylor series of functions in $\mathfrak{S}_f$. In subsection 8° this procedure is carried out for the partial sums P_n:

$$P_n g = \sum_{k=0}^{n} \hat{g}(k) z^k,\ \ g \in \mathcal{S}_f.$$

The problem of finding conditions on the "weight" f in order to make the sequence of projectors $\{P_n\}_{n\geqslant 0}$ bounded in $\mathfrak{S}_f$ is the known problem of the power basis $\{z^n\}_{n\geqslant 0}$ in a space with a weighted norm. This question was investigated by Hardy and Littlewood [137], Babenko [3], Gapoškin [17], Helson and Szegö [142], Rosenblum [162], Forelli [131], Plotkin [93], and Hunt, Muckenhoupt and Wheeden [144]. A necessary and sufficient condition in the case $\mathfrak{S} = H^2$ was found in [142]; it consisted of the fact that f must be an outer function and $|f|^2 = \exp(u + \tilde{v})$ on $\partial\mathbf{D}$, where u and v are bounded real functions, $\tilde{v}$ is the harmonic conjugate of v, and $\sup_{\partial\mathbf{D}}|v| < \pi/2$. The results in this section imply that this condition is also necessary and sufficient in the case $\mathfrak{E} = H^2(\mathfrak{M})$, $d\mathfrak{M} = \rho(r)drdl_1$.

In [144] a necessary and sufficient condition was found in a form which is more suitable for practical verification. Namely, the Helson-Szegö condition turns out to be equivalent to the fact that f is an outer function and

$$\sup_I \left(\frac{1}{|I|}\int_I |f|^2\,dl_1\right)\left(\frac{1}{|I|}\int_I |f|^{-2}\,dl_1\right) < +\infty,$$

where the supremum is taken over all intervals (arcs) I of the circle $\boldsymbol{\Gamma}$. Later we will introduce other conditions, which are only sufficient but are explicitly described, which guarantee that $\{z^n\}_{n\geqslant 0}$ is a basis in $\mathfrak{E}_f$. One version of these conditions in the case $\mathfrak{E} = H^2(l_2)$ was obtained in [93] by a different method.

A detailed investigation into the power basis problem leads far away from the basic goal of this chapter; but the arguments in this section show that boundary spaces are a convenient (and, hopefully, natural) means of describing basis problems. A similar observation applies to other questions which arise in the study of boundary spaces. For example, this is relevant to "Bernšteĭn inequalities" which connect the norms of a polynomial in $\mathfrak{E}$ and $\partial\mathfrak{E}$ and which become the classical Bernšteĭn inequality for a very special choice of the class $\mathfrak{E}$.

In conclusion we remark that uniform minimality of the powers $\{z^n\}_{n\geqslant 0}$ in a Hilbert space $\mathfrak{E}_f$ immediately leads to some estimates on the leading coefficients of polynomials which are orthogonal in the weighted metric of $\mathfrak{E}_f$. In fact, let $\{q_n\}_{n\geqslant 0}$ be polynomials of this kind and let π_n be orthoprojectors in $\mathfrak{E}_f$ onto $\mathfrak{L}(q_k\colon 0\leqslant k\leqslant n-1)$, $q_n = a_{nn}z^n + \cdots$. Then

$$1 = \operatorname{dist}(q_n, \mathscr{L}(q^k : 0\leqslant k\leqslant n-1)) = \|q_n - \pi_n q_n\|_f = |a_{nn}|\,\|z^n - \pi_n z^n\|_f,$$
$$\|z^n - \pi_n z^n\|_f = \operatorname{dist}(z^n, \mathscr{L}(z^k : 0\leqslant k\leqslant n-1)) \geqslant \operatorname{dist}(z^n, \mathscr{L}(z^k : k\neq n)).$$

Uniform minimality implies that this last distance is of the order $\|z^n\|_f$ as $n\longrightarrow\infty$ (in the sense $\asymp$). Therefore

$$|a_{nn}| \leqslant \frac{\text{const}}{\|z^n f\|_{\mathfrak{E}}}, \quad n\geqslant 0.$$

From the results of subsection 6° we then obtain estimates for $|a_{nn}|$ which are new even in the case $\mathfrak{E} = H^2(l_2)$ (see [96]).

We note again that the theorems on estimates of Taylor coefficients (subsection 3°) and power bases (8°) depend in an essential fashion on the description of boundary spaces given in 2° and 4°–6°. In 7° we present a table which implicitly contains a description of boundary spaces in many characteristic cases.

2°. *Boundary spaces.* We consider a Banach space $\mathfrak{E}$ consisting of analytic functions in $\mathbf{D}$ such that

a) the functionals $f\longrightarrow f(\zeta)$, $|\zeta| < 1$, are continuous on $\mathfrak{E}$;

b) $z^n\in\mathfrak{E}$, $n\geqslant 0$, and the sequence $\{z^n\}_{n\geqslant 0}$ is uniformly minimal in $\mathfrak{E}$; and

c) $\lim_n \|z^{n+1}\|/\|z^n\| = 1$.

From Lemma 2 it follows that all spaces with an invariant norm satisfy condition b); condition c) is not very troublesome if we exclude spaces which contain only regular functions in **D**. In fact, this last requirement gives us $\underline{\lim}_n \|z^n\|_{\mathfrak{S}}^{1/n} \leqslant 1$, since by condition b) $|\hat{f}(n)|^{1/n} \leqslant (\text{const}/\|z^n\|)^{1/n}$, and from condition a) it follows that $\underline{\lim}_n \|z^n\|^{1/n} \geqslant 1$. Therefore $\underline{\lim}_n \|z^n\|^{1/n} = 1$, and condition c) means only some "regularity" of the norm $\|\cdot\|_{\mathfrak{S}}$.

As basic examples of such spaces we consider (as in §2.1)

$$\mathcal{E} = H^p(\rho(r)\,dr\,dl_1), \quad \mathcal{E} = A(\lambda), \quad \mathcal{E} = l_A^p(w_n).$$

In the last one we require that w_n satisfy the condition $\lim_n w_{n+1}/w_n = 1$. We note also that an example of a Hilbert space $\mathfrak{S}$ which is invariant with respect to rotations is $l_A^2(w_n)$, where $w_n = \|z^n\|_{\mathfrak{S}}$, $n \geqslant 0$. In fact, invariance of the norm(22) implies invariance of the scalar product, and then

$$(z^n, z^k) = \int_{\partial D} (\alpha^n z^n, \alpha^k z^k)\, dl_1(\alpha) = \int_{\partial D} \alpha^{n-k}(z^n, z^k)\, dl_1(\alpha) = 0, \quad \text{if} \quad n \neq k.$$

We set

$$\partial\mathcal{E} = \left\{ f \in \mathcal{E} : \|f\|_{\partial\mathcal{E}} \overset{\text{def}}{=} \overline{\lim_n} \frac{\|z^n f\|_{\mathcal{E}}}{\|z^n\|_{\mathcal{E}}} < +\infty \right\}$$

and, as before, we let $\Phi_k \in \mathfrak{S}^*$ denote the functional in $\mathfrak{S}$ which takes a function g to its kth Taylor coefficient:

$$\Phi_k(g) = \hat{g}(k), \quad g \in \mathcal{E}. \tag{5}$$

LEMMA 3. *Suppose that the space $\mathfrak{S}$ satisfies conditions* a)–c) *and* $f \in \partial\mathfrak{S}$. *Then*

$$d\|f\|_{l_A^\infty} \leqslant \|f\|_{\partial\mathcal{E}} \leqslant \|f\|_{l_A^1},$$

where $d = d(\mathfrak{S}, \{z^n\})$ *is the minimality constant* (3).

PROOF. The left-hand inequality follows from b), (4), and c):

$$|\hat{f}(n)| = |\Phi_{n+k}(z^k f)| \leqslant \|z^k f\|_{\mathcal{E}} \|\Phi_{n+k}\| \leqslant \frac{\|z^k f\|_{\mathcal{E}}}{d\ \ \|z^{n+k}\|_{\mathcal{E}}},$$

$$|\hat{f}(n)| \leqslant \overline{\lim_k} \frac{\|z^k f\|_{\mathcal{E}}}{d\,\|z^{n+k}\|_{\mathcal{E}}} = \frac{1}{d}\|f\|_{\partial\mathcal{E}}.$$

From this inequality it follows that the space $\partial\mathfrak{S}$ satisfies condition a), and therefore the completion of $\partial\mathfrak{S}$ is realized naturally as a space of analytic functions in **D**. If $f \in l_A^1$, then $\Sigma_{n\geqslant 0}|\hat{f}(n)|\,\|z^n\|_{\partial\mathfrak{S}} < +\infty$; therefore $\Sigma_{n\geqslant 0}\hat{f}(n)z^n$ converges in the completion

(22) Without loss of generality, the norm $\|f\|_{\mathfrak{S}}$ can be replaced by the norm $(\int_{\partial D}\|f_\alpha\|_{\mathfrak{S}}^2\, dl_1)^{1/2}$. Therefore invariance of the norm with respect to rotations means (here and hereafter) that $\|f_\alpha\|_{\mathfrak{S}} = \|f\|_{\mathfrak{S}}$, $|\alpha| = 1$.

of $\partial\mathfrak{E}$, and from property a) it follows that its sum is f. This proves the right-hand inequality in Lemma 3. ●

If $|t| < 1$, then condition c) implies that $\Sigma_{n\geq 0}|t|^n\|z^n\|_{\mathfrak{E}} < \infty$, and so $(1 - tz)^{-1} \in \mathfrak{E}$. To each functional Φ in $\mathfrak{E}^*$ we associate its "Cauchy transform" $\mathfrak{K}\Phi$:

$$\mathscr{K}\Phi(t) = \Phi((1 - tz)^{-1}), \quad |t| < 1,$$

and we set

$$\mathscr{K}\mathscr{E}^* = \{\mathscr{K}\Phi : \Phi \in \mathscr{E}^*\}, \quad \|\mathscr{K}\Phi\|_{\mathscr{K}\mathscr{E}^*} = \|\Phi\|_{\mathscr{E}^*}.$$

Under the assumptions which we made at the start of this section we have $\widehat{\mathfrak{K}\Phi}(n) = \Phi(z^n)$, $n = 0, 1, \ldots$, and thus the natural duality between $\mathfrak{E}$ and $\mathfrak{K}\mathfrak{E}^*$ (for short, $\mathfrak{K}$-duality) is of the form

$$\langle g, \Psi\rangle = \sum_{n\geq 0} \hat{g}(n)\hat{\Psi}(n),$$

if $g \in \mathfrak{E}$, $\Psi \in \mathfrak{K}\mathfrak{E}^*$, and at least one of the functions g or Ψ is a polynomial.[23]

Let g be an analytic function in $\mathbf{D}$, and let $g^*\Phi_n$ be the functional in $\mathfrak{E}$ defined by the formula

$$g^*\Phi_n(f) = \widehat{gf}(n), \quad f \in \mathscr{E}.$$

It is clear that $g^*\Phi_n$ is a continuous functional for $n = 0, 1, \ldots$.

LEMMA 4. 1) $\mathfrak{K}\Phi_n = z^n$, $n = 0, 1, \ldots$;
2) $\mathfrak{K}g^*\Phi_n = \Sigma_{k=0}^n \hat{g}(n-k)z^k$, $n = 0, 1, \ldots$.

The proof is obvious. ●

We set

$$\partial^*\mathscr{K}\mathscr{E}^* = \left\{g : g \in A(\mathbf{D}), \quad \|g\|_{\partial^*\mathscr{K}\mathscr{E}^*} \overset{\text{def}}{=} \varlimsup_n \frac{\left\|\sum_{k=0}^n \hat{g}(n-k)z^k\right\|_{\mathscr{K}\mathscr{E}^*}}{\|z^n\|_{\mathscr{K}\mathscr{E}^*}} < +\infty\right\}.$$

The space $\partial^*\mathfrak{K}\mathfrak{E}^*$ is defined in a dual fashion with respect to $\partial\mathfrak{E}$; namely, in the definition gz^k is replaced by $g^*\Phi_k$.

3°. *Uniform minimality of powers.*

THEOREM 1. *Let $\mathfrak{E}$ be a space of analytic functions in $\mathbf{D}$ which satisfies conditions* a)–c) *in* 2°, *and let $f \in \mathfrak{E}$, $f(\zeta) \neq 0$, $|\zeta| < 1$. Let $X = \mathfrak{E}_f$ be the corresponding weight space. The following assertions are equivalent*:

(i) $\{z^n\}_{n\geq 0}$ *is a uniformly minimal sequence in $\mathfrak{E}_f$ (that is, $\mathfrak{E}_f \subset l_A^\infty(\|z^n\|_f)$).*

(ii) $f \in \partial\mathfrak{E}$ *and* $1/f \in \partial^*\mathfrak{K}\mathfrak{E}^*$.

PROOF. As we noted in subsection 1°, assertion (i) is equivalent to the following estimate on Taylor coefficients:

(23) The symbol $\langle\cdot,\cdot\rangle$ will always denote $\mathfrak{K}$-duality of spaces of analytic functions.

$$|\hat{g}(n)|\,\|z^n f\|_{\mathcal{E}} \leqslant \text{const}\,\|gf\|_{\mathcal{E}}, \quad g \in \mathcal{E}_f, \quad n \geqslant 0.$$

If we set $g = z^n/f$, we obtain

$$\|z^n f\|_{\mathcal{E}} \leqslant \text{const}\,\|z^n\|_{\mathcal{E}}, \quad n \geqslant 0,$$

and therefore $f \in \partial\mathfrak{E}$. On the other hand, condition b) in subsection 2° implies[24] that

$$|(z^n f)^{\wedge}(n)|\,\|z^n\|_{\mathcal{E}} \leqslant \text{const}\,\|z^n f\|_{\mathcal{E}};$$

that is,

$$\|z^n\|_{\mathcal{E}} \leqslant \text{const}\,\|z^n f\|_{\mathcal{E}}, \quad n \geqslant 0.$$

Therefore since $f \in \partial\mathfrak{E}$ we have the asymptotic "equation" $\|z^n\|_{\mathfrak{E}} \asymp \|z^n f\|_{\mathfrak{E}}$, which enables us to rewrite assertion (i) in the form

$$\left|\left(\frac{g}{f}\right)^{\wedge}(n)\right| \leqslant \text{const}\,\|g\|_{\mathcal{E}}\frac{1}{\|z^n\|_{\mathcal{E}}} \leqslant \text{const}\,\|g\|_{\mathcal{E}}\,\|z^n\|_{\mathcal{K}\mathcal{E}^*}, \quad n \geqslant 0.$$

It remains to note that the Cauchy transform of the functional $g \longrightarrow (g/f)^{\wedge}(n)$ is $\Sigma_{k=0}^{n}(1/f)^{\wedge}(k)z^{n-k}$. •

The next three subsections (4°–6°) are devoted to calculating the spaces $\partial\mathfrak{E}$ and $\partial^*\mathfrak{K}\mathfrak{E}^*$ for various classes $\mathfrak{E}$. As a result we obtain the table (in 7°) which lists the most characteristic cases of weighted spaces $X = \mathfrak{E}_f$ in which the powers $\{z^n\}_{n\geqslant 0}$ are uniformly minimal.

4°. *Calculation of* $\partial\mathfrak{E}$.

LEMMA 5. *Let* $\mathfrak{E}$ *be a Hilbert space which is invariant with respect to rotations, satisfies condition* c), *and for which*

$$\sup_{n\geqslant 0,\, k\geqslant 0}\frac{\|z^{n+k}\|_{\mathcal{E}}}{\|z^k\|_{\mathcal{E}}} < +\infty. \tag{6}$$

Then $\partial\mathfrak{E} = H^2$.

PROOF. As we already stated,

$$\mathcal{E} = l_A^2(\|z^n\|_{\mathcal{E}}) \text{ and } \|f\|_{\mathcal{E}}^2 = \sum_{n\geqslant 0}|\hat{f}(n)|^2\|z^n\|_{\mathcal{E}}^2.$$

If $f \in \partial\mathfrak{E}$, then

$$\sum_{n=0}^{N}|\hat{f}(n)|^2\|z^{n+k}\|_{\mathcal{E}}^2 \leqslant \|fz^k\|_{\mathcal{E}}^2 \leqslant \text{const}\,\|z^k\|_{\mathcal{E}}^2,$$

and therefore $\Sigma_{n=0}^{N}|\hat{f}(n)|^2 \leqslant \text{const}$ for arbitrary $N = 0, 1, \ldots$. Consequently $\partial\mathfrak{E} \subset H^2$.

---

(24) The notation const means, as before, a constant which does not depend on n and which is, in general, different from other constants indicated by const.

Conversely, if $f \in H^2$, then

$$\frac{\|z^k f\|_{\mathcal{E}}^2}{\|z^k\|_{\mathcal{E}}^2} = \sum_{n\geqslant 0} |\hat{f}(n)|^2 \frac{\|z^{n+k}\|_{\mathcal{E}}^2}{\|z^k\|_{\mathcal{E}}^2} \leqslant \text{const}\,\|f\|_{H^2}^2,$$

and $f \in \partial\mathfrak{E}$. ●

REMARK. Condition (6) is automatically satisfied if the operator S of multiplication by z (that is, $g \longrightarrow zg$) is a contraction in $\mathfrak{E}$ or[25] if $\sup_{n\geqslant 0}|S^n|_{\mathfrak{E}} < +\infty$. These requirements mean that $\mathfrak{E}$ consists of functions which "grow near the boundary". If the weight $\{\|z^n\|_{\mathfrak{E}}\}_{n\geqslant 0}$ is "irregular", then we can only guarantee that

$$\sum_{n\geqslant 0} |\hat{f}(n)|^2 \alpha_n < +\infty \Rightarrow f \in \partial\mathcal{E}, \quad \text{where } \alpha_n = \sup_{k\geqslant 0} \|z^{n+k}\|_{\mathcal{E}}^2 \|z^k\|_{\mathcal{E}}^{-2}.$$

LEMMA 6. *If in the hypothesis of Lemma* 5 *we replace* (6) *by*

$$\sup_{n,\,k\geqslant 0} \frac{\|z^{n+k}\|_{\mathcal{E}}}{\|z^n\|_{\mathcal{E}}\|z^k\|_{\mathcal{E}}} < +\infty, \tag{6'}$$

then $\partial\mathfrak{E} = \mathfrak{E}$ *and* $\|f\|_{\partial\mathfrak{E}} = \|f\|_{H^2}$.

PROOF. Let $f \in \mathfrak{E}$. Then

$$\|z^k f\|_{\mathcal{E}}^2 = \sum_{n\geqslant 0} |\hat{f}(n)|^2 \|z^{n+k}\|_{\mathcal{E}}^2 \leqslant \text{const}\,\|z^k\|_{\mathcal{E}}^2 \sum_{n\geqslant 0} |\hat{f}(n)|^2 \|z^n\|_{\mathcal{E}}^2,$$

$$\lim_k \frac{1}{\|z^k\|_{\mathcal{E}}^2} \sum_{n\geqslant 0} |\hat{f}(n)|^2 \|z^{n+k}\|_{\mathcal{E}}^2 = \|f\|_{H^2}^2. \quad ●$$

REMARK. By means of the same calculations for the space $\mathfrak{E} = l^p_A(w_n)$, $1 \leqslant p \leqslant \infty$, we obtain $\partial\mathfrak{E} = l^p_A$ in the case (6) and $\|f\|_{\partial\mathfrak{E}} = \|f\|_{l^p_A}$, $f \in \mathfrak{E}$, in the case (6′). An analogous result holds for $\mathfrak{E} = c_{0A}(w_n)$ and $\partial\mathfrak{E} = l^\infty_A$, if $w_n = o(1)$.

We let f_r denote the Abel mean of an analytic (in $\mathbf{D}$) function f: $f_r(\zeta) = f(r\zeta)$, $|\zeta| < 1$, $0 \leqslant r < 1$.

LEMMA 7. *Let F be a space of measurable functions on* $[0, 1)$ *with a monotone norm,*[26] *and let E be a space of analytic functions in* $\mathbf{D}$ *which satisfies conditions* a) *and* b), *where the operator of multiplication by z is an isometry. Let $\mathfrak{E} = F(E)$ be the collection of all functions f for which $f_r \in E$, $0 \leqslant r < 1$, and the function $n(r, f) \stackrel{\text{def}}{=} \|f_r\|_E$, $0 \leqslant r < 1$, belongs to F; $\|f\|_{\mathfrak{E}} \stackrel{\text{def}}{=} \|n(\cdot, f)\|_F$. Then the following assertions are true*:

1) *If $f \in \mathfrak{E}$ and $\sup_{0\leqslant r<1}\|f_r\|_E < +\infty$, then $f \in \partial\mathfrak{E}$.*

2) *If $f \in \mathfrak{E}$ and $\lim_{r\to 1-0}\|f_r\|_E = +\infty$, then $f \notin \partial\mathfrak{E}$.*

3) *If f is a polynomial or, more generally, $\lim_{r\to 1-0}\|f_r\|_E = \|f\|_E$, then $f \in \partial\mathfrak{E}$ and $\|f\|_{\partial\mathfrak{E}} = \|f\|_E/\|\mathbf{1}\|_E$.*

---

(25) The symbol $|F|_{\mathfrak{E}}$ denotes the operator norm in $\mathfrak{E}$.

(26) That is, a norm such that for arbitrary measurable functions g_1 and g_2 the conditions $|g_1| \leqslant |g_2|$ on $[0, 1)$ and $g_2 \in F$ imply that $g_1 \in F$ and $\|g_1\|_F \leqslant \|g_2\|_F$.

We always have

$$\varliminf_{r\to 1-0} n(r,\ f) \leqslant \|f\|_{\partial\mathfrak{E}}\|\mathbf{1}\|_E \leqslant \varlimsup_{r\to 1-0} n(r,\ f).$$

PROOF. 1) It is clear that $n(r, z^n f) = r^n n(r, f)$; monotonicity of the norm in F implies that $\|z^n f\|_{\mathfrak{E}} \leqslant \text{const}\|z^n\|_{\mathfrak{E}}$, $n \geqslant 0$, and $f \in \partial\mathfrak{E}$. We note that

$$\text{const} \leqslant \sup_r \|f_r\|_E/\|\mathbf{1}\|_E.$$

2) Let $t(r) \equiv r$ for $r \in [0, 1)$, and let χ_r be the characteristic function of the interval $(R, 1) \subset [0, 1)$. Then $\|t^n\|_F \sim \|t^n\chi_R\|_F$ as $n \longrightarrow \infty$, since

$$\|t^n(1-\chi_R)\|_F \leqslant \text{const}\, R^n = o((\sqrt{R})^n) \leqslant o(\|t^n\chi_{\sqrt{R}}\|_F) \leqslant o(\|t^n\|_F)$$

for an arbitrary R, $0 \leqslant R < 1$. Analogously

$$\|t^n n(\cdot,\ f)\|_F \sim \|t^n\chi_R n(\cdot,\ f)\|_F$$

as $n \longrightarrow \infty$. Consequently

$$\begin{aligned}\|z^n f\|_{\mathfrak{E}} = \|t^n n(\cdot,\ f)\|_F &\geqslant (1+o(1))\|t^n\chi_R\|_F \inf_{r\geqslant R} n(r,\ f)\\ &= (1+o(1))\|z^n\|_{\mathfrak{E}} \inf_{r\geqslant R} n(r,\ f).\|\mathbf{1}\|_E^{-1}\end{aligned} \tag{7}$$

for an arbitrary fixed R, $0 \leqslant R < 1$, as $n \longrightarrow \infty$. It follows that $f \notin \partial\mathfrak{E}$.

3) We estimate $\|z^n f\|_{\mathfrak{E}}$ from above as $n \longrightarrow \infty$:

$$\begin{aligned}\|z^n f\|_{\mathfrak{E}} &= \|t^n n(\cdot,\ f)\|_F = (1+o(1))\|t^n\chi_R n(\cdot,\ f)\|_F\\ &\leqslant (1+o(1))\|t^n\chi_R\|_F \sup_{r\geqslant R} n(r,\ f) = (1+o(1))\|z^n\|_{\mathfrak{E}} \sup_{r\geqslant R} n(r,\ f)\|\mathbf{1}\|_E^{-1}.\end{aligned}$$

If we compare this with (7) and use the condition in the lemma, we conclude that $f \in \partial\mathfrak{E}$ and $\|f\|_{\partial\mathfrak{E}} = \|f\|_E\|\mathbf{1}\|_E^{-1}$. ●

REMARK. If for an arbitrary function $f \in \mathfrak{E}$ the limit $\lim_{r\to 1-0}\|f_r\|_E$ exists, then $\partial\mathfrak{E}$ consists of precisely those f for which this limit is finite.

EXAMPLES. 1. Suppose that $E = H^p$, $1 \leqslant p \leqslant +\infty$; $F = L^q(\mathfrak{M})$, $1 \leqslant q \leqslant \infty$ and $\mathfrak{M}$ is a finite measure on $[0, 1)$ for which[27] $1 \in \operatorname{supp}\mathfrak{M}$. Then

$$\|f\|_{\mathfrak{E}} = \left\{\int_0^1 \left(\int_{\partial\mathbf{D}} |f(r\zeta)|^p\, dl_1(\zeta)\right)^{q/p} d\mathcal{M}(r)\right\}^{1/q} \tag{8}$$

and $\partial\mathfrak{E} = H^p$. The case $p = q = 2$ can also be treated by means of Lemma 5.

2. Let $E = H^p(\mathfrak{M}_1, \partial\mathbf{D})$ be the closure in $L^p(\mathfrak{M}_1, \partial\mathbf{D})$ of the linear span of the functions z^n, $n \geqslant 0$; let $F = L^q(\mathfrak{M})$, where $\mathfrak{M}$ is the measure in the preceding example. The norm in $\mathfrak{E} = F(E)$ is defined by analogy with (8) with l_1 replaced by $\mathfrak{M}_1$. In this case $f \in \partial\mathfrak{E}$ if $\sup_{0\leqslant r<1}\|f_r\|_{H^p(\mathfrak{M}_1)} < +\infty$. If $1 \leqslant p < \infty$ and the powers $\{z^n\}_{n\geqslant 0}$ form an Abel basis in $H^p(\mathfrak{M}_1, \partial\mathbf{D})$, then $\partial\mathfrak{E} = H^p(\mathfrak{M}_1)$. In this case we must have $d\mathfrak{M}_1 = |g|dl_1$, where g is an outer function in H^1; see [162]. Then $\partial\mathfrak{E} = g^{-1/p}H^p$.

---

(27) supp $\mathfrak{M}$ is the closed support of the measure $\mathfrak{M}$.

In the general case $\partial\mathcal{E}$ can be found as an "Abelian realization" of $H^p(\mathfrak{M}_1, \partial\mathbf{D})$.

3. Suppose that $E = H^p$, $1 \leqslant p \leqslant \infty$, and that $F = L_0^\infty(\lambda)$ is the collection of all (classes of) measurable functions on $[0, 1)$ with the norm

$$\|g\| = \sup_{0\leqslant r<1} |g(r)|/\lambda(r) < +\infty$$

such that $g(r) = o(\lambda(r))$, $r \longrightarrow 1$, and $\lim_{r\to 1}\lambda = \infty$. Then

$$\|f\|_{\mathcal{E}} = \sup_{0\leqslant r<1} \frac{1}{\lambda(r)} \left(\int_{\partial\mathbf{D}} |f(r\zeta)|^p \, dl_1(\zeta) \right)^{1/p}$$

and $\partial\mathcal{E} = H^p$. If $p = +\infty$ we obtain $\partial A(\lambda) = H^\infty$. As in Example 2, we can also consider the more general case $E = H^p(\mathfrak{M}_1, \partial\mathbf{D})$.

Another type of space for which boundary spaces can be calculated explicitly will now be described. Let E be a space of analytic functions, and let $\{\lambda_n\}_{n\geqslant 0}$ be a sequence of complex numbers such that $\overline{\lim}_n |\lambda_n|^{1/n} \leqslant 1$, $\lambda_n \neq 0$, $n \geqslant 0$. On the set $A(\mathbf{D})$ of all analytic functions in $\mathbf{D}$ we define an operator Λ by the equation

$$\Lambda f = \Lambda \Big(\sum_{n\geqslant 0} \hat{f}(n) z^n \Big) = \sum_{n\geqslant 0} \hat{f}(n) \lambda_n z^n$$

and we set $E_\Lambda = \Lambda^{-1}E$. If E is a normed space, we equip E_Λ with the "weighted" norm $\|f\|_\Lambda = \|\Lambda f\|_E$. It is clear that the properties of E_Λ depend on Λ and on the "quality" of the system of powers $\{z^n\}_{n\geqslant 0}$ with respect to E.

LEMMA 8. *Let E be the space in Lemma 7, and let $\{\lambda_n\}_{n\geqslant 0}$ be a sequence of positive numbers such that $\lim_n \lambda_{n+1}/\lambda_n = 1$. In each of the following cases $\partial E_\Lambda = E_*$, where E_* is the collection of all functions f in E_Λ for which the collection σf of means defined by the type of basis*[28] $\{z^n\}_{n\geqslant 0}$ *is bounded in E*:

1) $\{z^n\}_{n\geqslant 0}$ *is an unconditional basis in E, and* $\sup_{n,k\geqslant 0}\lambda_{n+k}\lambda_n^{-1} < +\infty$;

2) $\{z^n\}_{n\geqslant 0}$ *is a basis in E,* $\sup_{n\geqslant 0}\lambda_n < +\infty$, *and* $\{\lambda_{n+1}/\lambda_n\}_{n\geqslant 0}$ *is monotone for large n*;

3) $\{z^n\}_{n\geqslant 0}$ *is a Cesàro basis in E,* $\lambda_n \downarrow 0$, *and* $\{\lambda_n\}_{n\geqslant 0}$ *and* $\{\lambda_{n+1}/\lambda_n\}_{n\geqslant 0}$ *are convex for large n*;

4) $\{z^n\}_{n\geqslant 0}$ *is an Abel basis in E, and* $\{\lambda_n\}_{n\geqslant 0}$ *and* $\{1 - \lambda_{n+1}/\lambda_n\}_{n\geqslant 0}$ *are Hausdorff moment sequences.*

The proof makes use of the connection between an admissible summation method for power expansions in E and the supply of Hadamard multipliers (see [41] and [86]); this question is similar to one which will be considered in §3.3. In that section we shall show that if v_n, $n \geqslant 0$, are piecewise-continuous and bounded functions on $[0, 1)$ and $\lim_{t\to 1-0} v_n(t) = 1$, $n \geqslant 0$, then V-summability

$$f = \lim_{t\to 1-0} \sum_{n\geqslant 0} v_n(t) \hat{f}(n) z^n$$

(28) σf is the collection of all partial sums of the Taylor series of f in cases 1) and 2); it is the collection of Fejér sums (arithmetic means) in case 3); and it is the collection of Abel means in case 4).

of Taylor series in E implies that for an arbitrary finite measure $\mathfrak{M}$ on $[0, 1)$ the sequence of moments $\{\lambda_n\}_{n\geq 0}$, $\lambda_n = \int_0^1 v_n(t)d\mathfrak{M}(t)$, $n \geq 0$, defines a continuous operator Λ in E, and $|\Lambda|_{\mathcal{E}} \leqslant \text{const}\,\|\mathfrak{M}\|$.

Now we note that

$$\| z^n f \|_\Lambda = \| z^n \Lambda_n f \|_E = \| \Lambda_n f \|_E, \text{ where } \Lambda_n = \{\lambda_{n+k}\}_{k\geqslant 0}.$$

We assume that we have already established boundedness of the norms $\lambda_n^{-1}|\Lambda_n|_E$, $n = 0, 1, \ldots$. We refer to this assumption as "assumption (α)" and from it we deduce Lemma 8. First we establish the inclusion $\partial E_\Lambda \subset E_*$. Since $f \in \partial E_\Lambda$ implies that

$$\sup_{n\geqslant 0} \lambda_n^{-1} \| \Lambda_n f \|_E < +\infty,$$

it follows that the family of means $\sigma\lambda_n^{-1}\Lambda_n f = \lambda_n^{-1}\Lambda_n \sigma f$, $n \geqslant 0$, is uniformly bounded in E. By assumption (α) the family $\sigma f = \lim_n \lambda_n^{-1}\Lambda_n \sigma f$ is also bounded in E (taking the limit makes use of the Banach-Steinhaus theorem and the fact that $\sigma f \subset E$). Consequently $\partial E_\Lambda \subset E_*$. Conversely, if $f \in E_\Lambda$ and σf is bounded in E, then the families $\sigma\lambda_n^{-1}\Lambda_n f$, $n \geqslant 0$, are uniformly bounded. Later we will show that the operator $\Lambda_1\Lambda^{-1}$ is continuous in E. Since $z^n \Lambda_n \Lambda^{-1} = (z\Lambda_1\Lambda^{-1})^n$ and $\Lambda f \in E$, we then have $\lambda_n^{-1}\Lambda_n f \in E$. Therefore the norms of the functions $\lambda_n^{-1}\Lambda_n f = \lim_\sigma \sigma\lambda_n^{-1}\Lambda_n f$ are bounded by the constant $\sup_{\sigma,n}\|\sigma\lambda_n^{-1}\Lambda_n f\|_E$; that is, $f \in \partial E_\Lambda$.

Now in order to finish the proof it suffices to deduce assumption (α) from the hypothesis of the lemma; that is, to prove that the operators $\lambda_n^{-1}\Lambda_n$, $n \geqslant 0$, and $\Lambda_1\Lambda^{-1}$ are uniformly bounded in E. To fix ideas we will assume that in cases 2)–4) we have $\lambda_n \downarrow 0$ and $\lambda_{n+1}/\lambda_n \uparrow 1$ (other combinations of the symbols $\downarrow$ and $\uparrow$ require minor changes in the proof). Since the arguments for the operators $\lambda_n^{-1}\Lambda_n$ and $\Lambda_1\Lambda^{-1}$ are the same, it suffices to show that the sequences $\lambda_n^{-1}\Lambda_n$ are moment sequences (with respect to the corresponding V-method) and the variations of the corresponding measures are uniformly bounded. Only case 1) is an exception, but it follows immediately from the Lorch theorem on unconditional bases (see [24] or §3.3 below). Assertion 2) reduces to consideration of the functions $v_n(t) = \chi_{[1-1/(n+1),1)}$, $n \geqslant 0$; under certain assumptions the variation of the measure $\mathfrak{M}$ which is generated by $\lambda_n^{-1}\Lambda_n$ (that is, the measure such that $\lambda_n^{-1}\lambda_{n+k} = \mathfrak{M}([1 - 1/(k+1), 1))$, $k \geqslant 0$) equals 1. In assertion 4) we use the functions $v_n(t) = t^n$, $0 \leqslant t < 1$, $n \geqslant 0$. If $\lambda_k = \int_0^1 r^k \, d\mathfrak{M}$, $k \geqslant 0$, where $\mathfrak{M} \geqslant 0$, then

$$\lambda_n^{-1}\lambda_{n+k} = \lambda_n^{-1} \int_0^1 r^k r^n d\mathcal{M},$$

and the variation $\lambda_n^{-1}\int_0^1 r^n \, d\mathfrak{M}$ is again equal to 1. In case 3) the sequence $\{v_n\}_{n\geqslant 0}$ is defined by induction; we set $\sigma_0 = v_0 = 1$ and

$$\sigma_n \overset{\text{def}}{=} v_{n-1} - v_n = \frac{1}{n+1}\chi_{\left[1-\frac{1}{n+1},\, 1\right)}$$

for $n \geqslant 1$. Then the question of whether the sequence $\lambda_n^{-1}\Lambda_n = \{a_k\}_{k\geqslant 0}$ is a moment sequence reduces to the problem of finding a measure $\mathfrak{M}$ such that $b_k = \int_0^1 \sigma_k \, d\mathfrak{M}$, where $b_0 = a_0$ and $b_k = a_{k-1} - a_k$, $k \geq 1$. Consequently we seek $\mathfrak{M}$ by starting with the equations

$$a_0 = \int_0 d\mathcal{M} \text{ and } (k+1)b_k = \mathcal{M}\left(\left[1 - \frac{1}{k+1}, 1\right)\right), \quad k \geq 1.$$

For existence of such a measure it is sufficient (and necessary) that the variation Var of the sequence $\{(k+1)b_k\}_{k \geq 0}$ be finite (here we can assume that $\|\mathfrak{M}\| \leq \text{Var}$). If the sequence $\{a_k\}_{k\geq 0}$ is convex, then

$$\sum_{k=0}^{n} |(k+1)b_k - (k+2)b_{k+1}| \leq \sum_{k=0}^{n} (k+1)(b_k - b_{k+1}) + \sum_{k=0}^{n} b_{k+1} \leq 2\sum_{k=0}^{n+1} b_k - (n+1)b_{n+1},$$

and therefore $\text{Var} \leq 4a_0 = 4\lambda_n\lambda_n^{-1} = 4$. ●

REMARK. From the proof it is clear that a similar result is true for an arbitrary space E where the corresponding means σf are bounded for all $f \in E$. In particular, we can let $E = H^\infty$ or $E = l_A^\infty$.

EXAMPLE. For E we can take any of the spaces H^p, l_A^p $(1 \leq p \leq \infty)$, C_A, c_{0A} or $H^p(\mathfrak{M}_1, \partial\mathbf{D})$ in Example 2 following Lemma 7. In each of these, $\{z^n\}_{n\geq 0}$ is a basis or Cesàro basis (in the latter case this forces known restrictions on $\mathfrak{M}$ (see [162])), and thus for the weight $\{\lambda_n\}_{n\geq 0}$ we can use an arbitrary bounded sequence generated, for example, by a function λ in the Hardy skew field $\mathbf{H}$: $\lambda_n = \lambda(n)$, $n \geq 0$, as long as

$$\lim_n \lambda_{n+1}/\lambda_n = \lim_{t\to\infty} \exp\frac{\lambda'(t)}{\lambda(t)} = 1.$$

In particular, for $\lambda_n = n!/(n+m)!$, $n \geq 0$, $m \geq 1$, we arrive at the spaces E_{Jm} which are defined by the condition $J^m f \in E$, where $(Jf)(\zeta) = \int_0^\zeta f(t)\,dt$, $|\zeta| < 1$. We note that if $E = C_A$, then $\partial E_\Lambda = H^\infty$, and if $E = c_{0A}$, then $\partial E_\Lambda = l_A^\infty$.

LEMMA 9. *Suppose that the hypothesis of Lemma 8 is satisfied. Then, in each of the following cases,* $\partial E_\lambda = E_\lambda$ *and* $\|f\|_{\partial E_\Lambda} = \|f\|_E$:

1) $\{z^n\}_{n\geq 0}$ *is an unconditional basis in E and*

$$\sup_{n,k\geq 0} \lambda_{n+k}\lambda_n^{-1}\lambda_k^{-1} < \infty.$$

2) $\{z^n\}_{n\geq 0}$ *is a basis in E and* $\lambda_{n+1}/\lambda_n \downarrow 1$.

3) $\{z^n\}_{n\geq 0}$ *is a Cesàro basis in E and* $\{\lambda_{n+1}/\lambda_n\}_{n\geq 0}$ *is a convex decreasing sequence (for large n).*

The proof proceeds as in the previous lemma; we use the symbolism

$$\|z^n f\|_{E_\Lambda}/\|z^n\|_{E_\Lambda} = \left\|\sum_{k\geq 0} \hat{f}(k)\lambda_k \frac{\lambda_{n+k}}{\lambda_n\lambda_k} z^k\right\|_E. \quad ●$$

EXAMPLE. We consider spaces E as in the preceding example, replacing λ_n by $1/\lambda_n$. In particular, for $\lambda_n = n!/(n-m)!$, $n \geq m$, $\lambda_n = 1$, $0 \leq n < m$, we arrive at the spaces $E^{(m)}$ which are defined by the condition $f^{(m)} \in E$.

Concluding remarks on the norm in $\partial\mathfrak{E}$. 1. If $\mathfrak{E}$ is a Hilbert space and for each polynomial p the limit

$$\lim_n \| z^n p \|_{\mathcal{E}} / \| z^n \|_{\mathcal{E}} = \| p \|_{\partial\mathcal{E}}$$

exists, then $\| \cdot \|_{\partial\mathfrak{E}}$ is a Hilbert norm on the set of polynomials.

2. The operator of multiplication by z ($g \longrightarrow zg$) is an isometry in $\partial\mathfrak{E}$. In fact,

$$\| z^n (zf) \|_{\mathcal{E}} / \| z^n \|_{\mathcal{E}} = (\| z^{n+1} f \|_{\mathcal{E}} / \| z^{n+1} \|_{\mathcal{E}}) \, (\| z^{n+1} \|_{\mathcal{E}} / \| z^n \|_{\mathcal{E}})$$

and by condition c) $\|zf\|_{\partial\mathfrak{E}} = \|f\|_{\partial\mathfrak{E}}$.

3. If the powers $\{z^n\}_{n\geq 0}$ are complete and minimal(29) in $\partial\mathfrak{E}$ and the norm $\| \cdot \|_{\partial\mathfrak{E}}$ is a Hilbert norm, then the space $\partial\mathfrak{E}$ (or its completion $\widehat{\partial\mathfrak{E}}$) is isometrically isomorphic to H^2. In fact, in this case the operator $g \longrightarrow zg$ is an isometry and $\dim(\widehat{\partial\mathfrak{E}} \ominus z\widehat{\partial\mathfrak{E}}) = 1$. If $e \in \widehat{\partial\mathfrak{E}} \ominus z\widehat{\partial\mathfrak{E}}$ and $\|e\| = 1$, then from the Wald expansion (see [23]) it follows that $g \longrightarrow \Sigma_{n\geq 0}\hat{g}(n)z^n e$ is the aforementioned isomorphism of H^2 onto $\widehat{\partial\mathfrak{E}}$.

5°. *Calculation of* $\partial^*\mathfrak{K}\mathfrak{E}^*$. *A conjugate Bernšteĭn inequality.* Since the shift operator ($g \longrightarrow zg$) is an isometry in the space $\partial\mathfrak{E}$, we can imbed this space into the corresponding class of nonanalytic functions (or two-sided sequences if we are considering power series coefficients). Let $f = \Sigma_{k\in\mathbf{Z}}\hat{f}(k)z^k$, where $\hat{f}(k) = 0$ for $k < k(f)$, and suppose that $Pf \overset{\text{def}}{=} \Sigma_{k\geq 0}\hat{f}(k)z^k$ belongs to $\partial\mathfrak{E}$. By definition we set $\|f\|_{\mathfrak{D}\mathfrak{E}} = \|z^n f\|_{\partial\mathfrak{E}}$, where n is an arbitrary natural number which exceeds $k(f)$. The symbol $\mathfrak{D}\mathfrak{E}$ denotes the collection of all such functions f. The projector P is continuous in $\mathfrak{D}\mathfrak{E}$ (under the assumption that the polynomials are complete in $\partial\mathfrak{E}$) if and only if the system of powers $\{z^n\}_{n\geq 0}$ forms a basis of $\partial\mathfrak{E}$. We introduce the notation $P_n f = \Sigma_0^n \hat{f}(k)z^k$, which we will use in connection with other spaces. We define an involution $\mathcal{J}$ on the set $\mathfrak{P}$ of polynomials in z and z^{-1} by the equation

$$\mathcal{J}\left(\sum_k \hat{p}(k)\, z^k\right) = \sum_k \hat{p}(-k)\, z^k.$$

Theorems 1 and 2 enable us to find the space $\partial^*\mathfrak{K}\mathfrak{E}^*$ in practically all cases where $\partial\mathfrak{E} \neq \mathfrak{E}$; that is, where $\mathfrak{E}$ consists of "nonsmooth" functions. In the dual case of spaces of "smooth" functions in $\overline{\mathbf{D}}$ the operator ∂^* is of a different sort altogether; see Theorem 2, and the examples given later in the subsection. Among the assumptions in Theorem 3 we have inequality (9), which will be called a conjugate Bernšteĭn inequality. In the lemmas after Theorem 3 this inequality will be verified for all spaces $\mathfrak{E}$ which we need for studying weighted approximation problems. We also write a dual inequality for spaces of smooth functions, which in the case $\mathfrak{E} = C_A^{(1)} \overset{\text{def}}{=} \{f \in A(\mathbf{D})\colon f' \text{ is continuous in } \overline{\mathbf{D}}\}$ becomes the classical Bernšteĭn inequality (see [1]). We present many examples. We always assume, however, that the conjugate Bernšteĭn inequality is true under the condition that $\sup_{n\geq 0}|z^n|_{\mathfrak{E}} < +\infty$ (or some similar condition).

Let $\{w_n\}_{n\geq 0}$ be a sequence of positive numbers. We let $(\mathfrak{E}, l_A^\infty(w_n))$, or simply

(29) **For example, if they are minimal in $\mathfrak{E}$; see the proof of Theorem 3.**

$(\mathfrak{S}, w_n)$, denote the set of all analytic (in $\mathbf{D}$) functions (multipliers) f such that $g \in \mathfrak{S} \Rightarrow gf \in l_A^\infty(w_n)$.

THEOREM 2. *Let $\mathfrak{S}$ be a space of analytic functions in* $\mathbf{D}$ *which satisfies conditions* a)–c). *Then* $\partial^*\mathfrak{K}\mathfrak{S}^* = (\mathfrak{S}, \|z^n\|_{\mathfrak{S}})$.

PROOF. The inequality

$$\left\| \sum_{k=0}^{n} \hat{f}(k)\, z^{n-k} \right\|_{\mathscr{K}\mathscr{E}^*} \leqslant \text{const}\, \| z^n \|_{\mathscr{K}\mathscr{E}^*}, \quad n \geqslant 0,$$

means that

$$\left| \sum_{k=0}^{n} \hat{f}(k)\, \hat{g}(n-k) \right| \leqslant \text{const}\, \| g \|_{\mathscr{E}} \| z^n \|_{\mathscr{K}\mathscr{E}^*}, \quad n \geqslant 0,$$

for an arbitrary $g \in \mathfrak{S}$. By the closed-graph theorem functions f in $(\mathfrak{S}, \|z^n\|^{-1}_{\mathfrak{K}\mathfrak{S}^*})$ are also characterized by this property. Also, by condition b), $\|z^n\|_{\mathfrak{K}\mathfrak{S}^*} \asymp \|z^n\|^{-1}_{\mathfrak{S}}$. ●

LEMMA 10. *Suppose that $\mathfrak{S}$ is a Banach space which satisfies conditions* a)–c), $\|zf\|_{\mathfrak{S}} = \|f\|_{\mathfrak{S}}$ *for an arbitrary $f \in \mathfrak{S}$, and $(\mathfrak{D}\mathfrak{S})^*_+$ is the space of linear functionals Φ, defined on the set $\mathfrak{P}$ of all polynomials in z and z^{-1}, which are continuous in the metric of $\mathfrak{D}\mathfrak{S}$ and "analytic" in the sense that* $\Phi(z^{-k}) = 0$, $k \geqslant 1$. *Suppose that the involution $\mathcal{J}$ is continuous in $\mathfrak{D}\mathfrak{S}$. Then*

$$\partial^* \mathscr{K}\mathscr{E}^* = \mathscr{K}(\mathscr{D}\mathscr{E})^*_+.$$

REMARK. $(\mathfrak{D}\mathfrak{S})^*_+ \subset (\partial\mathfrak{S})^*$, but in general $(\mathfrak{D}\mathfrak{S})^*_+ \neq (\partial\mathfrak{S})^*$, even if the polynomials are dense in $\partial\mathfrak{S}$. For example, this is so for $\mathfrak{S} = C_A$: $(\mathfrak{D}\mathfrak{S})^*$ is the collection of all finite measures on $\partial\mathbf{D}$ and by the Riesz theorem (see [23]) we have $\mathfrak{K}(\mathfrak{D}\mathfrak{S})^*_+ = H^1$, while $(\partial\mathfrak{S})^* = C_A^*$ and $\mathfrak{K}C_A^*$ is the space of all functions which can be written as Cauchy-Stieltjes integrals. The equation $(\mathfrak{D}\mathfrak{S})^*_+ = (\partial\mathfrak{S})^*$ is valid if the projector P is continuous in $\mathfrak{D}\mathfrak{S}$.

COROLLARY. *If $\mathfrak{S}$ satisfies conditions* a)–c) *and the involution $\mathcal{J}$ is continuous in $\mathfrak{D}\mathfrak{S}$, then*

$$\partial^* \mathscr{K}\mathscr{E}^* \subset \mathscr{K}(\mathscr{D}\mathscr{E})^*_+.$$

The proof follows from assertion 1) in Theorem 3 (below) and Lemma 10, applied to the space $\partial\mathfrak{S}$. ●

PROOF OF LEMMA 10. By Theorem 2 we have $\partial^*\mathfrak{K}\mathfrak{S}^* = (\mathfrak{S}, 1)$. Let $g \in (\mathfrak{S}, 1)$. On the set of polynomials in z and z^{-1} the function g generates a linear functional Φ:

$$\Phi(f) = \lim_{n \to 0} (z^{-n} g z^n \mathcal{J} f)^\wedge(0), \quad f \in \mathscr{P}.$$

Continuity of Φ follows from the equation

$$(z^{-n} g z^n \mathcal{J} f)^\wedge(0) = (g z^n \mathcal{J} f)^\wedge(n),$$

continuity of the involution $\mathcal{J}$ and of the multiplication operator $p \longrightarrow gp$ from $\mathfrak{E}$ into l^∞_A, and the fact that the shift in $\mathfrak{E}$ is an isometry. Clearly $\Phi \in (\mathfrak{D}\mathfrak{E})^*_+$ and $\mathfrak{K}\Phi = g$.

Conversely, if $g = \mathfrak{K}\Phi$ and $\Phi \in (\mathfrak{D}\mathfrak{E})^*_+$, then for an arbitrary polynomial $p \in \mathfrak{E}$ we have

$$\widehat{gp}(n) = \sum_{k=0}^{n} \Phi(z^k)\,\hat{p}(n-k) = \sum_k \Phi(z^k)\,\hat{p}(n-k) = \Phi(\mathcal{J}z^{-n}p),$$

and therefore

$$\sup_{n\geqslant 0} |\widehat{gp}(n)| \leqslant \|\Phi\|\,\|\mathcal{J}\|\,\|p\|_{\mathcal{E}}. \quad \bullet$$

THEOREM 3. *Let $\mathfrak{E}$ be a Banach space which satisfies conditions* a)–c). *Then the following assertions are true*:

1) $\partial^*\mathfrak{K}\mathfrak{E}^* \subset \partial^*\mathfrak{K}(\partial\mathfrak{E})^*$.

2) *If the involution $\mathcal{J}$ is continuous in $\mathfrak{D}\mathfrak{E}$, if* $\sup_{n\geqslant 0}|P_n|_{\mathfrak{E}} < +\infty$, *and if*

$$\|p\|_{\partial\mathcal{E}}\,\|z^n\|_{\mathcal{E}} \leqslant \text{const}\,\|p\|_{\mathcal{E}} \tag{9}$$

for an arbitrary polynomial p of degree less than or equal to n, n = 0, 1, . . . , *with* const = $c(\mathfrak{E})$, *then*

$$\partial^*\mathcal{K}(\partial\mathcal{E})^* = \mathcal{K}(\partial\mathcal{E})^*$$

and

$$\partial^*\mathcal{K}\mathcal{E}^* = \mathcal{K}(\partial\mathcal{E})^*.$$

PROOF. 1) Let $g \in \partial^*\mathfrak{K}\mathfrak{E}^*$. Then

$$\left|\sum_{k=0}^{n} \hat{g}(k)\hat{f}(n-k)\right| \leqslant \text{const}\,\|f\|_{\mathcal{E}}\,\|z^n\|_{\mathcal{K}\mathcal{E}^*}$$

for arbitrary n and f with $n \geqslant 0$ and $f \in \mathfrak{E}$. Let $f_0 \in \partial\mathfrak{E}$, $0 \leqslant N \leqslant n$, and $f = z^{n-N}f_0$. Then

$$\left|\sum_{k=0}^{N} \hat{g}(k)\hat{f}_0(N-k)\right| \leqslant \text{const}\,\|z^{n-N}f_0\|_{\mathcal{E}}/\|z^n\|_{\mathcal{E}}$$

and by condition c)

$$\left|\sum_{k=0}^{N} \hat{g}(k)\hat{f}_0(N-k)\right| \leqslant \text{const}\,\|f_0\|_{\partial\mathcal{E}},$$

$$\left\|\sum_{k=0}^{N} \hat{g}(k)\,z^{N-k}\right\|_{\mathcal{K}(\partial\mathcal{E})^*} \leqslant \text{const}.$$

Now we just note that $\|z^n\|_{\mathfrak{K}(\partial\mathfrak{E})^*} \asymp 1$; that is, the sequence $\{z^n\}_{n\geqslant 0}$ is uniformly minimal in $\partial\mathfrak{E}$. In fact, let $n \geqslant N \geqslant 0$ and let q be a polynomial with $\hat{q}(N) = 0$. By condition b) we have (see (3) also)

$$\|z^{n-N}(z^N - q)\|_{\mathcal{E}} \geqslant d\,\|z^n\|_{\mathcal{E}},$$

and therefore $\|z^N - q\|_{\partial\mathfrak{E}} \geqslant d$.

2) Let $g \in \mathfrak{K}(\partial\mathfrak{S})^*$. By $\mathfrak{K}$-duality we know that for an arbitrary polynomial p of degree n

$$|\langle z^n \mathcal{J} p, g\rangle| \leqslant \|z^n \mathcal{J} p\|_{\partial\mathcal{E}} \|g\|_{\mathcal{K}(\partial\mathcal{E})^*} \leqslant \|\mathcal{J}\| \|p\|_{\partial\mathcal{E}} \|g\|_{\mathcal{K}(\partial\mathcal{E})^*}.$$

Using (9), we rewrite the preceding inequality in the form

$$\left|\sum_{k=0}^{n} \hat{g}(k)\hat{p}(n-k)\right| \leqslant \text{const}\, \|z^n\|_{\mathcal{K}\mathcal{E}^*} \|p\|_{\mathcal{E}},$$

where const does not depend on n or p. Since $\sup_{n\geqslant 0}|P_n|_{\mathfrak{S}} < \infty$, we have

$$\left\|\sum_{k=0}^{n} \hat{g}(k) z^{n-k}\right\|_{\mathcal{K}\mathcal{E}^*} \leqslant \text{const}\, \|z^n\|_{\mathcal{K}\mathcal{E}^*}.$$

Thus $\mathfrak{K}(\partial\mathfrak{S})^* \subset \partial^*\mathfrak{K}\mathfrak{S}^*$. By analogy we establish the inclusion $\partial^*\mathfrak{K}(\partial\mathfrak{S})^* \subset \mathfrak{K}(\partial\mathfrak{S})^*$. •

Now we can calculate ∂^*-spaces for all of the classes in Lemmas 5, 7 and 8.

LEMMA 5′. *Let $\mathfrak{S} = l^p_A(w_n)$, $1 \leqslant p < \infty$, where the numbers $w_n = \|z^n\|_{\mathfrak{S}}$ satisfy condition* c) *and* (6). *Then the conjugate Bernšteĭn inequality* (9) *holds, and therefore*

$$\partial^*\mathcal{K}\mathcal{E}^* = \mathcal{K}(\partial\mathcal{E})^* = l^{p'}_A, \quad \frac{1}{p}+\frac{1}{p'}=1.$$

PROOF. As we noted in the remark after Lemma 6, $\partial\mathfrak{S} = l^p_A$. Let g be a polynomial of degree less than or equal to n. Then

$$\|g\|^p_{\mathcal{E}} = \sum_{k=0}^{n} |\hat{g}(k)|^p \|z^k\|^p_{\mathcal{E}} \geqslant \frac{1}{\text{const}} \|z^n\|^p_{\mathcal{E}} \sum_{k=0}^{n} |\hat{g}(k)|^p = \frac{1}{\text{const}} \|z^n\|^p_{\mathcal{E}} \|g\|^p_{\partial\mathcal{E}},$$

where for const we can take the number on the left side of (6). •

REMARK. If $w_n \downarrow$, then (6) is automatically satisfied. If $p = 2$, then $\mathfrak{S}$ is a Hilbert space which is invariant with respect to rotations; here $\partial\mathfrak{S} = \partial^*\mathfrak{K}\mathfrak{S}^* = H^2$.

LEMMA 7′. *Let E, F, and $\mathfrak{S} = F(E)$ be the spaces in Lemma* 7. *Suppose that the involution $\mathcal{J}$ is continuous in $\mathfrak{D}E$ and $\sup_{0\leqslant r<1}\|g_r\|_{\mathfrak{D}E} < +\infty$ for an arbitrary $g \in \mathfrak{D}E$; $g_r = \Sigma_k r^{|k|}\hat{g}(k)z^k$. Then the conjugate Bernšteĭn inequality* (9) *is true. If the sequence of projectors $\{P_n\}_{n\geqslant 0}$ is bounded in $\mathfrak{S}$ (or, equivalently,*(30)*) in E), then*

$$\partial^*\mathcal{K}\mathcal{E}^* = \mathcal{K}(\partial\mathcal{E})^* = \mathcal{K}E^*.$$

In other cases

$$\partial^*\mathcal{K}\mathcal{E}^* = \mathcal{K}(\mathcal{D}E)^*_+.$$

PROOF. Let g be a polynomial of degree less than or equal to n, $n \geqslant 0$. Then for each r, $0 \leqslant r < 1$, we have

$$r^n g = z^n \mathcal{J}(z^n \mathcal{J} g_r)_r.$$

(30) See also Lemma 12 in subsection 8°.

Consequently $r^n\|g\|_E \leqslant \text{const}\|g_r\|_E$, where const does not depend on g or n. Since $\|t^n\|_F = \text{const}\|z^n\|_{\mathfrak{E}}$ and $\|g\|_E = \|g\|_{\partial\mathfrak{E}}$ (see Lemma 7), we obtain $\|z^n\|_{\mathfrak{E}}\|g\|_E \leqslant \text{const}\|g\|_{\mathfrak{E}}$ for an arbitrary polynomial g of degree less than or equal to n.

We still have to establish the inclusion $\mathfrak{K}(\mathfrak{D}\mathfrak{E})_+^* \subset \partial^*\mathfrak{K}\mathfrak{E}^*$, while the other assertions in the lemma are obtained from Theorem 3 and the corollary to Lemma 10. Let[31] $g \in \mathfrak{K}(\mathfrak{D}E)_+^*$ and $f \in \mathfrak{D}E$. Then

$$|\langle g, f_r\rangle| = |\langle g_r, f\rangle| \leqslant \|g\|\|f_r\|_{\mathfrak{D}E} \leqslant \text{const}\|g\|\|f\|_{\mathfrak{D}E}$$

and thus

$$\|g_r\|_{\mathcal{K}(\mathfrak{D}E)_+^*} \leqslant \text{const}\|g\|_{\mathcal{K}(\mathfrak{D}E)_+^*}.$$

By Lemma 10 the norm in $\mathfrak{K}(\mathfrak{D}E)_+^*$ is equivalent to the multiplier norm in $(E, \mathbf{1})$, and therefore

$$r^n|\widehat{fg}(n)| = |(f_r g_r)^\wedge(n)| \leqslant \text{const}\|g\|\|f_r\|_E$$

for an arbitrary $f \in \mathfrak{E}$. Thus

$$\|z^n\|_{\mathcal{E}}|\widehat{fg}(n)| \leqslant \text{const}\|g\|\|f\|_{\mathcal{E}}. \quad \bullet$$

REMARK. The constant const in the Bernšteĭn inequality does not exceed

$$\|1\|_E \|\mathcal{J}\|_E^2 M_E^2, \text{ where } M_E = \sup_{\|g\|_{\mathfrak{D}E}\leqslant 1}\ \sup_{0\leqslant r<1} \|g_r\|_{\mathfrak{D}E}.$$

EXAMPLES. Let h be a polynomial of degree less than or equal to n.

1. Let $E = H^p$ and $F = L^q(\mathfrak{M})$ be the spaces in Example 1 following Lemma 7. Then

$$\|h\|_{H^p}\left(\int_0^1 r^{nq}\,d\mathcal{M}\right)^{1/q} \leqslant \left(\int_0^1\left(\int_{\partial\mathbf{D}}|h(r\zeta)|^p\,dl_1(\zeta)\right)^{q/p} d\mathcal{M}\right)^{1/q},$$

and if $1 < p < \infty$, then $\partial^*\mathfrak{K}\mathfrak{E}^* = H^{p'}$, $1/p + 1/p' = 1$. The last assertion in the lemma says that $\partial^*\mathfrak{K}\mathfrak{E}^* = \mathfrak{K}(L^1)_+^* = H^\infty$ for $p = 1$ and $\partial^*\mathfrak{K}\mathfrak{E}^* = \mathfrak{K}(L^\infty)_+^* = H^1$ for $p = \infty$ (the Riesz brothers' theorem; see [23]).

2. Let $E = H^p(\mathfrak{M}_1, \partial\mathbf{D})$ and $F = L^q(\mathfrak{M})$ be the spaces in Example 2 following Lemma 7. Suppose that the measure $\mathfrak{M}_1$ is symmetric with respect to the real axis (that is, $\mathfrak{M}_1 e = \mathfrak{M}_1\bar{e}$, where the bar denotes complex conjugation, and $e \subset \partial\mathbf{D}$); let $\mathfrak{M}_1(\partial\mathbf{D}) = 1$; and suppose that the Rosenblum conditions (see [154]) are satisfied for some $M_E < +\infty$. If $p = q < +\infty$, then

$$\iint_{\mathbf{D}}\left|h\left(\frac{\zeta}{|\zeta|}\right)\right|^p|\zeta|^{np}\,d\mathcal{M}_1 d\mathcal{M} \leqslant M_E^p \iint_{\mathbf{D}}|h(\zeta)|^p\,d\mathcal{M}_1 d\mathcal{M},$$

if $p < +\infty$, then in the notation of Example 2 following Lemma 7 we have

$$\partial^*\mathcal{K}\mathcal{E}^* = H^{p'}(\mathcal{M}_1, \partial\mathbf{D}) = g^{\frac{1}{p'}}H^{p'}, \quad \frac{1}{p} + \frac{1}{p'} = 1,$$

for $p > 1$, and $\partial^*\mathfrak{K}\mathfrak{E}^* = gH^\infty$ for $p = 1$.

[31] Here we use the fact that $\mathfrak{D}E = \mathfrak{D}\mathfrak{E}$.

3. Let $E = H^p$ and $F = L_0^\infty(\lambda)$. Then

$$\|h\|_{H^p} \sup_{0\leqslant r<1} \frac{r^n}{\lambda(r)} \leqslant \sup_{0\leqslant r<1} \frac{1}{\lambda(r)} \left(\int_{\partial \mathbf{D}} |f(r\zeta)|^p \, dl_1(\zeta) \right)^{1/p},$$

and for $1 \leqslant p \leqslant \infty$ we have

$$\partial^* \mathcal{K} \mathcal{E}^* = H^{p'}, \quad \frac{1}{p} + \frac{1}{p'} = 1.$$

If $p = 1$ or ∞ (as in Example 1) we use the Riesz theorem. In particular, $\partial^* \mathfrak{K} A(\lambda)^* = H^1$.

REMARK. Theorems 2 and 3 and Lemmas 10, 5′, 7′, and 8′ (below) enable us to find $\partial^* \mathfrak{K} \mathfrak{E}^*$ without describing the conjugate space $\mathfrak{K} \mathfrak{E}^*$ of $\mathfrak{E}$. We note that such a description would require, in general, very sophisticated techniques. For example, in [37] it was shown that

$$\mathcal{K}(H^p(l_2))^* = \Lambda H^{p'}(l_2), \quad \frac{1}{p} + \frac{1}{p'} = 1, \quad 1 < p < \infty,$$

where $\Lambda = \{\lambda_n\}_{n\geqslant 0}$, $\lambda_n = 2n + 1$.

LEMMA 8′. *Suppose that the symbols E and Λ have the same meanings as in Lemma* 8, $\mathfrak{E} = E_\Lambda$, *and the involution $\mathcal{J}$ is continuous in $\mathfrak{D}\mathfrak{E} = \mathfrak{D}E$. Then in cases* 1) *and* 2) *of Lemma* 8, *and in case* 3) *under the assumption of convexity of the sequence* $\{1/\lambda_n\}_{n\geqslant 0}$, *we have the conjugate Bernšteĭn inequality*

$$\|g\|_E \lambda_n \leqslant \text{const} \|\Lambda g\|_E,$$

where g is a polynomial of degree less than or equal to n and const *does not depend on g or n. Here in the first and second cases we have* $\partial^* \mathfrak{K} \mathfrak{E}^* = \mathfrak{K} E^*$, *and in the third case* $\partial^* \mathfrak{K} \mathfrak{E}^* = \mathfrak{K}(\mathfrak{D}E)^*_+$.

The proof of this inequality follows from the equation $g = z^n \mathcal{J}(\Lambda^{-1} \Lambda z^n \mathcal{J}) g$ and an estimate on the operator $z^n \mathcal{J} \Lambda^{-1} z^n \mathcal{J}$ (that is, the operator $z^k \longrightarrow \lambda_{n-k}^{-1} z^k$, $0 \leqslant k \leqslant n$) by the method used in Lemma 8. In order to verify the second of the two equations,(32) it suffices to show that $\mathfrak{K}(\mathfrak{D}E)^*_+ \subset \partial^* \mathfrak{K} \mathfrak{E}^* = (\mathfrak{E}, \lambda_n)$. Let $h \in \mathfrak{K}(\mathfrak{D}\mathfrak{E})^*_+ = (E, 1)$. Then

$$(fh)^\frown(n) = \sum_{k=0}^{n} \hat{f}(k) \lambda_k (\hat{h}(n-k)/\lambda_k) = (fH)^\frown(n), \text{ where } H = \sum_{k=0}^{n} \hat{h}(k) z^k / \lambda_{n-k}.$$

Changing (if necessary) some number of factors $\lambda_0, \ldots, \lambda_N$ (depending only on the sequence $\{\lambda_k\}_{k\geqslant 0}$), we can see that the sequence $(1/\lambda_n, \ldots, 1/\lambda_0, 0, \ldots, 0)$ is convex for an arbitrary n. It is also clear that the powers $\{z^n\}_{n\geqslant 0}$ form a Cesaro basis in their linear span in the space $\mathfrak{K}(\mathfrak{D}E)^*_+$. From the estimates in Lemma 8 we obtain

$$\|H\|_{\mathcal{K}(\mathcal{D}E)^*_+} \leqslant \|h\|_{\mathcal{K}(\mathcal{D}E)^*_+} \lambda_n^{-1}$$

---

(32) The first equation takes into account the fact that $\mathfrak{K}E^* = \mathfrak{K}E^*_*$ (see the statement of Lemma 8), since the Cauchy transform depends only on the values of the functional on the closure of the set of polynomials in E.

and consequently

$$|(fh)^{\wedge}(n)| = |(fH)^{\wedge}(n)| \leqslant \|\Lambda f\|_E \|H\|_{(E,1)} \leqslant \text{const}\, \lambda_n^{-1} \|f\|_{E_\Lambda}. \quad \bullet$$

A remark similar to the one following Lemma 8 can also be made in this case.

EXAMPLE. Let E and $\{\lambda_n\}_{n\geqslant 0}$ be as in the example following Lemma 8, and let g be a polynomial of degree less than or equal to n. In the case $\lambda_n = n!/(n+m)!$, $n \geqslant 0$, $m \geqslant 1$, the conjugate Bernšteĭn inequality is of the form

$$\|g\|_{C_A} \frac{n!}{(n+m)!} \leqslant \|J^m g\|_{C_A},$$

$$\|g\|_{H^p} \frac{n!}{(n+m)!} \leqslant \|J^m g\|_{H^p},$$

where Jg is the primitive of g. If $E = H^p$, l_A^p, c_{0A}, or $H^p(\mathfrak{M}_1, \partial\mathbf{D})$, $1 < p < \infty$ (see the example following Lemma 8), then $\partial^*\mathfrak{K}E_\Lambda^* = \mathfrak{K}E^*$; if $E = C_A$ or $E = H^\infty$, then $\partial^*\mathfrak{K}E_\Lambda^* = H^1$; if $E = H^1$, then $\partial^*\mathfrak{K}E_\Lambda^* = H^\infty$.

6°. *Calculation of* $\partial^*\mathfrak{K}\mathfrak{E}^*$. *A Bernšteĭn inequality.* Here we make some observations on the space $\partial^*\mathfrak{K}\mathfrak{E}^*$ and a Bernšteĭn inequality for classes $\mathfrak{E}$ of "smooth" functions. First of all, from Theorem 2 and the corollary to Lemma 10 we see that, under the conditions in these assertions, $\partial^*\mathfrak{K}\mathfrak{E}^* = (\mathfrak{E}, \|z^n\|_{\mathfrak{E}})$ and $\partial^*\mathfrak{K}\mathfrak{E}^* \subset \mathfrak{K}(\mathfrak{D}\mathfrak{E})_+^*$, respectively. The first equation leads to the estimate $\hat{h}(n)\|z^n\|_{\mathfrak{E}} = O(1)$ if $h \in \partial^*\mathfrak{K}\mathfrak{E}^*$. Thus

$$\partial^*\mathcal{K}\mathcal{E}^* \subset \mathcal{K}(\mathcal{D}\mathcal{E})_+^* \cap l_A^\infty(\|z^n\|_{\mathcal{E}}). \tag{10}$$

LEMMA 11. *Let* $\mathfrak{E}$ *be a space of analytic functions which satisfies conditions* a)–c), *such that if* $f \in \mathfrak{E}$ *and* $h \in l_A^\infty(\|z^n\|_{\mathfrak{E}})$, *then* $fh \in l_A^\infty(\|z^n\|_{\mathfrak{E}})$. *Then*

$$\partial^*\mathcal{K}\mathcal{E}^* = l_A^\infty(\|z^n\|_{\mathcal{E}}).$$

In particular, this is so if $l_A^\infty(\|z^n\|_{\mathfrak{E}})$ *is an algebra with respect to pointwise multiplication of functions.*

PROOF. By condition b), $\mathfrak{E} \subset l_A^\infty(\|z^n\|_{\mathfrak{E}})$; the rest follows from Theorem 1. $\bullet$

LEMMA 9′. *Let* $\mathfrak{E}$ *and* Λ *be as in Lemma* 9. *Suppose that the involution* $\mathcal{J}$ *is continuous in* $\mathfrak{D}E$ *and* $\mathfrak{E} = E_\Lambda$. *If* $\lambda_n \leqslant \text{const}\, \lambda_k + \lambda_{n-k}$, $0 \leqslant k \leqslant n$, $n = 0, 1, \ldots,$ *and* $\mathfrak{E} \subset l_A^1$, *then*

$$\partial^*\mathcal{K}\mathcal{E}^* = \mathcal{K}(\mathcal{D}E)_+^* \cap l_A^\infty(\lambda_n).$$

PROOF. By virtue of earlier comments it suffices to show that

$$h \in \mathcal{K}(\mathcal{D}E)_+^* \cap l_A^\infty(\lambda_n) \Rightarrow h \in \partial^*\mathcal{K}\mathcal{E}^*.$$

Let $f \in \mathfrak{E}$. Then

$$[\Lambda(hf)]^{\wedge}(n) = [h\Lambda f]^{\wedge}(n) + [f\Lambda_n h]^{\wedge}(n),$$

where

$$\Lambda_n h = \sum_{k=0}^{n} (\lambda_n - \lambda_{n-k})\, \hat{h}(k)\, z^k.$$

Since $\Lambda f \in E$ and $h \in \mathfrak{K}(\mathfrak{D}E)_+^* = (E, 1)$, we have

$$|[h\Lambda f]^\wedge(n)| \leqslant \|f\|_{\mathcal{E}} \| h \|_{(E,1)}.$$

By hypothesis

$$|(\Lambda_n h)^\wedge(k)| \leqslant \text{const}\, |\hat{h}(k)|\, \lambda_k \leqslant \text{const}\, \|h\|_{l_A^\infty(\lambda_n)}.$$

Consequently

$$|[f\Lambda_n h]^\wedge(n)| \leqslant \text{const}\, \|f\|_{l_A^1} \|h\|_{l_A^\infty(\lambda_n)}$$

and

$$|(hf)^\wedge(n)| \leqslant \lambda_n^{-1}\left(\|f\|_{\mathcal{E}} \| h \|_{(E,1)} + \text{const}\, \|f\|_{l_A^1} \|h\|_{l_A^\infty(\lambda_n)}\right). \bullet$$

We make a few observations on Lemmas 11 and 9′. The conditions in these lemmas are in some sense complementary, although they do not exhaust all of the possibilities. The space $l_A^\infty(\|z^n\|_{\mathcal{E}})$ is a Banach algebra if and only if

$$\sum_{k=0}^{n} \|z^k\|_{\mathcal{E}}^{-1} \|z^{n-k}\|_{\mathcal{E}}^{-1} \leqslant \text{const}\, \|z^n\|_{\mathcal{E}}^{-1} \tag{11}$$

for $n \geqslant 0$ (for details see §3.2). Let $\|z^k\|_{\mathcal{E}}^{-1} = w(k)$, $k \geqslant 0$, where w is a function on $[0, +\infty)$ which decreases in a neighborhood of $+\infty$. Clearly the condition $w(2^n) \geqslant a^n$, where $a > 0$, is necessary for (11), and therefore w decreases "no faster" than some power $t^{-\alpha}$, $\alpha > 0$. On the other hand, it is easy to show that the conditions

$$\left|\frac{w'}{w}(t)\right| = O\left(\frac{1}{t}\right), \quad \int_1^\infty w(t)\,dt < +\infty$$

are sufficient for (11). For example, if w is a function in the Hardy field **H** (see §1.5), then the first condition is a result of the (necessary) condition

$$\overline{\lim_{t\to\infty}}\, (\log w(t))/\log t > -\infty,$$

and we just have to prove that w is integrable. If we let $\mathcal{E} = E_\Lambda$ and $\|z^k\|_{\mathcal{E}} = \lambda_k$, $k \geqslant 0$, then the space $l_A^\infty(\lambda_n)$ is an algebra in all of the following cases:

$$\lambda_n = n^{\alpha_1}(\log n)^{\alpha_2} \ldots (\log_k n)^{\alpha_{k+1}}, \quad n \geqslant n_0; \quad (\alpha_1, \ldots, \alpha_{k+1}) \succ (1, 1, \ldots, 1).$$

On the contrary, if $(\alpha_1, \ldots, \alpha_{k+1}) \preceq (1, \ldots, 1)$, the result does not follow.

We return to Lemma 9′, which includes the case of slowly increasing sequences $\{\lambda_n\}_{n\geqslant 0}$. From the hypothesis it follows that $\lambda_n \leqslant \text{const}\, n$. The inclusion $\mathcal{E} \subset l_A^1$ must therefore be proved not by means of the coarse estimate

$$\sum_{n\geqslant 0} |\hat{f}(n)| \leqslant \sum_{n\geqslant 0} \lambda_n^{-1} \quad (f \in \mathcal{E} = E_\Lambda),$$

but by means of the more subtle interaction of the norm $\|\cdot\|_E$ and the factors λ_n. It is not hard to show that $\lambda_n \leqslant \text{const}\, \lambda_k + \lambda_{n-k}$, $0 \leqslant k \leqslant n$, $n = 0, 1, \ldots$, if $\lambda_n = \lambda(n)$, where λ is an increasing function on $[0, +\infty)$ and $\lambda(t)/t$ is decreasing (in a neighborhood of $+\infty$). Example:

$$\lambda_n = n^{\alpha_1}(\log n)^{\alpha_2}\dots(\log_k n)^{\alpha_{k+1}} \text{ with } (\alpha_1,\dots,\alpha_{k+1}) \succ (1,0,\dots,0).$$

If we do not have $E_\Lambda \subset l^1_A$, then the description of $\partial^*\mathfrak{K}E^*_\Lambda$ becomes a complicated problem; apparently the best description of $\partial^*\mathfrak{K}E^*_\Lambda$ in this case is given by the "multiplier" equation $\partial^*\mathfrak{K}\mathfrak{E}^* = (\mathfrak{E}, \|z^n\|_{\mathfrak{E}})$; see also Example 3 (below).

EXAMPLE 1. Let E be one of the spaces H^p, l^p_A $(1 \leqslant p \leqslant \infty)$ or C_A, and let λ be a function on $[0, \infty)$ in the Hardy field $\mathbf{H}$ such that

$$\lim_{t\to+\infty} (\log \lambda(t))/\log t < +\infty \text{ and} \int_1^\infty \lambda(t)^{-1}\,dt < \infty.$$

Then for $\mathfrak{E} = E_\Lambda$ the boundary space $\partial^*\mathfrak{K}\mathfrak{E}^*$ consists of all analytic functions h in $\mathbf{D}$ for which

$$\hat{h}(n) = O\left(\frac{1}{\lambda_n}\right), \quad n \to \infty.$$

In particular, for $\lambda_n = n^\alpha$, $\alpha > 1$, E_Λ is the collection of all functions f whose derivatives of order α are in E.

EXAMPLE 2. Let E be as above; let $\lambda_n = n^\alpha$, $0 < \alpha \leqslant 1$, and $E_{\{n^\alpha\}} \overset{\text{def}}{=\!=} E_\alpha$.

a) $E = H^p$; by the theorem of Szász (see [174] and §3.3 below) we have $H^p_\alpha \subset l^1_A \Leftrightarrow \alpha > 1/p$ for $1 < p \leqslant 2$ or $\alpha > 1/2$ for $2 \leqslant p \leqslant \infty$. The case $p = 1$ is covered by a theorem of Hardy (see [23]): $H^1_1 \subset l^1_A$ (but of course $L^1_1 \not\subset l^1$). In each of these cases

$$\partial^*\mathcal{K}E^*_\alpha = \left\{f \in H^{p'}: \hat{f}(n) = O\left(\frac{1}{n^\alpha}\right)\right\},$$

where $1/p + 1/p' = 1$.

b) $E = C_A$; by a theorem of Bernšteĭn (see [39]) we have $C^{(\alpha)}_A \subset l^1_A \Leftrightarrow \alpha > 1/2$. Here

$$\partial^*\mathcal{K}(C^{(\alpha)}_A)^* = \{f \in H^1 : \hat{f}(n) = O(1/n^\alpha)\}.$$

The inclusion $E_\alpha \subset l^1_A$ for other classes of functions (which are symmetric in the sense of Semenov) is obtained as a corollary of eigenfunction expansions for isometric operators (see §3.3).

EXAMPLE 3. We consider in greater detail two cases which are not covered by Lemmas 11 and 9′ or by Examples 1 and 2.

a) Suppose that $E = l^p_A$, and that $\{\lambda_n\}_{n\geqslant 0}$ is an arbitrary sequence such that $\lim \lambda_{n+1}/\lambda_n = 1$, $\lambda_n \uparrow \infty$. Then (by the closed-graph theorem)

$$h \in \partial^*\mathcal{K}(E_\Lambda)^* \Leftrightarrow \left|\sum_{k=0}^{n} \hat{f}(k)\hat{h}(n-k)\right| \leqslant \frac{\text{const}\,\|f\|_{E_\Lambda}}{\lambda_n}, \ \forall f \in E_\Lambda$$

$$\Leftrightarrow \left(\sum_{k=0}^{n} |\hat{h}(n-k)|^{p'} \frac{1}{\lambda_k^{p'}}\right)^{1/p'} \leqslant \frac{\text{const}}{\lambda_n}, \quad n \geqslant 0.$$

In other words, $h \in \partial^*\mathfrak{K}(E_\Lambda)^*$ if and only if the function

$$H(\zeta) = \sum_{k\geqslant 0} |\hat{h}(k)|^{p'}\zeta^k, \quad |\zeta| < 1,$$

is a multiplier for the space $l_A^\infty(\lambda_n^{p'})$ if $1 < p \leqslant \infty$, or

$$\max_{0\leqslant k\leqslant n} \frac{|\hat{h}(k)|}{\lambda_{n-k}} \leqslant \frac{\text{const}}{\lambda_n},$$

if $p = 1$. Here there are two possible sufficient conditions which enable us to calculate $\partial^*\mathfrak{K}(E_\Lambda)^*$:

$$\left(\sum_{k=0}^{n} \lambda_{n-k}^{-p'}\lambda_k^{-p'}\right)^{1/p'} \leqslant \text{const}\, \lambda_n^{-1}, \quad n \geqslant 0, \tag{12}$$

and then $l_A^\infty(\lambda_n^{p'})$ is an algebra for $p > 1$ (for $p' = \infty$ the estimate (12) is equivalent to the fact that $l_A^1(\lambda_n)$ is an algebra) and the necessary condition $\hat{h}(k) = O(1/\lambda_k)$, $k \geqslant 0$, is also sufficient for $h \in \partial^*\mathfrak{K}(E_\Lambda)^*$; or

$$\Big(\sum_{n\geqslant 0} \lambda_n^{-p'}\Big)^{1/p'} < +\infty, \quad \lambda_n \leqslant \text{const}\,\lambda_k + \lambda_{n-k},$$

and then $l_A^p(\lambda_n) \subset l_A^1$, $\mathfrak{K}(\mathfrak{D}E)_+^* = l_A^{p'}$, and by Lemma 9′ $\partial^*\mathfrak{K}\mathfrak{E}^* = l_A^\infty(\lambda_n)$.

In the case of a unit weight ($\lambda_n \equiv 1$) Lemma 10 immediately gives us a more precise result: $\partial^*\mathfrak{K}(l_A^p)^* = l_A^{p'}$, $1 \leqslant p \leqslant \infty$.

b) Suppose that $E = l_A^2 \equiv H^2$, $\lambda_n = (n+1)^{1/2}$, $n \geqslant 0$, and $\mathfrak{E} = E_\Lambda$ is the class of all analytic functions in $\mathbf{D}$ with a finite Dirichlet integral

$$\int_{\mathbf{D}} |f'(\zeta)|^2\, dl_2 < +\infty.$$

We will show in this case that the boundary space $\partial^*\mathfrak{K}(E_\Lambda)^*$ is not given by the formulas in Lemmas 11 and 9′. In fact, the right side of the formula in Lemma 9′ gives us the space $H^2 \cap l_A^\infty(n^{1/2})$, and the inclusion (10) is of the form $\partial^*\mathfrak{K}\mathfrak{E}^* \subset H^2 \cap l_A^\infty(n^{1/2})$. From what we showed in part a) of this example, it suffices to find a function h such that $H(\zeta) \overset{\text{def}}{=} \sum_{k\geqslant 0} |\hat{h}(k)|^2 \zeta^k$ is not a multiplier in $l_A^\infty(n)$ but for which $h \in H^2 \cap l_A^\infty(n^{1/2})$. This inclusion is equivalent to the conditions $\sum_{k\geqslant 0} a_k < \infty$, $a_k = O(1/k)$, where $a_k = |\hat{h}(k)|^2$, $k \geqslant 0$. If $a_k \downarrow 0$, we have only succeeded in satisfying the first condition. Let $p_n = 2^{n^2}$, $p_{-1} = 0$, $m_n = p_n - p_{n-1}$, $n \geqslant 0$, and $a_k = \alpha_n/m_n$ for $p_{n-1} \leqslant k < p_n$, where $\alpha_n = n^{-3/2}$, $n \geqslant 1$. Then $\sum_{k\geqslant 0} a_k < +\infty$. On the other hand, the function H is not a multiplier in $l_A^\infty(n)$:

$$\sum_{k=0}^{p_n} \frac{a_k}{p_n - k + 1} \geqslant \text{const}\, a_{p_n-1} \log p_n = \text{const} \frac{\alpha_n}{m_n} n^2 \geqslant \text{const} \frac{\alpha_n n^2}{p_n} \neq O\Big(\frac{1}{p_n}\Big).$$

Returning to the foregoing necessary and sufficient condition for $h \in \partial^*\mathfrak{K}\mathfrak{E}^*$, $\mathfrak{E} = H^2_{1/2}$, we rewrite it in the form

$$\sum_{k=0}^{n} \frac{|\hat{h}(k)|^2}{n-k+1} \leqslant \frac{\text{const}}{n}, \quad n = 0, 1, \ldots.$$

We use the equation

$$(k+1)^{-1}(n-k+1)^{-1} = (n+2)^{-1}((k+1)^{-1} + (n-k+1)^{-1}),$$

and arrive at the condition

$$\frac{1}{n+2}\sum_{k=0}^{n} |\hat{h}(k)|^2 + \frac{1}{n+2}\sum_{k=0}^{n} \frac{|\hat{h}(k)|^2(k+1)}{n-k+1} \leqslant \frac{\text{const}}{n}, \quad n \geqslant 0.$$

Thus

$$h \in \partial^* \mathcal{K}(H^2_{1/2})^*$$

$$\Leftrightarrow (h \in H^2 \text{ and } g(\zeta)\log(1-\zeta) \in l^\infty_A, \text{where } g(\zeta) = \sum_{k\geqslant 0} |\hat{h}(k)|^2(k+1)\zeta^k, \ |\zeta|<1).$$

If we split the sum Σ_0^n into $\Sigma_0^{n/2}$ and $\Sigma_{n/2}^n$, it is not hard to show that either of the conditions:

(i) $|\hat{h}(k)|^2(k+1) = O(1/\log k)$ and $h \in H^2$, or

(ii) $\Sigma_{k\geqslant 0} |\hat{h}(k)|^{2p}(k+1)^p < +\infty$ for $1 \leqslant p < \infty$

guarantees that $h \in \partial^* \mathfrak{K}(H^2_{1/2})^*$.

To conclude this subsection we introduce several simple assertions on Bernšteĭn inequalities in spaces of "smooth" functions. Here we are considering inequalities of the form

$$\| p \|_{\mathcal{E}} \leqslant \text{const} \, \| z^n \|_{\mathcal{E}} \| p \|_{\partial\mathcal{E}},$$

which hold for an arbitrary polynomial of degree less than or equal to n.

1) If the powers $\{z^n\}_{n\geqslant 0}$ form a basis in $\mathfrak{E}$, then validity of a direct (conjugate) Bernšteĭn inequality in $\mathfrak{E}$ is equivalent to the validity of a conjugate (direct) Bernšteĭn inequality in $\mathfrak{K}\mathfrak{E}^*$.

2) If by a direct (conjugate) Bernšteĭn inequality for the space $\mathfrak{K}\mathfrak{E}^*$ instead of (9) we mean

$$\| p \|_{\mathcal{K}\mathcal{E}^*} \leqslant \text{const} \, \| z^n \|_{\mathcal{K}\mathcal{E}^*} \| p \|_{\mathcal{K}(\partial\mathcal{E})^*}$$

(respectively, the reverse inequality), then the preceding assertion remains valid if $\|z^n\|_{\mathfrak{E}} \asymp \|z^{2n}\|_{\mathfrak{E}}$, and the powers $\{z^n\}_{n\geqslant 0}$ form a Cesàro basis in $\mathfrak{E}$. We always have $\partial^*\mathfrak{K}\mathfrak{E}^* \subset \mathfrak{K}(\partial\mathfrak{E})^*$; the reverse inclusion is in general not true (for example, with $\mathfrak{E} = C_A$).

3) A Bernšteĭn inequality is valid for each of the spaces in Lemmas 6 and 9. The classical case $\lambda_n = n + 1$ has been analyzed by another method for polynomials which are not "analytic" (in [1], for example).

A general observation on the case where $\mathfrak{E}$ contains only "smooth" functions is that if $\mathfrak{E}$ is an algebra with respect to pointwise multiplication, whose space of maximal ideals is $\overline{\mathbf{D}}$, and $\partial^*\mathfrak{K}\mathfrak{E}^* \subset H^\infty$, then the conditions $f \in \partial\mathfrak{E}$ $(= \mathfrak{E})$ and $1/f \in \partial^*\mathfrak{K}\mathfrak{E}^*$ imply that $1/f \in \mathfrak{E}$. This applies to the classes $C_A^{(n)}$ and H^p_q.

Finally, we note that in Hilbert spaces (for example, of the type considered in the table with $p = q = 2$) we arrive at an estimate for the leading coefficients of orthogonal polynomials (see subsection 1°)

$$|a_{nn}| \leqslant \frac{\text{const}}{\| z^n \|_{\mathcal{E}}}, \quad n \geqslant 0.$$

7°. *Table.* The preceding subsections (3°–6°) were devoted to applications of Theorem 1. In Table 1 we express some of the conclusions in a more explicit form.

TABLE 1

Space	Necessary and sufficient condition for uniform minimality. Examples of weight functions. Comments.
$L^q(\mathfrak{M}, H^p)$ and, in particular, $H^p(d\mathfrak{M}\,dl_1)$, where $\mathfrak{M}$ is a finite measure on $(0, 1)$, $1 \in \operatorname{supp}\mathfrak{M}$, and $1 \leqslant p, q \leqslant \infty$.	$f \in H^p$, $1/f \in H^{p'}$. Here f is an outer function; the condition does not depend on q or $\mathfrak{M}$. Example: $f_\alpha(\zeta) = (1-\zeta)^\alpha$, $\lvert\zeta\rvert < 1$, for $-1/p < \alpha < 1/p'$ (for $p = 1$ or ∞ we set $\alpha = 0$).
H^p	The same condition.
$\{f: J^m f \in H^p\}$, where $Jf = \int_0^\zeta f\,dt$, $\lvert\zeta\rvert < 1$.	The same condition.
$A(\lambda)$, where $\lambda \uparrow +\infty$ as $r \uparrow 1$.	$f \in H^\infty$, $1/f \in H^1$. Here f is an outer function; the condition does not depend on λ. Example: f_α for $0 \leqslant \alpha < 1$.
C_A	$f \in C_A$, $1/f \in H^1$.
l^p_A	$f \in l^p_A$, $1/f \in l^{p'}_A$. Example: $f_\alpha(\zeta) = (1-\zeta)^\alpha$ for $1/p - 1 < \alpha < 1 - 1/p'$ (for $p = 1$ or ∞ we use the extreme values of α).
$C^{(n)}_A = \{f: f^{(n)} \in C_A\}$, $n \geqslant 1$.	$f \in C^{(n)}_A$, $1/f \in C^{(n)}_A$. (Since $1/f \in H^1 \cap l^\infty(k^n)$ leads to the condition $f(\zeta) \neq 0$ for all ζ, $\lvert\zeta\rvert \leqslant 1$, and $1/f \in C^{(n)}_A$).
$H^p_q = \{f: f^{(q)} \in H^p\}$, $1 < p \leqslant \infty$, $q \geqslant 0$.	$f \in H^p_q$, $1/f \in H^{p'} \cap l^\infty(n^q)$. For $p = q = 1$ the condition simplifies to f, $1/f \in H^1_1$, since here $1/f \in H^\infty$ and $(1/f)' = -f'/f^2 \in H^1$ together with $f' \in H^1$.
$H^2_{1/2}$, the space of functions with a finite Dirichlet integral.	$f \in H^2_{1/2}$, $\{\lvert(1/f)\hat{\ }(n)\rvert^2\}_{n\geqslant 0}$ is a multiplier in $l^\infty(n)$. Example: $f(\zeta) = 1/\log^k(1-\zeta)$, $\lvert\zeta\rvert < 1$; $k \geqslant 1$.

8°. *Power bases.* We consider the question of which spaces X have the property that for an arbitrary polynomial g

$$\| P_n g \|_X \leqslant \operatorname{const} \| g \|_X, \quad n \geqslant 0,$$

where $X = \mathfrak{S}_f$; here $\mathfrak{S}$ is a space of analytic functions where the analogous condition is assumed to be satisfied. In other words, we seek a characterization of those weights f for which the powers $\{z^n\}_{n\geqslant 0}$ form a basis[33] in the closure (with respect to $\mathfrak{S}_f$) of their linear span $\mathcal{P}^*_A$. The results which we obtain will make use of the transition to the boundary space $\partial\mathfrak{S}$, as in subsection 3°.

LEMMA 12. *Let $\mathfrak{S}$ be a Banach space of functions which contains the polynomials and satisfies conditions* a)–c) *of subsection* 2°.

(33) The word "basis" will be used in only this sense.

1) *If* $\{z^n\}_{n\geq 0}$ *is a basis in the metric of* $\mathfrak{S}$, *it is also a basis in the metric of* $\partial\mathfrak{S}$.

2) *If* $\{z^n\}_{n\geq 0}$ *is uniformly minimal in* $\mathfrak{S}$ *and in* $\mathfrak{S}_f$, *then* $\partial(\mathfrak{S}_f) = (\partial\mathfrak{S})_f$.

PROOF. 1) Since $z^nP_k = P_{n+k}z^n$ $(n,\ k \geq 0)$, from the estimate

$$\|P_kg\|_{\mathscr{E}} \leqslant \text{const}\,\|g\|_{\mathscr{E}},\quad g \in \mathscr{P}_A,$$

we have

$$\|P_kg\|_{\partial\mathscr{E}} = \varlimsup_n \frac{\|z^nP_kg\|_{\mathscr{E}}}{\|z^n\|_{\mathscr{E}}} \leqslant \text{const}\,\varlimsup_n \frac{\|z^ng\|_{\mathscr{E}}}{\|z^n\|_{\mathscr{E}}} = \text{const}\,\|g\|_{\partial\mathscr{E}}.$$

2) $$g \in \partial(\mathscr{E}_f) \Leftrightarrow \sup_{n\geqslant 0} \|z^ng\|_{\mathscr{E}_f}/\|z^n\|_{\mathscr{E}_f} < +\infty.$$

But since $\|z^n\|_{\mathfrak{S}_f} = \|z^nf\|_{\mathfrak{S}} \asymp \|z^n\|_{\mathfrak{S}}$ by hypothesis (and by Lemma 3), we have

$$g \in \partial(\mathscr{E}_f) \Leftrightarrow gf \in \partial\mathscr{E} \Leftrightarrow g \in (\partial\mathscr{E})_f. \ \bullet$$

COROLLARY. *If* $\|g\|_{\partial\mathfrak{S}} = \|g\|_{C_A}$ *or* $\|g\|_{\partial\mathfrak{S}} = \|g\|_{H^1(\mathfrak{M})}$, *where* $\mathfrak{M}$ *is a finite measure on* $\partial\mathbf{D}$, supp $\mathfrak{M}$ *is infinite, and g is an arbitrary polynomial, then there is no function* $f \in \mathfrak{S}$ *for which the powers* $\{z^n\}_{n\geq 0}$ *form a basis in* $\mathfrak{S}_f$.

In fact, uniform minimality of the sequence $\{z^n\}_{n\geq 0}$ is clearly necessary for it to be a basis, and we can use Lemma 12. The result for C_A also follows from classical theorems on Fourier series (see, for instance, [23]), and for $H^1(\mathfrak{M})$ it follows from a theorem of Forelli (see [131]). This theorem asserts that the Riesz projector P is continuous in $L^1(\mathfrak{M})$ under certain conditions on $\mathfrak{M}$. •

The particular case $\mathfrak{S} = H^1(l_2)$, $f \equiv 1$ is treated in [13]. As other examples we consider the spaces $\mathfrak{S} = A(\lambda)$ and $\mathfrak{S} = F(E)$ in Lemmas 7 and 7′ for $E = C_A, H^\infty, H^1$, or $H^1(\mathfrak{M})$; in particular,

$$\|g\|_{\mathscr{E}}^p = \int_0^1 \left(\int_{\partial\mathbf{D}} |g(r\zeta)|\, d\mathscr{M}(\zeta) \right)^p d\mathscr{M}_1,$$

where $\mathfrak{M}_1$ is a finite measure on $[0, 1)$.

THEOREM 4. *Let* $\mathfrak{S} = F(H^2)$, *where F is a space with a monotone norm (see Lemma* 7). *The powers* $\{z^n\}_{n\geq 0}$ *form a basis in* $\mathfrak{S}_f$ *if and only if f and* $1/f \in H^2$ *and*

$$|f|^2 = e^{u+\tilde{v}} \quad on\ \partial\mathbf{D}, \tag{13}$$

where u and v are bounded real functions on $\partial\mathbf{D}$, $\tilde{v}$ *is the harmonic conjugate of v, and* $\|v\|_{L^\infty} < \pi/2$.

REMARK. In this theorem the word "basis" can be understood as "basis in $\mathfrak{S}_f$" (and not just in the closure of the linear span of the z^n) if F has the property that the conditions $\varphi, \varphi_k \in F$, $|\varphi_k| \leqslant \varphi$ almost everywhere and $\lim_k \varphi_k = 0$ almost everywhere on $[0, 1)$ imply that $\lim_k \|\varphi_k\| = 0$. Example: $F = L^p(\mathfrak{M}_1)$, $1 \leqslant p < \infty$. We verify the fact that the polynomials are not dense in $\mathfrak{S}_f$. As we noted in the proof of Theorem 4, condition (13) implies that $\|f_rP_ng\|_{H^2} \leqslant \text{const} \cdot \|f_rg\|_{H^2}$ for arbitrary r, $0 < r < 1$,

$n \geqslant 0$, and $g \in \mathcal{P}_A$. Therefore

$$\|f_r(P_ng_r - g_r)\|_{H^2} \leqslant \text{const}\|f_rg_r\|_{H^2} \quad \text{and} \quad \lim_{n\to\infty}\|f_r(P_ng_r - g_r)\|_{H^2} = 0$$

for $0 \leqslant r < 1$. The condition on the norm $\|\cdot\|_F$ now gives us $\|P_ng - g\|_{\mathfrak{E}_f} \longrightarrow 0$ as $n \longrightarrow \infty$.

PROOF OF THEOREM 4. Lemma 12 and the Helson-Szegö theorem which we mentioned in subsection 1° imply the necessity of the conditions listed in the theorem. We now prove sufficiency. Let u_r and v_r denote the traces of the harmonic functions u and v on the circle $|\zeta| = r$. Clearly $\widetilde{v}_r = (\widetilde{v})_r$ and $\|v_r\|_\infty \leqslant \|v\|_\infty < \pi/2$. Since f is an outer function, we have $2 \log|f_r| = 2(\log|f|)_r = u_r + \widetilde{v}_r$, and by the Helson-Szegö theorem

$$\sup_{n\geqslant 0,\, 0\leqslant r<1} \|P_n\|_{H^2_{f_r}} < +\infty.$$

Let $g \in \mathcal{P}_A$. Then

$$\|f_rP_ng_r\|_{H^2} \leqslant \text{const}\,\|f_rg_r\|_{H^2} \quad \text{and} \quad \|fP_ng\|_{\mathcal{E}} \leqslant \text{const}\,\|fg\|_{\mathcal{E}},\ n = 0,\ 1,\ \ldots.$$

By the Banach-Steinhaus theorem this means that $\{z^n\}_{n\geqslant 0}$ is a basis in $\mathfrak{E}_f$. ●

To conclude this section we introduce an elementary sufficient condition for $\{z^n\}_{n\geqslant 0}$ to be a basis in the weighted class $\mathfrak{E}_f$. From Lemma 12 it follows that if the polynomials are dense in the boundary space $(\partial\mathfrak{E})_f$, then a necessary condition for having a basis is

$$\lim_n \left\|fP_n\frac{1}{f} - 1\right\|_{\partial\mathcal{E}} = 0,$$

or (using the notation $r_n(g) = g - P_ng$)

$$\lim_n \left\|fr_n\left(\frac{1}{f}\right)\right\|_{\partial\mathcal{E}} = 0.$$

The sufficient condition, on the one hand, uses the speed of convergence to zero of the sequence $\{\|fr_n(1/f)\|_{\partial\mathfrak{E}}\}_{n\geqslant 0}$, and on the other hand it uses some "regularity" of the norm in $\mathfrak{E}$ which takes place, for instance, in all spaces $\mathfrak{E} = F(E)$ where $\|f_r\|_E \uparrow$ as $r \uparrow 1$.

LEMMA 13. *Let $\mathfrak{E}$ be a space of analytic functions in which $\{z^n\}_{n\geqslant 0}$ is a basis and condition* c) *of* 2° *is satisfied, where $\|z^ng\|_{\mathfrak{E}} \leqslant \text{const}\cdot\|z^n\|_{\mathfrak{E}}\|g\|_{\partial\mathfrak{E}}$ for all $g \in \mathfrak{E}$ and $n \geqslant 0$. In the following cases the powers $\{z^n\}_{n\geqslant 0}$ form a basis in $\mathfrak{E}_f$*:

1) $\displaystyle\sum_{k\geqslant 0}\left\|fr_k\left(\frac{1}{f}\right)\right\|_{\partial\mathcal{E}} < +\infty$;

2) $\mathcal{E} = l^p_A(w_n)$, $w_n\downarrow$ *and* $K \overset{\text{def}}{=} \left(\displaystyle\sum_{k\geqslant 0}\left\|fr_k\left(\frac{1}{f}\right)\right\|^{p'}_{l^p_A}\right)^{1/p'} < +\infty,\ 1 \leqslant p \leqslant \infty$;

3) $\mathcal{E} = l^2_A(w_n)$, $w_n\downarrow$ *and* $f \in H^\infty$, $\dfrac{1}{f} \in H^2_{1/2}$.

REMARK. Assertion 3) is essentially a particular case of a theorem of Plotkin [93] involving the spaces $H^2(l_2)$ and $H^p(l_2)$, $p \geqslant 2$. The arguments of Plotkin show that under the assumptions in 3) the result in Lemma 13 holds if for $\mathfrak{E}$ we choose either $H^2(\rho drdl_1)$, as in the lemma, or $H^p(\rho drdl_1)$, $p \geqslant 2$. We recall that $H^2_{1/2}$ consists of functions with a finite Dirichlet integral.

PROOF OF THE LEMMA. 1) As we already noted, $z^k P_{n-k} = P_n z^k$ (we let $P_m = 0$ if $m < 0$), and thus for each $g \in \mathfrak{E}_f$ we have

$$fP_n \frac{g}{f} = fP_n \sum_{k\geqslant 0} \hat{g}(k)\, z^k \frac{1}{f} = f\sum_{k=0}^{n} \hat{g}(k)\, z^k P_{n-k}\frac{1}{f} = P_n g - f\sum_{k=0}^{n} \hat{g}(k)\, z^k r_{n-k}\left(\frac{1}{f}\right),$$

$$\begin{aligned} \left\| fP_n \frac{g}{f}\right\|_{\mathfrak{E}} &\leqslant \mathrm{const}\, \|g\|_{\mathfrak{E}} + \sum_{k=0}^{n} |\hat{g}(k)| \left\| z^k f r_{n-k}\left(\frac{1}{f}\right)\right\|_{\mathfrak{E}} \\ &\leqslant \mathrm{const}\, \|g\|_{\mathfrak{E}} + \mathrm{const} \sum_{k=0}^{n} |\hat{g}(k)|\, \|z^k\|_{\mathfrak{E}} \left\| f r_{n-k}\left(\frac{1}{f}\right)\right\|_{\partial\mathfrak{E}} \\ &\leqslant \mathrm{const}\, \|g\|_{\mathfrak{E}} \sum_{k=0}^{n} \left\| f r_k\left(\frac{1}{f}\right)\right\|_{\partial\mathfrak{E}}. \end{aligned} \tag{14}$$

Consequently $\sup_{n\geqslant 0} |P_n|_{\mathfrak{E}_f} < +\infty$.

2) Inequality (14) can be extended in a natural fashion:

$$\sum_{k=0}^{n} |\hat{g}(k)|\, \|z^k\|_{\mathfrak{E}} \left\| f r_{n-k}\left(\frac{1}{f}\right)\right\|_{\partial\mathfrak{E}} \leqslant K\|g\|_{\mathfrak{E}}.$$

3) In assertion 2) we set $p = 2$ and note that

$$\|1/f\|^2_{H^2_{1/2}} = \sum_{n\geqslant 0} |(1/f)^\frown(n)|^2 (n+1) = \sum_{k\geqslant 0} \|r_k(1/f)\|^2_{l^2_A}. \quad \bullet$$

REMARKS. 1. The functions f and $1/f$ play symmetric roles in Theorem 4. Thus in assertion 3) of Lemma 13 we can require that $f \in H^2_{1/2}$ and $1/f \in H^\infty$, at least if $\{w_n\}_{n\geqslant 0}$ is a Hausdorff moment sequence (that is, $\mathfrak{E} = H^2(\mathfrak{M}dl_1)$). Examples of functions f for assertion 3) in Lemma 13 can be found in [93].

2. As an example of a function f which satisfies the conditions in Theorem 4 we can use

$$f(\zeta) = \exp\left(\int_{\overline{\mathbf{D}}} \log(1-\zeta t)\, d\mathcal{M}(t)\right),$$

where $\mathfrak{M}$ is a real measure on $\overline{\mathbf{D}}$ such that $\int_{\overline{\mathbf{D}}} |d\mathfrak{M}| < ½$. In fact, we then have

$$\log|f|^2 = \tilde{v}, \qquad v = 2\int_{\overline{\mathbf{D}}} \arg(1-zt)\, d\mathcal{M}(t)$$

and

$$\|v\|_{L^\infty} \leqslant 2\int_{\overline{\mathbf{D}}} \|\arg(1-zt)\|_{L^\infty} |d\mathcal{M}(t)| < \frac{\pi}{2}.$$

In particular, we can take

$$f(\zeta) = \prod_{k\geqslant 1} (1-\zeta t_k)^{\mathcal{M}_k}, \quad \sum |\mathcal{M}_k| < \frac{1}{2}, \quad t_k \in \overline{\mathbf{D}}.$$

The case $\mathfrak{E} = H^2$, $f_\alpha(\zeta) = (1 - \zeta)^\alpha$, $-1/2 < \alpha < 1/2$, is contained in Babenko [3]. The index 1/2 is precise in the general case of Theorem 4, since $f_{-1/2} \notin H^2$ $(= \partial\mathfrak{E})$.

3. If $f \in \partial\mathfrak{E}$ and $1/f$ is a polynomial, then condition 1) in Lemma 13 is automatically satisfied. For example, if $f(\zeta) = (1 - \zeta)^{-1}$ and

$$\|g\|_{\mathfrak{E}} = \left(\int_0^1 \max_{n \geqslant 0} |\hat{g}(n)|^p r^{np} dr\right)^{1/p}, \quad 1 \leqslant p < \infty,$$

then the powers $\{z^n\}_{n \geqslant 0}$ form a basis in $\mathfrak{E}_f$ (here $\partial\mathfrak{E} = l_A^\infty$ and $\partial^*\mathcal{K}\mathfrak{E}^* = l_A^1$, in accord with Lemmas 7 and 7′).

4. New examples of spaces (of "smooth" functions) with a power basis can be obtained by considering the Cauchy transform $\mathcal{K}X^*$ of the conjugate space of $X = \mathfrak{E}_f$.

§2.4. Reduction to a problem of Bernšteĭn and auxiliary assertions

1°. Tests for weak invertibility which are more refined than those in §2.1 are based not on direct approximations by polynomials $p_n(\zeta) \longrightarrow 1/[f(\zeta)]$, $|\zeta| < 1$, so that $p_n f \longrightarrow \mathbf{1}$ in the corresponding space, but on the splitting of this problem into two easier problems, one of which is conveniently solved not in its original representation but in its "Fourier representation". A simple version of this approach consists of the following. Let X be a space of analytic (in $\mathbf{D}$) functions(34) $f \in X$, $f(\zeta) \neq 0$, $|\zeta| < 1$, where f satisfies the condition $e(t) \stackrel{\text{def}}{=} \exp t \log f \in X$ for all $t \in \overline{\mathbf{D}}$ (or only for $0 \leqslant t \leqslant 1$). Suppose that the set of functions of the form $e_\Phi = (e(\,\cdot\,), \Phi)$, $\Phi \in X^*$, is a quasi-analytic class in $\overline{\mathbf{D}}$ (that is, the equations $e_\Phi^{(n)}(t) = 0$, $n = 0, 1, \ldots,$ at any point t with $|t| \leqslant 1$ imply that $e_\Phi \equiv 0$). Now the problem of weak inversion of f reduces to the proof of the inclusion $(\log)^n f \in \mathfrak{L}(z^k f\colon k \geqslant 0) = E(f)$, $n = 0, 1, \ldots$. In fact, this inclusion implies that $e_\Phi^{(n)}(1) = 0$, $n \geqslant 0$, for an arbitrary functional Φ which is zero on $E(f)$. The assumption of quasi-analyticity gives us $e_\Phi \equiv 0$; in particular, $e_\Phi(0) = (\mathbf{1}, \Phi) = 0$. By the Hahn-Banach theorem this means that $\mathbf{1} \in E(f)$; that is, f is weakly invertible. Thus the original problem splits into two: (a) approximation of $(\log f)^n f$ by functions pf with $p \in \mathfrak{P}_A$; (b) proof of quasi-analyticity of the class $\{e_\Phi\colon \Phi \in X^*\}$. Problem (b) is usually easier to solve in terms of Taylor coefficients than is problem (a), using the estimates on $|f|$ which we obtained in Chapter 1.

In some cases this approach leads immediately to significant progress in the weak invertibility problem. We consider the following question, for example. Let λ and Λ be given weight functions on $[0, 1)$ with $\lambda < \Lambda$. How much faster than λ must Λ grow in order that each function $f \in A(\lambda)$, $f(\zeta) \neq 0$, $|\zeta| < 1$, will be weakly invertible in $A(\Lambda)$? A similar question can be asked for other kinds of spaces, but in this preliminary discussion we will only consider the one case which we stated. The estimates in

(34) All of the spaces in this section and succeeding sections satisfy conditions a) and b) in §2.3.2°; that is, a) the functionals $f \longrightarrow f(\zeta)$, $|\zeta| < 1$, are continuous in X; b) $\{z^n\}_{n \geqslant 0}$ is a uniformly minimal sequence in X. When we encounter a boundary space ∂X we also assume that we have c) $\lim_n \|z^{n+1}\|/\|z^n\| = 1$. The most important of these conditions, namely b), was treated in §2.3.

Chapter 1 make it easy to prove the following assertion.[35]

LEMMA 0. *Let* λ *and* Λ *be increasing positive functions on* $[0, 1)$, $M = \log \Lambda$, $\mu = \log \lambda$, $\mu(r) = \exp \psi(\log(1/(1-r)))$, $0 \leqslant r < 1$, *where* $l \stackrel{\text{def}}{=} \lim_{t\to\infty} \psi'(t) > 1$ *and* $\psi''\psi'^{-3/2}$ *is a monotone function. Suppose that* $M' > \mu'$ *and the function* $\mu(M'/\mu' - 1)$ *is monotone in a neighborhood of the point* 1. *Write* $M = \mu + h \circ \mu$, *where* h *is an increasing function on* $(0, +\infty)$, $h > 0$. *If*

$$\int^{\infty} \frac{h(t)}{t^{1+(1+\gamma)^{-1}}}\,dt = +\infty, \quad \gamma = \frac{2}{\pi}\arctan\frac{1}{c}, \quad c = \frac{2}{\pi} B\left(1+\frac{1}{l}\right)^{l+1}$$

(B *is the constant in Theorem* 2 *of* §1.3), *then each function* f *in the class* $A(\lambda)$, $f(\zeta) \neq 0$, $|\zeta| < 1$, *is weakly invertible in* $A(\Lambda)$.

PROOF. We estimate the derivatives of e_Φ, $\Phi \in A(\Lambda)^*$. In order to do this we first note that the estimates in Chapter 1 (Theorem 1 in §1.1 and Theorem 2 in §1.3) imply that for $t = x + iy$ and $x \geqslant 0$ we have

$$|e(t, \zeta)| = |f^t(\zeta)| = |f(\zeta)|^x e^{-y \operatorname{Im}(\log f(\zeta))} \leqslant |f|^x \lambda^{c|y|} \leqslant \lambda^{x+c|y|},$$
$$|e^{(n)}(t, \zeta)| = |\log f|^n |f^t(\zeta)| \leqslant \text{const}^n (\log \lambda)^n \lambda^{x+c|y|}(|\zeta|),$$

where $c = 2\pi^{-1}B(1 + 1/l)^{l+1}$.

Consequently for values of t in an angle of opening $\gamma\pi$, $\gamma = 2\pi^{-1} \arctan c^{-1}$, with a vertex at the point 1 and a bisector $(0, 1)$, we obtain

$$|e^{(n)}(t, \zeta)| \leqslant \text{const}^n (\log \lambda)^n \lambda(|\zeta|), \quad |\zeta| < 1.$$

Each linear functional Φ in $A(\Lambda)$ is, by the Riesz theorem, of the form

$$(f, \Phi) = \int_{\mathbf{D}} \frac{f(\zeta)}{\Lambda(|\zeta|)}\, d\mathcal{M}, \quad f \in A(\Lambda),$$

where $\mathfrak{M}$ is a finite measure in $\mathbf{D}$. Therefore

$$|e_\Phi^{(n)}(t)| \leqslant \text{const}^n \sup_{\zeta \in \mathbf{D}} \frac{(\log \lambda)^n \lambda(|\zeta|)}{\Lambda(|\zeta|)} \stackrel{\text{def}}{=} A_n, \quad n = 0, 1, \ldots,$$

for all values of t in the triangle $\Delta = \{t\colon \operatorname{Re} t \geqslant 0, |\arg(1-t)| \leqslant \gamma\pi/2\}$. It is not hard to show (see [60], Chapter V, §6) that under the conditions in the lemma we have $A_n < +\infty$, $n = 0, 1, \ldots$. Also

$$A_n = \sup_{\rho \geqslant \text{const}} \rho^n / T(\rho/\text{const}), \quad n \geqslant 0,$$

where $T(\rho) = \Lambda(\lambda^{-1}(e^\rho))e^{-\rho}$, $\rho \geqslant$ const; from the hypothesis it follows that the function $t \longrightarrow \log T(e^t)$ is convex in a neighborhood of $+\infty$ and

---

(35) We note that the restrictive "regularity" assumptions in Lemma 0 are satisfied in many cases; for example, this is so in any Hardy field (see §2.7).

$$\int^{\infty} \frac{\log T(\rho)}{\rho^{1+(1+\gamma)^{-1}}}\,d\rho = \int^{\infty} \frac{\log \Lambda(\lambda^{-1}(e^{\rho})) - \rho}{\rho^{1+(1+\gamma)^{-1}}}\,d\rho$$
$$= \int^{\infty} \frac{\mathrm{M}(r) - \mu(r)}{\mu^{1+(1+\gamma)^{-1}}}\,d\mu = \int^{\infty} \frac{h(t)}{t^{1+(1+\gamma)^{-1}}}\,dt = +\infty.$$

By a theorem of Salinas [166] the class $\{e_\Phi\colon \Phi \in A(\Lambda)^*\}$ is quasi-analytic in the triangle Δ. Consequently $\mathbf{1} \in \mathfrak{L}\{(\log f)^n f\colon n = 0, 1, \ldots\}$. But, as we already noted, $|(\log f)^n| \leqslant \text{const} \cdot (\log \lambda)^n$ and $(\log \lambda)^n \lambda = O(\Lambda)$ for an arbitrary $n = 0, 1, \ldots$, so that

$$\lim_k \| f\sigma_k - f(\log f)^n \|_{A(\Lambda)} = 0,$$

where the σ_k are the arithmetic means of the Taylor series of the function $(\log f)^n$. Therefore $(\log f)^n f \in E(f)$, $n = 0, 1, \ldots$. ●

COROLLARY. *Under these assumptions all nonvanishing functions in* $\mathbf{D}$ *in the space* $A(\lambda)$ *are weakly invertible in* $A(\Lambda)$, *where*

$$\Lambda = \lambda \exp\left((\log \lambda)^{(1+\gamma)^{-1}} \frac{1}{(\log\log \lambda) \ldots (\log_n \lambda)}\right), \quad n \geqslant 2.$$

REMARK. The constant B in the estimates in Chapter 1 is large enough to ensure that γ is sufficiently small, and we always have $(1+\gamma)^{-1} < 1$. But even with the best of all possible constants ($c = 1$) an argument of the kind in Lemma 0 cannot lead to the definitive order of the "component" $\log h(t) \sim \frac{1}{2}\log t$, which we will obtain in the following sections.

Thus the proposed method, although it enables us to strengthen significantly the theorems in §2.1, is too close to the idea of pushing f towards $\mathbf{1}$ by means of the powers f^t, on which §2.1 was based. It turns out to be more effective to split the weak invertibility problem as follows, in accord with Beurling's techniques in [118].[36] We introduce the following notation:

$$\theta_n(\zeta) = \sum_{k \geqslant 0} k^n \zeta^k, \quad |\zeta| < 1, \quad n = 0, 1, \ldots;$$

$$f * g(\zeta) = \sum_{k \geqslant 0} \hat{f}(k)\hat{g}(k)\zeta^k = \frac{1}{2\pi i} \int_{|t|=r} f\left(\frac{\zeta}{t}\right) g(t)\frac{dt}{t}, \quad |\zeta| < r < 1,$$

is the Hadamard convolution of functions f and g which are analytic in $\mathbf{D}$. The splitting into two problems consists of showing in succession that

(A) $\theta_n * f \in E(f)$ for all $n \geqslant 0$, and

(B) $\mathbf{1} \in \mathfrak{L}(\theta_n * f\colon n \geqslant 0)$.

First we point out the similarity of problems (a) and (A), and also (b) and (B). For problems (A) and (a) this is almost obvious; the first differs from the second in that the factor $(\log f)^n$ is replaced, roughly speaking, by $(\log f)^{(n)}$. More precisely,

---

(36) This work is discussed in detail in §§2.6 and 2.7.

$\theta_n * f = D^n f$, where $Df = zf'$; since D is a differentiation (that is, $Dfg = fDg + gDf$), we have $Dfg = f(Dg + gD\log f)$ for an arbitrary function g, or $D^n fg = fU^n g$, where the operator $U = U_f$ is defined by the equation $Uh = Dh + hD\log f$. Thus

$$\left.\begin{aligned} \theta_n * f = D^n f = f\cdot U_f^n 1, \\ U_f h = Dh + hD\log f,\ h\in A(\mathbf{D}). \end{aligned}\right\} \qquad (1)$$

Concerning problem (B), this problem and problem (b) can be reduced to the problem of quasi-analyticity of a certain class. Namely, to each functional $\Phi \in X^*$ which is analytic in the left half-plane $\{t\colon \operatorname{Re} t < 0\}$ we associate the function f_Φ:

$$f_\Phi(t) = \sum_{k\geqslant 0} \hat{f}(k)\,(\mathscr{K}\Phi)\hat{}\,(k)\,e^{kt}.$$

If the class of all such functions is quasi-analytic in $\operatorname{Re} t \leqslant 0$, then orthogonality of the functional Φ to all of the functions $\theta_n * f$ (that is, $f_\Phi^{(n)}(0) = 0$, $n = 0, 1, \ldots$) implies that $f_\Phi = 0$. In particular, $(\mathfrak{K}\Phi)\hat{}\,(0) = 0$ and $(\mathbf{1}, \Phi) = 0$. Therefore $\mathbf{1} \in \mathfrak{L}(\theta_n * f\colon n \geqslant 0)$.

Quasi-analyticity of the class $\{f_\Phi\colon \Phi \in X^*\}$ follows immediately from density of the polynomials $p|\mathbf{Z}_+ \overset{\text{def}}{=} \{p(n)\}_{n\geqslant 0}$, $p \in \mathfrak{P}_A$, in the weighted space of sequences $c_0(|\hat{f}(k)|\cdot \|z^k\|_X \epsilon_k^{-1})$, where $\Sigma_0^\infty \epsilon_k < +\infty$, $\epsilon_k > 0$. In order to see this, it suffices to find the derivatives $f_\Phi^{(n)}$ and then to approximate in this metric by the polynomials $p|\mathbf{Z}_+$, $p \in \mathfrak{P}_A$, the basis vectors $\{\delta_{nk}\}_{k=0}^\infty$ (δ_{nk} is the Kronecker delta). Now we use the following completeness criterion, which was essentially introduced by Bernšteĭn (see [60], Chapter V, §§4, 7).

LEMMA 1. *Let $\{\alpha_n\}_0^\infty$ be an increasing positive sequence such that $\{\log \alpha_n\}_0^\infty$ is convex with respect to* $\log n$. *If*

$$\sum_{n\geqslant 1} \frac{\log \alpha_n}{n^{3/2}} = +\infty,$$

then the polynomials $p|\mathbf{Z}_+ = \{p(n)\}_{n\geqslant 0}$, $p \in \mathfrak{P}_A$, are dense in the space $c_0(\alpha_n^{-1})$.

Thus we have the following result.

LEMMA 2. *Let X be a space of analytic functions in* $\mathbf{D}$ *satisfying conditions* a) *and* b). *Suppose that*

(i) $d_k > 0$, *and*

$$\{\log 1/d_k\,\|z^k\|_X\}_1^\infty$$

is a convex sequence with respect to $\log k$;

(ii) $f \in l_A^\infty(d_k^{-1})$, $\hat{f}(0) \neq 0$;

(iii) $\Sigma_1^\infty k^{-3/2}\log d_k \|z^k\|_X = -\infty$.

Then

$$\mathbf{1} \in \mathscr{L}_X(\theta_n * f : n \geqslant 0).$$

The proof is obtained from Lemma 1 and the following remarks. From (i) and (ii) it follows that

$$\lim_k \left(\log \frac{1}{d_k \| z^k \|_X}\right) \log^{-1} k = +\infty;$$

that is, $d_k < k^{-m} \|z^k\|^{-1}$ for $k > k(m)$ and for arbitrary $m > 0$. Therefore

$$\sum_1^\infty |\hat{f}(k)| \| z^k \|_X k^n \leqslant \text{const} \sum_1^\infty d_k k^n \| z^k \|_X < +\infty$$

and $\theta_n * f \in X$, $n = 0, 1, \ldots$. Then we just note that if the polynomials $\{p|\mathbf{Z}_+ : p \in \mathcal{P}_A\}$ are dense in the space $c_0(d_k \|z^k\|_X k^2)$, the sequence $(1, 0, \ldots)$ can be approximated by these polynomials in the space $c_0(|\hat{f}(k)| \cdot \|z^k\|_X k^2)$. ●

REMARK. In Lemma 2 we did not in fact use density of the polynomials, but only the fact that the sequence $\{\delta_{k0}\}_{k=0}^\infty = (1, 0, \ldots)$ could be approximated by the polynomials $p|\mathbf{Z}_+ = \{p(n)\}_0^\infty$, and therefore the condition of logarithmic convexity and condition (iii) need not be verified for the sequence $\{1/\|z^n\|_X d_n\}_0^\infty$, but just for an arbitrary smaller sequence which grows faster than an arbitrary power.

If a function f is not directly defined by its Taylor coefficients, but belongs to a space $X_1 \subset X$ satisfying condition b), we can use the estimate $|\hat{f}(n)| \cdot \|z^n\|_{X_1} \leqslant \text{const}$, $n \geqslant 0$. We consider two methods for finding X_1 in terms of the original space X.

I. Let F_1 be a space of measurable functions on $[0, 1)$ with a monotone norm, and let $X_1 = F_1(\partial X)$. Then for $f \in X_1$ we have $|\hat{f}(k)| \leqslant \text{const} \|t^k\|_{F_1}^{-1}$, $k \geqslant 0$. If $X = F(E)$ is a space of the type in Lemma 7 of §2.3, and if $F_1 \subset F$, then the conditions in Lemma 2 reduce to convexity of $\{\log \|t^k\|_{F_1} / \|t^k\|_F\}_1^\infty$ with respect to $\log k$, and

$$\sum_1^\infty k^{-3/2} \log \| t^k \|_{F_1} / \| t^k \|_F = +\infty.$$

Later we will often use the spaces $F = L^\infty(\lambda^{-1})$ and $F_1 = L^\infty(\lambda_1^{-1})$, where λ and λ_1 are increasing functions on $[0, 1)$ with $\lambda_1 \leqslant \lambda$. In this case

$$\| t^n \|_F^{-1} = \inf_{0 \leqslant r < 1} r^{-n} \lambda(r), \quad n \geqslant 0,$$

with the analogous relation for $\|t^n\|_{F_1}$.

II. Let

$$X_1 = \{f \in X : \sup_{n \geqslant 0} \| z^n f \|_X p_n^{-1} < +\infty\},$$

where $\{p_n\}_0^\infty$ is a given positive sequence. Here

$$|\hat{f}(k)| \cdot \| z^{n+k} \|_X \leqslant \text{const} \| z^n f \|_X \leqslant \text{const}\, p_n \quad \text{and} \quad |\hat{f}(k)| \leqslant \text{const} \inf_{n \geqslant 0} p_n \| z^{n+k} \|_X^{-1}.$$

In particular, the conditions in Lemma 2 for $f \in \partial X$ reduce to convexity of $\{\log 1/\|z^k\|_X\}_1^\infty$ with respect to $\log k$ and divergence of the series $\Sigma_1^\infty k^{-3/2} \log \|z^k\|_X$.

Finally, we note that by the results in §2.3 the same relations hold for the spaces X_g and $(X_1)_{g_1}$ if $g \in \partial X$, $1/g \in \partial^* \mathfrak{K} X^*$ and $g_1 \in \partial X_1$, $1/g_1 \in \partial^* \mathfrak{K} X_1^*$.

2°. The approximation problem (A) is significantly "rougher" than (B). The following simple considerations will suffice.

LEMMA 3. *Let X_1 and X be spaces which satisfy conditions* a) *and* b) *in* §2.3.2° *with $X_1 \subset X$; Let $\mathfrak{M}$ be a Banach space of multipliers from X_1 into X (that is, $g \in \mathfrak{M}$,*

$f \in X_1 \Rightarrow gf \in X$). Suppose that $\mathfrak{M}$ has the property of $(*)$-completeness of the polynomials (property (E) *or* (E*) *in* §2.1) *and that in case* (E*) *convergence in X does not depend on the boundary* $\partial\mathbf{D}$ *(property* (D) *in* §2.1). *If $f \in X_1$ and $U_f^n 1 \in \mathfrak{M}$, then $\theta_n * f \in E_X(f)$.*

PROOF. By the closed-graph theorem we have

$$\|gh\|_X \leqslant \text{const}\, \|g\|_{\mathfrak{M}} \cdot \|h\|_{X_1}, \quad h \in X_1, \quad g \in \mathfrak{M}.$$

If $f \in X_1$ and $U^n 1 \in \mathfrak{M}$ (see (1)), then there exists a sequence of polynomials $\{p_k\}_1^\infty$ such that

$$\sup_k \|p_k\|_{\mathfrak{M}} < +\infty, \quad \lim_k p_k(\zeta) = (U^n 1)(\zeta), \quad |\zeta| < 1,$$

(and in case (E) we also have $\lim_k \|p_k - U^n 1\|_{\mathfrak{M}} = 0$). Then

$$\theta_n * f(\zeta) = \lim_k p_k(\zeta) f(\zeta), \, |\zeta| < 1,$$

where $\|p_k f\|_X \leqslant \text{const} \cdot \|f\|_{X_1}$. Consequently $p_k f \xrightarrow{\text{w}.X} \theta_n * f$ and $\theta_n * f \in E_X(f)$. ●

We consider in more detail two cases corresponding to the classes I and II in 1°.

LEMMA 3a. *Let $X = F(E)$ and $X_1 = F_1(E)$ be spaces of type* I *in* 1° *(see also Lemma 7 in* §2.3). *If $\|gh\|_E \leqslant \text{const} \cdot \|g\|_E \cdot \|h\|_{H^\infty}$, $g \in E$, $h \in H^\infty$, and the function M_G, where $M_G(r) = \|G_r\|_{C_A}$, $0 \leqslant r < 1$, is a multiplier from F_1 into F, then G is a multiplier from X_1 into X.*

The proof is obvious. ●

The space of multipliers in this lemma has the property of circular symmetry, and therefore the polynomials are $(*)$-complete. Example: $F = L^p(\rho dr)$ and $E_1 = L^p(\rho_1 dr)$, where $1 \leqslant p \leqslant \infty$, and ρ and ρ_1 are positive functions on $[0, 1)$. Then the assumption in Lemma 3a on the function $G \in A(\mathbf{D})$ reduces to the estimate

$$\sup_r \rho(r) M_G(r)/\rho_1(r) < +\infty.$$

Additional results in the case $G = U^n\mathbf{1}$ (see (1)) can be obtained by using the estimates in Chapter 1. A positive continuous function α on $(0, +\infty)$ is said to be pseudomonotone if

$$\sup_{x>0,\, y>0} \alpha(x+y)/\alpha(x) < +\infty.$$

This condition means that there exist a decreasing function β and a constant $c > 0$ such that $c\beta \leqslant \alpha \leqslant \beta$.

LEMMA 4. *Suppose that λ is an increasing function on* $[0, 1)$,

$$\lambda(r) = \exp \varphi\left(\log \frac{1}{1-r}\right),$$

$0 \leqslant r \leqslant 1$, *and $\psi = \log \varphi$. Let $f \in A_\infty(\lambda)$; that is, $|f(\zeta)| \leqslant \text{const} \cdot \lambda(|\zeta|)$, $\zeta \in \mathbf{D}$, and $f(\zeta) \neq 0$, $|\zeta| < 1$.*

1) *If*

$$\varlimsup_{t\to\infty} \psi'(t) < +\infty \left(\textit{or} \quad 0 < \varliminf_{t\to\infty} \psi'(t) \leqslant \varlimsup_{t\to\infty} \psi'(t) < +\infty \right),$$

then

$$|\log f(\zeta)| \leqslant \text{const}\,(1-|\zeta|)^{-\text{const}}, \quad \zeta \in \mathbf{D},$$

or, respectively,

$$|\log f(\zeta)| \leqslant \text{const}\,(\log \lambda)^{\text{const}}, \quad \zeta \in \mathbf{D}.$$

2) *If* $\psi' \uparrow +\infty$ *and the function* $\psi''\psi'^{-3/2}$ *is pseudomonotone, then*

$$|\log f(\zeta)| \leqslant \text{const}\,(\log \lambda(|\zeta|))^{\text{const}}, \quad \zeta \in \mathbf{D},$$

PROOF. 1) By hypothesis $kt \leqslant \psi(t) \leqslant Kt$, where $t \geqslant t_0$ and $0 < k < K$, and therefore

$$\frac{1}{(1-r)^k} \leqslant \log \lambda(r) \leqslant \frac{1}{(1-r)^K}, \quad r_0 < r < 1.$$

From the inequality $\log|f| \leqslant \log\lambda + \text{const}$ and theorems of Cartwright (or Theorem 1 in §1.1 and Theorem 2 in §1.3) we obtain

$$|\log f(\zeta)| \leqslant \text{const}\,(1-|\zeta|)^{-\text{const}}, \quad \zeta \in \mathbf{D}.$$

2) By hypothesis there exists a function $\overline{\psi}$ on $(0, +\infty)$ such that $\overline{\psi}''\overline{\psi}'^{-3/2} \downarrow 0$ and $c\overline{\psi}''\overline{\psi}'^{-3/2} \leqslant \psi''\psi'^{-3/2} \leqslant \overline{\psi}''\overline{\psi}'^{-3/2}$, where $c > 0$. If we integrate (and, if necessary, change the constants) we obtain $k\overline{\psi} \leqslant \psi \leqslant K\overline{\psi}$ on $[t_0, +\infty)$. Since $\overline{\psi}'' > 0$, we have $\overline{\psi}' \uparrow +\infty$, and therefore $\overline{\psi}''\overline{\psi}'^{-2} \downarrow 0$. Thus

$$\log|f(\zeta)| \leqslant (\log\overline{\lambda}(|\zeta|))^{\text{const}}, \quad \zeta \in \mathbf{D},$$

where $\overline{\lambda} = \exp\overline{\varphi}(\log(1/(1-r)))$, $\overline{\varphi} = \exp\overline{\psi}$. Since the functions $\exp(\log\overline{\lambda})^{\text{const}}$ and $(\log\overline{\lambda})^{\text{const}}$ respectively satisfy the conditons in Theorem 2 of §1.3 and Theorem 1 of §1.1 in Chapter 1, we have

$$|\log f| \leqslant \text{const}(\log\overline{\lambda})^{\text{const}}.$$

From the inequality $k\overline{\psi} \leqslant \psi$ we now obtain assertion 2) of Lemma 4. ●

LEMMA 5. *Under the assumptions in Lemma* 4, *for each* $n = 0, 1, \ldots,$ *there exists a constant* const *such that in case* 1)

$$|U_f^n \mathbf{1}(\zeta)| \leqslant \text{const}\,(1-|\zeta|)^{-\text{const}}, \quad \zeta \in \mathbf{D},$$

or

$$|U_f^n \mathbf{1}(\zeta)| \leqslant \text{const}\,(\log\lambda)^{\text{const}};$$

and in case 2)

$$|U_f^n \mathbf{1}(\zeta)| \leqslant \text{const}\,(\log\lambda(|\zeta|))^{\text{const}}, \quad \zeta \in \mathbf{D}.$$

The proof of Lemma 5 will be given later. We now formulate a criterion for weak invertibility which is suitable for applications.

LEMMA 6. *Let λ_i be a function on* $[0, 1)$ *which satisfies the conditions in Lemma 5 and*

$$X_1 \subset A(\lambda_i) \subset \bigcup_{n=0}^{\infty} A((\log \lambda_i)^n \lambda_i) \subset X. \tag{2}$$

Suppose that the sequence $\{\log(\|z^n\|_{X_1}/\|z^n\|_X)\}_1^\infty$ *is convex with respect to* $\log n$ *and*

$$\sum_1^\infty n^{-3/2} \log \frac{\|z^n\|_{X_1}}{\|z^n\|_X} = +\infty. \tag{3}$$

Then each function f in X_1, $f(\zeta) \neq 0$, $|\zeta| < 1$, is weakly invertible in X.

The proof follows immediately from Lemmas 2, 3, and 5. ●

REMARK 1. The condition of logarithmic convexity and condition (3) may be verified not for the sequence $\{\|z^n\|_{X_1}/\|z^n\|_X\}_1^\infty$ but just for an arbitrary smaller sequence which grows faster than an arbitrary power (see the remark after Lemma 2).

REMARK 2. For the spaces $X_1 = F_1(E)$ and $X = F(E)$ in Lemma 3a the last inclusion in (2) can be replaced by the requirement that $(\log \lambda)^n$ be a multiplier from F_1 into F for each $n = 0, 1, \ldots$.

COROLLARY. *If $F = L^p(\rho\, dr)$ and $F_1 = L^p(\rho_1 dr)$, $1 \leqslant p \leqslant \infty$, and in the notation of Lemma 6*

$$(\log \lambda)^n = O(\rho_1/\rho) \quad as \quad r \to 1 - 0,$$

then $\theta_n * f \in E(f)$, $n = 0, 1, \ldots$.

The problem now is to translate conditions (2) and (3) into a common language and then verify their validity in concrete situations. This will be done in §§2.5–2.7.

We postpone the proof of Lemma 5 to the end of this section, and we now briefly consider subclasses $X_1 \subset X$ of type II in 1°. In this case, if we set $g = U_f^n 1$, we see that the condition $\Sigma_0^\infty |\hat{g}(k)| p_k < +\infty$ guarantees that $\theta_n * f \in E(f)$ (see (1)). Suppose that we already have estimates on the Taylor coefficients of $\log f$; that is, $|(\log f)\hat{}(n)| \leqslant \alpha_n$, where $\alpha_{n+k} \leqslant \text{const} \cdot \alpha_n \cdot \alpha_k$, $n, k \geqslant 0$ (see Lemma 7 below). Then

$$|U_f 1\hat{}(k)| \leqslant k\alpha_k \stackrel{\text{def}}{=} \beta_k$$

and by induction

$$|U_f^n 1\hat{}(k)| \leqslant \text{const}\, k^n \alpha_k + \{k\alpha_k\} * \ldots * \{k\alpha_k\} \leqslant \text{const}\, \{k\alpha_k\} * \ldots * \{k\alpha_k\};$$

here we have in mind the fact that $\{k\alpha_k\} * \cdots * \{k\alpha_k\} = \hat{h}^n(k)$, where $h = \Sigma_0^\infty k\alpha_k z^k$. We estimate the coefficients $\hat{h}^n(k)$ first of all in terms of $\max_{|\zeta|=r} |h(\zeta)|$:

$$|\hat{h}^n(k)| \leqslant \inf_{0 \leqslant r < 1} r^{-k} \max_{|\zeta|=r} |h^n(\zeta)|, \quad k = 0, 1, \ldots.$$

Sometimes this can be done immediately. For example, if $(\log f)\hat{}(k) = O(k^{\text{const}})$, then it is clear that $|U_f^n 1\hat{}(k)| = O(k^{\text{const}})$, with a different constant const. We have the following result.

COROLLARY. *If* $\|z^k f\|_X \leqslant \text{const}\, p_k$, $k \geqslant 0$, *and* $(\log f)^\frown(k) = O(k^{\text{const}})$, *then the condition*

$$\sum_1^\infty k^n p_k < +\infty, \quad n = 0, 1, \ldots,$$

implies that $\theta_n * f \in E(f)$, $n \geqslant 0$. *In particular, this is true for functions* f *in the boundary space* ∂X *with* $f(\zeta) \neq 0$, $|\zeta| < 1$.

Returning to Lemma 5, first we present three preliminary results which we will also use in later sections.

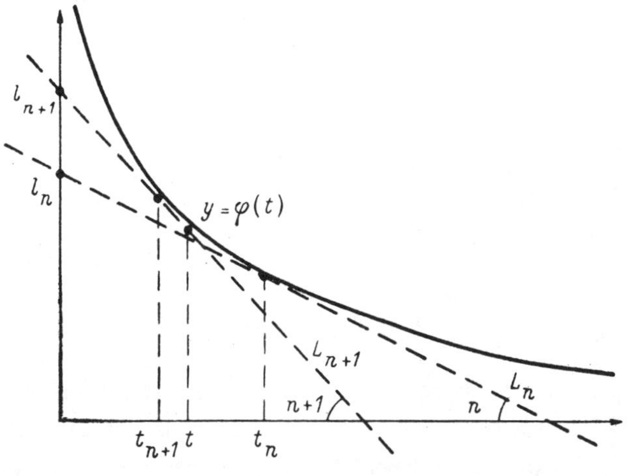

FIGURE 5

LEMMA 7. 1) *Let* λ *be an increasing positive function on* $[0, 1)$ *such that* $r \longrightarrow \log \lambda(r)$ *is a convex (downward) function of* $\log r$ (*that is,* $r\lambda'/\lambda \uparrow$). *Set*

$$d_n = \inf_{0 \leqslant r < 1} r^{-n}\lambda(r), \quad n \geqslant 0; \quad \lambda_0(r) = \sup_{n \geqslant 0} r^n d_n, \quad 0 \leqslant r < 1.$$

Then $\lambda_0 \leqslant \lambda$ *and* $\lambda \sim \lambda_0$ *as* $r \longrightarrow 1 - 0$.

2) *Suppose that* $\{d_n\}_0^\infty$ *is an increasing positive sequence and* $\{\log d_n\}_0^\infty$ *is convex upward. Then*

$$d_n^0 \overset{\text{def}}{=} \inf_{0 \leqslant r < 1} r^{-n}\lambda_0(r) = d_n, \quad n \geqslant 0,$$

where λ_0 *is defined as in the preceding assertion.*

PROOF. 1) We set $\varphi(t) = \log \lambda(e^{-t})$, $\varphi_0(t) = \log \lambda_0(e^{-t})$, $t > 0$, and $l_n = \log d_n$, $n \geqslant 0$. Then $l_n = \inf_{t \geqslant 0}(\varphi(t) + nt)$, $n \geqslant 0$, and thus l_n is the length of the segment intersected on the OY-axis by the tangent L_n to the graph of φ which has a slope equal to $-n$. Also $\varphi_0(t) = \sup_{n \geqslant 0}(l_n - nt)$, $t > 0$; that is, $\varphi_0(t)$ is the value at t of the function $y_n(t) = l_n - nt$ whose graph (a straight line) above this point is higher than all other such graphs (see Figure 5). Let t_n denote the point of intersection of L_n with the graph of φ. From what we have shown and from the convexity of φ it follows that on $[t_{n+1}, t_n]$ the graph of φ_0 is a broken line formed by segments of L_{n+1} and L_n. Therefore we always have $\varphi_0 \leqslant \varphi$, and $\varphi(t) - \varphi_0(t) = \varphi(t) - \varphi(t_n) + \varphi(t_n) - \varphi_0(t)$. If

t lies in the interval $[t_{n+1}, t_n]$ and $\varphi_0(t) = l_n - nt$, then

$$\varphi(t) - \varphi_0(t) = (t - t_n)\varphi'(\xi) + (t_n - t)\varphi'(t_n) = (t_n - t)(\varphi'(t_n) - \varphi'(\xi)),$$

where $t_{n+1} < \xi \leqslant t_n$. Convexity of φ implies that $\varphi(t_n) - \varphi'(\xi) \leqslant \varphi'(t_n) - \varphi'(t_{n+1}) = 1$, and therefore $\varphi(t) - \varphi_0(t) \leqslant t_n - t \leqslant t_n - t_{n+1}$. The same estimate holds for $\varphi_0(t) = l_{n+1} - (n+1)t$. Consequently $\varphi(t) - \varphi_0(t) = o(1)$ as $t \to 0$.

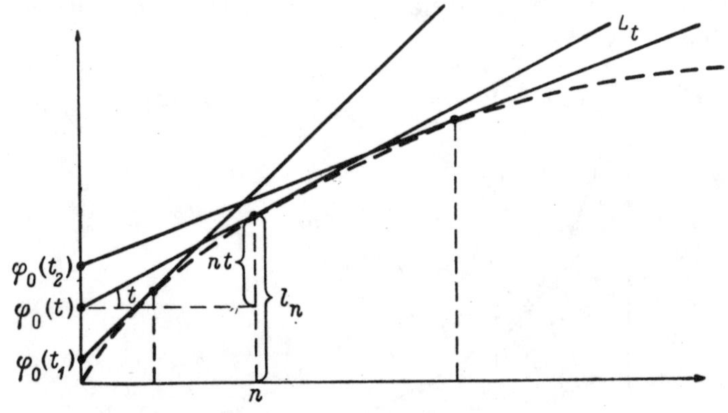

FIGURE 6

2) The equation $\varphi_0(t) = \sup_{n \geqslant 0}(l_n - nt)$ means that $\varphi_0(t)$ is the length of the interval intersected on the OY-axis by the tangent L_t to the broken line Λ which joins the points (k, l_k), $k = 0, 1, \dots$, having the slope t (see Figure 6). Also,

$$l_n^0 \stackrel{\text{def}}{=} \log d_n^0 = \inf_{t>0} (\varphi_0(t) + nt)$$

is the value at n of the function $y_t(s) = \varphi_0(t) + st$, $s > 0$, whose graph above this point is higher than all other such graphs. From convexity of Λ it follows that (by the definition of $\varphi_0(t)$) all such graphs lie above the point (n, l_n) except the tangent L_t which goes through this point. Consequently $l_n^0 = l_n$. ●

REMARK. The "regularizations" of the function λ and the sequence $\{d_n\}_0^\infty$ in Lemma 7 appear to be completely analogous to the constructions of Mandelbrojt in [60] for the case of a critical point $r = +\infty$. We note that if $\{d_n\}_0^\infty$ is logarithmically convex and $d_n \uparrow +\infty$, then $n^{-1}\log d_n \downarrow 0$, and if $d_n \uparrow \infty$ and $n^{-1}\log d_n \downarrow 0$, there exists a logarithmically convex sequence $\{d_n^{(1)}\}_0^\infty$ such that $d_n \leqslant d_n^{(1)} \leqslant d_n^2$, $n \geqslant 0$ (see [60]).

LEMMA 8. *Suppose that $r\lambda'/\lambda \uparrow +\infty$ as $r \to 1 - 0$ and $d_n = \inf_{0 \leqslant r<1} r^{-n}\lambda(r)$, $n = 0, 1, \dots$. Then $\{\log d_n\}_0^\infty$ is convex upward.*

PROOF. We rewrite the definition of d_n: $d_n = r_n^{-n}\lambda(r_n)$, where $r_n\lambda'(r_n)/\lambda(r_n) = n$. We set $\nu(r) = r\lambda'(r)/\lambda(r)$ and

$$d(n) = \frac{\lambda(\nu^{-1}(n))}{[\nu^{-1}(n)]^n}, \quad n = 0, 1, \dots.$$

If we define the function $d(t)$ to be linear on the intervals $(n, n+1)$, we obtain

$$\frac{d'}{d} \circ \nu = \frac{\lambda'}{\lambda\nu'} - \log r - \frac{\nu}{r\nu'} = -\log r \downarrow. \quad \bullet$$

LEMMA 9. *Let λ be a positive, unbounded, increasing function on $[0, 1)$, and let $\mu = \log \lambda$. Suppose that $k \geqslant 0$ and*

$$d_n(\lambda) = \inf_{0 \leqslant r < 1} r^{-n}\lambda(r); \quad \lambda_k(r) = \sup_{n \geqslant 0} n^k r^n d_n(\lambda).$$

1). *If the function $\mu''\mu'^{-2}$ is pseudomonotone and $\mu^{-\alpha}\mu'$ is unbounded in a neighborhood of 1 for some $\alpha \geqslant 1$, then*

$$\lambda_k \leqslant \text{const}\, \lambda (\log \lambda)^{k\alpha}.$$

2) *If μ/μ' is bounded in a neighborhood of 1, then*

$$d_n(\lambda(\log \lambda)^k) \leqslant \text{const}\, n^k d_n(\lambda), \quad n \geqslant 1.$$

PROOF. 1) By Lemma 8, $n \longrightarrow l(n) \overset{\text{def}}{=} \log d_n(\lambda)$ is a convex function, and therefore the maximum

$$\max_{n \geqslant 0} \{k \log n + n \log r + l(n)\}$$

is attained at the point $n = n(r)$ such that

$$\frac{k}{n} + \log r + l'(n) = 0.$$

This equation is to be interpreted to mean that there exists a unique n for which

$$l'(n+1) + \frac{k}{n+1} \leqslant \log \frac{1}{r} \leqslant l'(n) + \frac{k}{n}, \tag{4}$$

and the maximum is attained for $n(r) = n$ or $n(r) = n + 1$; here $l'(n)$ is the slope of the segment of the broken line with vertices $(n, l(n))$, $n = 1, 2, \ldots$. In a similar fashion we consider the equation

$$\log r + l'(m) = 0, \tag{5}$$

letting $m = m(r)$, where $\sup_{0 \leqslant r < 1} r^n d_n(\lambda) = r^m d_m(\lambda)$. We estimate the "central index" $n(r)$ in terms of $m(r)$. To calculate $m(r)$ we use a different method, which is contained in the proof of Lemma 7: we let $m(r) = m$ if $m \leqslant r\mu'(r) < m + 1$; that is, $m(r) = [r\mu'(r)]$. Since the function $n \longrightarrow l'(n) + k/n$ is decreasing, the inequality

$$\log \frac{1}{r} \geqslant l'(cm) + \frac{k}{cm} \tag{6}$$

implies that $n \leqslant cm$. We will show that there exists a constant c for which (6) is satisfied. We note that

$$l'(cm) \leqslant \log \frac{1}{\rho} \leqslant l'(cm-1) \quad \text{for} \quad cm - 1 < \rho\mu'(\rho) < cm. \tag{7}$$

Therefore (6) follows from the estimate

$$\log \frac{1}{\rho} + \frac{k}{\rho\mu'(\rho)} \leqslant \log \frac{1}{r}$$

or from

$$re^{\frac{k}{\rho\mu'(\rho)}} \leqslant \rho.$$

If we compare (4) and (5) and note that μ' is monotone, we see that the required inequality follows from

$$\mu'\left(re^{\frac{k}{cr\mu'(r)-(c+1)}}\right) \leqslant c\frac{r}{\rho}\mu'(r) - \frac{1}{\rho}(c+1).$$

Let $f = \mu'$ and $0 \leqslant x < x + \Delta < 1$. Then

$$\frac{1}{f(x)} - \frac{1}{f(x+\Delta)} = \Delta\frac{f'(\xi)}{f^2(\xi)}, \quad x \leqslant \xi \leqslant x + \Delta.$$

Since $f'f^{-2}$ is bounded, we obtain

$$0 \leqslant \frac{1}{f(x)} - \frac{1}{f(x+\Delta)} \leqslant \Delta\frac{f'(\xi)}{f^2(\xi)} \leqslant \Delta \sup\frac{f'(x)}{f^2(x)} \leqslant \alpha/f,$$

if $\Delta \leqslant (\alpha/f)(\sup f'/f^2)^{-1}$. Therefore (for $0 < \alpha < 1$) we have $f(x) \asymp f(x + \Delta)$ as $x \to 1 - 0$. Clearly

$$\Delta = r(\exp k/(cr\mu'(r) - (c+1)) - 1) \leqslant 1/2\mu'(r)$$

for large c, and (6) is satisfied for large m.

From this estimate it now follows that

$$\lambda_k(r) = n(r)^k r^{n(r)} d_{n(r)} \leqslant cm(r)^k r^{n(r)} d_{n(r)} \leqslant cm(r)^k \lambda_0(r) \leqslant c\mu'(r)^k \lambda(r) \leqslant c(\log\lambda)^{k\alpha}\lambda.$$

2) Let $d_n = r_n^{-n}\lambda(r_n)$ and $r_n\mu'(r_n) = n$. Clearly

$$\inf_{0\leqslant r<1} r^{-n}\lambda(r)\mu(r)^k \leqslant r_n^{-n}\lambda(r_n)\mu(r_n)^k \leqslant d_n \operatorname{const} n^k. \bullet$$

REMARK. Lemma 9 implies an analogue of the classical Wiman-Valiron inequality (see, for example, [94], Abschnitt IV, Aufgabe 56) for the central term of a power series: if $|f| \leqslant \lambda$, then under the assumptions in the lemma

$$\left|\sum_0^\infty \hat{f}(n)\zeta^n\right| \leqslant \sum_0^\infty |\hat{f}(n)||\zeta|^n \leqslant c\sup_{n\geqslant 0} n^2 d_n|\zeta^n| \leqslant \operatorname{const}\lambda(\log\lambda)^{2\alpha}.$$

If we use the Wiman-Valiron "comparison method" (comparing d_n and $\exp n^\alpha$), under the restrictive assumptions in Lemmas 4 and 5 we obtain

$$\sum_0^\infty r^n d_n \leqslant \operatorname{const}\lambda(\log\lambda)^{\frac{1}{2}+\varepsilon}$$

for an arbitrary $\epsilon > 0$.

PROOF OF LEMMA 5. In case 2) it suffices to show that the inequality $\log|g| \leqslant \text{const} \cdot \log\log\lambda$ implies that $\log|U_f g| \leqslant \text{const} \cdot \log\log\lambda$; that is,

$$\text{(i) } \log|Dg| \leqslant \text{const} \log\log\lambda,$$

$$\text{(ii) } \log|gD\log f| \leqslant \text{const} \log\log\lambda.$$

We note that by Lemma 4 we have $\log|\log f| \leqslant \text{const} \cdot \log\log\lambda$, and we just have to prove (i). We set

$$\underline{\lambda} = (\log\lambda)^c, \quad d_n = \inf_{0\leqslant r<1} r^{-n}\underline{\lambda}(r), \; n \geqslant 0.$$

Simple calculations show that for the function $\underline{\lambda}$ all of the conditions in part 1) of Lemma 9 are satisfied with $\alpha = 2$. Therefore $|g| \leqslant \underline{\lambda}$ implies that $|\hat{g}(k)| \leqslant d_k$, $k \geqslant 0$, and

$$|Dg(\zeta)| \leqslant \sum_0^\infty k|\hat{g}(k)||\zeta|^k \leqslant \text{const} \sup_{k\geqslant 0} k^3 d_k |\zeta|^k \leqslant \text{const}\, \underline{\lambda} (\log\underline{\lambda})^2 \leqslant \text{const}\, \underline{\lambda}^2.$$

In case 1) the proof of the lemma is obvious, since the inequalities

$$\log|\log f| \leqslant \text{const} \log\frac{1}{1-|\zeta|} \quad \text{and} \quad \log|g| \leqslant \text{const} \log\frac{1}{1-|\zeta|}$$

imply that

$$\log|Ug| \leqslant \text{const} \log\frac{1}{1-|\zeta|}.$$

For example, this can be seen immediately from the estimate on the convolution

$$|(\theta_1 * g)(\zeta)| = \frac{1}{2\pi}\left|\int_{\partial D} \theta_1(r_1 t)\, g(r_2\zeta/t)\, dl_1(t)\right|$$
$$\leqslant \inf_{r_1 r_2 = |\zeta|} \frac{1}{(1-r_1)^2} \cdot \frac{1}{(1-r_2)^c} \leqslant \frac{\text{const}}{(1-|\zeta|)^{c+2}}. \quad \bullet$$

§2.5. The fundamental theorem

Here we present the concluding calculations involving conditions (2) and (3) in Lemma 6 of §2.4 for spaces X of three types: $X = A(\lambda)$, $L^p(\rho dr; E)$, and $l^p_A(w_n)$. As we saw in §2.4, we must subject the weight functions λ, ρ and $\{w_n\}_{n\geqslant 0}$ to some restrictive requirements. Recall (see §2.0) that our goal is to achieve as narrow a gap as possible between the spaces X_1 and X with $X_1 \subset X$ while guaranteeing weak invertibility in X of all functions f in X_1 with $f(\zeta) = 0$, $|\zeta| < 1$. We note that the theorems presented below are sharp in a strict sense which will be discussed in §2.8. On the other hand, estimates "with a shift in the argument", which were mentioned in §1.3,(37) do not in general have the required precision (see Appendix III). Weak invertibility of slowly increasing functions is simplified by virtue of the fact that the necessary lower estimates are trivial (see §2.6). Examples for the theorems in this section are contained in §2.7.

(37) The estimates were obtained in [68] and [118] under less restrictive assumptions.

THEOREM 1a. *Let* λ *and* Λ *be positive increasing functions on* $[0, 1)$, *and let* $\mu = \log\lambda$, $\mathrm{M} = \log\Lambda$, $\mu(r) = \varphi(\log(1/(1-r)))$ *and* $\psi = \log\varphi$. *Suppose that the following conditions are satisfied*:

1) $0 < \underline{\lim}_{t\to\infty}\psi' \leqslant \overline{\lim}_{t\to\infty}\psi' < +\infty$ *or* $\psi' \uparrow \infty$ *and* $\psi''\psi'^{-3/2}$ *is pseudomonotone.*

2) *The function* M *is convex in a neighborhood of the point* 1, *and* $\mu \leqslant \mathrm{M}$.

3) *The function* $\mathrm{M}'(\mathrm{M}' - \mu')/\mathrm{M}''$ *is monotone in a neighborhood of the point* 1.

4) *The integral*

$$\int^{1} (\mathrm{M} - \mu)\frac{\mathrm{M}''}{\mathrm{M}'^{3/2}}\,dr \tag{1}$$

diverges.

Then each function f in the class $A(\lambda)$ *which does not vanish in* **D** *is weakly invertible in* $A(\Lambda)$.

PROOF. Let

$$d_n = \inf_r r^{-n}\lambda(r),\quad D_n = \inf_r r^{-n}\Lambda(r),\quad n \geqslant 0,$$

and

$$\lambda_0(r) = \sup_{n\geqslant 0} r^n d_n,\quad \Lambda_0(r) = \sup_n r^n D_n,\quad 0 \leqslant r < 1.$$

By Lemma 7 in §2.4 we have $\lim_{r\to 1-0}\Lambda/\Lambda_0 = 1$; by Lemma 8 in §2.4 we have $D_n/D_{n+1}\uparrow$. Consequently (see [94], Abschnitt IV, Aufgabe 42) for each natural number n there exists a number r_n, $0 \leqslant r_n < 1$, such that $\Lambda_0(r_n) = r_n^n D_n$. It is clear from the proof of Lemma 7 that $r_n\mathrm{M}'(r_n) = n$. Also

$$\lambda_0(r_n) = \sup_k r_n^k d_k \geqslant r_n^n d_n$$

and therefore

$$D_n/d_n \geqslant \Lambda_0(r_n)/\lambda_0(r_n),\quad n = 1, 2, \ldots.$$

From conditions 3) and 4) it follows that $\mathrm{M}' > \mu'$ for values of the argument near 1; thus if we let θ denote the inverse of the function M', we obtain

$$\begin{gathered}\sum_1^\infty [\mathrm{M}(\theta(n)) - \mu(\theta(n))]\frac{1}{n^{3/2}} \leqslant \mathrm{const}\sum_1^\infty \frac{1}{n^{3/2}}\log\frac{D_n}{d_n},\\ \int_{\theta(1)}^{1} (\mathrm{M} - \mu)\frac{\mathrm{M}''}{\mathrm{M}'^{3/2}}\,dr = \int_1^\infty (\mathrm{M}\circ\theta - \mu\circ\theta)\frac{dt}{t^{3/2}} \leqslant \mathrm{const}\sum_1^\infty \frac{1}{n^{3/2}}\log\frac{D_n}{d_n}.\end{gathered} \tag{2}$$

In order to use Lemma 6 in §2.4 and the first remark which follows it, we just have to note that (i) $n^k = O(\Lambda\circ\theta(n)/\lambda\circ\theta(n))$ for each $k \geqslant 1$; (ii) the sequence $\{\mathrm{M}\circ\theta(n) - \mu\circ\theta(n)\}_1^\infty$ is convex with respect to $\log n$; (iii) $(\log\lambda)^k\lambda = O(\Lambda)$ for each $k \geqslant 1$. Assertion (ii) is clear, since

$$(\mathrm{M}\circ\theta(e^t) - \mu\circ\theta(e^t))' = (\mathrm{M}'\circ\theta(e^t) - \mu'\circ\theta(e^t))\,e^t/\mathrm{M}''\circ\theta(e^t)$$

is a monotone function in a neighborhood of 1. And then we also obtain assertion (i):

$$\{M \circ \theta(n) - \mu \circ \theta(n) - k \log n\}_1^\infty$$

is a convex, and therefore increasing, sequence (otherwise we would contradict the divergence of the series on the left side of (2)). From conditions 3) and 4) it follows that the function $M'(M' - \mu')/M''$ grows without bound; therefore $\log M' = o(M - \mu)$ and $\log \mu' = o(M - \mu)$. Condition 1) gives us $\mu \leqslant \text{const}\, \mu'$, and therefore $\log \mu = o(M - \mu)$, which proves assertion (iii). ●

THEOREM 1b. *Let E_1 and E_2 be spaces of analytic functions in* D *which satisfy conditions* a) *and* b) *in* §2.2.2°, *and suppose that* $0 < \inf_n \|z^n\|_{E_k} \leqslant \sup_n \|z^n\|_{E_k} < +\infty$, $k = 1, 2$. *Let λ and Λ satisfy the conditions in Theorem* 1a, *and let* $X_{\text{int}} = L^p_w(E_1)$ *and* $X = L^q_W(E_2)$, *where* $w = 1/\lambda$, $W = 1/\Lambda$ *and* $1 \leqslant p, q \leqslant \infty$. *Then each function f in the class X_{int} which does not vanish in* D *is weakly invertible in X.*

PROOF: We estimate the norms of powers:[38]

$$\|t^n\|_{L^1_w} \leqslant \|t^n\|_{L^p_w} \leqslant \|t^n\|_{L^\infty_w} = \max_r r^n w(r) \overset{\text{def}}{=} r_n^n w(r_n),$$

$$\|t^n\|_{L^1_w} = \int_0^1 r^n w(r)\, dr \geqslant \int_0^{r_n} r^n w(r)\, dr \geqslant w(r_n) \int_0^{r_n} r^n dr = \frac{r_n}{n+1} \|t^n\|_{L^\infty_w}.$$

Since $r_n \longrightarrow 1$, for large n we have

$$\frac{1}{2(n+1)} \|t^n\|_{L^\infty_w} \leqslant \|t^n\|_{L^p_w} \leqslant \|t^n\|_{L^\infty_w} = 1/d_n(\lambda). \tag{3}$$

If $f \in L^p_w(E_1)$, then by Lemma 9 in §2.4 we have[39]

$$|f(\zeta)| \leqslant \sum_0^\infty |\hat f(n)|\, |\zeta|^n \leqslant \text{const} \sum_0^\infty \frac{|\zeta|^n}{\|t^n\|_{L^p_w}} \leqslant \text{const} \sum_0^\infty |\zeta|^n (n+1)\, d_n(\lambda) \tag{4}$$
$$\leqslant \text{const}\, \lambda (\log \lambda)^2 \leqslant \text{const}\, \lambda^2(|\zeta|).$$

Since the majorant λ^2 satisfies the hypothesis of Theorem 1a along with λ, we find from Lemmas 5 and 6 in §2.4 that $\theta_n * f \in E_{A^\infty}(f)$, where $E_{A^\infty}(f)$ is the invariant subspace with respect to S which is generated by the function f and which is closed in $A^\infty = \bigcup_{k=0}^\infty A(\lambda(\log \lambda)^k)$. We show that $A^\infty \subset X$. In fact, if $g \in A^\infty$, then $g \in A(\lambda(\log \lambda)^k)$ for some $k \geqslant 0$, and thus by Lemma 9 in §2.4 we have

$$|\hat g(n)| \leqslant \text{const}\, d_n(\lambda(\log \lambda)^k) \leqslant \text{const}\, n^k d_n(\lambda),$$

---

(38) Here we only use monotonicity of the weight function w; and here $t(r) \equiv r$.

(39) Hereafter we only consider the second version of condition 1) in Theorem 1a. In the first case

$$\underline{\lambda} = \exp(1-r)^{-\alpha} \leqslant \lambda \leqslant \exp(1-r)^{-\beta} = \bar\lambda$$

for some positive numbers α and β, and therefore

$$|f(\zeta)| \leqslant \text{const}\, \bar\lambda (\log \bar\lambda)^2 \text{ and } \log \bar\lambda \leqslant (\log \lambda)^{\text{const}}$$

which does not change later calculations.

and thus by the same lemma[40]

$$\|g_r\|_{l^1_A} \leqslant \text{const} \sum_0^\infty |\hat{g}(n)| r^n \leqslant \text{const} \sup_{n\geqslant 0} r^n n^{k+2} d_n(\lambda) \leqslant \text{const}\, \lambda (\log \lambda)^{2(k+2)}.$$

Consequently

$$W(r) \|g_r\|_{E_2} \leqslant \text{const} \|g_r\|_{l^1_A} / \Lambda(r) \leqslant \text{const}\, \lambda (\log \lambda)^{2(k+2)} / \Lambda,$$

and, as we noted in the proof of Theorem 1a, the right side of the inequality is bounded in a neighborhood of 1. This implies that $g \in L^\infty_W(E_2) \subset L^q_W(E_2)$. Thus $\theta_n * f \in E_X(f)$, $n = 0, 1, \ldots$. The rest follows from Lemma 6 in §2.4, the remark which follows it, the estimate (2) in the proof of Theorem 1a, and the inequality

$$\text{const} \frac{\|z^n\|_{X_{int}}}{\|z^n\|_X} \geqslant \frac{\|t^n\|_{L^\infty_w}}{2(n+1)\|t^n\|_{L^\infty_W}} = \frac{D_n}{2(n+1)d_n}. \quad \bullet$$

REMARK 1. In the proof of Theorem 1b we only used the fact that $X_{\text{int}} = F_{\text{int}}(E_1)$ and $X = F(E_2)$, where the spaces F_{int} and F have a monotone norm, and $L^\infty_w \subset F_{\text{int}} \subset L^1_w$ and $L^\infty_W \subset F$. The spaces $\partial X_{\text{int}} = E_1$ and $\partial X = E_2$ have no effect on the problem of weak invertibility in this case; on the contrary, here the important thing is the "weight" which is concentrated in **D** and which reconstructs the norm $\|g\|_X$ from $\|g_r\|_{\partial X}$.

REMARK 2. We can change the form of the requirements on λ and Λ in Theorems 1a and 1b (this applies only to conditions 2) and 3); typical requirement: $\mu''/M'' \downarrow$) so that if they are satisfied by a function Λ they are also satisfied by $\Lambda_\epsilon = \lambda(\Lambda/\lambda)^\epsilon$, $\epsilon > 0$. Then Theorems 1a and 1b give us a test for weak invertibility of functions in $X = A(\lambda)$ or $L^p_{1/\lambda}(E)$ in the projective limits

$$X_{+0} = \bigcap_{\varepsilon>0} A(\lambda^{1-\varepsilon}\Lambda^\varepsilon), \quad \bigcap_{\varepsilon>0} L^p_{1/\lambda^{1-\varepsilon}\Lambda^\varepsilon}(E)$$

respectively; for details see §2.7.

THEOREM 1c. *Let $X_{\text{int}} = l^p_A(w_n)$ and $X = l^q_A(W_n)$, where $\{w_n\}_0^\infty$ is a moment sequence of a function w, satisfying the conditions in Theorem* 1b (*in the sense that $w_n = \|t^n\|_{L^s_w}$, $n \geqslant 0$, for some s with $1 \leqslant s \leqslant \infty$*), *the sequence $\{\log w_n/W_n\}_0^\infty$ is convex with respect to* $\log n$, *and*

$$\sum_{n\geqslant 1} n^{-3/2} \log \frac{w_n}{W_n} = +\infty.$$

Then each function f in X_{int} which does not vanish in **D** *is weakly invertible in X.*

---

[40] For the first version of condition 1) in Theorem 1a it is more convenient to use another method. It was shown in Theorem 1a that the terms in the minorant series (2) form a convex sequence with respect to $\log n$, and this series diverges. Therefore $-k \log n + (M(\theta(n)) - \mu(\theta(n)))$ grows without bound and $\lim_n n^{-k} D_n/d_n = +\infty$ for each $k > 0$. Therefore

$$\|g\|_X \leqslant \|g\|_{L^\infty_W(l^1_A)} = \sup_r W(r) \sum_0^\infty |\hat{g}(n)| r^n \leqslant \sum_0^\infty |\hat{g}(n)|/D_n \leqslant \sum_0^\infty n^k d_n(\lambda)/D_n < +\infty.$$

The proof is completely analogous to the proof of the preceding theorem. We obtain $\theta_n * f \in E_{A^\infty}(f)$ as before, and for the proof of the inclusion $A^\infty \subset X$ we use the arguments in the preceding remark.

§2.6. Weak invertibility of slowly growing functions

1°. In this section we establish criteria for weak invertibility of "slowly growing" functions; that is, functions f for which

$$\int^1 (\log M_f(r)/1-r)^{1/2}\,dr < +\infty.$$

A necessary condition for weak invertibility will be deduced from the aforementioned theorem of Beurling [118]; a precise formulation will be given later. Beurling's proof was based on using the (bounded) function $\exp((z+1)/(z-1))$, whose importance in questions of weighted approximation (in spaces of "slowly increasing" functions) had already been pointed out by Carleman (see [18] and [115]), and was used by Keldyš (see [45] and [70]) and then by many others (see [171], [87], [112] and [149]). The abstract idea of the Keldyš method will be described in §2.8. This method seems to explain more clearly the reasons for a lack of weighted approximability (weak invertibility) than do the Beurling arguments, but it requires several additional assumptions. The approach in §2.8 enables us to obtain an independent proof of a criterion for weak invertibility of slowly increasing functions. This criterion, roughly speaking, consists of the equivalence of weak invertibility of such functions to divergence of the series $\Sigma_1^\infty n^{-3/2}\log\|z^n\|_X$. Numerical examples for the theorems in this section will be given in §2.7.

2°. *Necessary conditions.* In [118] Beurling proved the following theorem.

THEOREM 0. *Let* $\{w_n\}_0^\infty$ *be a sequence of positive numbers such that* $w_k^2 \leqslant w_n$ *for* $n \leqslant 2k$ *and* $\Sigma_0^\infty w_n^c < +\infty$ *for some* $c > 0$. *Let* $X = \bigcup_{c>0} l_A^\infty(w_n^c)$.

1) *If* $\exp((z+1)/(z-1))$ *is weakly invertible in* X, *then*

$$\sum_1^\infty n^{-3/2}\log w_n = -\infty. \tag{1}$$

2) *If* $\lim_n w_n^{1/n} = 1$, $w_n \leqslant n^{-c}$ *for* $n = 1, 2, \ldots$ *and some* $c > 0$, *and if* $\{\log w_n^{-1}\}_0^\infty$ *is convex with respect to* $\log n$, *and* (1) *holds, then in* X *individual spectral analysis is possible*; *that is, each nonvanishing function is weakly invertible.*

We note that the space X in Theorem 0 equipped with the natural topology of an inductive limit (see [43]) is a topological algebra with respect to ordinary multiplication of functions or with respect to the Hadamard product. This fact is essential to [118]. A detailed comparison of Theorem 0 with the results in this chapter will be made in §2.7. From Theorem 0 we immediately obtain the following result.

THEOREM 1. *Let* X *be a space of analytic functions in* $\mathbf{D}$ *satisfying conditions* a) *and* b) *in* §2.2.2°. *If* $\|z^k\|_X^2 \leqslant \|z^n\|_X$ *for* $n \leqslant 2k$ *and* $\Sigma_{n\geqslant 0}\|z^n\|_X^c < +\infty$ *for some* $c >$ 0, *and if the function* $\exp((z+1)/(z-1))$ *belongs to* X *and is weakly invertible there, then*

$$\sum_{1}^{\infty} n^{-3/2} \log \| z^n \|_X = -\infty. \tag{2}$$

We note that $f_0 = \exp((z+1)/(z-1))$ is a bounded inner function (see [23]) and $\hat{f}_0(n) = O(n^{-3/4})$, but $\hat{f}_0(n) \neq o(n^{-3/4})$ (see [158]). Consequently $f_0 \in X$ if the number c in Theorem 1 is less than 4. Theorem 1 provides an easily verifiable necessary condition for weak invertibility of "slowly increasing" functions in spaces of the type $X = l^p_A(w_n)$, since here $\|z^n\|_X = w_n$, $n \geqslant 0$. For spaces with uniform or integral norms the following result is useful.

LEMMA 1. *Suppose that λ is a positive function on $[0, 1)$, $\lim_{r\to 1-0} \lambda = +\infty$, and $d_n = \inf_{0\leqslant r<1} r^{-n}\lambda(r)$, $n = 0, 1, \ldots$. Then*

$$\sum_{1}^{\infty} \frac{\log d_n}{n^{1+\varepsilon}} \leqslant c(\varepsilon) \int_0^1 \left(\frac{\log \lambda(r)}{1-r}\right)^{1-\varepsilon} dr, \quad 0 < \varepsilon < 1.$$

PROOF. We set $\lambda_0(r) = \sup_{n\geqslant 0} d_n r^n$, $0 \leqslant r < 1$. From the proof of Lemma 7 in §2.4 it follows that $\{\log d_n\}_0^\infty$ is convex upward and $\lambda_0 \leqslant \lambda$. Therefore $d_n \uparrow +\infty$ and $n^{-1}\log d_n \downarrow 0$. We set $k(n) = n^{-1}\log d_n$, $n = 1, 2, \ldots$, and we extend the definition of k linearly as a decreasing function on $(1, \infty)$. Let $N(\sigma) = k^{-1}(2\sigma)$, $0 < \sigma < \sigma_0$. Then

$$\lambda_0(e^{-\sigma}) \geqslant e^{-\sigma([N(\sigma)]+1)} d_{[N(\sigma)]+1} = e^{\sigma([N(\sigma)]+1)} \geqslant e^{\sigma N(\sigma)},$$

$$\int_0 N(\sigma)^{1-\varepsilon}\, d\sigma \leqslant \int_0 \left(\frac{\log \lambda_0(e^{-\sigma})}{\sigma}\right)^{1-\varepsilon} d\sigma \leqslant \text{const} \int^1 \left(\frac{\log \lambda(r)}{1-r}\right)^{1-\varepsilon} dr,$$

$$\int_0 N(\sigma)^{1-\varepsilon} d\sigma = \frac{1}{2} \int^\infty k\left(t^{\frac{1}{1-\varepsilon}}\right) dt = \frac{1-\varepsilon}{2} \int^\infty k(u)\, u^{-\varepsilon} du \geqslant \text{const} \sum_1^\infty \frac{\log d_n}{n^{1+\varepsilon}}. \bullet$$

THEOREM 1a. *If*

$$\lim_{r\to 1-0} \log \lambda(r) \Big/ \log \frac{1}{1-r} > 0$$

and the function $f_0 = \exp((z+1)/(z-1))$ is weakly invertible in $A(\lambda)$, then

$$\int^1 \left(\frac{\log \lambda(r)}{1-r}\right)^{1/2} dr = +\infty. \tag{3}$$

If also $\lambda \uparrow +\infty$, then condition (3) is necessary for weak invertibility of f_0 in the space $L^p_w(E)$, $w = 1/\lambda$, $1 \leqslant p \leqslant \infty$; here E is a space of analytic functions which satisfies conditions a) *and* b) *in* §2.3.2° *and in which the operator of multiplication by z is an isometry.*

PROOF. Let $d_n = \inf_{0\leqslant r<1} r^{-n}\lambda(r)$, $n \geqslant 0$. Then

$$d_n \geqslant \inf_{0\leqslant r<1} r^{-n}(1-r)^{-\text{const}} \geqslant n^{\text{const}} \quad \text{and} \quad \sum_1^\infty \| z^n \|^c_{A(\lambda)} = \sum_1^\infty d_n^{-c} < +\infty$$

for an appropriate choice of a constant $c > 0$. The other condition in Theorem 1 is also clear:

$$\| z^k \|^2_{A(\lambda)} = \sup_{0 \leqslant r < 1} r^{2k}/\lambda(r)^2 \leqslant \sup_{0 \leqslant r < 1} r^n/\lambda(r)^2 < \| z^n \|_{A(\lambda)} \tag{4}$$

for $n \leqslant 2k$ and $n \geqslant n_0$ (since $\lim_{r \to 1-0} \lambda = +\infty$). Now (3) follows from Lemma 1.

The inequalities (3) in §2.5 show that for the space $X = L^p_w(E)$ it suffices to verify a condition of the type (4):

$$\| 1 \|_E^{-2} \| z^k \|^2_{L^p_w(E)} = \| t^k \|^2_{L^p_w} \leqslant \left(\int_0^1 r^{\frac{np}{2}} / \lambda^p dr \right)^{2/p} \leqslant \| t^n \|_{L^p_w} \left(\int_0^1 \lambda^{-p} dr \right)^{1/p}$$

for $n \leqslant 2k$ since, without loss of generality, we can assume that the last integral is less than 1, and thus the required condition is satisfied. ●

REMARK. If we have existence of the limit

$$l = \lim_{r \to 1-0} \log \lambda(r) / \log \frac{1}{1-r},$$

then (3) implies that $l = +\infty$, and the first condition in Theorem 1a is not needed.

3°. A sufficient condition for weak invertibility of "slowly increasing" functions essentially coincides with (2) and (3), but we need additional assumptions in order to prove it. We note also that the use of the Beurling theorem is very problematical here.

THEOREM 2. *Let X_1 and X be spaces of analytic functions which satisfy conditions* a) *and* b) *of* §2.3.2°, *with $X_1 \subset X$. Suppose that*

1) $\overline{\lim}_{n \to \infty} (\log \log \|z^n\|^{-1}_{X_1}) / \log n < 1$;
2) $0 \leqslant \overline{\lim}_{n \to \infty} (\log \|z^n\|^{-1}_{X_1} / \log \|z^n\|^{-1}_X) < 1$;
3) $\{\log \|z^n\|^{-1}_X\}_0^\infty$ *is a convex sequence with respect to* $\log n$ *and* (2) *holds.*

Then each nonvanishing function $f \in X_1$ is weakly invertible in X.

PROOF. By 1) we have

$$|\hat{f}(n)| \leqslant \frac{\text{const}}{\| z^n \|_{X_1}} \leqslant \text{const} \exp n^c,$$

where $0 \leqslant c < 1$, and therefore

$$|f(\zeta)| \leqslant \text{const} \exp (1 - |\zeta|)^{-\text{const}}.$$

From Lemma 5 in §2.4 it follows that

$$| U^n_f 1 (\zeta) | \leqslant \text{const} (1 - |\zeta|)^{-\text{const}}$$

for $n = 0, 1, \ldots$ (the constant depends on n). In order to verify the existence of the approximation $\theta_n * f \in E_X(f)$, we use Lemma 3 in §2.4 and conclude that

$$f \in X_1, \; g \in A\left(\frac{1}{(1-r)^{\text{const}}}\right) \Rightarrow fg \in X. \tag{5}$$

We prove (5):

$$|(fg)^\wedge(n)| \leqslant \sum_{k=0}^{n} |\hat{f}(k)|\,|\hat{g}(n-k)| \leqslant \text{const}\cdot n^{\text{const}}\cdot \max_{0\leqslant k\leqslant n} \|z^k\|_{X_1}^{-1},$$

$$\sum_1^\infty |(fg)^\wedge(n)|\,\|z^n\|_X \leqslant \text{const}\sum_1^\infty n^{\text{const}}\|z^n\|_X\cdot \max_{0\leqslant k\leqslant n}\|z^k\|_{X_1}^{-1} < +\infty,$$

since by 2)

$$\|z^k\|_X\cdot\|z^k\|_{X_1}^{-1} \leqslant \text{const}\,\|z^k\|_X^\alpha,\quad 1>\alpha>0,$$

and by 3) $\|z^n\|_X \uparrow +\infty$ and $n^c = O(\|z^n\|_X^{-1})$ as $n \to \infty$ with $c > 0$ arbitrary. Now we use Lemma 2 in §2.4 and the remark which follows it, with a minorant sequence

$$\{\|z^n\|_X^{-\alpha}\}_0^\infty:\ \|z^n\|_X^{-\alpha} \leqslant \text{const}\,\|z^n\|_{X_1}/\|z^n\|_X,\quad 0<\alpha<1\,. \bullet$$

REMARK. Conditions 1) and 2) do not significantly reduce the generality of the theorem if

$$\sum_1^\infty n^{-3/2}\log\|z^n\|_{X_1}^{-1} < +\infty. \tag{6}$$

Convergence of this series (under minor assumptions on the sequence $\{\|z^n\|_{X_1}\}_0^\infty$) is equivalent to convergence of the integral

$$\int_0^1 (\log\lambda_0(r)/1-r)^{1/2}\,dr,\quad \lambda_0(r) = \sup_{n\geqslant 0} r^n/\|z^n\|_{X_1}$$

(see Lemmas 1 and 2), and this is just the case of "slowly increasing" functions to which this section is devoted. Condition (6) shows that the requirements 1) and 2) in Theorem 2 are essentially requirements of "regularity of variation" of the sequence $\{\|z^n\|_{X_1}\}_0^\infty$. This theorem (as well as Theorem 1) can be used for spaces of the type $l_A^p(w_n)$. An estimate on weighted sup-norms is contained in the following lemma.

LEMMA 2. *Let* λ *be a positive function on* $[0, 1)$ *and let* $r \to \log\lambda(r)$ *be a function of* $\log r$ *which is convex downwards* (*that is,* $r\lambda'/\lambda\uparrow$). *Set* $d_n = \inf_{0\leqslant r<1} r^{-n}\lambda(r)$, $n = 0, 1, \ldots$. *Then*

$$\int_0^1 \left(\frac{\log\lambda(r)}{1-r}\right)^{1-\varepsilon} dr \leqslant c(\varepsilon)\sum_1^\infty \frac{\log d_n}{n^{1+\varepsilon}},\quad 0<\varepsilon<1.$$

PROOF. Let $\lambda_0(r) = \sup_{n\geqslant 0} r^n d_n$, $0 \leqslant r < 1$. From Lemma 7 in §2.4 (and its proof) it is clear that $\lambda(r) \sim \lambda_0(r)$ as $r \to 1-0$ and $\{\log d_n\}_0^\infty$ is a convex sequence. Therefore $d_n \uparrow +\infty$ and $n^{-1}\log d_n \downarrow 0$. As in the proof of Lemma 1 we set $k(n) = n^{-1}\log d_n$ and extend k by linearity, and we set $N(\sigma) = k^{-1}(\sigma/2)$, $0 < \sigma < \sigma_0$. Then

$$\lambda_0(e^{-\sigma}) = \sup_{n\geqslant 0} e^{-\sigma n} d_n = e^{-\sigma n(\sigma)} d_{n(\sigma)} \leqslant d_{n(\sigma)} \leqslant d_{[N(\sigma)]+1}$$

$$= e^{k([N(\sigma)]+1)([N(\sigma)]+1)} \leqslant e^{\frac{\sigma}{2}(N(\sigma)+1)},$$

$$\int_0 \left(\frac{\log \lambda_0(e^{-\sigma})}{\sigma}\right)^{1-\varepsilon} d\sigma \leqslant c \int_0 N(\sigma)^{1-\varepsilon} d\sigma = c\frac{1-\varepsilon}{2}\int^{\infty} k(u) u^{-\varepsilon} du \leqslant c(\varepsilon) \sum_1^{\infty} \frac{\log d_n}{n^{1+\varepsilon}}. \bullet$$

THEOREM 2a. *Let λ and Λ be positive increasing functions on* $[0, 1)$ *such that*

1a)
$$\varlimsup_{r\to 1-0} \frac{\log\log\lambda(r)}{\log\frac{1}{1-r}} < +\infty;$$

2a) $\lambda(r^\alpha) = O(\Lambda^\beta(r))$ *as* $r \to 1-0$ *for some α and β,* $0 < \alpha, \beta < 1$;

3a) $(1-r)\Lambda'(r)/\Lambda(r)$ *is a monotone function in a neighborhood of* 1, *and*

$$\int^1 \left(\frac{\log \Lambda(r)}{1-r}\right)^{1/2} dr = +\infty. \tag{7}$$

Also let $w = 1/\lambda$ and $W = 1/\Lambda$, and let E_1 and E_2 be spaces of analytic functions which satisfy conditions a) *and* b) *of* §2.3.2°, *in which*

$$0 < \inf_n \|z^n\|_{E_k} \leqslant \sup_n \|z^n\|_{E_k} < +\infty \ (k=1,\ 2).$$

Then each nonvanishing function in the spaces $A(\lambda)$ or $L^p_w(E_1)$, $1 \leqslant p \leqslant \infty$, is weakly invertible in $A(\Lambda)$ or $L^p_W(E_2)$.

PROOF. We show that conditions 1a)–3a) imply conditions 1) and 2) of Theorem 2 if for X_1 we choose one of the spaces $A(\lambda)$ or $L^p_w(E_1)$, and for X we choose $A(\Lambda)$ or $L^p_W(E_2)$. Condition 1a immediately gives us 1) (see also (3) in §2.5):

$$\|z^n\|_{X_1} \geqslant \text{const}\,\|t^n\|_{L^1_w} = \text{const}\int_0^1 r^n w(r)\,dr \geqslant \text{const}\int_0^1 r^n e^{-(1-r)^{-\text{const}}} dr$$

$$\geqslant \text{const}\frac{1}{n+1}\cdot\sup_{0\leqslant r<1} r^n \exp(-(1-r)^{-\text{const}}) \geqslant \text{const}\exp(-n^{\text{const}}).$$

Analogously we obtain 2):

$$\|z^n\|_{X_1} \geqslant \text{const}\frac{1}{n+1}\sup_{0\leqslant r<1} r^n/\lambda(r) = \text{const}\frac{1}{n+1}\sup_{0\leqslant r<1} r^{\alpha n}/\lambda(r^\alpha)$$

$$\geqslant \text{const}\frac{1}{n+1}\sup_{0\leqslant r<1} r^{\alpha n}/\Lambda^\beta(r) \geqslant \text{const}\frac{1}{n+1}\left(\sup_{0\leqslant r<1} r^{\frac{\alpha n}{\beta}}\Big/\Lambda(r)\right)^\beta \geqslant \text{const}\frac{1}{n+1}\|z^n\|^\beta_X,$$

where in the last inequality we use the fact that $\alpha < \beta$ (without loss of generality).

Concerning condition 3) in Theorem 2, it is easier to replace this condition by convexity with respect to $\log n$ of the minorant sequence $1/\|z^n\|_{L^\infty_W} \leqslant \text{const}/\|z^n\|_X$ (see the last assertion in the proof of Theorem 2). We define a function L on $(0, +\infty)$ by the formula

$$L(\alpha) = \inf_{0\leqslant r<1}\left(\log\Lambda(r) + \alpha\log\frac{1}{r}\right), \quad \alpha > 0.$$

Clearly

$$L(n) = \log\|z^n\|^{-1}_{L^\infty_W} \text{ and } L(\alpha) = \log\Lambda(\nu^{-1}(\alpha)) + \alpha\log 1/\nu^{-1}(\alpha),$$

where ν^{-1} is the inverse of $\nu(r) = r\Lambda'(r)/\Lambda(r)$, $0 \leq r < 1$. Therefore convexity of L with respect to $\log\alpha$ is equivalent to monotonicity of the function $\alpha L'(\alpha)$ or (since ν is obviously monotone) of the function

$$\nu\cdot L' \circ \nu = \nu\log 1/r = r \log 1/r \; (\Lambda'(r)/\Lambda(r)).$$

This last function increases as $r \uparrow 1$, since $(1-r)\Lambda'(r)/\Lambda(r)$ increases:

$$\left(r\log\frac{1}{r}\cdot\frac{\Lambda'}{\Lambda}\right)' = \left(\frac{\Lambda'}{\Lambda}\right)' r\log\frac{1}{r} + \frac{\Lambda'}{\Lambda}\log\frac{1}{r} - \frac{\Lambda'}{\Lambda}$$

$$= \left(1-\log\frac{1}{r}\right)\left[\left(\frac{\Lambda'}{\Lambda}\right)'\left(1-\log\frac{1}{r}\right)^{-1} r\log\frac{1}{r} - \frac{\Lambda'}{\Lambda}\right]$$

$$> \left(1-\log\frac{1}{r}\right)\left[(1-r)\left(\frac{\Lambda'}{\Lambda}\right)' - \frac{\Lambda'}{\Lambda}\right] > 0$$

for $1/2 < r < 1$ ($\log(1/r) > 1 - r$ for $1/2 < r < 1$). The rest follows from Lemma 2. •

REMARK. As in Theorem 1b of §2.5, here we can replace the spaces L^p_w and L^p_W by arbitrary spaces F_1 and F with monotone norms such that $L^1_w \supset F_1 \supset L^\infty_w$ and $L^1_W \supset F \supset L^\infty_W$. Conditions 1a) and 2a) do not impose any significant new conditions on the growth of λ if we restrict ourselves to the case of "slowly increasing" majorants such that $\int_0^1 (\log\lambda(r)/(1-r))^{1/2}\,dr < +\infty$ as in the theorem which we proved. Thus if we abstract from the requirements([41]) of "regularity of growth" 1a) and 2a), we can say that each slowly increasing function f, $f(\zeta) \neq 0$, $|\zeta| < 1$, is weakly invertible in the spaces $A(\Lambda)$ and $L^p_{1/\Lambda}(E)$ if the majorant Λ satisfies the conditions of Theorem 2a.

§2.7. Analysis of conditions for weak invertibility. Inductive and projective limits. Examples

1°. In this section we present some commentary, applications, and examples for the theorems in §§2.5 and 2.6. We will analyze the conditions on "regularity of growth" imposed on the spaces in §§2.5 and 2.6, and we will show that the theorem in §2.5 turns into the criteria in §2.6 in the case of slowly increasing functions (see subsection 2°). We also solve the following problem: given a space of analytic functions, find the closest possible spaces X_{int} and X_{ext} with $X_{\text{int}} \subset X \subset X_{\text{ext}}$ in which individual spectral analysis is possible; that is, in which all nonvanishing functions are weakly invertible (equivalent to an analogue of the Tauberian theorem of Wiener; see §2.0). It is shown in §2.8 that our result is essentially sharp; in subsection 3° of that section it will be compared with the aforementioned (see §2.6) theorem of Beurling. The present section ends with numerical examples which illustrate the theorems in §§2.5 and 2.6.

2°. *Analysis of the conditions in the theorems in* §§2.5 *and* 2.6. The conditions in Theorems 1a and 1b in §2.5 appear to be hard to verify, and thus it is useful to note right away that they are satisfied if λ and Λ–or φ and Φ;

([41]) All of these requirements are satisfied for "slowly increasing" majorants λ which belong to a Hardy field which is closed with respect to the operations exp and log and contains the constants (for instance, the field H; see §2.7.2°).

$$\log\lambda(r)=\varphi\left(\log\frac{1}{1-r}\right),\quad \log\Lambda(r)=\Phi\left(\log\frac{1}{1-r}\right)\Big)$$

–belong to some Hardy field $\mathfrak{R}$ which is closed with respect to the operations exp and log and which contains the constants. In fact, all of these conditions are formulated in terms of monotonicity of certain rational functions of λ, λ', λ'' and $\log\lambda$ (and the corresponding functions for Λ). But each such function belongs to the same Hardy field $\mathfrak{R}$ and therefore is monotone in a neighborhood of the point 1 (if λ and Λ are given directly in terms of φ and Φ, then the rational functions which we mentioned also depend on the exponential e^t, $0 < t < +\infty$, which also belongs to $\mathfrak{R}$). In particular, if we are dealing with a criterion for weak invertibility of slowly increasing functions (Theorem 2a in §2.6), we can restrict our attention to the Hardy field **H** (of "elementary" functions). Now we introduce a basic (quantitative) condition for weak invertibility (divergence of the integral (1) in §2.5) in a form which is suitable for applications. Since the case of "slowly increasing" majorants was analyzed in detail in §2.6, here we will only consider the case

$$\int^1\left(\frac{\log\lambda(r)}{1-r}\right)^{1/2}dr=+\infty.$$

This condition is assumed to be satisfied throughout the present section (unless we specifically say otherwise; see the first remark after Lemma 1).

LEMMA 1. *Let λ and Λ be positive increasing functions on* $[0, 1)$, *and let h be an increasing function on* $(0, +\infty)$.

1) *If* $\log\lambda \overset{\text{def}}{=} \mu = \log\Lambda - h(\log\Lambda)$ *and α is a function on* $(0, +\infty)$ *for which* $M = \alpha(M')$, $M = \log\Lambda$, *then*

$$\int^1 (M-\mu)\frac{M''}{M'^{3/2}}\,dr=\int^\infty\frac{h(\alpha(t))}{t^{3/2}}\,dt.$$

2) *If* $M = \mu + h\circ\mu$, *and the function μ satisfies condition* 1) *in Theorem* 1a *in* §2.5, *and in the case*

$$\sup_r\,(1-r)\frac{\mu'(r)}{\mu(r)}<+\infty$$

(*that is,* $\overline{\lim}_{t\to\infty}\psi'(t) < +\infty$) *it also satisfies the condition* $\mu'^2 = O(\mu\mu'')$, *and if for the function h we have* $h' = O(1)$ *and* $h''(t) = o(t^{-1})$ *as* $t\to+\infty$, *then the integrals*

$$\int^1 (M-\mu)\frac{M''}{M'^{3/2}}\,dr \quad\textit{and}\quad \int^\infty\frac{h(\alpha(t))\,dt}{t^{3/2}} \tag{1}$$

both diverge or both converge (*here α is a function on* $(0, +\infty)$ *for which* $\mu = \alpha\circ\mu'$).

The proof of assertion 1) is clear. In case 2) we have

$$M'=\mu'+(h'\circ\mu)\mu'\asymp\mu' \quad\text{and}\quad M''=\mu''+(h''\circ\mu)\mu'^2+(h'\circ\mu)\mu''\asymp\mu'',$$

since $(h''\circ\mu)\mu'^2 = o(\mu'^2/\mu) = o(\mu'')$ in the first version of condition 1) in Theorem 1a in §2.5; the second version of this condition also leads to the desired result: the equation

$$(\mu/\mu')' = 1 - \frac{\mu\mu''}{\mu'^2} = \left\{\left(\frac{1}{\psi'}\right)' - \frac{1}{\psi'}\right\} \circ \left(\log\frac{1}{1-r}\right) = o(1)$$

implies that $\mu'^2/\mu = O(\mu'')$ and $(h'' \circ \mu)\mu'^2 = o(\mu'')$. Consequently

$$(\mathrm{M} - \mu)\frac{\mathrm{M}''}{\mathrm{M}'^{3/2}} \asymp (h \circ \mu)\frac{\mu''}{\mu'^{3/2}}. \bullet$$

COROLLARY 1. *Under the hypothesis of Lemma* 1 *a necessary condition for divergence of the integrals in* (1) *is*

$$\int^{\infty} \frac{h(t)}{t^{3/2}}\, dt = +\infty.$$

In fact,

$$\inf_r (1 - r)\frac{\mu'(r)}{\mu(r)} > 0,$$

and therefore $\mu \leqslant \text{const}\, \mu'$; that is, $\alpha(t) \leqslant \text{const} \cdot t$, $0 < t < +\infty$. $\bullet$

COROLLARY 2. *If under the hypothesis of Lemma* 1 *we have a convex function* κ *such that* $\mu' \leqslant \kappa(\mu)$, *then for divergence of the integrals in* (1) *it suffices to have one of the following integrals diverge*:

$$\int^{\infty} \frac{h(t)}{\kappa(t)^{3/2}} \kappa'(t)\, dt \quad \text{or} \quad \int^{\infty} \frac{h(t)}{t\kappa(t)^{1/2}}\, dt. \tag{2}$$

In particular, if $\beta \downarrow 0$ *and* $\beta \circ \mu$ *is convex downward, we can take* $\kappa = 1/|\beta'|$.

In fact, by the hypothesis of the corollary we have $\kappa^{-1} \leqslant \alpha$ and

$$\int^{\infty} \frac{h \circ \alpha}{t^{3/2}}\, dt \geqslant \int^{\infty} \frac{h \circ \kappa^{-1}}{t^{3/2}}\, dt = \int^{\infty} \frac{h(u)}{\kappa(u)^{3/2}} \kappa'(u)\, du.$$

Since $\lim_{r \to 1-0} \mu'/\mu = +\infty$ (see the proof of Corollary 1), we have $t = o(\kappa(t))$, and therefore $\kappa' \uparrow +\infty$. Thus $u\kappa'(u) \geqslant \kappa(u)$ for $u \geqslant u_0$, and

$$\int^{\infty} \frac{h(u)}{\kappa(u)^{3/2}} \kappa'(u)\, du \geqslant \int^{\infty} \frac{h(u)}{\kappa(u)^{1/2} u}\, du. \bullet$$

REMARKS. 1. Under the assumptions in the second part of the lemma we have

$$\int^{\infty} \frac{h(\alpha(t))}{t^{3/2}}\, dt \leqslant \text{const} \int^{\infty} \frac{\alpha(t)}{t^{3/2}}\, dt \leqslant \text{const} \int^{\infty} \frac{\mu\mu''}{\mu'^{3/2}}\, dr,$$

so that the last integral diverges, if we wish to establish weak invertibility of functions in $A(\lambda)$ or $L^p_{1/\lambda}(E)$ in the space $A(\Lambda)$, or respectively $L^p_{1/\Lambda}(E)$, using Theorems 1a and 1b in §2.5. Later, in Lemma 2, we will show that convergence of the first integral implies convergence of the integral $\int^1 (\mu/(1-r))^{1/2}\, dr$. Weak invertibility of functions with such majorants was considered in §2.6.

2. If in the conditions of Lemma 1 we write $M = \mu + h_1 \circ \mu'$ ($\mu = M - h_1 \circ M'$) and require that for h_1 we have $h_1 = O(t)$ and $h_1'/h_1 = O(1/t)$, then the first integral in (1) converges simultaneously with

$$\int^{\infty} \frac{h_1(t)}{t^{3/2}}\, dt.$$

This follows from the fact that under our assumptions we have $M' \asymp \mu'$ and $h_1(\mu') \asymp h_1(M')$, since

$$\log \frac{h_1(c\mu')}{h_1(\mu')} = \frac{h_1'(\xi)}{h_1(\xi)}(c-1)\mu',$$

where $\mu' \leqslant \xi \leqslant c\mu'$ and

$$1 \leqslant \frac{h_1(c\mu')}{h_1(\mu')} \leqslant \text{const}.$$

3. The second integral in (2) clearly indicates the necessary order of growth of h for the possibility of weak inversion of functions in $A(e^{\mu})$ or $L^p_w(E)$, $w = e^{-\mu}$, in the space $A(e^{\mu + h\circ\mu})$, or respectively $L^p_W(E)$, $W = e^{-\mu - h\circ\mu}$. The behavior of the function κ in the case $\mu' = \kappa(\mu)$ can be described qualitatively by saying that it grows more slowly the more rapidly $\lambda = \exp \mu$ grows, and we always have $\kappa(t) \geqslant t$ and

$$\int^{\infty} \frac{dt}{\varkappa(t)} = \int^{1} \frac{\mu'}{\varkappa(\mu)}\, dr = \int^{1} dr < +\infty.$$

All of this is also true (with μ replaced by M) for the case $\mu = M - h \circ M$.

4. We list some cases (see also subsection 3° below) in which it is easy to describe functions h for which the integrals in (1) and (2) are infinite. We assume that the conditions in Lemma 1 are satisfied.

(i) $\mu = \log \lambda$ is of order $\alpha > 1$ with respect to $(1-r)^{-1}$:

$$\alpha = \lim_{r \to 1-0} \frac{\log \mu(r)}{\log (1-r)^{-1}};$$

then

$$(1-r)\mu'/\mu \asymp \text{const}, \quad \mu(r) \geqslant (1-r)^{-\alpha+\delta}$$

for an arbitrary $\delta > 0$ and

$$\mu' \leqslant C(1-r)^{-1}\mu \leqslant C\mu^{1+\frac{1}{\alpha}+\delta};$$

it suffices to take $h(t) \geqslant t^{1/2+\alpha/2+\delta}$ for some $\delta > 0$.

(ii) If

$$\lim_{r \to 1-0} \frac{\log \mu(r)}{\log(1-r)^{-1}} = +\infty,$$

then $h(t) = t^{1/2+\delta}$, $\delta > 0$, gives us divergence of the integrals in (1) and (2).

(iii) If $\int^1 (\mu'/\mu)^{1/2}\, dr = +\infty$, then $h(t) = t^{1/2}$ gives us divergence of the integrals in (1); here

$$\int_0^1 \frac{h(\mu)\mu''}{\mu'^{3/2}}\,dr \geqslant \frac{1}{2}\int_0^1 \frac{\mu^{1/2}\mu''}{\mu'^{3/2}}\,dr = -\left(\frac{\mu}{\mu'}\right)^{1/2}\Big|_0^1 + \frac{1}{2}\int_0^1 \left(\frac{\mu'}{\mu}\right)^{1/2} dr = +\infty;$$

for example, this is so for $\mu = \exp(1-r)^{-1}$.

(iv) Downward convexity of the function $\beta \circ \mu$ with $\beta(t) = 1/\log_n t$, $n \geqslant 1$, corresponds to growth of μ such that $\mu(r) \geqslant \exp \cdots \exp \mathrm{const}(1-r)^{-1}$. Here a sufficient condition for divergence of the integrals in (2) is

$$\int^{\infty} \frac{h(t)}{t^{3/2}(\log t \ldots \log_{n-1} t)^{1/2} \log_n t}\,dt = +\infty.$$

We can often draw the reverse ("necessary") conclusion about the speed of growth of h which guarantees divergence of the integrals in (1) and (2). Suppose, for instance, that the conditions in the lemma are satisfied and

$$\psi(t) \leqslant \gamma t,\ t \geqslant t_0,\ \gamma \geqslant 1 \left(\mu(r) = \exp\psi\left(\log\frac{1}{1-r}\right)\right).$$

Then

$$\mu'(r) = \mu(r)\psi'\left(\log\frac{1}{1-r}\right)\frac{1}{1-r} \geqslant \mu(r)^{1+1/\gamma}\,\psi'\left(\log\frac{1}{1-r}\right) \geqslant \mathrm{const}\,\mu(r)^{1+\frac{1}{\gamma}},$$

and $\alpha(t) \leqslant \mathrm{const}\cdot t^{\gamma/(1+\gamma)}$. Therefore

$$\int^{\infty} \frac{h(\alpha(t))}{t^{3/2}}\,dt \leqslant \int^{\infty} \frac{h(t^{\gamma/(1+\gamma)})}{t^{3/2}}\,dt = c\int^{\infty} \frac{h(u)}{u^{3/2+1/2\gamma}}\,du.$$

In particular, if $\gamma = 1$ (that is, $\mu(r) \leqslant (1-r)^{-1}$ and the majorant λ is almost "slowly increasing"), then h cannot grow significantly more slowly than $h(u) = u(\log u \cdots \log_n u)^{-1}$.

5. Remark 4 shows that the least order of growth of h for which in the space $A(e^{\mu + h\circ\mu})$, respectively $L^p_w(E)$, we have weak invertibility of all functions f, $f(\zeta) \neq 0$, $|\zeta| < 1$, in $A(e^{\mu})$, respectively $L^p_w(E)$, does not exceed $1/2 + 1/2\alpha$, where α $(0 < \alpha \leqslant +\infty)$ is the order of $\mu = \log\lambda$ with respect to $(1-r)^{-1}$. For majorants of a special form we will prove the sharpness of this estimate in §2.8.

Now we show that Theorems 1a and 1b in §2.5 become Theorem 2a in §2.6 if we restrict our attention to slowly increasing and sufficiently "regular" majorants λ.

LEMMA 2. *If μ is a function which grows without bound,* $\sup \mu'/\mu^2 < +\infty$, *and the functions μ'/μ and μ''/μ' are comparable to* $(1-r)^{-1}$, *then the integrals*

$$\int^1 \left(\frac{\mu}{1-r}\right)^{1/2} dr \quad \textit{and} \quad \int^1 \frac{\mu\mu''}{\mu'^{3/2}}\,dr$$

both converge or both diverge.

PROOF. First we note that the conditions $\mu'/\mu \asymp (1-r)^{-1}$ and $\mu''/\mu' \asymp (1-r)^{-1}$ imply that the integrals $\int^1 \mu\mu''(\mu')^{-3/2}\,dr$ and $\int^1 \mu^{1/2}(1-r)^{-1/2}\,dr$ converge or diverge together. After an integration, the case $\mu''/\mu' = o((1-r)^{-1})$ gives us the estimate $\mu \leqslant \text{const}(1-r)^{1-\epsilon}$, $\epsilon > 0$; that is, the function μ is bounded and therefore can be ignored. If $(1-r)^{-1} = o(\mu''/\mu')$, then

$$\mu \geqslant \text{const}\,(1-r)^{-n}$$

for an arbitrary $n > 0$ and for sufficiently large r, and therefore $\int^1 (\mu/(1-r))^{1/2}\,dr = \infty$. The second integral in this case satisfies the estimate

$$\int^1 \frac{\mu\mu''}{\mu'^{3/2}}\,dr \geqslant \text{const} \int^1 \frac{\mu}{\mu'^{1/2}(1-r)}\,dr \geqslant \text{const} \int^1 \frac{dr}{1-r}.$$

The cases $\mu''/\mu' \asymp (1-r)^{-1}$ and $\mu'/\mu = o((1-r)^{-1})$ or $(1-r)^{-1} = o(\mu'/\mu)$ are treated analogously. ●

Thus the condition

$$\int^1 \frac{MM''}{M'^{3/2}}\,dr = +\infty,$$

which comes from Theorems 1a and 1b in §2.5 and which is sufficient for weak invertibility of each bounded function f in $\mathbf{D}$ for which $f(\zeta) \neq 0$, $|\zeta| < 1$, coincides with the necessary and sufficient condition

$$\int^1 \left(\frac{M}{1-r}\right)^{1/2} dr = +\infty,$$

in Theorem 2a of §2.6.

3°. *Spectral analysis of the shift operator in inductive and projective limits.* Let X be a Banach space of analytic functions in $\mathbf{D}$; we construct the closest possible "approximating" spaces X_i and X_e for X with $X_i \subset X \subset X_e$ in which each function which does not vanish in $\mathbf{D}$ is weakly invertible. To do this, we set(42) $X_e = \bigcup_{n=0}^{\infty} X_n$, $X_0 = X$, $X_n \subset X_{n+1}$, $n \geqslant 0$, and $X_i = \bigcap_{n=0}^{\infty} X_n$, $X_0 = X$, $X_n \supset X_{n+1}$, $n \geqslant 0$, and we equip these sets with the topologies of inductive and projective limits respectively (see [43]). The spaces X_n will be constructed so that: a) in X_e all functions $f \in X_n$ with $f(\zeta) \neq 0$, $|\zeta| < 1$, will be weakly invertible in X_{n+1}; b) in X_i all functions $f \in X_{n+1}$ with $f(\zeta) \neq 0$, $|\zeta| < 1$, will be weakly invertible in X_n. This can be done on the basis of results in §§2.4 and 2.5. We will do this for two types of spaces which we considered in §2.5, namely, $A(\lambda)$ and $L^p_w(E)$.

THEOREM 1. *Let λ be an increasing positive function on* $[0, 1)$ *which belongs to a Hardy field $\mathfrak{R}$ which contains the constants and is closed with respect to the operations* log *and* exp. *Let h be an increasing function on* $(0, +\infty)$ *in the Hardy field* $\mathbf{H}$ *for which* $h'(t) = O(1)$ *and one of the integrals in* (1) *or* (2) *diverges. Then in the space*

$$X = \bigcup_{\varepsilon < 1} A(\lambda \exp \varepsilon h \circ \mu) \tag{3}$$

each nonvanishing function is weakly invertible.

---

(42) Later we will also consider constructions of a different sort, where $X_e = \bigcap_1^{\infty} X_n$, $X \subset X_{n+1} \subset X_n$, $n \geqslant 1$, and $X_i = \bigcup_1^{\infty} X_n$, $X_n \subset X_{n+1} \subset X$, $n \geqslant 1$.

PROOF. Let $\mu_\epsilon = \mu + \epsilon h \circ \mu$, $\epsilon < 1$. Since $\mathbf{H} \circ \mathfrak{R} \subset \mathfrak{R}$ (see [12], Chapter V, Appendix, §4), it follows that all of the "regularity" conditions in subsection 2° are satisfied, and in order to apply Lemma 1 or Corollary 2 we have to show that:

(i) for an arbitrary $\epsilon < 1$ there exist $\delta > 0$ and $\epsilon_1 < 1$ such that $\mu_\epsilon + \delta h \circ \mu_\epsilon \leqslant \mu_{\epsilon_1}$;

(ii) $\int^1 h(\mu_\epsilon)\mu_\epsilon'' \mu_\epsilon'^{-3/2}\, dr$ diverges for an arbitrary ϵ, $0 < \epsilon < 1$, together with the integrals in (1) and (2). The estimate (i) follows from the inequality $h \circ \mu_\epsilon \leqslant \text{const} \cdot h \circ \mu$, which is fairly obvious:

$$h(\mu + \varepsilon h \circ \mu) = h \circ \mu + \varepsilon h'(\xi)\, h \circ \mu \leqslant \text{const}\, h \circ \mu.$$

Since congruences for functions in the same Hardy field can be differentiated, and since $\mu_\epsilon \asymp \mu$, we have $\mu_\epsilon' \asymp \mu'$, $(\mu_\epsilon')^{-1/2} \asymp (\mu')^{-1/2}$, and finally $\mu_\epsilon''(\mu_\epsilon')^{-3/2} \asymp \mu''(\mu')^{-3/2}$. This proves assertion (ii). ●

THEOREM 2. *Under the conditions imposed on* λ *and* h *in Theorem* 1, *in the space*

$$X = \bigcup_{\varepsilon<1} L^p_{w_\varepsilon}(E), \quad \frac{1}{w_\varepsilon} = \lambda e^{\varepsilon h \circ \mu},$$

each nonvanishing function is weakly invertible. Here E is an arbitrary space of analytic functions which satisfies conditions a) *and* b) *in* §2.3.2°, *and*

$$0 < \inf_n \| z^n \|_E \leqslant \sup_n \| z^n \|_E < +\infty.$$

REMARK. Results similar to Theorems 1 and 2 also hold for spaces which are constructed by "reduction" of the function λ. Namely, under the assumptions in Theorems 1 and 2 the same conclusions follow for the spaces

$$\bigcup_{\varepsilon>0} A(\lambda e^{-\varepsilon h \circ \mu}) \quad \text{and} \quad \bigcup_{\varepsilon>0} L^p_{w_{-\varepsilon}}(E), \quad w_\varepsilon^{-1} = \lambda e^{\varepsilon h \circ \mu}.$$

The proofs are the same, except that instead of the second assertion in Lemma 1 we use the first.

For spaces of the type $l^p_A(w_n)$ similar results can also be established (on the basis of Theorem 1c in §2.5), but the conditions on the weight sequences turn out to be very complicated. We turn to a comparison of Theorems 1 and 2 with the Beurling theorem (Theorem 0 in §2.6).

LEMMA 3. *Let* $\{w_n\}_0^\infty$ *be a sequence of positive numbers such that* $w_n \leqslant n^{-C}$, $n = 1, 2, \ldots$, *for some* $C > 0$ *and* $\{\log d_n\}_0^\infty$, *where* $d_n = 1/w_n$ *is a convex sequence. Set*

$$\lambda(r) = \sup_{n\geqslant 0} r^n d_n, \quad \lambda_c(r) = \lambda(r^{1/c})^c, \quad 0 \leqslant r < 1, \quad c > 0.$$

Then $\bigcup_{c<\infty} l^\infty_A(w_n^c) = \bigcup_{c<\infty} A(\lambda_c)$.

The proof is essentially contained in Lemma 7 of §2.4. If $f \in A(\lambda_c)$, $c > 0$, then by this lemma

$$|\hat{f}(n)| \leqslant \inf_{0 \leqslant r < 1} r^{-n}\lambda_c(r) = \left(\inf_{0 \leqslant r < 1} r^{-\frac{n}{c}}\lambda\left(r^{\frac{1}{c}}\right)\right)^c = d_n^c,$$

and $f \in l_A^\infty(w_n^c)$. Conversely, if $f \in l_A^\infty(w_n^c)$, then

$$|f(\zeta)| \leqslant \sum_0^\infty |\hat{f}(n)|\,|\zeta|^n \leqslant \text{const} \sup_{n \geqslant 0} |\zeta|^n n^2 |\hat{f}(n)|$$

$$\leqslant \text{const} \sup_{n \geqslant 0} |\zeta|^n d_n^{c_1} = \text{const}\,\lambda_{c_1}(|\zeta|),$$

and $f \in A(\lambda_{2c_1})$. ●

Thus the Beurling result (assertion 2) in Theorem 0 of §2.6) is equivalent to the possibility of individual spectral analysis in the space $\bigcup_{c>0} A(\lambda_c)$ under the condition(43) (see Lemmas 1 and 2 in §2.6)

$$\int^1 \left(\frac{\log \lambda(r)}{1-r}\right)^{1/2} dr = \infty$$

together with mild conditions on λ (of the type described in the hypothesis of Theorem 2a in §2.6). Under the more restrictive conditions (which are satisfied by all majorants which are composed of elementary functions) of Theorem 1 we obtain a more precise assertion: "analysis" is also possible in the space (3). The Beurling scale $\{A(\lambda_c)\}_{c>0}$ is significantly "rougher" than the scale $\{A(\lambda^{1/c})\}_{c>0}$ (consider the functions $\lambda(r) = \exp \cdots \exp(1/(1-r))$; see also Appendix III), and the latter scale for exponential order 1/2 is "rougher" than the scale (3); loosely speaking, we can say that the transition to the power scale $\{A(\lambda^{1/c})\}_{c>0}$ corresponds to changing the function $h(t) = \sqrt{t}$ to $h(t) = t$.

Now we note that the possibility of localization of principal invariant subspaces for the ("finite"!) power scale

$$X = \bigcup_{c<c_0} A(\lambda^c), \quad 0 < c_0 \leqslant +\infty, \tag{4}$$

follows from the results in §2.1 and the estimate in Chapter 1. We have the following result.

THEOREM 3. *Let $\lambda = \exp\varphi(\log(1/(1-r)))$, $0 \leqslant r < 1$, be an increasing function on $[0, 1)$, and suppose that for $\psi = \log\varphi$ we have the conditions $1 < \lim_{t\to\infty}\psi'(t)$, $\psi' \uparrow$, and $\psi''\psi'^{-3/2}$ is a monotone function. Then in the space (4) individual spectral analysis is possible.*

In fact, together with λ the function λ^c, $c > 0$, satisfies the same conditions. By Theorem 2 in §1.3, if $f \in X$ and $f(\zeta) \neq 0$, $|\zeta| < 1$, then for each $\delta > 0$ there exists an $\epsilon > 0$ such that $1/f^\epsilon \in A(\lambda^\delta)$. And it is clear that f and the pair of spaces X and $\mathfrak{M} = A(\lambda^\delta)$ for small $\delta > 0$ satisfy the hypotheses of Theorems 1 and 1′ in §2.1. Consequently f is weakly invertible in X. ●

From (4) in §2.5 it follows that, under the same conditions on λ,

(43) We note that this condition does not change if we replace λ by λ_c.

$$\bigcup_{c<+\infty} A(\lambda^c) = \bigcup_{c<\infty} L^p_{1/\lambda^c}(E), \tag{5}$$

where E is the space in Theorem 2.

COROLLARY. *Let λ be the function in Theorem 3, and let E be the space in Theorem 2. Then in the inductive limit (5) each nonvanishing function in* **D** *is weakly invertible.*

We note that if $\lim_{t\to\infty}\psi'(t) = 1$, the assertions in Theorems 1 and 2 are significantly stronger than the possible analogues of Theorem 3 and its corollary which could be obtained directly by means of the results in §2.1. The reason for this seems to be that, under the assumptions which we made, the (best possible) lower estimate for functions f in $A(\lambda)$ is significantly weaker than the requirement $|f| \geqslant 1/\lambda^{\text{const}}$ (see §1.3).

For projective limits we have theorems which are analogous to Theorems 1–3. We will state analogues of Theorems 1 and 3 and will omit their proofs, which essentially coincide with the proofs just given.

THEOREM 1$'$. *Under the conditions in Theorem 1 on λ and h, in the space*

$$X = \bigcap_{\varepsilon>0} A(\lambda e^{\varepsilon h\circ\mu})$$

each nonvanishing function in **D** *is weakly invertible.*

THEOREM 3$'$. *Under the conditions in Theorem 3 on λ and E, each nonvanishing function in the space*

$$X = \bigcap_{c>c_0} L^p_{1/\lambda^c}(E), \quad 0 \leqslant c_0 < +\infty,$$

is weakly invertible.

REMARK. Theorems 1, 3, 1$'$ and 3$'$ enable us to solve the problem stated at the start of the present subsection. Suppose, for instance, that $X = A(\lambda)$ and the functions λ and h satisfy the conditions in Theorems 1 and 1$'$. We set

$$X_{\text{int}} = \bigcup_{\varepsilon>0} A(\lambda e^{-\varepsilon h\circ\mu}), \quad X_{\text{ext}} = \bigcap_{\varepsilon>0} A(\lambda e^{\varepsilon h\circ\mu}).$$

Then $X_{\text{int}} \subset X \subset X_{\text{ext}}$ and in the spaces X_{int} and X_{ext} individual spectral analysis is possible. As we already mentioned, in §2.8 we will prove that the order of growth of the function h which is admissible in Theorems 1 and 1$'$ cannot be improved.

4°. *Examples.* For calculations with specific majorants λ it is often more convenient to use Remark 2 after Lemma 1 rather than the lemma itself and its corollaries. We carry out such calculations in two simple cases.

1. Let

$$\lambda(r) = \exp(1-r)^{-\alpha} \prod_{k=1}^{n} \log_k^{\alpha_k} \frac{1}{1-r}, \quad r_0 \leqslant r < 1,$$

where $\alpha_k \in \mathbf{R}$, $\alpha > 0$, and $\log_k = \log\cdots\log\,(k \text{ times})$.

If

$$(\alpha, \alpha_1, \ldots, \alpha_n) \succ (1, -2, -2, \ldots, -2),$$

then

$$\int^{1}\left(\frac{\log\lambda}{1-r}\right)^{1/2} dr<+\infty$$

and we use Theorem 2a in §2.6. If Λ is another majorant which satisfies the conditions of this theorem (for instance, if $\Lambda\in\mathbf{H}$) and for which $\int^1(\log\Lambda/(1-r))^{1/2}\,dr=+\infty$, then all functions f, $f(\zeta)\neq 0$, in $A(\lambda)$ are invertible in the space $A(\Lambda)$; an analogous assertion is true for spaces of the type $L^p_w(E)$. For example, we can let

$$\Lambda(r)=\exp(1-r)^{-1}\left(\prod_{k=1}^{n}\log_k^{-2}\frac{1}{1-r}\right).$$

If

$$(\alpha,\ \alpha_1,\ \ldots,\ \alpha_n)\gneqq(1,\ -2,\ -2,\ \ldots,\ -2),$$

then we can use the results in this section and Theorems 1a and 1b in §2.5. We use the following notation:

$$\mu=\log\lambda,\quad \Pi=\prod_{k=1}^{n}\log_k^{\alpha_k}\frac{1}{1-r},\quad r_0<r<1.$$

Then

$$\mu'=\mu\left\{\frac{\alpha}{1-r}+\sum_{k=1}^{n}\frac{\alpha_k}{(1-r)\log\frac{1}{1-r}\cdots\log_k\frac{1}{1-r}}\right\}.$$

For the function h_1 in Remark 2 after Lemma 1 we take

$$h_1(t)=t^{1/2}(\log t\ldots\log_N t)^{-1},\ N\geqslant n.$$

Then

$$h_1(\mu')\asymp h_1\left(\frac{\mu(r)}{1-r}\right)\asymp(1-r)^{-\frac{\alpha}{2}-\frac{1}{2}}\prod_{k=1}^{n}\log_k^{\frac{\alpha_k}{2}-1}\frac{1}{1-r}\cdot\prod_{k=n+1}^{N}\log_k^{-1}\frac{1}{1-r}.$$

Now we can draw the same conclusion as in the previous case if we let

$$\log\Lambda(r)=\log\lambda(r)+(1-r)^{-\frac{\alpha}{2}-\frac{1}{2}}\prod_{k=1}^{n}\log_k^{\frac{\alpha_k}{2}-1}\frac{1}{1-r}\cdot\prod_{k=n+1}^{N}\log_k^{-1}\frac{1}{1-r}.$$

Also, by Theorems 1, 2 and 1$'$, in the spaces

$$\bigcup_{\varepsilon<\varepsilon_0}A(\lambda(\Lambda/\lambda)^{\varepsilon})=\bigcup_{\varepsilon<\varepsilon_0}A(e^{(1-\varepsilon)\log\lambda+\varepsilon\log\Lambda}),$$

$$\bigcup_{\varepsilon>0}A(\lambda(\Lambda/\lambda)^{-\varepsilon}),\quad \bigcap_{\varepsilon>0}A(\lambda(\Lambda/\lambda)^{\varepsilon})$$

and in the corresponding $L^p_{1/\lambda}(E)$-spaces individual spectral analysis is possible.

In particular, for $\alpha=1$ and $\alpha_k\equiv 0$ all functions f, $f(\zeta)\neq 0$, in $A(\exp(1/(1-r)))$ are weakly invertible in the space

$$A\left(\exp\left(\frac{1}{1-r}+\frac{\varepsilon}{1-r}\prod_{k=1}^{N}\log_k^{-1}\frac{1}{1-r}\right)\right)$$

for arbitrary $\epsilon > 0$ and $N \geqslant 0$, and even (Theorem 1′) in the projective limit (intersection) of these spaces as $\epsilon \to 0$. The results in §2.1 and the estimates in Chapter 1 give us a less refined result; namely, the same conclusions hold for

$$\log\Lambda(r)=\frac{1}{1-r}\log^2\frac{1}{1-r},$$

and localizability of principal ideals is guaranteed only in the space

$$\bigcup_{n=1}^{\infty} A\left(\exp\frac{1}{1-r}\log^n\frac{1}{1-r}\right).$$

2. Let

$$\lambda(r)=\underbrace{\exp\ldots\exp}_{n+1}\frac{1}{(1-r)^\alpha},\quad 0\leqslant r<1,$$

where $n \geqslant 1$ and $\alpha > 0$. Here

$$\mu'=\mu\frac{\alpha}{1-r}\prod_{k=1}^{n-1}\log_k\mu;$$

if we choose h_1 as in Example 1, we obtain

$$h_1(\mu')\asymp\mu^{1/2}\frac{1}{(1-r)^{1/2}}\prod_{k=1}^{n}\log_k^{-1/2}\mu\cdot\prod_{k=n+1}^{N}\log_k^{-1}\mu\overset{\text{def}}{=}\Delta(r).$$

Analogues of all of the results in the previous example hold for the pair of spaces $A(\lambda)$ and $A(\Lambda)$, $\log\Lambda = \log\lambda + \Delta$. In particular, for $n = 1$ we obtain a majorant Λ with

$$\log\Lambda(r)=\exp\frac{1}{(1-r)^\alpha}+\frac{1}{(1-r)^{1/2+\alpha/2}}\prod_{k=1}^{N-1}\log_k^{-1}\frac{1}{1-r}\cdot\exp\frac{1}{2(1-r)^\alpha}.$$

§2.8. Abstract form of the Keldyš method. Counterexamples

1°. We return to the topic mentioned in §2.0 and in §2.1.1° concerning "negative" results; that is, the problem of constructing functions in a given space X which are not weakly invertible. Concerning spaces of "slowly increasing" functions, one such problem was considered in §2.6 in connection with the Beurling theorem; namely, it was shown that (under certain "regularity" assumptions on the norm $\|\cdot\|_X$) the inner function

$$f_0=\exp\frac{z+1}{z-1},$$

is weakly invertible if and only if $\Sigma_1^\infty n^{-3/2}\log\|z^n\|_X = -\infty$. In this section we present another method for obtaining similar results, developing the method which Keldyš used in [45] for showing that f_0 was not weakly invertible in $H^2(l_2)$. This method is based on an analysis of the properties of inner functions which prevent weak invertibility of these functions in spaces with slowly growing majorants. The general method for showing noninvertibility (see Lemma 1 below) is applicable to almost any space X, but its realization requires the construction of functions which have simultaneously extremal growth (with respect to X) on some curve and extremal decrease on another curve (see §2.1.1°). The existence of such functions is nontrivial even in spaces $A(\lambda)$ with the very simple weights $\lambda_\alpha(r) = \exp(1-r)^{-\alpha}$, $\alpha \geqslant 1$ (see subsection 4°). A precise result for these spaces is the fact that for an arbitrary $\alpha > 1$ there exist functions f in $A(\lambda_\alpha)$ with $f(\zeta) \neq 0$, $|\zeta| < 1$, which are not weakly invertible in $A(\lambda_\alpha \lambda_{(1+\alpha)/2-\epsilon}^{1/\epsilon})$ for any $\epsilon > 0$. We recall that by the fundamental theorem (see §2.7.4°) each such function is weakly invertible in the space $A(\Lambda)$ with

$$\Lambda(r) = \lambda_\alpha(r) \exp\left(\frac{1}{(1-r)^{\frac{\alpha+1}{2}} \log \frac{1}{1-r} \cdots \log_n \frac{1}{1-r}}\right), \quad n \geqslant 1.$$

On the basis of Lemma 1 similar examples can be constructed for majorants λ which have an infinite exponential order of growth (see subsection 5°). Here the functions λ are of a special form (they are of the form M_f for some f), but for an arbitrary comparatively regular majorant λ_1 we can construct a function λ of the form described in subsection 5° such that $\log\log\lambda_1 \sim \log\log\lambda$. Thus there is no "speed of growth" for a majorant λ which makes it possible to localize the ideals in spaces of the type $A(\lambda)$ or to lower the power order 1/2 for h which guarantees weak invertibility of functions in $A(\lambda)$ in the space $A(\lambda e^{h(\log\lambda)})$.

2°. *The general procedure.* Let X be a Banach space of analytic functions in $\mathbf{D}$ which contains $\mathbf{1}$ and satisfies condition a) of §2.3.2°; that is, a space such that the functionals ψ_ζ, $\psi_\zeta(f) = f(\zeta)$, $f \in X$, are continuous in X for all ζ with $|\zeta| < 1$. Let Ω be a simply connected domain in $\mathbf{C}$ with $\Omega \neq \mathbf{C}$, and let $G \in A(\Omega)$. The function G is said to be outer (in Ω) if $G \circ \omega$ is an outer function in $\mathbf{D}$ and ω is a conformal mapping of $\mathbf{D}$ onto Ω. By an outer function in $\mathbf{D}$ we mean (conflicting somewhat with normal usage; see [23] and [117]) a function F with a bounded characteristic–that is,

$$\sup_{0\leqslant r<1} \int_{\partial\mathbf{D}} |\log|F(r\zeta)|| \, dl_1(\zeta) < +\infty$$

–for which there are no inner factors in the standard factorization (see [23]).

LEMMA 1. *Let γ be a closed Jordan curve, $\gamma \subset \overline{\mathbf{D}}$, which is symmetric with respect to the real axis and which intersects the boundary of the disk only at the point $\zeta = 1$ (that is, $\gamma \cap \partial\mathbf{D} = \{1\}$). Suppose that $f \in X$ and G is an outer function in the interior* Int γ *of the curve γ such that*

1) $\|\psi_\zeta\|/|f(\zeta)| \leqslant |G(\zeta)|$, $\zeta \in \gamma$;
2) $f(x) = o(1/G(x))$, $x \to 1-0$.

Then $E_X(f) \neq X$; that is, f is not weakly invertible in X.

PROOF. If $\mathbf{1} = \lim_n p_n f$, $p_n \in \mathcal{P}_A$, then

$$|p_n(\zeta) f(\zeta) - 1| \leqslant \|\psi_\zeta\| \cdot \|p_n f - 1\|_X.$$

Consequently $|p_n(\zeta)| \leqslant \text{const} \cdot \|\psi_\zeta\|/|f(\zeta)|$ (since $\|\psi_\zeta\| \geqslant \|\mathbf{1}\|_X^{-1} > 0$), and by condition 1) we have $|p_n(\zeta)| \leqslant \text{const} \cdot |G(\zeta)|$, $\zeta \in \gamma$. Since G is an outer function, we have $|p_n(\zeta)| \leqslant \text{const} \cdot |G(\zeta)|$ in Int γ, and therefore

$$|1/f(x)| = \lim |p_n(x)| \leqslant \text{const}\, |G(x)|,\ 0 \leqslant x_0 \leqslant x < 1,$$

which contradicts condition 2). ●

REMARK. A function f which satisfies the conditions in the lemma reminds one of an inner function [23] (and in the cases $X = H^\infty$ or $X = C_A$ there are no other functions which satisfy 1) and 2)). Condition 1) means that f "supports" the norm $\|\psi_\zeta\|$ well; this norm must almost be attained by the function f for all $\zeta \in \gamma$; the optimal case $\|\psi_\zeta\| \leqslant \text{const} \cdot |f(\zeta)|$, $\zeta \in \gamma$, produces a simplified form of condition 2): $f(x) = o(1)$ as $x \longrightarrow 1 - 0$. On the other hand, the second requirement on f means that its speed of decrease on the radius Re $z = 0$ is extremal among functions in the class X, and f "extinguishes" an arbitrary outer function in Int γ. For example, the inner function $f_0 = \exp((z + 1)/(z - 1))$ satisfies the conditions in the lemma for the curve γ: $|\zeta - 1/2| = 1/2$ and the outer function $G = \mathbf{1}$ (the case $X = H^\infty$; if $X = C_A$ we take $f = (z - 1)f_0$; see [18]). For the space $X = H^2(l_2)$ or even $X = H^2(\rho dl_2)$, where $\rho(r) \geqslant \text{const} \cdot (1 - r)^{\text{const}}$, $r_0 \geqslant r > 1$, we can use the same curve γ and the function $G = (1 - z)^{-\text{const}}$ (with a positive constant in the exponent, possibly different from the previous one) since here $\|\psi_\zeta\| \leqslant \text{const} \cdot (1 - |\zeta|)^{-\text{const}}$. The first such example was the aforementioned example of Keldyš.

After the author's work [87] appeared, containing many of the results in this section, the work of Kelleher and Taylor [149] appeared, and it made use of a similar construction for investigating closed z-invariant subspaces (ideals) in algebras of entire functions of the type $\bigcup_{c<\infty} A(\lambda^c)$ with a majorant λ which has a finite exponential order of growth.

2a°. *Illustration: the noninvertibility of* $\exp((z + 1)/(z - 1))$. The fact that the inner function $f_0 = \exp((z + 1)/(z - 1))$ is not weakly invertible in spaces $A(\lambda)$ with a "slowly increasing" majorant was deduced in §2.6 from a theorem of Beurling. Now we will show that we can use the procedure in subsection 2° in order to obtain a more direct proof of this fact. This proof seems more intuitive and elementary than the first one, although it requires new assumptions on λ.

THEOREM 1. *Suppose that λ is an increasing positive function on* $[0, 1)$, $\log \lambda(r) = \varphi(\log(1/(1 - r)))$, $\psi = \log \varphi$ *and ψ is convex in a neighborhood of* $+\infty$. *If*

$$\int^1 \left(\frac{\log \lambda(r)}{1 - r}\right)^{1/2} dr < +\infty, \tag{1}$$

then $E_{A(\lambda)}(f_0) \neq A(\lambda)$.

PROOF. We will show that there exists a smooth curve γ of the type required by Lemma 1 which touches $\partial \mathbf{D}$ and for which

$$\int_\gamma \log \frac{\lambda(|\zeta|)}{|f_0(\zeta)|} |d\zeta| = \int_\gamma \log \lambda(|\zeta|) |d\zeta| + \int_\gamma \log \frac{1}{|f_0(\zeta)|} |d\zeta| < +\infty. \tag{2}$$

Since $|f_0| \leqslant 1$ in $\mathbf{D}$, the integrals must approach each other. If $y = y(x)$, $0 \leqslant x < 1$, is the equation of the upper half $\gamma_+ = \gamma \cap \{\zeta\colon \operatorname{Im}\zeta \geqslant 0\}$ of the curve γ and $\zeta = x + iy$, then[44]

$$\int\limits_{\gamma_+} \log \frac{1}{|f_0(\zeta)|} |d\zeta| = \int\limits_{\gamma_+} \frac{1-x}{(1-x)^2 + y(x)^2} ds = \int\limits_{\gamma_+} \frac{(1-x)\,ds}{y^2(x)\left(\left(\frac{1-x}{y(x)}\right)^2 + 1\right)} \asymp \int\limits_{\gamma_+} \frac{1-x}{y^2(x)} ds,$$

since γ_+ touches $\partial\mathbf{D}$ $(\lim_{x \to 1-0}(1-x)/y(x) = 0)$. If we write the equation of γ_+ in polar coordinates: $r = r(t)$, $0 < t \leqslant \pi/2$, we switch to the problem of finding a sufficiently smooth function r for which

$$\int\limits_0 \frac{1 - r\cos t}{r^2 \sin^2 t} \sqrt{r^2(t) + (r'(t))^2}\, dt < +\infty,$$

$$\int\limits_0 \log \lambda(r) \sqrt{r^2(t) + (r'(t))^2}\, dt < +\infty.$$

Since the expression

$$\frac{1-r^2}{t} = \frac{1-(x^2+y^2)}{r\sin t} \cdot \frac{r \sin t}{t} = \left| \frac{(1-x)(1+x)}{y} - y \right| r \frac{\sin t}{t}$$

converges to zero as $t \longrightarrow 0$ (that is, $y \longrightarrow 0$), we have $\lim_{t \to 0} r'(t) = 0$ and $\sqrt{r^2 + r'(t)^2} \sim 1$ as $t \longrightarrow 0$. Thus we have finite integrals

$$\int\limits_0 \log \lambda(r(t))\, dt$$

and

$$\int\limits_0 \frac{1 - r\cos t}{r^2 \sin^2 t} dt = \int\limits_0 \frac{1 - r(1 - 2\sin^2 t/2)}{r^2 \sin^2 t} dt \asymp \int\limits_0 \frac{1-r}{t^2} dt.$$

We set $(1 - r(t))^{-1} \log \lambda(r(t)) = 1/t^2$. Then

$$t^2(r) = \frac{1-r}{\log \lambda(r)}, \quad t'(r) = -\frac{1}{2}\left(\frac{1-r}{\log \lambda(r)}\right)^{-1/2} \frac{\lambda(r) \log \lambda(r) + \lambda'(r)(1-r)}{\lambda(r) \log^2 \lambda(r)},$$

and therefore

$$\int\limits_0 \frac{1-r}{t^2} dt = \int\limits_0 \log \lambda(r(t))\, dt = \int^1 t'(r) \log \lambda(r)\, dr$$

$$= \int^1 \frac{\log^{1/2} \lambda(r)}{(1-r)^{1/2}} dr + \int^1 \frac{\lambda'(r)(1-r)^{1/2}}{\lambda(r) \log^{1/2} \lambda(r)} dr < +\infty,$$

since the first of these integrals converges by condition (1) and the second converges to it by virtue of the estimate $\sup \psi' < +\infty$:

$$\frac{\lambda'(r)(1-r)^{1/2}}{\lambda(r)\log^{1/2}\lambda(r)} = \frac{\lambda'(r)}{\lambda(r)} \cdot \frac{(1-r)}{\log \lambda(r)} \cdot \frac{\log^{1/2}\lambda(r)}{(1-r)^{1/2}} \leqslant \text{const}\, \frac{\log^{1/2}\lambda(r)}{(1-r)^{1/2}}.$$

---

(44) The symbol $\asymp$ temporarily indicates that the integrals both converge or both diverge.

Now we note that condition (1) and the convexity of ψ imply that $\psi'_\infty \overset{\text{def}}{=} \lim_{t\to\infty}\psi' \leqslant 1$. If $\psi'_\infty < 1$, then in the following arguments we replace the majorant λ by the larger function $\exp(1-r)^{-\alpha}$, $\psi'_\infty < \alpha < 1$. In that case, as in the case $\psi'_\infty = 1$, it is not hard to show that γ is a convex curve. For the construction of the function G which is outer in $\operatorname{Int}\gamma$ and satisfies $|G(\zeta)| = \lambda(|\zeta|)/|f_0(\zeta)|$, $\zeta \in \gamma$, we have to show (see [95]) that

$$\int_\gamma \log|G(t)|\, d\mu_t(\cdot,\ \zeta) < +\infty, \tag{3}$$

where $\mu(e, \zeta)$ is the harmonic measure of a set $e \subset \gamma$ at some point $\zeta \in \operatorname{Int}\gamma$ with respect to $\operatorname{Int}\gamma$. Let Δ be an arc on γ; let $0 \leqslant r < 1$, and let $\beta(\Delta, r)$ be the angle through which Δ is seen from r; let $|\Delta|$ be the length of Δ. Since γ is convex and touches the line $\operatorname{Im}\zeta = 1$, we can show(45) that

$$\mu(\Delta,\ r) \leqslant \frac{1}{\pi}\beta(\Delta,\ r) \leqslant \operatorname{const}\frac{1}{1-r}|\Delta|. \tag{4}$$

The inequality on the left constitutes the well-known Lindelöf theorem. Thus conditions (2) and (3) coincide, and in order to apply Lemma 1 we have to verify condition 2) in that lemma. Here it suffices to have

$$\log G(r) = o\left(\frac{1}{1-r}\right), \quad r \to 1-0.$$

But clearly $\lim_{r\to 1-0}\mu(\Delta, r) = 0$ if the closed arc Δ does not contain the point 1. The required relation now follows from the Poisson integral representation

$$\log G(r) = \int_\gamma \log|G(t)|\, d\mu_t(\cdot\ ,\ r),$$

inequality (4), and the Lebesgue limit theorem. ●

3°. *A converse of the fundamental theorem in the case* $\lambda(r) = \exp(1-r)^{-\alpha}$, $\alpha > 1$.

THEOREM 2. *Let* $\lambda_\alpha(r) = \exp(1-r)^{-\alpha}$, $\alpha > 1$. *There exists a function* $f \in A(\lambda_\alpha)$ *which is not weakly invertible in the space*

$$A\left(\lambda_\alpha \exp\frac{1/\varepsilon}{(1-r)^{\frac{\alpha+1}{2}-\varepsilon}}\right)$$

for any $\epsilon > 0$.

The proof of Theorem 2 is based on the following elementary lemma.

LEMMA 2. *Let* $\alpha > 1$ *and*

$$p(\zeta) = \sum_{k=0}^{[\alpha]} a_k \frac{1}{(1-\zeta)^{\alpha-k}}, \quad |\zeta| < 1,$$

where the a_k *are real numbers.*

1) *The function* M,

---

(45) By comparing the sides of some triangles in the plane.

$$M(r)=\max_{|\zeta|=r}\operatorname{Re}p(\zeta),\quad 0\leqslant r<1, \tag{5}$$

can be written in the form

$$M(r)=\sum_{k=0}^{[\alpha]}b_k\frac{1}{(1-r)^{\alpha-k}}+o\left(\frac{1}{(1-r)^{\alpha-[\alpha]}}\right)+O(1),\quad r\to 1-0, \tag{6}$$

where the b_k are real and $b_0>0$ if $a_0\neq 0$. Here the maximum in (5) *is attained on a curve γ which satisfies the conditions in Lemma* 1 *and the condition*

$$\lim_{\substack{|\zeta|\to 1\\ \zeta\in\gamma}}|\arg(1-\zeta)|=\frac{\pi}{\alpha+1}.$$

2) *For arbitrary $b_0,\ldots,b_{[\alpha]}$ with $b_0>0$ there exist coefficients $a_0,\ldots,a_{[\alpha]}$ with $a_0<0$ such that* (6) *is satisfied.*

We will deduce the theorem from Lemma 2. To do this we set $b_0=1$ and $b_k=0$ for $1\leqslant k\leqslant[\alpha]$ and let p be the "polynomial" in $1/(1-\zeta)$ with leading coefficient $a_0<0$ whose existence is guaranteed by Lemma 2.([46]) Suppose that

$$f(\zeta)=\exp\left(p(\zeta)-\frac{\text{const}}{(1-\zeta)^{\alpha-[\alpha]}}\right),\quad |\zeta|<1;$$

here the constant const is positive, and thus

$$|f(\zeta)|\leqslant\exp\left(\frac{1}{(1-|\zeta|)^{\alpha}}+O\left(\frac{1}{(1-|\zeta|)^{\alpha-[\alpha]}}\right)-\frac{\text{const}}{(1-|\zeta|)^{\alpha-[\alpha]}}\right)\leqslant\text{const}\,\lambda_\alpha(|\zeta|).$$

It is also clear that

$$\max_{|\zeta|=r}|f(\zeta)|\geqslant\exp\left(M(r)-\frac{\text{const}}{(1-r)^{\alpha-[\alpha]}}\right)=\exp\left(\frac{1}{(1-r)^{\alpha}}+O\left(\frac{1}{(1-r)^{\alpha-[\alpha]}}\right)\right),$$

where the maximum is attained on a curve γ_f which asymptotically approaches the curve γ in Lemma 2. In particular, for an arbitrary $\delta>0$ the curve γ_f (asymptotically) lies in the angle

$$\Gamma_\delta=\left\{\zeta:|\arg(1-\zeta)|<\frac{\pi}{\alpha+1}+\delta\right\}.$$

Now in order to use Lemma 1 it suffices to find an outer function G defined in the angle([47]) Γ_δ such that for $\zeta\in\gamma_f$ we have

$$\frac{\lambda_\alpha(|\zeta|)\exp\left\{\frac{1}{\varepsilon}(1-|\zeta|)^{\varepsilon-\frac{\alpha+1}{2}}\right\}}{|f(\zeta)|}\leqslant\exp\left\{\text{const}\,(1-|\zeta|)^{\varepsilon-\frac{\alpha+1}{2}}\right\}\leqslant|G(\zeta)|,$$

and as $x\to 1-0$ we have $\log|G(x)|=o(\log 1/f(x))$; that is,

$$\log|G(x)|=o\left(\frac{1}{(1-x)^{\alpha}}\right).$$

([46]) In the simple case $\alpha=2$ all of the constants can be found explicitly: $a_0=-4$, $a_1=3$.

([47]) Such a function is also outer in the smaller domain $\operatorname{Int}\gamma\cap\Gamma_\delta$.

The required outer function is obtained if on the sides of the angle we set

$$\log|G(\zeta)| = \text{const}\,(1-|\zeta|)^{-\frac{\alpha+1}{2}+\varepsilon}.$$

If δ is sufficiently small, this can be done since the condition on the modulus of an outer function in the angle Γ_δ is the relation

$$-\infty < \int\limits_{\partial\Gamma_\delta} \log|G(\zeta)|\cdot(1-|\zeta|)^{\frac{\alpha+1}{2}-1-\delta'}|d\zeta| < +\infty,$$

where

$$\delta' = \left(\frac{2}{\alpha+1}+\delta\right)^{-1} - \frac{\alpha+1}{2}$$

is small together with δ. ●

PROOF OF LEMMA 2. 1) We set $w = (1-\zeta)^{-1}$ and we note that under this change of variable the circle $\{\zeta\colon |\zeta| = r\}$ becomes the circle C_r in the right half-plane which is given by the parametric equation

$$w = \frac{1}{1-r^2} + \frac{r}{1-r^2}e^{it}, \quad 0 \leqslant t < 2\pi,$$

or, with the notation $\rho = 1/(1-r)$, it is given by the equation

$$w = \rho\frac{\rho+(\rho-1)e^{it}}{2\rho-1}, \quad 0 \leqslant t < 2\pi.$$

First we will show that the function

$$M\left(\frac{\rho-1}{\rho}\right) = \max_{w\in C_r} \operatorname{Re} p_1(w), \quad \rho \geqslant 1, \tag{7}$$

is of the form

$$\sum_{k=0}^{[\alpha]} b_k\rho^{\alpha-k} + o(\rho^{[\alpha]-\alpha}) + O(1);$$

here $p_1(w) = \Sigma_0^{[\alpha]} a_k w^{\alpha-k}$, $\operatorname{Re} w \geqslant 0$. Since

$$w = w_\rho = \rho\frac{1+e^{it}}{2}\left[1 + \frac{1}{\rho\left(1-\frac{1}{2\rho}\right)} - \frac{e^{it}}{(1+e^{it})\rho\left(1-\frac{1}{2\rho}\right)}\right],$$

it suffices to show that there exists a function $c_+ = c_+(\rho)$ which depends regularly on $1/\rho$ in a neighborhood of $\rho = \infty$ and is such that $e^{it} = c_+(\rho)$ furnishes the maximum for the expression in (7). The generalized polynomial p_1 can be written in the form

$$p_1(w_\rho) = a_0\rho^\alpha\left(\frac{1+e^{it}}{2}\right)^\alpha [1 + O(\rho^{-1})], \tag{8}$$

where $O(\rho^{-1})$ depends regularly on ρ^{-1} in a neighborhood of $\rho = \infty$, has an order of growth not exceeding ρ^{-1} uniformly on C_r as $\rho = 1/(1-r) \longrightarrow \infty$, and maintains these properties after differentiation with respect to t. The extremal point $e^{it} = c_+(\rho)$ is determined by the equation

$$\frac{\partial}{\partial t}\operatorname{Re}\left(\frac{1+e^{it}}{2}\right)^{\alpha}[1+O(\rho^{-1})]=0,$$

which for $\rho = +\infty$ becomes

$$\operatorname{Re}\frac{\partial}{\partial t}\left(\frac{1+e^{it}}{2}\right)^{\alpha}=\alpha 2^{-\alpha}\operatorname{Im} e^{it}(1+e^{it})^{\alpha-1}=0.$$

This last equation has a simple root e^{it_0} with the least positive argument $t_0 = 2\pi/(\alpha+1)$. By the implicit function theorem it follows that the necessary function(48) c_+, $c_+(\infty) = e^{it_0}$, exists and is regular in a neighborhood of $\rho = \infty$. It is clear that the same extremum is attained on the symmetric curve $c_- = c_-(\rho)$ with $c_-(\infty) = e^{-it_0}$. From geometric considerations it is clear that $|\arg w_\rho| \sim \frac{1}{2}|\arg c_\pm(\rho)|$ as $\rho \longrightarrow +\infty$, and therefore on the curve of maxima γ we have

$$|\arg(1-\zeta)| \sim \pi(\alpha+1)^{-1}$$

as $\zeta \longrightarrow 1$, $\zeta \in \gamma$.

2) From (8) it is clear that the leading coefficient b_0 in (6) runs through the half-line $[0, \infty)$ when a_0 runs through $(-\infty, 0]$. Since a change in a_m only changes the coefficients b_k for which $k \geqslant m$ and since the $b_1, \ldots, b_{[\alpha]}$ depend continuously on $a_1, \ldots, a_{[\alpha]}$, it suffices to show that b_m takes on arbitrarily large and arbitrarily small values for the appropriate choices of a_m. But it is clear(49) that for $e^{it} = c_+(\rho)$ the mth term (of the expansion in powers of $1/\rho$) in the function $O(\rho^{-1})$ in (8) is of the form

$$(A+a_m)\frac{1}{\rho^m}(1+O(\rho^{-1})),$$

where the constant A depends only on $a_0, \ldots, a_{m-1}$. Thus the contribution of this term to the corresponding maximum depends asymptotically linearly on a_m. ●

4°. *A converse of the fundamental theorem for special majorants.* In order to apply Lemma 1 to spaces of the type $A(\lambda)$ we have to construct functions $f \in A(\lambda)$, $f(\zeta) \neq 0$, $|\zeta| < 1$, for which the maximum modulus M_f is close to the natural limit. Here the proximity of order $\log M_f \sim \log \lambda$ is clearly insufficient for proving theorems like Theorem 1 in the previous part, but finding more precise estimates is very difficult. The results to be introduced below tell us that for any majorant λ which satisfies the conditions in §§2.4–2.7 there exists a majorant λ_* such that $\log \lambda_* \sim \log \lambda$ (or only $\log\log \lambda_* \sim \log\log \lambda$), and there exists a function $f \in A(\lambda_*)$, $f(\zeta) \neq 0$, $|\zeta| < 1$, which is not weakly invertible in some space $A(\lambda_* e^{h(\mu_*)})$, where $\mu_* = \log \lambda_*$, and the function h is so large that

$$\varliminf_{t\to\infty}\frac{\log h(t)}{\log t} \geqslant \frac{1}{2};$$

(48) Here and hereafter, without loss of generality, we assume that $a_0 < 0$. In the case $a_0 > 0$ (which is not needed for the proof of the theorem) we should take the first root to be $e^{it_0} = 1$: We note also that positivity of the required maximum follows from the condition $\alpha > 1$.

(49) The asymptotics of $c_+(\rho)$ do not depend on a_m, $m > 0$.

from Remark 2 after Lemma 1 in §2.7 it follows that a similar assertion is not possible for

$$\varliminf_{t\to\infty}\frac{\log h(t)}{\log t}>\frac{1}{2}.$$

First we consider the case of "rapidly increasing" majorants; that is, majorants for which

$$\lim_{r\to 1}\frac{\log\log\lambda(r)}{\log\dfrac{1}{1-r}}=+\infty.$$

It suffices to show that we cannot invert the corresponding function in some space $A(\lambda_* e^{h(\log\lambda_*)})$ with $\varliminf_{t\to\infty}\log h(t)/\log t \geqslant 1/2$, since by the results in §2.7 a function h with power order greater than $1/2$ gives us weak invertibility of all functions in $A(\lambda_*)$ in this case. We will say that a function N defined on $[0, +\infty)$ belongs to the class K if 1) $N\uparrow +\infty$ and the functions N and $N^{-\delta}$ are convex in a neighborhood of $+\infty$ for some δ, $0<\delta<1$, and 2) $N''/N'^{3/2}=o(1)$ in a neighborhood of $+\infty$. Similar conditions were analyzed in detail in §1.5.

THEOREM 3. *Let λ be a function on* $[0, 1)$, *let* $\mu=\log\lambda$, *and let*

$$\log\mu(r)=\psi\left(\log\frac{1}{1-r}\right),\quad 0\leqslant r<1.$$

Suppose that the function N, $N(t)=\psi'(\log t)t^{-1}$, belongs to the class K. Then the following assertions are true:

1) $\log\mu\sim\log\mu_N$, *where*

$$\mu_N(r)=\sqrt{2\pi}\left[N'\left(\frac{1}{1-r}\right)\right]^{1/2}\exp\left(\int_0^{(1-r)^{-1}}N(t)\,dt\right),\quad 0\leqslant r<1.$$

2) *There exist functions λ_* and f such that* a) $f\in A(\lambda_*)$, $f(\zeta)\neq 0$, $|\zeta|<1$; b) $\log\lambda_*\sim\mu_N$; c) *the function f is not weakly invertible in some space $A(\lambda_* e^{h(\log\lambda_*)})$, where*

$$\varliminf_{t\to\infty}\frac{\log h(t)}{\log t}\geqslant\frac{1}{2}. \tag{9}$$

PROOF. We set $H(u)=\int_0^u n(t)\,dt$, where $n=N^{-1}$ is the inverse of N, and

$$F(w)=\int_0^\infty e^{uw-H(u)}du. \tag{10}$$

The function F is regular in the whole complex plane, is bounded in each half-plane $\operatorname{Re} w<C$ and is real on the axis $\operatorname{Im} w=0$. If we use the Laplace method, we can show that

$$\max_{|\zeta|=r}\operatorname{Re}F\left(\frac{1}{1-\zeta}\right)\sim\log\lambda(r),\quad r\to 1-0, \tag{11}$$

and

$$\min_{|\zeta|=r}\operatorname{Re}F\left(\frac{1}{1-\zeta}\right)\sim -\log\lambda(r),\quad r\to 1-0. \tag{12}$$

To do this, it suffices to show that the hypothesis of the following theorem of Egorov [36] is satisfied:

Suppose that $H \in C^{(2)}_{[0,\infty)}$ and

a) $\lim_{x\to\infty} x^2 H''(x) = \lim_{x\to\infty} H'(x) = +\infty$;

b) for some $A > 0$ we have $H''(u) \sim H''(x)$ when $x \longrightarrow \infty$ and $|u - x| \leqslant A(H''(x))^{-1/2}$.

Then the function $x \longrightarrow xt - H(x)$ for large values of t has a unique maximum at $x = c(t)$, and the integral $F(t) = \int_0^\infty e^{xt-H(x)}\,dx$ asymptotically equals the contribution of this point:

$$V_c(t) = \sqrt{2\pi}\,\rho_{c(t)} \exp(-H(c(t)) + tc(t))$$

(here $\rho_{c(t)} = [H''(c(t))]^{-1/2}$ is the radius of influence of the maximum point $c(t)$).

In our case $H'(u) = n(u)$ and $H''(u) = 1/N'(n(u))$, so that the first condition in Egorov's theorem reduces to condition 1) for functions N in the definition of the class K. Since $H''(u) = 1/N'(n(u)) = n'(u)$ is a monotone function, in order to verify condition b) we have to show that

$$\lim_{x\to\infty} \frac{H''(x)}{H''(x \pm A\rho_x)} = 1$$

for an arbitrary $A > 0$. If we let $g(x) = H''(x)$, we obtain

$$\left|\frac{g(x + A\rho_x)}{g(x)} - 1\right| = \left|\frac{g'(\xi)A\rho_x}{g(x)}\right| = \frac{A\,|g'(\xi)|}{g(x)^{3/2}} \leqslant A\frac{|g'(\xi)|}{g(\xi)^{3/2}} = A\frac{|N''(n(\xi))|}{N'(n(\xi))^{3/2}},$$

if H'' is decreasing (that is, N is convex downward). Analogous estimates for

$$\left|\frac{g(x)}{g(x + A\rho_x)} - 1\right|$$

when N is convex upward, together with condition 2) in the definition of the class K, give us the required result. In the same way we consider the quotient $g(x - A\rho_x)/g(x)$.

If we apply Egorov's theorem (with $c(t) = N(t)$ and $(H''(c(t)))^{-1/2} = (N'(t))^{1/2}$), we obtain an asymptotic equation similar to (11):

$$\max_y \operatorname{Re} F(x + iy) = \int_0^\infty e^{ux - H(u)}\,du \sim \sqrt{2\pi}\,(N'(x))^{1/2} \exp(-H(N(x)) + xN(x))$$

$$= \sqrt{2\pi}\,(N'(x))^{1/2} \exp\left(\int_0^x N(s)\,ds\right). \text{ (50)}$$

Since the mapping $\zeta \longrightarrow 1/(1-\zeta)$ takes the circle $\{\zeta\colon |\zeta| = r\}$, $r < 1$, to the circle C_r with diameter $[1/(1+r), 1/(1-r)]$, relation (11) is proved.

---

(50) In fact

$$H(N(x)) = \int_0^{N(x)} n(t)\,dt = \int_0^x sN'(s)\,ds = sN(s)\Big|_0^x - \int_0^x N(s)\,ds = xN(x) - \int_0^x N(s)\,ds.$$

In order to verify (12) we first obtain an asymptotic expression for

$$\min_y \operatorname{Re} F(x+iy),$$

and then we go to the circle C_r. By the definition of the contribution of the maximum point (see Russian p. 20 of [36]) in the integral (10) we have

$$F(x) \sim V_c(x) = \int\limits_{c(x)-\rho_c(x)\tau(x)}^{c(x)+\rho_c(x)\tau(x)} e^{xu-H(u)}\,du,$$

where the function τ, $\tau(x) \uparrow +\infty$, grows arbitrarily slowly.

We denote the lower limit of integration by $a(x)$ and the upper limit by $b(x)$. Then

$$\frac{a(x)}{b(x)} = \frac{N(x) - \tau(x) N'(x)^{1/2}}{N(x) + \tau(x) N'(x)^{1/2}} = 1 - 2\tau(x)\frac{N'(x)^{1/2}}{N(x) + \tau(x) N'(x)^{1/2}} = 1 - o(1),$$

if we choose τ such that $\tau(x)N'(x)^{1/2}/N(x) = o(1)$ [see condition 1) in the definition of the class K]. Consequently

$$1 > \frac{a(x)}{b(x)} = 1 - o(1),$$

and if we set

$$y(x) = \frac{\pi(1-\varepsilon)}{a(x)}, \quad \frac{1-\varepsilon}{1+\varepsilon} = \frac{a(x)}{b(x)}, \tag{13}$$

we obtain

$$a(x) \leqslant u \leqslant b(x) \Rightarrow \pi(1-\varepsilon) \leqslant yu \leqslant \pi(1+\varepsilon),$$

and therefore

$$-F(x) \leqslant \min_y \operatorname{Re} F(x+iy) \leqslant \operatorname{Re} F(x+iy(x)) = \int\limits_0^\infty e^{ux-H(u)} \cos uy(x)\,du$$

$$= \int\limits_{a(x)}^{b(x)} e^{ux-H(u)} \cos uy(x)\,du + o(V_c(x)) \leqslant \cos\pi(1-\varepsilon) \int\limits_{a(x)}^{b(x)} e^{ux-H(u)}\,du + o(V_c(x))$$

$$= V_c(x)\cos\pi(1-\varepsilon) + o(V_c(x)) \sim (-V_c(x)),$$

since $\lim_{x\to\infty} \epsilon(x) = 0$ (see (13)). Thus

$$\min_y \operatorname{Re} F(x+iy) \sim (-V_c(x)) \sim (-F(x)).$$

In order to deduce (12) from this relation, we first note that for all $w \in \operatorname{Int} C_r$ we have

$$\operatorname{Re} F(w) \geqslant -F(\operatorname{Re}(w)) \geqslant -F\left(\frac{1}{1-r}\right) \sim -V_c\left(\frac{1}{1-r}\right).$$

Therefore it suffices to find a point $w_r \in \operatorname{Int} C_r$ such that

$$\operatorname{Re} F(w_r) \sim -V_c\left(\frac{1}{1-r}\right).$$

LEMMA 3. *We have $N'(u) \sim N'(v)$ when $u \to +\infty$ and $0 < u - v < AN(u)^{-1}$, where $A > 0$ is arbitrary.*

LEMMA 4. *If*

$$x_r = \frac{1}{1-r} - AN\left(\frac{1}{1-r}\right)^{-1}, \quad y_r = BN(x)^{-1},$$

then $w_r = x_r + iy_r \in \operatorname{Int} C_r$ for r sufficiently close to 1. Here A and B are arbitrary positive constants.

The proof of Lemma 3 is obtained from the definition of the class K in the same way that we obtained condition b) in Egorov's theorem, and therefore the proof will be omitted.

PROOF OF LEMMA 4. We have to show that

$$\left|\frac{1}{1-r^2} - w_r\right|^2 \leqslant \frac{r^2}{(1-r^2)^2}$$

for large r. We have

$$\left|\frac{1}{1-r^2} - w_r\right|^2 = \left(x_r - \frac{1}{1-r^2}\right)^2 + y_r^2 = \left(\frac{r}{1-r^2} - \frac{A}{N\left(\frac{1}{1-r}\right)}\right)^2 + \frac{B^2}{N^2\left(\frac{1}{1-r} - \ldots\right)}$$

$$= \frac{r^2}{(1-r^2)^2} - \frac{2Ar}{(1-r^2)N\left(\frac{1}{1-r}\right)} + \frac{A^2}{N^2\left(\frac{1}{1-r}\right)} + \frac{B^2}{N^2\left(\frac{1}{1-r} - \ldots\right)}.$$

It remains to show that

$$\frac{2Ar}{(1-r^2)N\left(\frac{1}{1-r}\right)} > \frac{A^2}{N^2\left(\frac{1}{1-r}\right)} + \frac{B^2}{N^2\left(\frac{1}{1-r} - \ldots\right)}$$

as $r \to 1 - 0$, and for this it suffices to show that

$$N\left(\frac{1}{1-r}\right)^{-1} > N\left(\frac{1}{1-r} - \frac{A}{N\left(\frac{1}{1-r}\right)}\right)^{-2} \quad \text{as} \quad r \to 1 - 0.$$

From Lemma 3 we have

$$N\left(\frac{1}{1-r}\right) - N\left(\frac{1}{1-r} - \frac{A}{N\left(\frac{1}{1-r}\right)}\right) = N'(\xi)\frac{A}{N\left(\frac{1}{1-r}\right)} = N'\left(\frac{1}{1-r}\right)\frac{A(1+o(1))}{N\left(\frac{1}{1-r}\right)}$$

and therefore

$$N\left(\frac{1}{1-r} - \frac{A}{N\left(\frac{1}{1-r}\right)}\right) = N\left(\frac{1}{1-r}\right)\left\{1 - \frac{N'\left(\frac{1}{1-r}\right)}{N^2\left(\frac{1}{1-r}\right)}A(1+o(1))\right\}$$

$$= N\left(\frac{1}{1-r}\right)(1+o(1)),$$

by virtue of condition 1) in the definition of the class K. ●

In accord with Lemma 4 we choose $A(r)$ such that the point

$$w_r = x_r + i\pi N(x_r)^{-1}, \quad x_r = \frac{1}{1-r} - A_r N\left(\frac{1}{1-r}\right)^{-1},$$

belongs to Int C_r for $r_0 \leqslant r < 1$ and $\lim_{r \to 1-0} A_r = 0$. Then[51]

$$\operatorname{Re} F(w_r) = -V_c(x_r)(1 + o(1)) = -\sqrt{2\pi N'(x_r)}^{1/2} e^{\int_0^{x_r} N(s)\,ds}(1 + o(1))$$

$$= -\sqrt{2\pi} N'\left(\frac{1}{1-r}\right)^{1/2} \exp\left(\int_0^{(1-r)^{-1}} N(s)\,ds\right)\left[\frac{N'(x_r)}{N'\left(\frac{1}{1-r}\right)}\right]^{1/2}$$

$$\times\left(\exp \int_{x_r}^{(1-r)^{-1}} N(s)\,ds\right)(1 + o(1)) = -V_c\left(\frac{1}{1-r}\right)(1 + o(1)),$$

since $N'(x_r) \sim N'(1/(1-r))$ by virtue of Lemma 3 and

$$1 \leqslant \exp\left(\int_{(1-r)^{-1} - AN\left(\frac{1}{1-r}\right)^{-1}}^{(1-r)^{-1}} N(s)\,ds\right) \leqslant \exp\left(AN\left(\frac{1}{1-r}\right)^{-1} N\left(\frac{1}{1-r}\right)\right) = \exp A_r \to 1$$

as $r \longrightarrow 1 - 0$. Thus the asymptotic equations (11) and (12) are proved. Now we set

$$\lambda_*(r) = \max_{|\zeta| \leqslant r} \left|\exp\left(-F\left(\frac{1}{1-\zeta}\right)\right)\right| = \exp\left(-\min_{|\zeta| \leqslant r} \operatorname{Re} F\left(\frac{1}{1-\zeta}\right)\right),$$

$$f(\zeta) = (1-\zeta)\exp\left(-F\left(\frac{1}{1-\zeta}\right)\right), \quad |\zeta| < 1.$$

Since $N \in K$, we have

$$\log \mu_N \sim \int_0^{(1-r)^{-1}} N(t)\,dt = \psi\left(\frac{1}{1-r}\right) + \text{const} = \log\log \lambda(r) + \text{const},$$

and thus assertion 1) in the theorem is proved.

Turning to assertion 2), we let γ denote a curve which is symmetric with respect to $[0, 1)$, at points of which the maximum is obtained in the definition of λ_*:

$$\lambda_*(r) = \left|\exp\left(-F\left(\frac{1}{1-\zeta_r}\right)\right)\right|, \quad |\zeta_r| = r, \quad \zeta_r \in \gamma.$$

After a conformal transformation into a half-plane $\zeta \longrightarrow w = (1-\zeta)^{-1}$ the curve γ becomes the curve Γ whose points $w_r = w(\zeta_r)$ satisfy the conditions

$$w_r = x_r + iy_r, \quad \frac{1}{1-r} - A_r N\left(\frac{1}{1-r}\right)^{-1} \leqslant x_r \leqslant \frac{1}{1-r}, \quad y_r \sim \frac{\pi}{N(x_r)}.$$

It is easy to show that under these conditions there exists a function α, $\alpha(x) \downarrow 1$ as

---

(51) It is not hard to see that if instead of using (13) we set $y(x) = -\pi N(x)^{-1}$, then $a(x)\pi/N(x) = \pi(1 + o(1))$ and $b(x)\pi/N(x) = \pi(1 + o(1))$, and therefore $\operatorname{Re} F(x + iy(x)) \sim (-V_c(x))$.

$x \to +\infty$, such that $y_r \leqslant \pi\alpha(x_r)N(x_r)^{-1}$ and the function $s = \pi\alpha/N$ satisfies the conditions in Warschawski's theorem on conformal transformations[52] (see Appendix I). Suppose that ω is a conformal mapping of the strip

$$\operatorname{Int}\tilde{\Gamma} \overset{\text{def}}{=} \{x+iy : |y| < s(x)\}$$

onto the strip $\Pi = \{z\colon \operatorname{Im} z < \pi/2\}$; G_Π is an outer function in the strip Π such that $\log|G_\Pi(x+iy)| \geqslant x^{-2}e^x$, $x > 0$. Then $g \overset{\text{def}}{=} G_\Pi \circ \omega$ is an outer function in $\operatorname{Int}\tilde{\Gamma}$, and therefore in $\operatorname{Int}\Gamma$, and

$$\log|g(x+iy)| \geqslant (\operatorname{Re}\omega)^{-2} \exp \operatorname{Re}\omega,$$

$$\operatorname{Re}\omega(x+iy) = \frac{\pi}{2}\int_0^x \frac{dt}{s(t)} + o(1).$$

Consequently

$$\log|g(x+iy)| \geqslant \text{const}\left(\int_0^x N(t)\,dt\right)^{-2} \exp\left(\frac{1}{2}\int_0^x \frac{N(t)}{\alpha(t)}\,dt\right),$$

and by using an estimate for x_r and the monotonicity of N we obtain

$$\log|g(x_r+iy_r)| \geqslant \text{const}\left(\int_0^{(1-r)^{-1}} N\,dt\right)^{-2} \exp\left(\frac{1}{2}\int_0^{(1-r)^{-1}} \frac{N}{\alpha}\,dt\right).$$

We compare the right side of this inequality with the function $\mu_* = \log\lambda_*$:

$$\log\mu_* \geqslant \log\mu_* - \frac{1}{2}\log N' \geqslant \int_0^u N\,dt + \text{const}, \quad \int_0^u N\,dt = \int_0^u \frac{N}{N'}\,dN$$

$$\geqslant \text{const}\int_0^u N N^{-1-\delta}\,dN = \text{const}\, N^{1-\delta} \geqslant \text{const}\,(N')^{\frac{1-\delta}{1+\delta}}.$$

Therefore

$$\log|g(x_r+iy_r)| \geqslant \text{const}\,\frac{1}{(\log\mu_*(r))^2}[\mu(r)]^{\frac{1}{2}+o(1)} N'\left(\frac{1}{1-r}\right)^{-\frac{1}{4}+o(1)}$$

$$\geqslant \text{const}\,(\log\mu_*(r))^{-2-\frac{1}{4}\frac{1+\delta}{1-\delta}} \mu_*(r)^{\frac{1}{2}+o(1)}.$$

Now we set $\Lambda(r) = \lambda_*(r)|G(\zeta_r)|$, where $G(\zeta) = g(1/(1-\zeta))$, $|\zeta| < 1$. Then the functions f and G satisfy conditions 1) and 2) in Lemma 1 (with λ replaced by Λ), and

$$\mathrm{M}(r) \equiv \log\Lambda(r) = \mu_*(r) + \log|g(w_r)| = \mu_* + h \circ \mu_*.$$

Clearly (9) holds for the function h. ●

---

(52) From condition 1) for the function N it follows that $N'/N^{1+\delta}$ is decreasing and $\int^\infty N'^2/N^3\,dx \leqslant \int^\infty N^{-2+\delta}\,dN < +\infty$, and the decrease in N'/N^2 implies that $\int^\infty |(1/N)''| < +\infty$. Thus for the function $s_0 = \pi/N$ the conditions in Warschawski's theorem are clearly satisfied, and therefore we can find the required function α.

In a similar fashion, but with the use of many conformal mappings, we can prove an analogous theorem for majorants λ with finite exponential order

$$\lim_{r\to 1-0} \frac{\log\log\lambda(r)}{\log\frac{1}{1-r}} < +\infty.$$

We state such a theorem.

THEOREM 4. *Suppose that* $\mu = \log\lambda$, $\log\mu(r) = \psi(\log(1/(1-r)))$, $0 \leqslant r < 1$, *and the function* ψ *satisfies the conditions in Theorem* 2 *in* §1.4:

$$1 \leqslant \psi'_\infty = \lim_{t\to\infty} \psi'(t) < +\infty;\quad \int^1 \left(\frac{\log\lambda}{1-r}\right)^{1/2} dr = +\infty;$$

the function a, $a(\sigma) = \psi(\sigma) - \sigma$, $\sigma > 0$, *is convex in a neighborhood of* $+\infty$ *and* (*in the case* $a' < 0$) *is comparable with* σ^α *for some* α *with* $0 < \alpha \leqslant 1/3$. *Let* F *be the function constructed in Theorem* 2 *of* §1.4, $g = (1-z)F$, *and* $\lambda_* = M_F$. *Then* 1) $\log\lambda \sim \log\lambda_*$; 2) $g \in A(\lambda_*)$, $g(\zeta) \neq 0$, $|\zeta| < 1$; 3) *there exists a function* h *on* $(0, +\infty)$ *such that* $\underline{\lim}_{t\to\infty} \log h(t)/\log t \geqslant 1/2$, *but* g *is not weakly invertible in the space* $A(\lambda_* e^{h(\mu')})$.

CHAPTER 3

ANALYSIS AND SYNTHESIS ON THE POINT AND UNITARY SPECTRA

§3.0. Problems and results

1°. The questions in the theory of linear operators which we consider in this chapter involve the general analysis-synthesis problem which we discussed in the introductory Chapter 0. We recall that this problem consists of describing the linear operators T on a space X that satisfy the following conditions:

(A) Each subspace $E \subset X$ which is invariant with respect to T (that is, $TE \subset E$) contains at least one eigenvector ("analysis").

(B) Each part of the operator T (that is, each restriction $T|E$ where $TE \subset E$) has a complete set of eigenvectors (or root vectors) ("synthesis").

The analysis-synthesis problem is studied in this chapter for extreme cases; namely, when the spectrum $\sigma(T)$ contains only one point or one limit point, and when $\sigma(T)$ lies on the unit circle $\Gamma = \partial \mathbf{D} \stackrel{\text{def}}{=} \{\zeta \in \mathbf{C}: |\zeta| = 1\}$ and T is in some sense close to a unitary operator. The analysis problem also gets involved with questions on the multiplicity of the unitary spectrum, and we will analyze such questions in this chapter.

On the whole the results in Chapter 3 can be described by saying that the possibility of spectral analysis (and all the more for synthesis) requires significant analytic "regularity" of the resolvent of the operator T and its powers T^n, $n \geqslant 0$, "smoothness" of the operator, and so forth; without these restrictions we not only lose the analysis-synthesis properties, but operators with a complete set of eigenvectors (or root vectors) are indistinguishable from operators in "general position", which do not have any particular structure. The analysis-synthesis problem in each of these two cases ($\sigma(T) \sim \{0\}$ and $\sigma(T) \subset \Gamma$) generates its own specific questions. Comparing and constrasting these questions might turn out to be useful for "general spectral theory".

2°. *Description of results.* We turn to a brief description of the sections in this chapter; first we consider some terminology and notation which will be used in all of the sections. The word "subspace" will always mean a closed linear subset of the original linear space. By an operator we mean a continuous linear mapping from one normed space to another; here I will be the identity operator in the corresponding space. The symbol $E_\lambda = \operatorname{Ker}(T - \lambda I)$ denotes the subspace of all eigenvectors of T corresponding to the number $\lambda \in \mathbf{C}$; $E_\lambda = \{x \in X: (T - \lambda I)x = 0\}$. Following the usual terminology, we will say that an operator T in the space X is complete if the collection of its eigensubspaces (or root subspaces, $\stackrel{\text{def}}{=} \bigcup_{n \geqslant 1} \operatorname{Ker}(T - \lambda I)^n$) is complete in X. Complete operators are similar to finite-dimensional operators, which have very "good" properties with respect to the analysis-synthesis problem; namely, an arbitrary finite-dimensional operator admits spectral synthesis ("the fundamental theorem of algebra").

In §3.1 we show that completeness of an operator T in an infinite-dimensional (Banach or Hilbert) space, even together with complete continuity, does not guarantee that the operator will not have "very bad parts" $T|E$, $TE \subset E$, and for such operators spectral synthesis is impossible. More precisely, in §3.1 we show that an arbitrary completely continuous operator $T: X \longrightarrow X$ has a complete, completely continuous extension $\widetilde{T}$; that is, $\widetilde{T}$ is an operator in a larger space Y, $X \subset Y$, and $\widetilde{T}X \subset X$ with $\widetilde{T}|X = T$. The proof is based on a uniqueness theorem for entire functions which is proved in the same section. This method is also applied in §3.3.

The impossibility of synthesis for general completely continuous operators forces us to turn to subclasses of this class, choosing those which are most closely connected with other problems in mathematical analysis. For many cases of this type investigations of the analysis-synthesis problem were carried out in the works of Hamburger [134], Wermer [177], Brodskiĭ [10], Markus [62], and others. In §3.2 we consider weighted shift operators $T = S\Lambda$ which are defined on elements of a basis in X by the equations $Te_n = \lambda_n e_{n+1}$, $n \geqslant 0$. In this section, for $X = l^p$, $1 \leqslant p \leqslant \infty$, the set $\{e_n\}_{n \geqslant 0}$ is the natural basis of this space. Actually the analysis-synthesis problems are formulated for the (complete) adjoint operator $T^* = \Lambda^* S^*$, for which the vectors e_n, $n \geqslant 0$, are root vectors. In the case considered in §3.2 we have $\sigma(T) = \{0\}$, and the operator has no other root vectors. The possibility of spectral analysis in the language of the operator T (conjugate to T^*) means that each noncyclic vector (1) $x = \{x_n\}_{n \geqslant 0}$, $X \in l^p$, can be zero only in one of the first coordinates, say, $x_0 = 0$; and the possibility of synthesis means that each T-invariant subspace coincides with one of the "standard" subspaces l^p_n:

$$l^p_n = \{x \in l^p:\ \ x = \{x_k\}_{k \geqslant 0},\ x_k = 0,\ 0 \leqslant k < n\}.$$

Here we note that the possibility of spectral synthesis for T^* is equivalent to the condition that all of its invariant subspaces form a linearly ordered chain. Such operators are said to be unicellular, and their important role in the general theory of linear operators is well known (see [9] and [25]).

In §3.2 we make use of an algebraic point of view; the problem of unicellularity of operators $S\Lambda$ is connected with the study of weighted Banach algebras of the form $l^p(w_n)$, $w_n = \lambda_0 \cdots \lambda_{n-1}$, $w_0 = 1$. The resulting criteria for unicellularity involve a relatively subtle effect of the "regularity" properties of the variation of the moduli $|\lambda_n|$ upon the properties of the operator $S\Lambda$. Here monotonicity of the moduli on each of the subsets P and $\mathfrak{N}\backslash P$, where $\mathfrak{N}$ is the sequence of natural numbers and P is an arithmetic progression, implies unicellularity, and if we assume that the set P is sparser than a progression ($P = \{n_k: k \geqslant 1\}$, $\sup_{k \geqslant 0}(n_{k+1} - n_k) = +\infty$), then this implication does not hold. The results in §3.2 enable us to solve a problem of Šilov

(1) We recall that a cyclic (or reproducing, or weakly invertible) vector of T is defined by the property $\mathfrak{L}(T^n x: n \geqslant 0) = X$, where $\mathfrak{L}(\ldots)$ denotes the closed linear span of the set in parentheses.

on the existence of nonstandard ideals in prime algebras of the type $l^1(w_n)$.

In §3.3 we consider the analysis-synthesis problem for complete operators whose point spectra $\sigma_p(T)$(2) lie on the circle Γ and which are close to isometries in a given sense. Operators such that $\mathfrak{L}(E_\lambda: \lambda \in \Gamma) = X$ are said to be Γ-complete; their closeness to isometric operators is described by the conditions of strong ergodicity

$$\sup_{n \geq 0} \| T^n \|_X < +\infty,$$

C-ergodicity

$$\sup_{\lambda \in \Gamma,\, n \geq 0} \left\| \frac{1}{n+1} \sum_{k=0}^{n} T^k \lambda^{-k} \right\|_X < +\infty,$$

A-ergodicity

$$\sup_{|\lambda|>1} \| R(\lambda, T) \|_X (|\lambda| - 1) < +\infty,$$

where $R(\lambda, T)$ is the resolvent of T, and others. Each of these conditions enables us to apply an ergodic theorem (see [30] and [108]) which guarantees the existence of a "spectral projector" P_λ onto the eigensubspace E_λ which is perpendicular to all other such subspaces: $P_\lambda P_\mu = \mathbf{0}$ for $\lambda, \mu \in \Gamma$, $\lambda \neq \mu$. This enables us to associate to each vector $x \in X$ its formal Fourier series

$$x \sim \sum_{\lambda \in \sigma_p(T)} P_\lambda x. \tag{1}$$

Inevitably the study of Γ-complete operators leads to an analogy with rotation operators (group of rotations on Γ) in spaces of classical mathematical analysis (L^p, l^p, Orlicz spaces, and so forth). This analogy forces us to vary some of the terminology used in other sections. In addition to previous definitions, we will now say that an A-ergodic operator T admits *harmonic* analysis if

$$P_\lambda x \in \mathcal{L}(T^n x : n \geq 0) \stackrel{\text{def}}{=} E(x), \quad \lambda \in \Gamma,$$

for an arbitrary $x \in X$, and admits *harmonic* synthesis if $E(x) = \mathfrak{L}(P_\lambda x: \lambda \in \Gamma)$. It turns out that among the classes of operators listed above, only the class of strongly ergodic operators admits harmonic synthesis, while "analysis" is possible for A-ergodic operators. Operators of the first type also have the following stronger property: to each subgroup of Γ there corresponds a "spectral projector", and these spectral projectors are uniformly bounded. The structure of complete C-ergodic and A-ergodic operators which do not admit spectral synthesis constitutes the most technical part of §3.3. There we show that many operators with a spectrum contained in $\mathbf{D} = \{\zeta: \zeta \in \mathbf{C}, |\zeta| < 1\}$ have C-ergodic Γ-complete extensions.

We note that summability of the series (1) by any linear method leads to harmonic (and therefore spectral) synthesis. In connection with this, in §3.3 we verify analogues of classical theorems of Lebesgue, Dini-Lipschitz, Szász, and Bernšteĭn on convergence (and summability methods) of Fourier series in eigenfunctions of strongly ergodic operators.

---

(2) $\sigma_p(T) \stackrel{\text{def}}{=} \{\lambda: \lambda \in \mathbf{C}, E_\lambda \neq \{\mathbf{0}\}\}$.

In §3.3 we also consider the question of localization of the unitary spectrum; that is, the question of which Γ-complete operators have the property $\overline{\sigma_p(T)} = \sigma(T)$. We show that under "two-sided" conditions on T (imposed on the resolvent $R(\lambda, T)$ for $|\lambda| < 1$ and $|\lambda| > 1$ or on powers T^n for $n \geqslant 0$ and $n \leqslant 0$) a necessary and sufficient condition for localization of the point spectrum (for all operators in the class) is "non-quasi-analyticity", and under "one-sided" conditions the requirement is close to strong ergodicity. In the same section we consider localization of the spectrum for multipliers and we consider the "Wiener-Pitt phenomenon".

This last problem, which we solve in §3.3, consists in determining the possible massiveness of the point spectrum $\sigma_p(T) \cap \Gamma$ in dependence on the growth of the powers $\| T^n \|$, $n \geqslant 0$. At the beginning of §3.3 it is pointed out that $\sigma_p(T) \cap \Gamma$ is no bigger than a countable set for an arbitrary A-ergodic operator T in a separable Banach space. We demonstrate the connection between γ-capacity of the set $\sigma_p(T) \cap \Gamma$ (γ is some convex integrable function) and the growth of the powers $\| T^n \|$. This connection is investigated for the case of Hilbert spaces, but the method of proof is also suitable for Banach spaces which have an orthogonal basis. In the limiting case, when the γ-capacity coincides with the Lebesgue measure l_1 (γ is the Dirac δ-function), this connection consists of the fact that

$$l_1(\sigma_p(T) \cap \Gamma) > 0 \Rightarrow \sum_{n \geqslant 0} \| T^n \|^{-2} < +\infty,$$

and, conversely, for numbers w_n with $1 \leqslant w_{n+k} \leqslant w_n w_k$ and $\Sigma_{n \geqslant 0} w_n^{-2} < \infty$, there exists an operator T in a Hilbert space with the properties that $\|T^n\| \leqslant w_n$ for $n \geqslant 0$, and $\sigma_p(T) \supset \Gamma$.

§3.4 is devoted to the question of existence of cyclic vectors for a class of operators which, roughly speaking, have an empty point spectrum. Here we can only speak of a known analogy with the analysis-synthesis problem. This analogy consists of the fact that there is a relation between a vector being cyclic and the existence of nonzero projections onto each portion of the spectrum of the operator being considered. More precisely, in §3.4 we consider two-sided shift operators S in some spaces X of generalized functions on Γ. The problem is to calculate the multiplicity of the spectrum of $T = S \times S$ in $X \times X$; that is, to find the numbers

$$m_1(T) \overset{\text{def}}{=} \inf \{\text{card}\, K \colon \mathscr{L}(T^n K \colon n \geqslant 0) = X \times X\},$$

$$m_2(T) \overset{\text{def}}{=} \inf \{\text{card}\, K \colon \mathscr{L}(T^n K \colon n = 0, \pm 1, \ldots) = X \times X\},$$

where K is a subset of the direct product $X \times X$ and card K is the cardinality of the set K. It is easy to see that density of the polynomials in X implies that $m_2(S) = 1$ and $m_2(S \times S) \leqslant 2$, and if the operator is not quasi-analytic, then $m_1(S) = 1$ and $m_1(S \times S) \leqslant 2$. The problem reduces to clarifying the degree of "independence" of the components of the direct product $S \times S$ for which $m_i(S \times S) = 1$, $i = 1, 2$. If we omit the details, we can state the solution as follows: if the space X has a natural realization as a space of measurable functions on Γ, then $m_i(S \times S) = 2$; and if not, then $m_i(S \times S) = 1$. As an example, for the spaces $X = l^p_{\mathbf{Z}}$ the first case holds for $1 \leqslant p \leqslant 2$, and the second case holds for $p > 2$. From this criterion we deduce the simplicity of the spectrum of a countably-multiple shift in spaces $X(E)$ satisfying the condition $m_i^X(S \times S) = 1$, and we make applications to l^p-spaces, Orlicz and Lorentz spaces, and so forth.

3°. *Notation and terminology.* We preserve the basic notation in Chapter 2. Use of the letter λ as a spectral parameter is new. We present a list of frequently encountered symbols:

$R(\zeta, T)$ is the resolvent of the operator T; that is, the inverse of $(\zeta I - T)$ for $\zeta \in \mathbf{C}$.

$\sigma(T)$ and $\sigma_p(T)$ are the spectrum and point spectrum respectively of T.

$\mathfrak{S}_p$ is the Schatten-von Neumann class (see [24]); here $\mathfrak{S}_1$ is the class of all nuclear operators and $\mathfrak{S}_\infty$ is the class of all completely continuous operators.

$\Gamma \stackrel{\text{def}}{=} \partial \mathbf{D} = \{\zeta: \zeta \in \mathbf{C}, |\zeta| = 1\}$.

$\mathbf{Z}$ is the set (group) of all integers.

S is the common symbol for a shift operator in various classes of (formal) Fourier series ($S(\Sigma_n \hat{f}(n) z^n) = \Sigma_n \hat{f}(n) z^{n+1}$) and in sequence spaces ($S\{x_n\} = \{y_n\}$, where $y_n = x_{n-1}$).

l^p is the class of all complex sequences on the set $\mathfrak{N}$ of natural numbers (or on $\mathfrak{N} \cup \{0\}$) which are pth power summable; $l^p_{\mathbf{Z}}$ is the same class for the set $\mathbf{Z}$.

$E_\lambda = \operatorname{Ker}(T - \lambda I)$ is the eigensubspace corresponding to a number $\lambda \in \sigma_p(T)$.

$E_X(f) = E_X^T(f) = \mathfrak{L}(T^n f\colon n \geqslant 0)$ has the same meaning as in the previous chapter.

We recall, finally, that orthogonality of vectors x and y (in a Banach space) means that $\|x + \lambda y\| \geqslant \|x\|$ and $\|y + \lambda x\| \geqslant \|y\|$ for all $\lambda \in \mathbf{C}$; minimality of a sequence of vectors $\{x_n\}_{n \geqslant 1}$ in a space X means that $\operatorname{dist}(x_n, \mathfrak{L}(x_k; k \neq n)) > 0$, $n = 1, 2, \ldots$; uniform minimality means that $\inf_n \operatorname{dist}(x_n/\|x_n\|, \mathfrak{L}(x_k\colon k \neq n)) > 0$.

§3.1. Complete extensions of completely continuous operators

1°. Let T be a completely continuous linear operator acting in a Banach space X. In this section we will say that T is a complete operator if its root subspaces which correspond to nonzero points in the spectrum are complete in X. In the recent literature there have been many theorems which guarantee completeness of an operator $T \in \mathfrak{S}_\infty$ (see [24]). But the structure of complete operators has not been studied in sufficient detail. This remark applies primarily to the conditions under which the property of completeness is inherited when we pass from an operator T to its restriction to certain invariant subspaces. Fundamental results in this area were achieved by Wermer in [177] and his proteges in [122], [163] (the case of normal operators in a Hilbert space) and by Hamburger in [134]. The latter work is important since it contained the first examples which showed that complete operators in $\mathfrak{S}_\infty$ have, in general, nonzero Volterra parts. We recall that by a part of an operator T we mean an operator T_0 of the form $T_0 = T/E$ (the restriction of T to an invariant subspace E).

It was not clear how general the situation was which Hamburger had discovered.

In this section we show that an arbitrary Volterra operator can be a part of a completely continuous operator.(3)

In the case of a Hilbert space X, it follows that an arbitrary operator in $\mathfrak{S}_\infty$ is a part of a complete operator. In order to see this, it suffices to extend $(I-P_0)T(I-P_0)$ to a complete Volterra operator, where P_0 is the orthoprojector onto the closed linear span of the root vectors of T.

Another important problem concerning the structure of complete operators consists of describing the structure of the invariant subspaces of these operators. A natural conjecture is that each T-invariant subspace E can be recovered from the root vectors of T which are contained in E. If this is indeed the case, then we say that for the operator T spectral synthesis is possible.

It is known (see [177]) that spectral synthesis is not always possible even for normal operators in a Hilbert space. From the basic results in this section it follows that for general complete, completely continuous operators not only is spectral synthesis not possible, but also the structure of invariant subspaces can be arbitrarily complicated (that is, as complicated as the analogous structure for an arbitrary Volterra operator).(4) At the same time, for operators in $\mathfrak{S}_\infty$ distinguished by a certain "analytic" (resolvent) completeness criterion, spectral synthesis is usually possible (see [62]).

The extension method which we describe in this section will be used in §3.3 for constructing examples of operators having resolvents with a preassigned growth which do not admit synthesis and do not have the property of localization of the spectrum.

We will use the following terminology and notation.

Let X be a separable Banach space and let T be a Volterra operator in X (that is, $T \in \mathfrak{S}_\infty$ and the spectrum $\sigma(T)$ consists of the point zero). By an extension of T we mean an arbitrary operator $\widetilde{T}$ acting in a space $\widetilde{X}$ such that $\widetilde{X} \supset X$, $\widetilde{T}X \subset X$ and $\widetilde{T}|X = T$.

If $\lambda \in \sigma(A)$, where $\lambda \neq 0$ and $A \in \mathfrak{S}_\infty$, then the spectral projector P_λ onto the corresponding root subspace is defined by the formula

$$P_\lambda = \frac{1}{2\pi i} \int\limits_{|\zeta-\lambda|=\epsilon} R(\zeta, A)\, d\zeta,$$

where $\epsilon > 0$ is sufficiently small and $R(\zeta, A)$ is the resolvent of A.

Let X_1 and X_2 be Banach spaces. We let $X_1 \oplus X_2$ denote the direct sum of X_1 and X_2 equipped with any of the usual norms. The spaces $\{0\} \oplus X_2$ and $X_1 \oplus \{0\}$ will be identified with X_2 and X_1 respectively.

2°. *An extension theorem.*

THEOREM. *Let X be a separable Banach space and let T be a Volterra operator in X. There exist a space Y and an operator A acting in Y such that*

1) *$Y = H \oplus X$, where H is a Hilbert space;*
2) *$AX \subset X$ and $A|X = T$;*

---

(3) The treatment follows the author's work [80].

(4) We note that a normal operator for which spectral synthesis is not possible is automatically not completely continuous.

3) *A is a complete operator, and $A \in \mathfrak{S}_\infty$;*

4) $\bigcap_{0\neq\lambda\in\sigma(A)}(I - P_\lambda)Y = X$, *where P_λ is the spectral projector for A.*

The proof of this theorem depends on an elementary lemma concerning entire functions. This lemma has the flavor of a uniqueness theorem for a certain class of functions and is connected with the "simultaneous approximation" problem (see §3.4).

LEMMA. *Suppose that λ is a strictly increasing function defined on $[0, \infty)$, $\lim_{r\to\infty}\lambda(r) = \infty$, and $A_\infty(\lambda, \mathbf{C}) = \{f\colon f$ is an entire analytic function such that $f(0) = 0$ and $|f(\zeta)| \leqslant C_f\lambda(|\zeta|)$ for all complex numbers ζ and some constant $C_f\}$. Then there exist sequences $\{\gamma_n\}_1^\infty$ and $\{\zeta_n\}_1^\infty$ of complex numbers such that $\sum_n|\gamma_n|^2 < \infty$ and $\lim_n|\zeta_n| = +\infty$, and*

$$f \in A_\infty(\lambda, \mathbf{C}) \quad \textit{and} \quad \sum_n |\gamma_n|^2 |f(\zeta_n)|^2 < \infty \Rightarrow f \equiv 0.$$

The proof of this lemma will be given below; first we use the lemma to prove the theorem.

PROOF OF THE THEOREM. Suppose that $Y = H \oplus X$, where H is a Hilbert space and $\dim H = \infty$. For simplicity (5) we assume that T is a cyclic operator; that is, that there exists a vector $g \in X$ for which the system $\{T^n g\}_0^\infty$ is complete in X. Suppose that $\{e_n\}_1^\infty$ is an orthonormal basis in H. The operator A with the required properties will be sought in the following form:

$$\begin{array}{ccc} H & \oplus & X \\ \Lambda\downarrow & \overset{\Gamma}{\searrow} & \downarrow T \\ H & \oplus & X \end{array}$$

that is, $A|X = T$, $P_H A|H = \Lambda$ and $P_X A|H = \Gamma$, where P_H and P_X are the natural projectors from Y onto H and X.

Here the operator Λ is determined by the equations $\Lambda e_n = \lambda_n e_n$, $n = 1, 2, \ldots,$

$$\lim_n \lambda_n = 0, \tag{1}$$

and Γ will be defined later. Completeness of the operator A and properties 2) and 4) (in the theorem) are implied by the following conditions:

1. $A\varphi_n = \lambda_n\varphi_n$, $n = 1, 2, \ldots$.
2. $\varphi_n = e_n + f_n$, where $f_n \in X$, $n = 1, 2, \ldots$.
3. The system $\{\varphi_n\}_1^\infty$ is complete in Y.

The first of these conditions gives us the equations

$$\lambda_n e_n + \Gamma e_n + T f_n = \lambda_n e_n + \lambda_n f_n, \quad n \geqslant 1;$$

that is, $\Gamma e_n = (\lambda_n I - T)f_n$, $n \geqslant 1$. Now we set $f_n = \gamma_n(\lambda_n I - T)^{-1}g$, $n \geqslant 1$, where g is a cyclic vector of T and $\{\gamma_n\}_1^\infty$ is a sequence of complex numbers. Then $\Gamma e_n = \gamma_n g$, $n \geqslant 1$. Consequently the condition for continuity (coinciding with the condition for complete continuity since Γ is a one-dimensional operator) of Γ can be written in the form

---

(5) Later we will indicate the changes which are necessary in order to prove the theorem in the general case.

$$\sum_n |\gamma_n|^2 < \infty. \tag{2}$$

We write out the condition for completeness of the system $\{\varphi_n\}_1^\infty$ in more detail. The orthogonality condition $x \perp \varphi_n$, $n = 1, 2, \ldots, x \in Y^*$, implies that $x = \mathbf{0}$ if and only if

$$\sum_n |(h, f_n)|^2 = \infty \tag{3}$$

for $h \in X^*$, $h \neq \mathbf{0}$. In fact, if $x = x_0 + h$, $x_0 \in H$, $h \in X^*$, then the orthogonality condition $x \perp \varphi_n$, $n \geqslant 1$, can be written as the equation $(x_0, e_n) = -(h, f_n)$. The rest follows from the Riesz-Fischer theorem.

Thus, in order to establish the existence of the operator A, we have to construct sequences $\{\lambda_n\}_1^\infty$, $\{\gamma_n\}_1^\infty$ and $\{f_n\}_1^\infty$ satisfying conditions (1)–(3). By virtue of the definition of the sequence $\{f_n\}_1^\infty$, condition (3) can be written in the form

$$\sum_n |\gamma_n|^2 |(h, (\lambda_n I - T)^{-1} g)|^2 < \infty \Rightarrow h = \mathbf{0} \quad (h \in X^*). \tag{4}$$

We set $R(\zeta) = (\zeta^{-1} I - T)^{-1}$, where ζ is a complex number, and we let $\lambda(r) = \max_{|\zeta|=r} \|R(\xi)\|$. Then each function F_h, $F_h(\zeta) = (R(\zeta)g, h)$, belongs to the space $A_\infty(\lambda, \mathbf{C})$ which was introduced in the lemma, and existence of the required sequences $\{\lambda_n\}_1^\infty$ and $\{\gamma_n\}_1^\infty$ follows from the lemma if we note that the equations $F_h = \mathbf{0}$ and $h = \mathbf{0}$ are equivalent (cyclicity of g).

Now we make a few remarks about the construction of A in the general case. In X we choose a complete system $\{g_j\}_{j \in J}$, where J is some (no larger than countable) index set,(6) and we partition the sequence of natural numbers $\mathfrak{N}$ into J disjoint infinite subsets N_j, $j \in J$. On $\mathfrak{L}(e_k\colon k \in N_j)$,(7) using the basis $\{e_k\}_{k \in N_j}$ and the vector g_j in X, we construct, as before, operators Λ_j and Γ_j (with sequences $\{\lambda_n^{(j)}\}_{n=1}^\infty$ and $\{\gamma_n^{(j)}\}_{n=1}^\infty$) such that conditions (1) and (2) are satisfied and the left side of the implication (4) implies that $F_h(\zeta) = (h, R(\zeta)g_j) \equiv 0$; that is,

$$h \perp T^n g_j, \quad n \geqslant 0. \tag{5}$$

Without loss of generality we can also assume that

$$\sup_n |\lambda_n^{(j)}| < \frac{1}{j}, \quad \sum_{n=1}^\infty |\gamma_n^{(j)}|^2 \leqslant 2^{-j}, \quad j \in J.$$

If we set $\Lambda = \Sigma_j \oplus \Lambda_j$ and $\Gamma = \Sigma_j \oplus \Gamma_j$, then Λ and Γ are completely continuous operators, and the condition for completeness of the system $\{\varphi_n\}_1^\infty$ (obtained by uniting the systems $\{\varphi_k^{(j)}\}_{k=1}^\infty$, $j \in J$) can be written in the form

$$\sum_{j \in J} \sum_{n=1}^\infty |\gamma_n^{(j)}|^2 |(h, (\lambda_n^{(j)} I - T)^{-1} g_j)|^2 < \infty \Rightarrow h = \mathbf{0}.$$

Relation (5) and completeness of the system $\{g_j\}_{j \in J}$ imply that the above condition is indeed satisfied. ●

(6) We will assume that J is either the sequence of natural numbers or some finite segment of this sequence.

(7) Recall that $\mathfrak{L}(e_k\colon k \in N_j)$ is the closed linear span of the vectors e_k, $k \in N_j$.

3°. *Proof of the lemma.* Let $\{r_n\}_1^\infty$ and $\{\epsilon_n\}_1^\infty$ be monotone sequences of positive numbers with $\lim_n r_n = \infty$ and $\lim_n \epsilon_n = 0$, where the numbers $2\pi r_n$ and $1/\epsilon_n$ are integers for all $n \geq 1$. We partition the circle $\{\zeta: |\zeta| = r_n\}$ in the complex plane by means of the points $\zeta_1^{(n)}, \ldots, \zeta_{p_n}^{(n)}$ into $p_n = 2\pi r_n/\epsilon_n$ equal parts. We list all of the points $\zeta_1^{(n)}, \ldots, \zeta_{p_n}^{(n)}, n = 1, 2, \ldots$, in a sequence $\{\zeta_k\}_1^\infty$ (in the order of growth of the modulus $|\zeta_k|$, and in the order of growth of the argument on a given circle).

The sequence $\{\gamma_k\}_1^\infty$ is defined by the equations $\gamma_k = \epsilon_n$ if $|\zeta_k| = r_n$. First we show that the numbers ϵ_n, $n \geq 1$, can be chosen so small that

$$\sum_{i=1}^{p_n} \varepsilon_n |f(\zeta_i^{(n)})|^2 - \int_{|\zeta|=r_n} |f(\zeta)|^2 |d\zeta| = o(1) \tag{6}$$

as $n \longrightarrow \infty$ for all $f \in A_\infty(\lambda, \mathbf{C})$.

We have the inequality[8]

$$\left|\sum_{i=1}^{p_n} \varepsilon_n |f(\zeta_i^{(n)})|^2 - \int_{|\zeta|=r_n} |f(\zeta)|^2 |d\zeta|\right| \leqslant \sum_{i=1}^{p_n}\left|\int_{\Delta_i} (|f(\zeta_i^{(n)})|^2 - |f(\zeta)|^2) |d\zeta|\right|$$

$$\leqslant \sum_{i=1}^{p_n} \varepsilon_n \max_{\zeta\in\Delta_i} \left||f(\zeta_i^{(n)})|^2 - |f(\zeta)|^2\right| \leqslant 2C_f\lambda(r_n) \sum_{i=1}^{p_n} \varepsilon_n \max_{\zeta\in\Delta_i} |f(\zeta_i^{(n)}) - f(\zeta)|$$

$$\leqslant 2C_f\lambda(r_n) \sum_{i=1}^{p_n} \varepsilon_n C_f\lambda(r_n+1)(r_n+1)\varepsilon_n = 2C_f^2\lambda(r_n)\lambda(r_n+1)(r_n+1)2\pi r_n\varepsilon_n.$$

Consequently condition (6) is satisfied if

$$\lambda^2(r_n+1)(r_n+1)^2\varepsilon_n = o(1), \quad n\to\infty. \tag{7}$$

From (6) we obtain

$$\sum_k |\gamma_k|^2 |f(\zeta_k)|^2 = \sum_{n=1}^\infty \varepsilon_n \sum_{i=1}^{p_n} \varepsilon_n |f(\zeta_i^{(n)})|^2 = \sum_n \varepsilon_n \left(\int_{|\zeta|=r_n} |f(\zeta)|^2 |d\zeta| + o(1)\right)$$

$$= \sum_n \varepsilon_n \left(r_n \int_0^{2\pi} |f(r_n e^{i\theta})|^2 d\theta + o(1)\right) \geqslant K \sum_{n=1}^\infty \varepsilon_n r_n^3,$$

since

$$\int_0^{2\pi} |f(r_n e^{i\theta})|^2 d\theta = r_n^2 \int_0^{2\pi} \left|\frac{f(r_n e^{i\theta})}{r_n e^{i\theta}}\right|^2 d\theta,$$

and the integrals on the right do not decrease as $r_n \longrightarrow \infty$ (assuming that $f \not\equiv 0$).

Thus the lemma will be proved if we can find sequences $\{r_n\}_1^\infty$ and $\{\epsilon_n\}_1^\infty$ such that

a) $\lim_n r_n = \infty$; b) $\sum_n \varepsilon_n r_n < \infty$;

c) $\lambda^2(r_n+1)(r_n+1)^2\varepsilon_n = o(1)$, $n\to\infty$; d) $\sum_n \varepsilon_n r_n^3 = \infty$.

We rewrite condition c) in the form $\epsilon_n = \delta_n/r_n^2\lambda^2(r_n+1)$, where $\lim \delta_n = 0$, and we use the notation $\lambda_0(r) = r\lambda^2(r+1)$, $r \geq 0$, $\psi = \lambda_0^{-1}$ and $x_n = \lambda_0(r_n)$, $n = 1, 2, \ldots$. Then conditions a)–d) can be written as follows:

---

(8) Here Δ_i is the arc of the circle $\{\zeta: |\zeta| = r_n\}$ which joins the points $\zeta_i^{(n)}$ and $\zeta_{i+1}^{(n)}$.

a) $\lim_n x_n = \infty$; b) $\sum_n \frac{\delta_n}{x_n} < \infty$;

c) $\lim_n \delta_n = 0$; d) $\sum_n \frac{\delta_n \psi^2(x_n)}{x_n} = \infty$.

Now it is clear that it suffices to find an increasing sequence $\{x_n\}_1^\infty$ such that $\Sigma_n x_n^{-1} < \infty$ and $\Sigma_n \psi^2(x_n)/x_n = \infty$.

But if we assume that no such sequence exists, then convergence of the series $\Sigma_n x_n^{-1}$ would imply that $\Sigma_n \psi(x_n)/x_n < \infty$, and, by virtue of the monotone decrease of the general term

$$\frac{\psi(x_n)}{x_n} = \frac{r_n}{\lambda_0(r_n)} = \frac{r_n}{r_n \lambda^2(r_n+1)} = \frac{1}{\lambda^2(r_n+1)},$$

that

$$\frac{\psi(x_n)}{x_n} = o\left(\frac{1}{n}\right), \quad n \to \infty;$$

that is, $1/x_n = o(1/n\psi(x_n))$ as $n \to \infty$. But it is easy to produce an example which contradicts the above conclusion (if the n_k are integers such that $n_k \uparrow \infty$ and $\Sigma_1^\infty 1/\psi(n_k) < \infty$, then $\{x_i\}_1^\infty$, where $x_i = n_k \psi(n_k)$, $n_{k-1} < i \leqslant n_k$, is the desired sequence). ●

§3.2. Nonstandard ideals, unicellularity, and algebras connected with the shift operator

1°. In this section we investigate the structure of invariant subspaces of a special but sufficiently broad class of operators; namely, weighted shift operators. Our main concern is the question of whether spectral analysis-synthesis is possible for these operators. Chapter 2 was devoted to shift operators with $\sigma(T) \neq \{0\}$; certain aspects of this problem will be treated in §3.3. Here we consider the case where $\sigma(T) = \{0\}$; that is, essentially the operator of multiplication by z in spaces of formal power series in z. We make use of an "algebraic" approach to the problems of analysis-synthesis and unicellularity (see below); we establish sufficient conditions for unicellularity which are almost necessary. The exposition follows the treatment in the author's work [83].

Let S denote the (right) shift operator in the space(9) l^p of all sequences of complex numbers which are pth power summable, $1 \leqslant p \leqslant \infty$:

$$Se_n = e_{n+1}, \quad n = 0, 1, \ldots;$$

here $e_n = \{\delta_{nk}\}_{k=0}^\infty$ is the natural basis in l^p (where δ_{nk} is the Kronecker delta). Suppose that $\{\lambda_n\}_0^\infty \in l^\infty$, $\Lambda e_n = \lambda_n e_n$, $n \geqslant 0$, and $T = S\Lambda$ is a weighted shift operator in l^p. We will be interested in conditions on the sequence $\{\lambda_n\}_0^\infty$ under which the operator $T = S\Lambda$ turns out to be unicellular. We recall that an operator T is said to be unicellular (see [9] or [25]) if the collection of all of its invariant subspaces is linearly ordered by inclusion; that is, if the condition $TE_i \subset E_i$, where E_i is a (closed)

(9) The same symbol S will also be used to denote a shift operator in other sequence spaces when this does not lead to confusion.

subspace, $i = 1, 2$, implies that $E_1 \subset E_2$ or $E_2 \subset E_1$. In the unicellular case the structure of the invariant subspaces of $T = S\Lambda$ turns out to be very simple; each T-invariant subspace coincides with one of the *standard* invariant subspaces l^p_n, $n \geqslant 0$:

$$l^p_n = \{x : x = \{x_i\}_{i=0}^{\infty},\ x_i = 0 \ \text{ for } \ 0 \leqslant i < n\}.$$

We note that an operator T is unicellular if and only if the adjoint operator T^* is. But in the case $\sigma(T) = \{0\}$ the left shift T^* has as root vectors the basis vectors e_n, $n = 0, 1, \ldots$. Therefore unicellularity for T^* is equivalent to the possibility of spectral synthesis. In other words, each proper T^*-invariant subspace is finite-dimensional.

In the study of unicellularity of operators $S\Lambda$ it is convenient to introduce the concept of a basis operator (see [78]). If the necessary condition for unicellularity

$$r(S\Lambda) \overset{\text{def}}{=} \lim_n |\lambda_0 \ldots \lambda_{n-1}|^{1/n} = 0 \tag{1}$$

is satisfied and $S\Lambda$ is a basis operator, then it is unicellular and the system $\{T^n x\}_0^\infty$, $x \in l^p$, is "close" to the natural basis of l^p (precise definitions will be given at the start of subsection 2°). In order to state the fundamental results of this section, we first replace the operator $T = S\Lambda$ by a shift in the weighted space $l^p(w_n)$.

LEMMA 0. *Let* $w_n = \lambda_0, \ldots, \lambda_{n-1}$, $n \geqslant 1$, $w_0 = 1$, $\lambda_k \neq 0$, $k \geqslant 0$, *and*

$$l^p(w_n) = \left\{x:\ x = \{x_i\}_{i=0}^{\infty},\ \sum_0^{\infty} |x_i|^p\, |w_i|^p < \infty\right\}.$$

Then the operator $T = S\Lambda$ *in* l^p *is unitarily equivalent to the shift* S *in* $l^p(w_n)$.

This (obvious) assertion enables us to consider the operator S in $l^p(w_n)$ instead of $S\Lambda$ in l^p. In subsection 2° we establish necessary and sufficient conditions for an operator to be a basis operator; it is necessary that each of the spaces $l^p(w_{n+j})$, $n \geqslant 0$, $j = 0, 1, \ldots$, be a Banach algebra with convolution as the multiplication (if to each element $x = \{x_i\}_0^\infty$, $x \in l^p(w_n)$, we associate the formal power series $x \longrightarrow \hat{x}(Z) = \Sigma_0^\infty x_i Z^i$, then the convolution $x * y$ corresponds to the product $\hat{x}(Z)\hat{y}(Z)$).

Then we introduce various sufficient conditions, expressed explicitly in terms of the weight sequence $\{w_n\}_0^\infty$; they are definitive for $p = 1$ or ∞. In conjunction with the results in subsection 4° (below) these assertions give us a satisfactory answer to the question of whether $S\Lambda$ is a basis operator (and sometimes whether it is unicellular); see also the remarks at the end of 2°.

If $l^p(w_n)$ is a Banach algebra (with respect to convolution) and condition (1) is satisfied, then the unique maximal ideal of this algebra is $l^p_1(w_n) = \{x: x = \{x_i\}_0^\infty, x_0 = 0\}$. For such (prime) algebras our main interest is in the problem of describing all closed ideals, and, in particular, the question [10] of whether each such ideal coincides with one of the standard ideals $l^p_k(w_n)$, $k \geqslant 1$:

$$l^p_k(w_n) = \{x : x = \{x_i\}_{i=0}^{\infty},\ x_i = 0,\ 0 \leqslant i < k\}.$$

---

[10] For $p = 1$ this question was posed by Šilov around 1940. This question and its connection with unicellularity were studied by E. A. Gorin.

We should note that the absence of nonstandard ideals in the algebra $l^p(w_n)$ (equivalent to unicellularity of $S\Lambda$) indicates that a certain approximation problem is solvable. Namely, it is possible to find arbitrarily precise approximations of an arbitrary element in l^p_k, $k \geqslant 0$, in terms of polynomials with respect to the system

$$\{T^n x\}_{n\geqslant 0}, \tag{2}$$

where $x \in l^p_k$, $x = \{x_i\}_0^\infty$ and $x_k \neq 0$. An analogous connection between "nonstandard ideals" and approximation problems can be observed in other (related) algebras. For example, Domar (see [126]) showed that each ideal is "standard" in (two-sided) algebras of the type $l^1_{\mathbf{Z}}(\rho_n)$ under certain conditions (like smoothness and convexity) on the weight $\{\rho_n\}_{n\in\mathbf{Z}}$, and from this he deduced a criterion for solvability of a problem of Bernšteĭn on weighted approximation by means of polynomials (see also [71]). On the other hand, Koosis (see [150]), considering Domar's problem, constructed examples of algebras (with "irregular" weights) in which there are nonstandard ideals. We note finally that a "one-sided" analytic analogue of Domar's problem (that is, the question of existence of prime ideals in the algebras $l^p(w_n)$ with $r(S\Lambda) > 0$) was considered by Gurariĭ in [27], [28], Korenbljum in [51], [52], and the author in [82]. In such algebras there are continua of "standard" ideals, and a description of them is closely connected with the problem of spectral synthesis. We note that Korenbljum (see [51], [52]) obtained a complete description of the ideals in algebras of smooth functions of the type $l^2_A(n^k)$ and $C^{(n)}_A = \{f\colon f \in A(\mathbf{D}), f^{(n)} \in C_A\}$; these results resemble theorems of Beurling in [117] and Rudin in [164].

The results in subsection 2° give us various sufficiently general criteria for determining whether each ideal in $l^p(w_n)$ coincides with one of the standard ideals. These criteria involve (logarithmic) convexity of the sequence $\{w_n\}_0^\infty$ and are in some sense similar to the requirement of monotonicity of $|\lambda_n| = |w_{n+1}/w_n|$, $n \geqslant 0$.

In subsection 4° we prove a general theorem, already stated in [78], on nonunicellularity of $S\Lambda$. It shows that unicellularity depends on the monotone decrease of $\{|\lambda_n|\}_0^\infty$ on arithmetic progressions and is lost if we pass to sequences with "large" gaps (that is, sufficiently rapid decrease of $|\lambda_n|$, $n \geqslant 0$, for $n = n_k$, $\sup_k(n_{k-1} - n_k) = \infty$, and on the complementary sequence, leads to the existence of a large collection of nonstandard invariant subspaces; for details, see subsection 4°).

By using the constructions in 4° , in the last subsection we construct examples of algebras $l^p(w_n)$ with nonstandard ideals.

Almost all works on invariant subspaces of shift operators (see, for example, [83], [115], [117], [127], [128], [130], [132], [133], [140], [160]) take off from the fundamental investigation of Beurling [117] (isometric shifts) or another limiting case treated by Donoghue [127] which is connected with unicellularity of such operators. In order to illustrate the difficulties which arise in the study of general weighted shift operators, we note that a description of just the cyclic(11) vectors of the operator $T = S\Lambda$, $\lambda_n = (1 + 1/(n + ½))^{½}$, $n \geqslant 0$, in the space l^2 is equivalent to solving the known function-

(11) A vector x is said to be cyclic (reproducing) for an operator T if the system (2) is complete. An operator T is said to be cyclic if it has at least one reproducing vector.

theory problem (see [100]) on density of the polynomials in the weighted space $H^2(|\varphi|^2 dl_2)$ with an analytic weight φ. Some idea of the problems involved here is given by the examples and theorems in §2.1.

2°. *Criteria for an operator to be a basis operator.* As we already mentioned, an operator $T = S\Lambda$ is unicellular in l^p if and only if the system (2) is complete in l^p_n, where $n = \min\{i\colon x_i \neq 0\}$ and $x = \{x_i\}_0^\infty$ is an arbitrary nonzero element of l^p. The condition for being a basis operator (technically more convenient) imposes more restrictions on T.

DEFINITION (see [76] and [78]). An operator T acting in l^p is said to be a *basis operator* if it is cyclic and for each $x \in l^p$, $x \neq \mathbf{0}$, there exist a (linear) isomorphism V of l^p onto itself, an integer $i > 0$, and a sequence of complex numbers $\{t_n\}_0^\infty$, $t_n = t_n(x)$, such that $Ve_{n+i} = t_n T^n x$, $n \geqslant 0$ (recall that $\{e_n\}_0^\infty$ is the natural basis in l^p).

REMARK. Here and hereafter, $1 \leqslant p \leqslant \infty$. In the case $p = \infty$, unicellularity, the basis property, and other such concepts which depend on the type of convergence in the underlying space will always be considered for the weak topology $\sigma(l^\infty, l^1)$, which is generated on l^∞ by the functionals in l^1. For $p = \infty$ we will often consider the restrictions of operators to the (separable) subspace

$$c_0 = \{x\colon\ x \in l^\infty;\ \ x = \{x_i\}_0^\infty;\ \ \lim_i x_i = 0\}.$$

The analogous meaning is given to the symbol $c_0(w_n)$.

THEOREM 1. *Let $T = S\Lambda$ be a weighted shift operator in l^p; let $\lambda_n \neq 0$, $n \geqslant 0$, and let $r(T) = 0$. The following assertions are equivalent*:

1) *T is a basis operator.*

2) *$l^p(w_{n+j})_{n\geqslant 0}$ is a Banach algebra for each $j = 0, 1, \ldots$.*

For the proof we need an assertion which was established in [78].

LEMMA 1. 1) *If $T = S\Lambda$ is unicellular in l^p, then* a) $\lambda_n \neq 0$, $n = 0, 1, \ldots$, *and* b) $r(T) = 0$.

2) *Under the hypothesis of Theorem 1 the operator $T = S\Lambda$ is a basis operator in l^p if and only if the operator $V(x)$ (for an arbitrary $x \neq \mathbf{0}$)*

$$V(x)\, e_n \overset{\text{def}}{=} \frac{w_i}{x_i w_n} T^{n-i} x;\ \ n \geqslant i = i(x) \overset{\text{def}}{=} \min\{k\colon x_k \neq 0\},$$

can be extended to an isomorphism of l^p_i onto itself.

3) *If T is a basis operator and $r(T) = 0$, then*

$$\| V(e_i + x) \| \leqslant \text{const} \| x \|_p + 1,\ \ x \in l^p_{i+1}, \tag{3}$$

for each $i = 0, 1, \ldots$ (the constant depends on i).

Concerning condition 1b), we note that if $r(T) > 0$, then the formal series $\hat{x}(\zeta) = \Sigma_0^\infty x_k \zeta^k$, $\{x_k\}_0^\infty \in l^p(w_n)$, converge for $|\zeta| < r(T)$, and T is unitarily equivalent to the operator of multiplication by z in the space $l^p_A(w_n)$. Thus we have invariant subspaces $\{f\colon f(\lambda) = 0, f \in l^p_A(w_n)\}$ for arbitrary λ, $|\lambda| < r(T)$, and this contradicts the unicellularity of T.

PROOF OF THEOREM 1. 1) $\Rightarrow$ 2). Using Lemma 0 in subsection 1°, we pass from the operator $T = S\Lambda$ to the shift S in $l^p(w_n)$. Then S is a basis operator(12) in $l^p(w_n)$, and for each $x \in l^p_{i+1}(w_n)$ the sequence $\{S^n(e_i + x)\}_{n\geqslant 0}$ is equivalent to the natural basis $\{e_{n+i}\}_{n\geqslant 0}$ of the space $l^p_i(w_n)$ (see the definition of a basis operator and part 2) of Lemma 1). From the estimate (3) for the norm of the isomorphism which connects these equivalent bases we obtain (for each $y \in l^p_i(w^p_n)$)

$$\Big\| \sum_{n\geqslant 0} y_{n+i} S^n (e_i + x) \Big\| \leqslant \| y \| (1 + \mathrm{const} \, \| x \|)$$

and

$$\Big\| \sum_{n\geqslant 0} y_{n+i} S^n (\lambda e_i + x) \Big\| \leqslant \| y \| (|\lambda| + \mathrm{const} \, \| x \|).$$

If in $l^p_i(w_n)$ we introduce a multiplication by means of the rule $e_k \odot e_m = e_{k+m-i}$, $k, m \geqslant i$, the preceding inequality can be written in the form $\|y \odot t\| \leqslant \mathrm{const}\, \|y\| \cdot \|t\|$ for arbitrary y and t in $l^p_i(w_n)$. Thus the space $l^p(w_{n+i})_{n\geqslant 0}$ becomes a Banach algebra (with respect to the usual multiplication of formal power series).

2) $\Rightarrow$ 1). We consider the algebra $l^p_i(w_n)$ with the multiplication just introduced: $e_k \odot e_m = e_{k+m-i}$, $k, m \geqslant i$. Since $r(T) = 0$, we have $\underline{\lim}_n |w_{n+i}|^{1/n} = 0$, and therefore $l^p_i(w_n)$ is a prime algebra (if Φ is a multiplicative functional on $l^p_i(w^p_n)$, then $\Phi(e_k) = \zeta^{k-i}$, $k \geqslant i$, where $\zeta = \Phi(e_{i+1})$; and since $\{\Phi(e_k)\}_{k=i}^{\infty} \in l^p(w_{n+i}))^*$, we have $\zeta = 0$, and $\Phi(\Sigma_i^\infty c_k e_k) = c_i$). Now let $x \in l^p_i$, $x_i \neq 0$. We let $Ux = a$ denote the image of an element x in $l^p(w_n)$ under the natural isomorphism $l^p \longrightarrow l^p(w_n)$. Then $a = \{a_k\}_0^\infty \in l^p_i(w_n)$ and $a_i \neq 0$. Consequently a is an invertible element in the algebra $l^p_i(w_n)$, and the mapping

$$b \xrightarrow{\psi} b \odot a, \quad b \in l^p_i(w_n),$$

is an isomorphism. But then $V = U^{-1}\psi U$ is also an isomorphism (of l^p_i onto itself), and

$$V e_n = U^{-1}\psi U e_n = \frac{1}{w_n} U^{-1}\psi e_n = \frac{1}{w_n} U^{-1}(e_n \odot a)$$

$$= \frac{1}{w_n} U^{-1}(S^{n-i}a) = \frac{1}{w_n} U^{-1}S^{n-i}Ux = \frac{1}{w_n} T^{n-i}x,$$

where $n \geqslant i$. Now by definition T is a basis operator. ●

REMARKS. 1. It is not hard to show (for example, by Lemma 1; see also [78]) that if the hypothesis of Theorem 1 is satisfied and $T = S\Lambda$ is a basis operator, then it is unicellular. Ralph Gellar has noted (private communication) that if T is a basis operator, then $r(T) = 0$.

2. It is possible to have $r(T) = 0$ with $\sigma(T) \neq \{0\}$ (here σ is the spectrum). But if T is a basis operator, then $\sigma(T) = \{0\}$ whenever $r(T) = 0$ (here $l^p(w_n)$ is an algebra and $\lambda e_0 + e_1$ is an invertible element for $\lambda \neq 0$).

3. It is not known whether there exist unicellular weighted shift operators which are not basis operators. A variation on this question is the following: does there exist

(12) The concept of a basis operator is transferred from l^p to $l^p(w_n)$ by means of the natural isomorphism U: $l^p \leftrightarrow l^p(w_n)$.

an operator $T = S\Lambda$ which is unicellular in l^p, $1 \leqslant p \leqslant \infty$, under the uniqueness condition $|\lambda_n| \downarrow 0$, $n \longrightarrow \infty$? From [74] or Theorem 1 it follows that for $p = 1$ this is not so.

2°. *Sufficient conditions.* Here we introduce various conditions(13) which guarantee that an operator is a basis operator; these conditions follow from Theorem 1 and are expressed (as explicitly as possible) in terms of the sequence $\{\lambda_n\}_0^\infty$.

We let $\mathbf{R}$ denote the real line, $\mathbf{R} = (-\infty, \infty)$, and we let $\mathbf{R}_+$ denote the positive half-line, $\mathbf{R}_+ = (0, \infty)$; we let $\mathbf{R}^m = \mathbf{R} \times \cdots \times \mathbf{R}$ and $\mathbf{R}_+^m = \mathbf{R}_+ \times \cdots \times \mathbf{R}_+$ (m times). Also, we let $x = (x_0, \ldots, x_m) \in \mathbf{R}_+^{m+1}$ and $\alpha = (\alpha^{(0)}, \ldots, \alpha^{(m)}) \in \mathbf{R}^{m+1}$. As usual, $x^\alpha \overset{\text{def}}{=\!=} x_0^{\alpha^{(0)}} \cdots x_m^{\alpha^{(m)}}$. By a *generalized polynomial* we mean a function P defined on $\mathbf{R}_+^{m+1}$ by the formula

$$P(x) = \sum_{i=0}^{N} a_i x^{p_i}, \quad x \in \mathbf{R}_+^{m+1},$$

where $p_i \in \mathbf{R}^{m+1}$ and a_i is a complex number, $1 \leqslant i \leqslant N = N(P)$.

THEOREM 2. 1) *Let $T = S\Lambda$, $\lambda_n \neq 0, n \geqslant 0$, and $r(T) \geqslant 0$. Then a sufficient (and for $p = 1$ or ∞ a necessary) condition for T to be a basis operator is the relation*

$$\sup_m \left(\sum_{k=0}^{m} \left| \frac{w_{m+j}}{w_{k+j} w_{m-k+j}} \right|^{p'} \right)^{1/p'} < \infty \quad (j = 0, 1, \ldots), \tag{4}$$

where $1/p + 1/p' = 1$.

2) *Each of the following conditions guarantees that $T = S\Lambda$ is a basis operator in the scale $\{l^p\}_{1 \leqslant p \leqslant \infty}$*:(14)

a) $|\lambda_n| \downarrow 0 (\lambda_n \neq 0)$; $\{\lambda_n\}_{n=0}^\infty \in \bigcup_{r<\infty} l^r$ (*or, equivalently,* $\lambda_n = O(n^{-\epsilon})$, $\epsilon > 0$).

b) $|\lambda_n| = |P(x_n)/Q(x_n)|$ *for large n and* $\lim_n \lambda_n = 0$; *here P and Q are generalized polynomials on* $\mathbf{R}_+^{m+1}$ *and* $x_n = (n, \log n, \ldots, \log_m n)$ *for large n, while* $\log_m n = \log_{m-1}(\log n)$ *and* $\log_1 n = \log n$.

3) *If* $|\lambda_n| \downarrow 0$ *and the conditions* $|\widetilde{\lambda}_n| \downarrow 0$ *and* $|\widetilde{\lambda}_n| \leqslant |\lambda_n|$, $n \geqslant 0$, *imply that* $\widetilde{T} = S\widetilde{\Lambda}$ *is a basis operator in* c_0, *then (compare with assertion* 2a))

$$\{\lambda_n\}_{n=0}^\infty \in \bigcup_{r<\infty} l^r. \tag{5}$$

REMARKS. 1. The sufficient condition (4) depends monotonically on p; if it holds for a given p, then it holds for all p_1 such that $1 \leqslant p_1 \leqslant p$.

Recently Kerlin and Lambert [180] showed that condition (4) is also necessary for $S\Lambda$ to be a basis operator.

2. Assertions 2a) and 3) of Theorem 2 show that as a necessary condition for $T = S\Lambda$ to be a basis operator in the scale $\{l^p\}_{1 \leqslant p \leqslant \infty}$, expressed just in terms of the speed of decrease of the (monotone) sequence $\{|\lambda_n|\}_0^\infty$, we have condition (5). At the same time a "regularly" varying sequence $\{|\lambda_n|\}_0^\infty$ can decrease much more slowly (while keeping $T = S\Lambda$ as a basis operator in the scale $\{l^p\}_{1 \leqslant p \leqslant \infty}$); for example, it can decrease like $\{1/\log_m n\}_{n \geqslant N}$ for an arbitrary $m \geqslant 1$ (see assertion 2b) in Theorem 2).

---

(13) Some of these conditions were presented in [76].

(14) That is, in each of the spaces of the scale; for $p = \infty$ we consider the space c_0.

THEOREM 3. 1) *If for some* $n \geqslant 1$ *the operator* T^n $(T = S\Lambda)$ *is completely continuous, then*

a) *if* $|\lambda_{kn+i}| \geqslant |\lambda_{(kn+1)n+i}|$, $i = 0, 1, \ldots, n-1$, $k \geqslant 0$, *then* T *is a basis operator in the space* l^1; *and*

b) *if one of the operators* $T\Lambda_i$, $\Lambda_i = \{\lambda_{kn+i}\}_{k=0}^{\infty}$, $i = 0, 1, \ldots, n-1$, *is a basis operator in the scale* $\{l^p\}_{1 \leqslant p \leqslant \infty}$ *and* $|\lambda_{kn+i}| \geqslant |\lambda_{(k+1)n+i}|$, $i = 0, \ldots, n-1$, $k \geqslant 0$, *then* $T = S\Lambda$ *is a basis operator in the scale* $\{l^p\}_{1 \leqslant p \leqslant \infty}$.

2) *If* $T = S\Lambda$ *is a basis operator in the space* l^p *and* $|\widetilde{\lambda}_n| \leqslant |\lambda_n|$, $\Sigma_n \log |\lambda_n/\widetilde{\lambda}_n| < \infty$, *then* $\widetilde{T} = S\widetilde{\Lambda}$ *is also a basis operator in* l^p.

3) *If* $T_1 = S\Lambda^{(1)}$ *is a basis operator in the scale* $\{l^p\}_{1 \leqslant p \leqslant \infty}$, *and the sequence* $\{|\lambda_n^{(i)}|\}_{n=0}^{\infty}$, $2 \leqslant i \leqslant N$, *is not increasing, then* $T = S\Lambda$, $\lambda_n = \Pi_{i=1}^N \lambda_n^{(i)}$, $n \geqslant 0$, *is also a basis operator in* $\{l^p\}_{1 \leqslant p \leqslant \infty}$.

4) *Suppose* $\{\lambda_n\}_0^{\infty}$ *and* $\{\mu_n\}_0^{\infty}$ *are sequences such that*

a) $|\lambda_{n+1}| \leqslant |\lambda_n|$, $|\mu_{n+1}| \leqslant |\mu_n|$, $|\mu_n| < |\lambda_n|$, $n \geqslant 0$;

b) *both satisfy* (4) *for* $p = \infty$, *and*

c) $|\mu_{n+1}/\lambda_{n+1}| \leqslant |\mu_n/\lambda_n|$ *for large* n.

Then each of the sequences $\{\gamma_n^{(i)}\}_{n=0}^{\infty}$, $i = 1, 2, 3$, $|\gamma_n^{(1)}|^{-1} = |\mu_n|^{-1} + |\lambda_n|^{-1}$, $|\gamma_n^{(2)}|^{-1} = |\mu_n|^{-1} - |\lambda_n|^{-1}$, $|\gamma_n^{(3)}|^{-2} = |\mu_n|^{-2} + |\lambda_n|^{-2}$, *satisfies* (4) *for* $p = \infty$ *(that is, they generate basis operators in the scale* $\{l^p\}_{1 \leqslant p \leqslant \infty}$).

REMARKS. 1. Assertion 1) in Theorem 3 implies that $T = S\Lambda$ with $\lambda_{2n+1} = 1$, $n \geqslant 0$, $|\lambda_{2n}| \downarrow 0$, $\{\lambda_{2n}\}_{n=0}^{\infty} \in \bigcup_{r<\infty} l^r$, is a basis operator in $\{l^p\}_{1 \leqslant p \leqslant \infty}$. Such criteria (of unicellularity), connected with "regularity" of the sequence $\{\lambda_n\}_0^{\infty}$ on arithmetic subprogressions, were considered in detail in [74]; see also the remarks after Theorem 4 in subsection 4°.

2. Under the conditions in assertion 2) of Theorem 3 we can establish similarity of T and $\widetilde{T}$ (similar operators are simultaneously both basis operators or not). More general theorems on similarity of operators of this type were obtained in [26].

3. Assertions 3) and 4) in Theorem 3 enable us to construct new basis operators from old ones, using the rules in 3) and 4); see also assertion 3) in Theorem 2.

4. As concrete examples of basis (and therefore unicellular) operators in l^p, $1 \leqslant p \leqslant \infty$, we can use operators which satisfy condition 1) or 2) in Theorem 2 or 1)–4) in Theorem 3. We pick out two cases (following [78]) which are of interest for the theory of functions:

$$\lambda_n = q^n, \quad n \geqslant 0, \quad |q| < 1$$

(the operator $f(\zeta) \to \zeta f(q\zeta)$, $|\zeta| < 1$, in the Hardy space H^2; this operator was studied by Donoghue in [127], the author in [76], and Kaz′min in [42]); and

$$\lambda_n = (n+1)^{-\alpha}, \quad \alpha > 0, \quad n \geqslant 0,$$

the Weyl operator J_α of fractional integration in H^2; in particular, $(J_1 f)(\zeta) = \int_0^\zeta f(t)\,dt$, $|\zeta| < 1$, $f \in H^2$; this case was studied by Fišman and Sas′ko in [103], the author in [76], and others (see [79] and [83]).

LEMMA 2. *Let j be an integer, $j \geqslant 0$, and let $\{\lambda_n\}_0^\infty$ be a sequence of complex numbers satisfying the following conditions*:

1) $|\lambda_n| \geqslant |\lambda_{n+1}| > 0$, $n \geqslant 0$.

2) *There exist sequences of natural numbers $\{\alpha_n\}_0^\infty$ and $\{\beta_n\}_0^\infty$ such that*

a) $0 < \alpha_n < \beta_n \leqslant \frac{n}{2}$, $n = 1, 2, \ldots$; $\frac{n}{\beta_n} = O(1)$.

b) $1 - \dfrac{|\lambda_{n-\alpha_n+j}|}{|\lambda_{\alpha_n+j-1}|} \geqslant \delta > 0$ *for* $n \geqslant n(j)$;

c) $\left|\dfrac{\lambda_{n-\beta_n+j}}{\lambda_{\beta_n+j-1}}\right|^{\alpha_n+1}\left(1 - \left|\dfrac{\lambda_{n-\beta_n+j}}{\lambda_{\beta_n+j-1}}\right|\right)^{-1} = O(1)$, $n \to \infty$.

Then condition (4) *is satisfied for a given j and for $p = \infty$.*

PROOF. We denote the general term in the sum (4) by $\Delta_{k,m}$ (or $\Delta_{k,m,j}$ if we need to include the dependence on j):

$$\Delta_{k,m} = \left|\frac{w_{m+j}}{w_{k+j}w_{m-k+j}}\right| = \frac{1}{|\lambda_0 \ldots \lambda_{j-1}|}\left|\frac{\lambda_{m-k+j} \ldots \lambda_{m+j-1}}{\lambda_j \ldots \lambda_{k+j-1}}\right|.$$

We must show that $\Sigma_{k=0}^m \Delta_{k,m} = O(1)$ as $m \to \infty$. First we note that $\Delta_{k,m} = \Delta_{m-k,m}$ for each k, $0 \leqslant k \leqslant m$, and $\Delta_{k+1,m} \leqslant \Delta_{k,m}$ for $0 \leqslant k < [m/2]$(15) (monotonicity of the sequence $\{|\lambda_i|\}_0^\infty$). Therefore

$$\sum_{k=0}^{m} \Delta_{k,m} \leqslant \left(\left[\frac{m}{\beta_m}\right] + 1\right)\sum_{k=0}^{\beta_m} \Delta_{k,m} = O\left(\sum_{k=0}^{\beta_m} \Delta_{k,m}\right)$$

by virtue of condition 2a). Also,

$$\sum_{k=0}^{\beta_m} \Delta_{k,m} = \sum_{k=0}^{\alpha_m} \Delta_{k,m} + \sum_{k=\alpha_m+1}^{\beta_m} \Delta_{k,m} \leqslant$$

(by monotonicity of the sequence $\{|\lambda_i|\}_0^\infty$)

$$\leqslant \frac{1}{|\lambda_0 \ldots \lambda_{j-1}|}\sum_{k=0}^{\alpha_m}\left|\frac{\lambda_{m+j-k}}{\lambda_{k+j-1}}\right|^k + \frac{1}{|\lambda_0 \ldots \lambda_{j-1}|}\sum_{k=\alpha_m+1}^{\beta_m}\left|\frac{\lambda_{m+j-k}}{\lambda_{k+j-1}}\right|^k \leqslant$$

(for the same reason)

$$\leqslant \frac{1}{|\lambda_0 \ldots \lambda_{j-1}|}\left(\sum_{k=0}^{\alpha_m}\left|\frac{\lambda_{m+j-\alpha_m}}{\lambda_{\alpha_m+j-1}}\right|^k + \sum_{k=\alpha_m+1}^{\beta_m}\left|\frac{\lambda_{m+j-\beta_m}}{\lambda_{\beta_m+j-1}}\right|^k\right)$$

$$\leqslant \frac{1}{|\lambda_0 \ldots \lambda_{j-1}|}\left\{\left(1 - \left|\frac{\lambda_{m+j-\alpha_m}}{\lambda_{\alpha_m+j-1}}\right|\right)^{-1} + \left|\frac{\lambda_{m+j-\beta_m}}{\lambda_{\beta_m+j-1}}\right|^{\alpha_m+1}\left(1 - \left|\frac{\lambda_{m+j-\beta_m}}{\lambda_{\beta_m+j-1}}\right|\right)^{-1}\right\} = O(1)$$

as $m \to \infty$ by virtue of conditions(16) 2b) and 2c). ●

LEMMA 3. 1) *If $\lambda_n \neq 0$ and $\lambda_n = 1/\log_m n$ for large $n \geqslant 0$, then the sequence $\{\lambda_n\}_0^\infty$ satisfies all of the conditions in Lemma 2 (with $\alpha_n = [\epsilon(\log n)^2]$, $n \geqslant 1$, and $\beta_n = [\beta n]$, $n \geqslant 0$, for arbitrary $\epsilon > 0$ and $0 < \beta < ½$).*

---

(15) $[x]$ denotes the integer part of a number x.

(16) It is clear from the conditions in Lemma 2 that we can always assume that $\lim_m \alpha_m = \infty$ (since if $\inf|\lambda_n| > 0$ and $\sup \alpha_n < \infty$ simultaneously, then condition 2c) would not be satisfied).

2) *If the sequence* $\{\lambda_n\}_0^\infty$ *satisfies all of the conditions in Lemma* 2 *with* $\alpha_n = [t\alpha_n']$ *and* $\{\beta_n\}_0^\infty$ *for large* $t > 0$ *and some* $\{\alpha_n'\}_0^\infty$ *such that* $\alpha_n' = o(\beta_n)$, *then the sequence* $\{|\lambda_n|^\epsilon\}_0^\infty$, $\epsilon > 0$, *satisfies the same conditions.*

PROOF. 1) We have to establish the existence of constants δ, C and $n(j)$ (depending on ϵ, β and m, but not on n) such that

$$\text{a)}\ \left|\frac{\lambda_{n+j-\alpha_n}}{\lambda_{\alpha_n+j-1}}\right| \leqslant 1-\delta,\quad n \geqslant n(j);$$

$$\text{b)}\ \left|\frac{\lambda_{n+j-\beta_n}}{\lambda_{\beta_n+j-1}}\right|^{\alpha_n+1} \leqslant C\left(1-\left|\frac{\lambda_{n+j-\beta_n}}{\lambda_{\beta_n+j-1}}\right|\right),\quad n \geqslant n(j),$$

where $\alpha_n = [\epsilon(\log n)^2]$ and $\beta_n = [\beta_n]$, $\epsilon > 0$, $0 < \beta < ½$.

First we prove a):

$$\frac{\log_m([\varepsilon(\log n)^2]+j-1)}{\log_m(n+j-[\varepsilon(\log n)^2])} \leqslant \frac{\log_m 2\varepsilon(\log n)^2}{\log_m n/2}$$
$$\leqslant \frac{(2\varepsilon+1)\{\log_{m-1}(2\log\log n)\}}{\log_m n/2} \leqslant \frac{2(2\varepsilon+1)\{\log_{m+1} n\}}{\log_m n/2} \leqslant \frac{1}{2}$$

for $n \geqslant n(j, \epsilon, m)$, where the second inequality follows from $\log \epsilon x \leqslant (\epsilon+1)\log x$ for $\epsilon > 0$ and $x > n(\epsilon)$, and $\log \epsilon x \leqslant \epsilon \log x$ for $\epsilon \geqslant 1$.

In order to prove b) we first note that for $0 < \beta < 1$ we have

$$\log_m(\beta n + A_n) - \log_m n = \frac{\log\beta}{\log n\cdot(\log\log n)\ldots(\log_{m-1} n)}(1+o(1)), \tag{6}$$

if $n \to \infty$ and $\sup_n |A_n| < \infty$. In fact, for $m = 1$ relation (6) is obvious $(\log(\beta n + A_n) - \log n = \log\beta + O(1/n))$, and for $m > 1$ it follows from elementary calculations and the use of induction:

$$\log_m(\beta n + A_n) - \log_m n = \log(1+\rho_n) = \rho_n + O(\rho_n^2)$$
$$= \left\{\frac{\log\beta}{\log n\ldots(\log_{m-2} n)} + o\left(\frac{1}{\log n\ldots(\log_{m-2} n)}\right)\right\}\frac{1}{\log_{m-1} n} + O(\rho_n^2)$$
$$= \frac{\log\beta}{\log n\ldots(\log_{m-1} n)}(1+o(1)),\qquad \text{since}\quad \rho_n^2 = o(\log n\ldots(\log_{m-1} n))^{-1}$$

as $n \to \infty$, where $\rho_n = (\log_{m-1} n)^{-1}(\log_{m-1}(\beta_n + A_n) - \log_{m-1} n)$ and $\lim_n \rho_n = 0$ by the induction hypothesis.

Now we rewrite the left side of b) in the form

$$\left|\frac{\lambda_{n+j-\beta_n}}{\lambda_{\beta_n+j-1}}\right|^{[\varepsilon(\log n)^2]+1} = (1+\sigma_n)^{[\varepsilon(\log n)^2]+1},$$

where

$$\sigma_n = \frac{\log_m(\beta n + A_n') - \log_m((1-\beta)n + A_n'')}{\log_m((1-\beta)n + A_n'')}\quad \text{and}\quad \sup_n |A_n'| \leqslant j,\ \sup_n |A_n''| \leqslant j+1.$$

We then have

$$\left|\frac{\lambda_{n+j-\beta_n}}{\lambda_{\beta_n+j-1}}\right|^{[\varepsilon(\log n)^2]+1} = \exp\{([\varepsilon(\log n)^2]+1)\log(1+\sigma_n)\}$$

$$= \exp\{\varepsilon(\log n)^2\sigma_n(1+o(1))\} = \exp\left\{\frac{\varepsilon\log n\cdot\log\frac{\beta}{1-\beta}}{\log\log n\ldots(\log_{m-1}n)(\log_m(1-\beta)n)}(1+o(1))\right\},$$

where the last equality is a consequence of (6). On the other hand,

$$1-\left|\frac{\lambda_{n+j-\beta_n}}{\lambda_{\beta_n+j-1}}\right| = -\sigma_n = \frac{\log\frac{1-\beta}{\beta}(1+o(1))}{\log n\ldots(\log_{m-1}n)(\log_m(1-\beta)n)}$$

as $n \longrightarrow \infty$. If we divide the left side of b) by the right side, we find that

$$\left|\frac{\lambda_{n+j-\beta_n}}{\lambda_{\beta_n+j-1}}\right|^{\alpha_n+1}\left(1-\left|\frac{\lambda_{n+j-\beta_n}}{\lambda_{\beta_n+j-1}}\right|\right)^{-1} = \exp\left\{\frac{-\varepsilon\log n\cdot\log\frac{1-\beta}{\beta}}{\log\log n\ldots(\log_m n)}(1+o(1))\right\}$$

$$\times\frac{\log n\ldots(\log_m n)}{\log\frac{1-\beta}{\beta}} \leqslant \text{const}\exp\left(-\varepsilon\log\frac{1-\beta}{\beta}(\log n)^{1-\delta}\right)\cdot(\log n)^{1+\delta},$$

where $\delta > 0$ is sufficiently small. Since $0 < \beta < ½$, this expression converges to zero as $n \longrightarrow \infty$, and therefore it is less than ½ for all $n > n(j, \epsilon, \beta, m)$. Thus inequalities a) and b) are proved, and therefore assertion 1) of Lemma 3 is proved.

2) If $\epsilon \geqslant 1$, then the required assertion follows immediately from conditions 2b) and 2c) of Lemma 2 on the original sequence $\{\lambda_n\}_0^\infty$ and from the following two elementary facts:

A) The function f, $f(y) = (1-y)/(1-y^\sigma)$, $y \in [0, 1)$, is bounded on $[0, 1)$ if $\sigma > 0$.

B) $$\left|\frac{\lambda_{n+j-\beta_n}}{\lambda_{\beta_n+j-1}}\right|^{\varepsilon(\alpha_n+1)} \leqslant \left|\frac{\lambda_{n+j-\beta_n}}{\lambda_{\beta_n+j-1}}\right|^{\alpha_n+1}$$

(monotonicity of $\{|\lambda_i|\}_0^\infty$).

If $0 < \epsilon < 1$, then condition 2b) of Lemma 2 is verified for $\{|\lambda_n|^\epsilon\}_0^\infty$ with the help of A) once again. In order to verify condition 2c) we can use the inequalities

$$\left|\frac{\lambda_{n-\beta_n+j}}{\lambda_{\beta_n+j-1}}\right|^{\varepsilon(\alpha_n+1)}\left(1-\left|\frac{\lambda_{n-\beta_n+j}}{\lambda_{\beta_n+j-1}}\right|^{\varepsilon}\right)^{-1} \leqslant \left|\frac{\lambda_{n-\beta_n+j}}{\lambda_{\beta_n+j-1}}\right|^{\left[\frac{\varepsilon}{2}\alpha_n'\right]+1}\left(1-\left|\frac{\lambda_{n-\beta_n+j}}{\lambda_{\beta_n+j-1}}\right|^{\varepsilon}\right)^{-1}$$

$$\leqslant \text{const}\left|\frac{\lambda_{n-\beta_n+j}}{\lambda_{\beta_n+j-1}}\right|^{\left[\frac{\varepsilon}{2}\alpha_n'\right]+1}\left(1-\left|\frac{\lambda_{n-\beta_n+j}}{\lambda_{\beta_n+j-1}}\right|\right)^{-1} = O(1),$$

as $n \longrightarrow \infty$ by virtue of condition 2c) of Lemma 2 for the sequence $\{\lambda_n\}_0^\infty$ (the first inequality is valid because we can assume, as in the proof of Lemma 2, that $\lim_n \alpha_n' = \infty$; for the second inequality see assertion A)). ●

PROOF OF THEOREM 3. 1a) Using the notation in the proof of Lemma 2, we show that $\sup_{k,m}\Delta_{k,m} < \infty$ for each $j = 0, 1, \ldots$ (see part 1) of Theorem 2). Since

$\Delta_{k,m} = \Delta_{m-k,m}$, $0 \leqslant k \leqslant m$, we can assume that $k \geqslant m/2$. If $k \leqslant m-n$, then

$$\Delta_{k,m} = \frac{1}{|\lambda_0 \dots \lambda_{j-1}|}\left|\frac{1}{\lambda_0 \dots \lambda_{j-s-1}} \cdot \frac{\lambda_{m-k+j} \dots \lambda_{m+j-s-1}}{\lambda_{j-s} \dots \lambda_{k+j-1}} \lambda_{m+j-s} \dots \lambda_{m+j-1}\right| \leqslant \text{const} = \text{const}\,(n,\ j)$$

(here $0 \leqslant s \leqslant n-1$ and $(m-k-s)$ is divisible by n). If $k > m-n$, then

$$\Delta_{k,m} \leqslant \frac{\text{const}\,(n,\ j)}{|\lambda_j \dots \lambda_{j+s}|} \leqslant \text{const};$$

here, as before, $0 \leqslant s \leqslant n-1$. Consequently $\sup_{k,m} \Delta_{k,m} < \infty$.

1b) Again it suffices to verify (4) (with $p=\infty$) for the sequence $\{\lambda_k\}_0^\infty$. Since for one of the sequences $\{\lambda_{nk+i}\}_{k=0}^\infty$ (for example, with $i=0$) condition (4) is satisfied and the other sequences are monotone, if in (4) we sum with respect to subprogressions with a difference equal to the "period" n, we obtain

$$\sum_{k=0}^{m} \Delta_{k,m,j} = \sum_{r=0}^{n-1} \sum_{\substack{0 \leqslant k \leqslant m \\ k=sn+r}} \Delta_{k,m,j}$$

$$\leqslant \sum_{r=0}^{n-1} C_r \sum_{0 \leqslant s \leqslant \left[\frac{m-r}{n}\right]} \Delta^0_{s, \left[\frac{m-2r}{n}\right] \pm {}^0_1, \left[\frac{j+r}{n}\right]} \leqslant \text{const}\,(j,\ \{\lambda_i\}_0^\infty).$$

(To see the first inequality note that $\Delta^0_{s,\alpha,\beta}$ is a term in the corresponding sum (4) for the sequence $\{\lambda_{kn}\}_{k=0}^\infty$. The middle index α equals

$$\left[\frac{m+j-r-1}{n}\right] - \left[\frac{j+r-1}{n}\right] = \left[\frac{m-2r}{n}\right] \pm {}^0_1,$$

where the choice of $+0$ or -1 depends on the indices m, j, r and n.)

2) We set $a_0 = 1$, $a_n = \Pi_{k=1}^{n-1} \widetilde{\lambda}_k/\lambda_k$, $n \geqslant 1$, and $A\{x_i\}_0^\infty = \{a_i x_i\}_0^\infty$, $\{x_i\}_0^\infty \in l^p$. Then $A(S\Lambda)A^{-1} = S\widetilde{\Lambda}$, and the assertion is proved.

3) By hypothesis $|\lambda_n/\lambda_n^{(1)}| \geqslant |\lambda_{n+1}/\lambda_{n+1}^{(1)}|$, $n \geqslant 0$, and therefore

$$\left|\frac{\lambda_{n+s}^{(1)}}{\lambda_n^{(1)}}\right| \geqslant \left|\frac{\lambda_{n+s}}{\lambda_n}\right|$$

for all $n \geqslant 0$ and $s \geqslant 0$.

Consequently $\Delta_{k,m}(\lambda) \leqslant \Delta_{k,m}(\lambda^{(1)})$ for all k and m, and the relation (4) for the sequence $\{\lambda_n\}_0^\infty$ follows from the same relation for $\{\lambda_n^{(1)}\}_0^\infty$. Since $r(T) = 0$, we are done (see part 1) of Theorem 2).

4) This final assertion in Theorem 3 follows from the previous one if we set $\lambda_n^{(1)} = \mu_n$ and $\lambda_n^{(2)} = \gamma_n^{(2)}/\mu_n$ for the sequence $\{\gamma_n^{(2)}\}_0^\infty$, and $\lambda_n^{(1)} = \lambda_n$ and $\lambda_n^{(2)} = \gamma_n^{(i)}/\lambda_n$ for the sequences $\{\gamma_n^{(i)}\}_0^\infty$, $i = 1, 3$. ●

PROOF OF THEOREM 2. 1) Let $\alpha_n = |w_{n+j}|$, $n \geqslant 0$. We will show that condition (4) for a fixed j is sufficient to make $l^p(\alpha_n)$ a Banach algebra. In fact,

$$\sum_{m\geqslant 0}\alpha_m^p\Big|\sum_{k+i=m}x_ky_i\Big|^p\leqslant\sum_{m\geqslant 0}\Big(\sum_{k+i=m}|x_k|\alpha_k|y_i|\alpha_i\frac{\alpha_m}{\alpha_k\alpha_i}\Big)^p$$

$$\leqslant\sum_{m\geqslant 0}\Big(\sum_{k+i=m}|x_k|^p\alpha_k^p|y_i|^p\alpha_i^p\Big)\Big(\sum_{k+i=m}\Big(\frac{\alpha_m}{\alpha_k\alpha_i}\Big)^{p'}\Big)^{p/p'}$$

$$\leqslant c_j^p\sum_{m\geqslant 0}\sum_{k+i=m}|x_k|^p\alpha_k^p|y_i|^p\alpha_i^p=c_j^p\|x\|_{l^p(\alpha_n)}^p\|y\|_{l^p(\alpha_n)}^p,$$

where c_j is the number in the left side of (4). For $p = 1$ it is easy to prove the converse. Thus, if $l^1(\alpha_n)$ is an algebra, we have $\|e_n * e_m\| \leqslant \text{const}\, \|e_n\| \cdot \|e_m\|$; that is, $\alpha_{n+m} \leqslant \text{const}\, \alpha_n\alpha_m$, $n, m \geqslant 0$. If $p = \infty$, then as a "test" function we choose $x = \{\alpha_n^{-1}\}_0^\infty$ (or a segment of it $(\alpha_0^{-1}, \alpha_1^{-1}, \dots, \alpha_r^{-1}, 0, 0, \dots)$ if we are working in $c_0(\alpha_n)$). Then

$$x^2=\Big\{\sum_{k=0}^{n}\alpha_k^{-1}\alpha_{n-k}^{-1}\Big\}_{n=0}^{\infty},$$

and the inequality $\|x^2\| \leqslant c\,\|x\|^2$ gives us

$$\sup_{n\geqslant 0}\sum_{k=0}^{n}\frac{\alpha_n}{\alpha_k\alpha_{n-k}}<\infty.$$

2) Each of the conditions a) and b) guarantees that $r(T) = 0$, and thus it suffices to verify (4) for $p = \infty$.

a) In this case $\lambda_n = O(n^{-\epsilon})$, $n \to \infty$, $\epsilon > 0$, and $|\lambda_n| \downarrow 0$. We have

$$\sum_{k=0}^{m}\Big|\frac{w_n}{w_kw_{m-k}}\Big|\leqslant 2\Big(2+\sum_{k=1}^{[m/2]}\Big|\frac{w_m}{w_kw_{m-k}}\Big|\Big)\leqslant C_N+2\sum_{k=N}^{[m/2]}\Big|\frac{\lambda_{m-k}\dots\lambda_{m-1}}{\lambda_0\dots\lambda_{k-1}}\Big|$$

$$\leqslant C_N+\frac{2}{|\lambda_0\dots\lambda_{N-1}|}\sum_{k=N}^{[m/2]}\Big|\frac{\lambda_{m-k}\dots\lambda_{m-N-1}}{\lambda_N\dots\lambda_{k-1}}\Big|\cdot|\lambda_{m-N}\dots\lambda_{m-1}|$$

$$\leqslant C_N+C_N'\sum_{k=N}^{[m/2]}|\lambda_{m-N}\dots\lambda_{m-1}|\leqslant C_N+C_N''\Big(\Big[\frac{m}{2}\Big]-N\Big)\Big(\frac{1}{m-N}\Big)^{\epsilon N}.$$

The last expression is bounded as $m \to \infty$ if $\epsilon N > 1$.

b) Let

$$P(x)=\sum_{i=0}^{N}a_ix^{p_i},\quad Q(x)=\sum_{i=0}^{M}b_ix^{q_i}$$

be the generalized polynomials which are mentioned in the hypothesis of the theorem. Without loss of generality we can assume that the sequences $p_0, p_1, \dots, p_N$ and $q_0, q_1, \dots, q_M$ of points in $\mathbf{R}^{m+1}$ are lexicographically increasing (recall that $x \succ y$ in the sense of lexicographical order in $\mathbf{R}^{m+1}$ means the existence of a number s, $0 \leqslant s \leqslant m$, such that $x^i = y^i$, $0 \leqslant i < s$, and $x^s > y^s$; here $x = (x^0, \dots, x^m)$ and $y = (y^0, \dots, y^m)$), and all of the coefficients a_i and b_i are nonzero. Under these conditions we will write $Q \succ P$ if $q_0 \succ p_0$. We define a function x with values in $\mathbf{R}_+^{m+1}$ by setting

$$t \to x(t) = (t, \log t, \ldots, \log_m t)$$

for sufficiently large t ($t \geqslant n(m)$).

First we note that if $Q \succ P$, then $|P(x(t))/Q(x(t))| \downarrow 0$ as $t \longrightarrow +\infty$, $t \geqslant t_0$. The converse is also true, but obvious, because $x^p \prec x^q$ (that is, $p \prec q$) is equivalent to

$$\lim_{t\to+\infty} \frac{(x(t))^p}{(x(t))^q} = 0.$$

'For real polynomials P and Q the desired result follows from membership of $P(x(\cdot))/Q(x(\cdot))$ in the Hardy field $\mathbf{H}$, and in the general case

$$\left|\frac{P}{Q}\right|^2 = \left(\operatorname{Re}\frac{P}{Q}\right)^2 + \left(\operatorname{Im}\frac{P}{Q}\right)^2 = \left(\frac{\operatorname{Re} P\bar{Q}}{|Q|^2}\right)^2 + \left(\frac{\operatorname{Im} P\bar{Q}}{|Q|^2}\right)^2,$$

where $|Q|^2 \succ P\bar{Q}$ and thus $|Q|^2 \succ \operatorname{Re} P\bar{Q}$ and $|Q|^2 \succ \operatorname{Im} P\bar{Q}$.

Also, since $Q \succ P$, we have $q_0 \succ p_0$; that is, $q_0 = (q_0^{(0)}, \ldots, q_0^{(m)})$, $p_0 = (p_0^{(0)}, \ldots, p_0^{(m)})$ and there exists a number s, $0 \leqslant s \leqslant m$, such that $p_0^{(i)} = q_0^{(i)}$, $0 \leqslant i < s$, and $q_0^{(s)} > p_0^{(s)}$. We choose $\epsilon > 0$ so small that $q_0^{(s)} > p_0^{(s)} + \epsilon$. Then $Q \succ \zeta_s^\epsilon P$, where ζ_ϵ is a monomial, $\zeta_\epsilon(x) = x_s$, if $x = (x_0, \ldots, x_m) \in \mathbf{R}_+^{m+1}$. We set $\lambda_n^{(2)} = (\zeta_s^\epsilon P)(x_n)/Q(x_n)$ and $\lambda_n^{(1)} = \zeta_s^{-\epsilon}(x_n) = (\log_s n)^{-\epsilon}$, $n \geqslant n(m)$. From what we have already shown we have $|\lambda_n^{(2)}| \downarrow 0$ as $n \longrightarrow \infty$. To the sequences $\{\lambda_n^{(1)}\}_0^\infty$ and $\{\lambda_n^{(2)}\}_0^\infty$ we apply assertion 3) of Theorem 3 and Lemma 3 (or the already proven part 2a) of Theorem 2 if $s = 0$) and we conclude that $\{\lambda_n\}_0^\infty$ satisfies condition (4) for $p = \infty$.

3) Suppose that $|\lambda_n| \downarrow 0$ as $n \longrightarrow \infty$, and $\{\lambda_n\}_0^\infty \notin \bigcup_{r<\infty} l^r$. We construct a sequence $\{\tilde{\lambda}_n\}_0^\infty$, $|\tilde{\lambda}_n| \downarrow 0$, $|\tilde{\lambda}_n| \leqslant |\lambda_n|$, such that (4) is not satisfied for $p = \infty$. Since the sequence $\{\lambda_n\}_0^\infty$ is decreasing, the condition which we imposed on it means that

$$\sup_{n\geqslant\infty} |\lambda_{2n}|^r n = \infty$$

for each $r > 0$. Therefore we can construct inductively an increasing sequence of natural numbers $\{n_k\}_1^\infty$ such that

$$|\lambda_{2n_{k+1}}|^{2n_k} n_{k+1} > 2k\,|\lambda_0\lambda_1 \ldots \lambda_{2n_k-1}|,\quad n_{k+1} > 4n_k,\quad k = 1, 2, \ldots.$$

We set $\tilde{\lambda}_m = \lambda_{2n_{k+1}}$ if $2n_k \leqslant m < 2n_{k+1}$. Clearly $|\tilde{\lambda}_m| \downarrow 0$ as $m \longrightarrow \infty$, and $|\tilde{\lambda}_m| \leqslant |\lambda_m|$, $m \geqslant 0$. Also,

$$\sum_{s=0}^{2n_{k+1}} \left|\frac{\tilde{\lambda}_{2n_{k+1}-s} \ldots \tilde{\lambda}_{2n_{k+1}-1}}{\tilde{\lambda}_0 \ldots \tilde{\lambda}_{s-1}}\right| \geqslant \sum_{s=2n_k+1}^{n_{k+1}} \left|\frac{\tilde{\lambda}_{2n_{k+1}-s} \ldots \tilde{\lambda}_{2n_{k+1}-2n_k} \ldots \tilde{\lambda}_{2n_{k+1}-1}}{\tilde{\lambda}_0 \ldots \tilde{\lambda}_{2n_k-1} \ldots \tilde{\lambda}_{s-1}}\right|$$

$$= \sum_{s=2n_k+1}^{n_{k+1}} \left|\frac{\tilde{\lambda}_{2n_{k+1}-2n_k} \ldots \tilde{\lambda}_{2n_{k+1}-1}}{\tilde{\lambda}_0 \ldots \tilde{\lambda}_{2n_k-1}}\right| = \frac{|\lambda_{2n_{k+1}}|^{2n_k}}{|\tilde{\lambda}_0 \ldots \tilde{\lambda}_{2n_k-1}|}(n_{k+1} - 2n_k)$$

$$\geqslant \frac{|\lambda_{2n_{k+1}}|^{2n_k}}{|\lambda_0 \ldots \lambda_{2n_k-1}|}(n_{k+1} - 2n_k) > \frac{1}{2}\frac{|\lambda_{2n_{k+1}}|^{2n_k}}{|\lambda_0 \ldots \lambda_{2n_k-1}|} n_{k+1} > k$$

for $k = 1, 2, \ldots$, where the equality follows from $\widetilde{\lambda}_\alpha = \widetilde{\lambda}_{2n_{k+1}-\beta} = \lambda_{2n_{k+1}}$ for $2n_k \leqslant \alpha < 2n_{k+1}$ and $0 < \beta \leqslant 2n_{k+1} - 2n_k$ (and $n_{k+1} < 2n_{k+1} - 2n_k$).

4°. *Criteria for nonunicellularity*; This subsection is very closely connected with the second part of [74], and it contains the proof of the theorem on nonunicellularity which we mentioned in subsection 1°. As we have already noted (see also [74], [76] and [78]), the results of this subsection (together with Theorems 1–3) show that "regularity" (of the monotone decrease type) of the sequence $\{|\lambda_n|\}_0^\infty$ on arithmetic subprogressions is one of the fundamental features which causes $S\Lambda$ to be unicellular (and to be a basis operator). Our method of proof will be, as before (see [74] and [76]), the "method of similar systems" which is based on the following version of the Paley-Wiener theorem.

LEMMA 4. *Suppose that A is a subset of the sequence of natural numbers $\mathfrak{N}$, $\{x_\alpha\}_{\alpha\in A}$ is a collection of vectors in l^p (or c_0, if $p = \infty$), $M = \mathfrak{L}(x_\alpha)_{\alpha \in A}$ is the closed linear span of the elements x_α, $\alpha \in A$, and $l^p(A) = \mathfrak{L}(e_\alpha\colon \alpha \in A)$ ($\{e_n\}_0^\infty$ is the standard basis in l^p). If*

$$\Big\| \sum_{\alpha\in A} c_\alpha (x_\alpha - e_\alpha) \Big\| \leqslant \theta \Big(\sum_\alpha |c_\alpha|^p \Big)^{1/p} \tag{7}$$

for some θ, $0 \leqslant \theta < 1$, and for arbitrary complex numbers c_α, $\alpha \in A$, then there exists a continuous (and linear) operator $\Gamma\colon l^p(A) \longrightarrow l^p(\mathfrak{N}\backslash A)$ such that

$$M = l^p(A) \oplus \Gamma l^p(A) = \{x + \Gamma x\colon\ x \in l^p(A)\}.$$

In particular, the codimensions of M and $l^p(A)$ in l^p coincide, and the system $\{x_\alpha\}_{\alpha\in A}$ is complete in l^p if $A = N$.

REMARK. Condition (7) follows, for example, from the inequality

$$\Big(\sum_{\alpha\in A} \| x_\alpha - e_\alpha \|_p^{p'} \| \Big)^{1/p'} < 1. \tag{8}$$

The proof of Lemma 4 is elementary and essentially repeats the proof of Lemma 5 in [74]; therefore it will be omitted.

THEOREM 4. *Suppose that $\{\lambda_n\}_0^\infty$ is a bounded sequence of complex numbers, $\lambda_n \neq 0$, $n \geqslant 0$, and $P = \{n_k\}_1^\infty$ is an increasing sequence of natural numbers such that*

$$\sup_{k\geqslant 1} (n_{k+1} - n_k) = \infty.$$

Then there exists a sequence $\{\mu_k\}_1^\infty$, $\mu_k > 0$, $\mu_k \downarrow 0$ as $k \longrightarrow \infty$, for which an arbitrary weighted shift operator $\widetilde{T} = S\widetilde{\Lambda}$ with

$$\widetilde{\lambda}_n = \begin{cases} \lambda_n, & n \notin P, \\ \widetilde{\mu}_k, & n = n_k,\ k \geqslant 1 \ \text{and}\ |\widetilde{\mu}_k| \leqslant \mu_k, \end{cases}$$

is not unicellular.

REMARKS. 1. The conditions on the sequence $\{\mu_k\}_1^\infty$ (depending on $\{\lambda_n\}_0^\infty$) which occur in the proof of Theorem 4 can, in specific situations, be reduced "to numbers". The calculations become especially simple if we also assume that $|\lambda_n| \downarrow 0$ and P is a sparse (for example, lacunary) sequence. Thus in the case $\lambda = 2^{-n}$, $P = \{2^k + 1\}_1^\infty$ we can take $\mu_k = 2^{-2^{(k+1)2}}$, $k \geqslant 1$.

2. As a corollary to Theorems 3 and 4 we have the following assertion. Let $P = \{n_k\}_1^\infty$ be a sequence of natural numbers, and let $K_1(P)$ be the class of all operators $S\Lambda$ for which $\{|\lambda_n|\}$ is not increasing on $\mathfrak{N}\backslash P$ and decreases to zero on P. In order for each operator in $K_1(P)$ to be unicellular in l^1 it is sufficient that $n_{k+1} - n_k = \text{const}$, $k \geqslant k_0$, and necessary that $\sup_k(n_{k+1} - n_k) < \infty$.

An analogous assertion is true in the whole scale $\{l^p\}_{1 \leqslant p \leqslant \infty}$, but for the class $K_\infty(P)$ of all operators with the additional requirement (for instance) that $\{\lambda_n\}_{n \in P} \in \bigcup_{r<\infty} l^r$.

3. Finally, we introduce a "naive" explanation of Theorem 4 which provides the basis of the formal proof of this theorem. If an operator $T = S\Lambda$ is not one-to-one; that is, if $\lambda_n = 0$ for some $n \geqslant 0$, then it is not unicellular, since T "locks" the basis vectors $e_0, \ldots, e_n$ into the invariant subspace $\{x: x = \{x_i\}_0^\infty, x_i = 0 \text{ for } i > n\}$, which clearly does not coincide with any of the l_k^p, $k \geqslant 0$. This example is based on linear algebra considerations and is trivial. But from it we can obtain the needed construction for Theorem 4; namely, we choose λ_n very small ("almost zero") for the numbers $n = n_k$, $k \geqslant 1$. Then the operator $T^{n_k - n_{k-1}}$, $T = S\Lambda$, almost "extinguishes" the basis vectors e_s, $n_{k-1} \leqslant s \leqslant n_k$, while at the same time $T^{n_k - n_{k-1}} e_s$ for $s > n_k$ is very large if $n_{k+1} - n_k \gg n_k - n_{k-1}$. Then we quantify this difference in order to establish the lack of completeness of the system $\{T^n x\}_{n \geqslant 0}$ in each of the subspaces l_m^p, $m \geqslant 0$, for all elements $x = \{x_i\}_0^\infty$ which are "sensitive" to the oscillations of the quantities $\|T^n e_s\|$, $n \geqslant 0$, $s \geqslant 0$.

PROOF OF THEOREM 4. For simplicity we will write n_k as $n(k)$, $k \geqslant 1$, in this proof. We consider the function

$$m(\beta) = \sup_{k \leqslant \beta} \{n(k+1) - n(k)\}, \quad \beta \in \mathfrak{N}.$$

Let $\beta_1, \beta_2, \ldots$ be growth points of the function m; that is, points $\beta \in \mathfrak{N}$ such that $m(\beta) > m(\beta - 1)$ and $\beta_0 = 1$. It is clear that for each $i \in \mathfrak{N}, i < \beta_k$, we have $m(\beta_k) > n(i + 1) - n(i)$; also $m(\beta_k) \uparrow \infty$ by hypothesis. Consequently $\mathfrak{N} = \bigcup_1^\infty I_k$, where $I_k = [m(\beta_{k-1}), m(\beta_k))$ and $m(\beta_0) = 0$, and the intervals $\{I_k\}$ do not intersect. We choose an element $x \in l^p$ satisfying the following conditions:[17]

1) $x = \{x_i\}_0^\infty$, $x_i = 0$ if $i \neq n(\beta_k) + 1$, $k \geqslant 1$; $x_{n(\beta_k)+1} \neq 0$, $k \geqslant 1$, and $|x_i| \leqslant 1$ for all $i \in \mathfrak{N}$.

2) $$\left(\sum_{j=n(\beta_k)+2}^{\infty} \left| \frac{x_j \lambda_j \cdots \lambda_{j+s-1}}{x_{n(\beta_k)+1} \lambda_{n(\beta_k)+1} \cdots \lambda_{n(\beta_k)+s}} \right|^p \right)^{1/p} = \delta_{k,s} < \infty$$

for all k and s ($s \in I_k$) where $\sum_{k=1}^\infty \sum_{s \in I_k} \delta_{k,s} \leqslant 1/2$.

---

(17) In order to avoid (inessential) requirements, we will first assume that $|\lambda_k| \leqslant 1$, $k \geqslant 0$, and $1 \leqslant p < \infty$. The case of the space c_0 is treated analogously (changing the l^p-norm to the l^∞-norm). The coordinates of the required element x (for $n(k) + 1$, $k \geqslant 1$) can be found by an inductive construction or by an explicit calculation (see the proof of Theorem 5).

We introduce the notation

$$\sigma_l = \sum_{n(l)\leqslant m(\beta_{k-1})<n(l+1)} \sum_{s\in I_k} \frac{(k-1)^{1/p}}{|x_{n(\beta_k)+1}\lambda_{n(\beta_k)+1}\cdots\lambda_{n(\beta_k)+s}|}, \quad l=1,2,\ldots,$$

and we choose a sequence $\{\mu_k\}_1^\infty$ such that

$$\text{(i)}\qquad \sigma_k\mu_k < 2^{-k-2},\ k\geqslant 1;$$

$$\text{(ii)}\ 0<\mu_{k+1}<\mu_k<|\lambda_k|,\ k\geqslant 1.$$

We consider the operator $\widetilde{T} = S\widetilde{\Lambda}$ which is defined in the hypothesis of the theorem. If we show that the closed linear span $\mathfrak{L}(\widetilde{T}^s x\colon s\geqslant 0)$ of the elements $\{\widetilde{T}^s x\}_{s\geqslant 0}$ does not coincide with $l^p_{n(\beta_1)+1}$, the theorem will be proved.

Let $s\geqslant 0$ be an integer. There exists a unique $k=k(s)\geqslant 1$ such that $s\in I_k$. We set

$$d_s = \left\| \frac{1}{x_{n(\beta_k)+1}\widetilde{\lambda}_{n(\beta_k)+1}\cdots\widetilde{\lambda}_{n(\beta_k)+s}}\widetilde{T}^s x - e_{n(\beta_k)+s+1}\right\|_p$$

for all $s\geqslant 0$ and $k=k(s)$. If we can show that

$$\Big(\sum_{s\geqslant 0} d_s^{p'}\Big)^{1/p'} < 1, \tag{9}$$

then (8) will be satisfied with the role of A being played by the set[18] $A=\{\alpha\in\mathfrak{N}\colon$ there exists $k\geqslant 0$ such that $n(\beta_k)+1+n(\beta_{k-1}+1)-n(\beta_{k-1})\leqslant\alpha\leqslant n(\beta_k+1)\}$. Since the set $\mathfrak{N}\backslash A$ is infinite ($n(\beta_k+1)+1<n(\beta_{k+1})+1+n(\beta_k+1)-n(\beta_k)$ for each $k\geqslant 1$), by virtue of Lemma 4 and the remark which follows it we have $\mathfrak{L}(\widetilde{T}^s x\colon s\geqslant 0) = l^p(A)\oplus\Gamma l^p(A)\neq l^p_{n(\beta_1)+1}$.

We still have to prove (9). We will show that $\Sigma_s d_s<1$ and therefore the operator $\widetilde{T}=S\widetilde{\Lambda}$ is unicellular in all of the spaces l^p, $1\leqslant p\leqslant\infty$. We have

$$d_s = \Big(\sum_{j=0}^{n(\beta_k)+s} |z_j^{(s)}|^p + \sum_{j=n(\beta_k)+s+2}^{\infty} |z_j^{(s)}|^p\Big)^{1/p} \leqslant \Big(\sum_{j=0}^{n(\beta_k)+s}|z_j^{(s)}|^p\Big)^{1/p} + \Big(\sum_{j=n(\beta_k)+s+2}^{\infty}|z_j^{(s)}|^p\Big)^{1/p},$$

where $\{z_j^{(s)}\}_{j=0}^\infty = (x_{n(\beta_k)+1}\widetilde{\lambda}_{n(\beta_k)+1}\cdots\widetilde{\lambda}_{n(\beta_k)+s})^{-1}\widetilde{T}^s x$, $k=k(s)$, and therefore

$$\sum_s d_s \leqslant \sum_s\Big(\sum_{j=0}^{n(\beta_k)+s}|z_j^{(s)}|^p\Big)^{1/p} + \sum_s\Big(\sum_{j=n(\beta_k)+s+2}^{\infty}|z_j^{(s)}|^p\Big)^{1/p}.$$

We estimate the term on the right side of this inequality:

$$\text{(A)}\ \sum_{s=0}^{\infty}\Big(\sum_{j=0}^{n(\beta_k)+s}|z_j^{(s)}|^p\Big)^{1/p} \leqslant \sum_{l\geqslant 1}\ \sum_{\substack{k\\ (n(l)\leqslant m(\beta_{k-1})<n(l+1))}}\ \sum_{s\in I_k}\Big(\sum_{j=0}^{n(\beta_k)+s}|z_j^{(s)}|^p\Big)^{1/p}$$

$$= \sum_{l\geqslant 1}\sum_k\sum_{s\in I_k}\Big(\sum_{j=0}^{n(\beta_k)}\Big|\frac{x_j\widetilde{\lambda}_j\cdots\widetilde{\lambda}_{j+s-1}}{x_{n(\beta_k)+1}\widetilde{\lambda}_{n(\beta_k)+1}\cdots\widetilde{\lambda}_{n(\beta_k)+s}}\Big|^p\Big)^{1/p} \overset{\text{def}}{=} \Sigma_1.$$

---

(18) For $k=0$ we set $n(\beta_{-1}+1)-n(\beta_{-1})=0$ by definition. The fact that A is not empty follows from the inequality $m(\beta_k)>n(\beta_{k-1}+1)-n(\beta_{k-1})$.

Now we note that $s \in I_k$ implies that $s \geqslant m(\beta_{k-1})$, and thus for $m(\beta_{k-1}) \geqslant n(l)$ we have $s \geqslant n(l)$. Consequently for $j = n(\beta_i) + 1$, $i < k$, the numbers $j, j+1, \ldots, j+s-1$ include a number of the form $n(m)$ with $m \geqslant l$ (in fact, $n(\beta_i + 1)$ occurs among these numbers since $n(\beta_i) + 1 \leqslant n(\beta_i + 1) = n(\beta_i) + (n(\beta_i + 1) - n(\beta_i)) \leqslant n(\beta_i) + m(\beta_{k-1}) \leqslant j + s - 1$; if $n(l) < n(\beta_i)$, then $\beta_i + 1 > l$, and otherwise $n(\beta_i) + 1 \leqslant n(l) \leqslant j + s - 1$, so we are done). From condition 1) in the definition of x and condition (ii) in the definition of $\{\mu_k\}_1^\infty$ we obtain

$$|x_j \tilde{\lambda}_j \ldots \lambda_{j+s-1}| \leqslant \mu_l,$$

where $j = n(\beta_i) + 1$, $1 \leqslant i \leqslant k-1$, $s \in I_k$, and $n(l) \leqslant m(\beta_{k-1}) < n(l+1)$. If we extend the estimate of the sum (A), we can write(19)

$$\sum\nolimits_1 \leqslant \sum_{l \geqslant 1} \sum_k \sum_{s \in I_k} \frac{(k-1)^{1/p} \mu_l}{|x_{n(\beta_k)+1} \lambda_{n(\beta_k)+1} \ldots \lambda_{n(\beta_k)+s}|} = \sum_{l \geqslant 1} \sigma_l \mu_l < \frac{1}{4}$$

by virtue of condition (i) for the sequence $\{\mu_k\}_1^\infty$.

(B) $$\sum_{s \geqslant 0} \Bigl(\sum_{j=n(\beta_k)+s+2} |z_j^{(s)}|^p \Bigr)^{1/p}$$

$$= \sum_{k \geqslant 1} \sum_{s \in I_k} \Bigl(\sum_{j=n(\beta_k)+2}^{\infty} \Bigl| \frac{x_j \tilde{\lambda}_j \ldots \tilde{\lambda}_{j+s-1}}{x_{n(\beta_k)+1} \lambda_{n(\beta_k)+1} \ldots \lambda_{n(\beta_k)+s}} \Bigr|^p \Bigr)^{1/p}$$

$$\leqslant \sum_{k \geqslant 1} \sum_{s \in I_k} \Bigl(\sum_{j=n(\beta_k)+2}^{\infty} \Bigl| \frac{x_j \lambda_j \ldots \lambda_{j+s-1}}{x_{n(\beta_k)+1} \lambda_{n(\beta_k)+1} \ldots \lambda_{n(\beta_k)+s}} \Bigr|^p \Bigr)^{1/p} = \sum_{k \geqslant 1} \sum_{s \in I_k} \delta_{k,s} < \frac{1}{2}.$$

If we combine the estimates (A) and (B), we see that

$$\sum_{s \geqslant 0} d_s < \frac{3}{4} < 1.$$

Thus inequality (9) is proved. ●

COROLLARY 1. *If $L = \mathfrak{L}(\tilde{T}^s x\colon s \geqslant 0)$ ($\tilde{T}$ and x are the same as in Theorem 4), then $L = l^p(A) \oplus \Gamma l^p(A)$, Γ being a continuous operator from $l^p(A)$ into $l^p(\mathfrak{N} \setminus A)$. Therefore the operator $\tilde{T}$ is "strongly" nonunicellular: to each point ζ, $|\zeta| = 1$, we can associate a T-invariant subspace L_ζ, where $L_{\zeta_1} \not\subset L_{\zeta_2}$ if $\zeta_1 \neq \zeta_2$.*

PROOF. First we note that $l^p(A) \oplus \Gamma l^p(A) = l^p(A) \oplus \Gamma' l^p(A)$ if and only if $\Gamma = \Gamma'$. If $x = \{x_i\}_0^\infty$ is the element of l^p which was constructed in the proof of Theorem 4, then there exists an n such that $x_n \neq 0$ and $n \notin A$ (see the definition of A in the proof of Theorem 4). We set $x(\zeta) = \zeta x_n e_n + \Sigma_{i \neq n} x_i e_i$, $|\zeta| = 1$. Since the condition on x which were imposed in the proof of Theorem 4 depend only on the moduli $|x_i|$, $i \geqslant 0$, we have $L_\zeta = \mathfrak{L}(\tilde{T}^s x(\zeta)\colon s \geqslant 0) = l^p(A_\zeta) \oplus \Gamma_\zeta l^p(A_\zeta)$, where $A_\zeta = A$ for all ζ, $|\zeta| = 1$. Let j be the smallest integer for which $x_j \neq 0$ ($j = n_1 + 1$ in the notation of Theorem 4). Then $j < n$ (since $j \in A$), and $\Gamma_\zeta e_j = \zeta x_n e_n + \Sigma_{i > j, i \neq n} x_i e_i$. Consequently $\Gamma_{\zeta_2} \neq \Gamma_{\zeta_2}$ tor $\zeta_1 \neq \zeta_2$, and so $L_{\zeta_1} \not\subset L_{\zeta_2}$ and $L_{\zeta_2} \not\subset L_{\zeta_1}$. ●

COROLLARY 2. *If $L = \mathfrak{L}(\tilde{T}^s x\colon s \geqslant 0)$ ($\tilde{T}$ and x are the same as in Theorem 4), then $\tilde{T}/L$ is similar to a weighted shift operator in l^p. In particular,* $\dim (L/\tilde{T}L) = 1$.

(19) Here we use the fact that $\tilde{\lambda}_{n(\beta_k)+i} = \lambda_{n(\beta_k)+i}$ if $1 \leqslant i \leqslant s$, $s \in I_k$ and $k = k(s)$.

The corollary follows from the equivalence (proved in Theorem 4) of the natural basis in $l^p(A)$ and the sequence $\{t_s\widetilde{T}^s x\}_{s\geq 0}$, where the t_s are normalizing factors. ●

REMARK. By means of special choices of the sequences $\{\lambda_n\}_0^\infty$, $\{\mu_k\}_1^\infty$, and $\{x_i\}_0^\infty$ we can produce various features of the operator $\widetilde{T}|L$ (where $\widetilde{T}$ and L are the same as in Theorem 4). For example, we can produce the following cases: $\widetilde{T}|L$ is unicellular; $\widetilde{T}|L$ is similar to the shift S in l^p; $\widetilde{T}|L$ is similar to $\widetilde{T}$ (here the structure of the invariant subspaces of $\widetilde{T}$ becomes very complicated; the operator $\widetilde{T}$ repeats itself on the subspace L, and so forth).

5°. *Nonstandard ideals.* Here we construct examples of algebras $l^p(w_n)$ with $r(S\Lambda) = 0$ and with a large collection of nonstandard ideals. First we mention a connection between Theorems 1 and 4 and a question considered by Šilov. If $\underline{\lim}_n |w_n|^{1/n} = 0$ and each of the spaces $l^p(w_{n+j})_{n\geq 0}$, $j = 0, 1, \ldots$, is an algebra when we let convolution be the multiplication, then $T = S\Lambda$ is not only unicellular but also a basis operator (see Theorem 1). It is not hard to find examples of sequences $\{\lambda_n\}_0^\infty$ for which $l^p(w_n)$ is an algebra and $l^p(w_{n+1})$ is not[20] (for $p = 1$ this property is possessed by each sequence $\{\lambda_n\}_0^\infty$ such that $w_{i+k} \leq w_i w_k$, $\lambda_0 \cdots \lambda_{n_k-1} = \lambda_{n_k} \cdots \lambda_{2n_k-1}$ and $\lim \lambda_{n_k} = 0$; see also Remark 2 after Theorem 5). In such a space each of the sequences $\{T^n x\}_{n\geq 0}$ (where $x = \{x_i\}_0^\infty$, $x_0 \neq 0$ (after normalization), is a basis in l^p) is equivalent to the natural basis $\{e_n\}_0^\infty$, but for vectors x in l_1^p with $x_1 \neq 0$ the analogous assertion is not true. It is rather difficult to combine the possibility of convolution in $l^p(w_n)$ (that is, $\|x * y\|_p \leq \text{const}\, \|x\|_p \cdot \|y\|_p$) with the constructions in Theorem 4.

THEOREM 5. *For each p, $1 \leq p \leq \infty$, there exist prime algebras $l^p(w_n)$ (or $c_0(w_n)$ for $p = \infty$) containing continua of nonstandard ideals.*

The proof uses the results of Theorem 4 for a special choice of the sequence $\{\lambda_n\}_0^\infty$. In the notation of Theorem 4 we set $n_k = 2^k$, $k \geq 1$, and $x_3 = \lambda_0 = \lambda_1 = 1$. Then $m(\beta) = 2^\beta$, $\beta \geq 1$; $\beta_k = k + 1$, $k \geq 0$, and $I_k = [2^k, 2^{k+1})$. Now the sequences $\{\lambda_n\}_{n\neq 2^k}$, $\{\lambda_{2^k} = \mu_k\}_1^\infty$ and $\{x_i\}_0^\infty$ are constructed inductively (and simultaneously) by means of the following rules:[21]

a) $\lambda_{2^k+s} = \lambda_s/2$, $1 \leq s \leq 2^k - 1$, $k \geq 1$.

b) We use the notation

$$c_{k,s} = \frac{k}{x_{2^k+1}\lambda_{2^k+1}\cdots\lambda_{2^k+s}}, \quad s \in I_k, \quad C_k = \sum_{s\in I_k} c_{k,s}$$

and $\Delta_k = 2^{-(k+1)}C_k^{-1}$, and we choose $0 < x_{2^{k+1}+1} < x_{2^k+1}/2$ and $x_{2^{k+1}+1} < \Delta_k/2$.

c) $C_k\mu_k < 2^{-2^k-1}$, $0 < \mu_k < \mu_{k-1}$.

We will show that under these conditions $l^p(w_n)$ is a prime algebra for each p, $1 \leq p \leq \infty$, and conditions 1) and 2) and (i) and (ii) in the proof of Theorem 4 are

---

(20) In the third remark after Theorem 1 we posed the problem of whether unicellular and basis operators are the same (for $r(S\Lambda) = 0$). If the answer turns out to be (unexpectedly) affirmative, then some elementary considerations already presented in this text would answer Šilov's question.

(21) Here we assume that $\lambda_0, \ldots, \lambda_{2^k-1}, \mu_1, \ldots, \mu_{k-1}$ and $x_{2+1}, x_{2^2+1}, \ldots, x_{2^k+1}$ have already been choosen. As in Theorem 4 we let $x_i = 0$ if $i \neq 2^k + 1$, $k \geq 1$. The method used in part a) is essentially due to Gorin.

satisfied, which showed that $S\widetilde{\Lambda}$ ($S\Lambda$ in the present notation) was not unicellular.

First we show that

$$\sup_m \sum_{i=0}^{m} \frac{w_m}{w_i w_{m-i}} = \sup_m \sum_{i=0}^{m} \frac{\lambda_{m-i} \cdots \lambda_{m-1}}{\lambda_0 \cdots \lambda_{i-1}} < \infty, \tag{10}$$

which will imply the possibility of convolution in each of the spaces $l^p(w_n)$, $1 \leqslant p \leqslant \infty$ [see the proof of part 1) of Theorem 2]. We note right away that $l^p(w_n)$ is primary (that is, $r(S\Lambda) = 0$) by virtue of the obvious relation $\lim_n \lambda_n = 0$. We estimate

$$\sum_{i=0}^{m} \frac{w_m}{w_i w_{m-i}}, \quad 2^k < m \leqslant 2^{k+1}.$$

in terms of the analogous sum corresponding to the index $m' = m - 2^k$. We have

$$\begin{aligned}
\sum_{i=1}^{m-1} \frac{\lambda_{m-i} \cdots \lambda_{m-1}}{\lambda_0 \cdots \lambda_{i-1}} &= \sum_{0<i<m-2^k} \frac{\lambda_{m-i} \cdots \lambda_{m-1}}{\lambda_0 \cdots \lambda_{i-1}} + \sum_{m-2^k \leqslant i < m} \frac{\lambda_{m-i} \cdots \lambda_{m-1}}{\lambda_0 \cdots \lambda_{i-1}} \\
&\leqslant \frac{1}{2} \sum_{0<i<m-2^k} \frac{\lambda_{m-2^k-i} \cdots \lambda_{m-2^k-1}}{\lambda_0 \cdots \lambda_{i-1}} + \sum_{m-2^k \leqslant i \leqslant 2^k} + \sum_{2^k < i < m} \\
&= \frac{1}{2} \sum_{0<i<m-2^k} \frac{\lambda_{m-2^k-i} \cdots \lambda_{m-2^k-1}}{\lambda_0 \cdots \lambda_{i-1}} + \sum_{m-2^k \leqslant i \leqslant 2^k} + \sum_{2^k < i < m} \frac{\lambda_i \cdots \lambda_{m-1}}{\lambda_0 \cdots \lambda_{m-i-1}} \\
&\leqslant \frac{1}{2} \sum_{0<i<m-2^k} \frac{\lambda_{m-2^k-i} \cdots \lambda_{m-2^k-1}}{\lambda_0 \cdots \lambda_{i-1}} + \sum_{m-2^k \leqslant i \leqslant 2^k} + \frac{1}{2} \sum_{2^k < i < m} \frac{\lambda_{i-2^k} \cdots \lambda_{m-2^k-1}}{\lambda_0 \cdots \lambda_{m-i-1}} \\
&= \left(\frac{1}{2} + \frac{1}{2}\right) \sum_{0<i<m-2^k} \frac{\lambda_{m-2^k-i} \cdots \lambda_{m-2^k-1}}{\lambda_0 \cdots \lambda_{i-1}} + \sum_{m-2^k \leqslant i \leqslant 2^k} \leqslant \sum_{0<i<m-2^k} \\
&\quad + (2^{k+1} - m + 1) \frac{\mu_k}{\lambda_0 \cdots \lambda_{2^k-1}} < \sum_{0<i<m-2^k} + 2^{-k-1},
\end{aligned}$$

where the last inequality follows from condition c) and the fact that $x_{2^{k+1}} < 1$. It follows that for an arbitrary $m > 2$

$$\sum_{i=0}^{m} \frac{w_m}{w_i w_{m-i}} < 2 + 3 + \frac{1}{2} + \frac{1}{4} + \cdots = 6,$$

which proves (10).

Condition 1) in the proof of Theorem 4 is satisfied. And condition 2) is also satisfied:

$$\begin{aligned}
\sum_{k=1}^{\infty} \sum_{s \in I_k} \delta_{k,s} &\leqslant \sum_{k=1}^{\infty} \sum_{s=2^k}^{2^{k+1}-1} \frac{1}{x_{2^k+1} \lambda_{2^k+1} \cdots \lambda_{2^k+s}} \sum_{j=2^k+2}^{\infty} |x_j| \\
&\leqslant \sum_{k=1}^{\infty} C_k \sum_{n=k+1}^{\infty} x_{2^n+1} < \sum_{k=1}^{\infty} C_k x_{2^{k+1}+1} 2 < \sum_{k=1}^{\infty} \Delta_k C_k < \frac{1}{2},
\end{aligned}$$

where condition b) is used beginning with the third inequality.

Conditions (i) and (ii) in the proof of Theorem 4 follow from condition c), since $\sigma_k = C_k$ for our choice of $\{n_k\}_1^\infty$.

Now we use the constructions in Theorem 4 and Corollary 1 of that theorem. ●

REMARKS. 1. In addition to the concluding part of the proof of Theorem 4, the hypothesis of Theorem 5 also enables us to keep Corollaries 1 and 2 in subsection 4° and the remark following them.

2. If we only consider the algebras $l^1(w_n)$, then the construction of the sequence $\{\lambda_n\}_0^\infty$ is simplified and can be described as follows. As before, we set $\lambda_0 = \lambda_1 = 1$, and the desired sequence is of the form

$$\begin{array}{l}(\lambda_0)\,\lambda_1,\ \lambda_2,\ \lambda_3,\ \lambda_4,\ \ldots,\ \lambda_8,\ \ldots,\ \lambda_{16},\ \ldots\\ (1),\ 1,\ \mu_1,\ 1,\ \mu_2,\ \ldots,\ \mu_3,\ 1,\ \ldots,\ \mu_4\ldots.\end{array}$$

New numbers μ_k arise in the 2^kth position with $k \geqslant 1$, after which the sequence repeats itself (from λ_1 to λ_{2^k-1}), and so forth. The numbers μ_k, $k \geqslant 1$, must decrease sufficiently rapidly.

There is another possible method for constructing $\{\lambda_n\}_0^\infty$ which preserves the monotonicity of $\{\lambda_{n_k}\}_{k\geqslant 1}$ and $\{\lambda_n\}_{n\neq n_k}$. We describe the corresponding sequence (for $p = 1$):

a) $n_0 = 0$, and $n_{k+1} = 2n_k + 2$, $k \geqslant 0$

b) $\lambda_{n_k} = \mu_k$, and $\lambda_i = \gamma_k$ for $n_k < i < n_{k+1}$, where $\gamma_k^{n_k+1} = \lambda_0\lambda_1 \cdots \lambda_{n_k-1}\mu_k$. The sequences $\{\mu_k\}_1^\infty$ and $\{\gamma_k\}_1^\infty$ are constructed inductively:

c) $2^{k^2}(k-1)\gamma_k^{n_1+2}(\lambda_0 \cdots \lambda_{n_1}) < 1/8$, $k \geqslant 0$.

d) $2^{k^2}(k-1)\gamma_k\mu_{k-1} < 1/8$, $k \geqslant 0$.

e) $\mu_k < \gamma_k < \gamma_j$ if $j < k$; $\gamma_k < \mu_i$ if $i \leqslant k-1$, and $\mu_k < \mu_{k-1}$, $k \geqslant 1$.

Now (with the element $x = \{x_i\}_0^\infty$, $x_i = 0$, $i \neq n_k + 1$, $x_{n_{k+1}} = 2^{-k^2}$, $k \geqslant 1$) we can carry out calculations which are analogous to the ones which we used in Theorems 4 and 5.

3. We emphasize once again that nonunicellularity of an operator $T = S\Lambda$ in Theorem 5 arises from a relatively subtle "irregularity" of the sequence $\{\lambda_n\}_0^\infty$, since, for example, each element $x = \{x_i\}_0^\infty$ with $x_0 \neq 0$ is cyclic in this case ($\{x_iw_i\}_0^\infty$ does not belong to the unique maximal ideal in $l^p(w_n)$).

§3.3. Spectral analysis on the unitary point spectrum

1°. In this section we solve some spectral analysis problems for operators which have a complete family of eigenvectors for which the corresponding eigenvalues lie on the unit circle $\Gamma = \{\lambda \in \mathbf{C}: |\lambda| = 1\}$. If we restrict our attention to unitary operators in a Hilbert space, the answers to all of our questions are given by the spectral theorem. But going outside of this class of operators is unnatural, as is the restriction to the Hilbert space case. On the other hand, if we assume that the point spectrum is contained in the circle Γ (or some other set), then we must subject our operator to certain conditions which "connect" it with the circle (or the other set); this is necessary in order to stay within the realm of good theory. We will make use of the following conditions:

Strong ergodicity of T:

$$\sup_{n\geqslant 0}\|T^n\| < +\infty.$$

C-ergodicity:

$$\sup_{|\lambda|=1,\, n\geqslant 0} \frac{1}{n+1}\left\| \sum_{k=0}^{n} T^k \lambda^k \right\| < +\infty .$$

A-ergodicity:

$$\sup_{|\lambda|>1} \| R(\lambda, T) \| (|\lambda| - 1) < +\infty .$$

Before we treat the main topic we make a few preliminary remarks. The classical examples of operators with a unitary point spectrum are rotations of a circle in the Lebesgue spaces $L^p = L^p(\Gamma, l_1)$ or in the space of almost-periodic sequences (functions on the group $\mathbf{Z}$ of all integers). We also encounter this situation in the spectral theory of finite-dimensional operators with algebraically simple spectra lying in Γ (see [19]). The role of almost-periodicity and Fourier analysis in similar situations became clear after the appearance of the important investigations of Lorch [155], deLeeuw and Glichsberg [152], and Ljubič [58], [59]. The fundamental ideal here is to divide a vector x into "harmonics" $P_\lambda x$ corresponding to frequencies $\lambda \in \Gamma$, and then to "assemble" x by means of these harmonics. In the problems of this section the possibility of finding harmonics follows from an ergodic theorem (see [108]). In order to preserve the analogy with classical analysis on the group Γ, we supplement the terminology in §§3.1 and 3.2 as follows. We will say that an operator T acting in a space X admits *harmonic analysis* if $P_\lambda x \in E^T(x) = \mathfrak{L}(T^n x\colon n \geqslant 0)$ for arbitrary $\lambda \in \Gamma$ and $x \in X$. The possibility of *harmonic synthesis* means that $E^T(x) = \mathfrak{L}(P_\lambda x\colon \lambda \in \Gamma)$, $x \in X$. If we can perform harmonic analysis, this does not guarantee spectral analysis, but harmonic synthesis does guarantee spectral synthesis.

In subsection 2° we show that spectral analysis is possible for a maximal class of operators, where the projectors P_λ are defined for A-ergodic operators. In the strongly ergodic case we construct "spectral" projectors P_γ on subsets $\gamma \subset \Gamma$ which are subgroups of Γ. We note that in [165], which is not connected with analysis-synthesis problems, Rudin gave an example of a strongly ergodic operator which has the aforementioned projections only onto subsets γ which are finite unions of subgroups to within a finite set of elements.

In subsection 3° we show that the only one of the above classes of operators which admits harmonic (or only spectral) synthesis is the class of strongly ergodic operators. We show that an arbitrary completely continuous operator whose spectrum lies in the disk $\mathbf{D}$ can be extended to a Γ-complete (that is, $\mathfrak{L}(E_\lambda\colon \lambda \in \Gamma) = X$, where $E_\lambda \overset{\text{def}}{=} \operatorname{Ker}(T - \lambda I)$) and C-ergodic operator. This is achieved by the method used in §3.1. The other counterexamples use specific features of certain spaces of analytic functions in $\mathbf{D}$.

Subsection 4° contains analogues of classical theorems on convergence of Fourier series under various "smoothness" conditions. By smoothness we mean smoothness with respect to a given isometric operator T, and the sense of convergence varies from absolute convergence ($\Sigma_{\lambda\in\Gamma} \|P_\lambda x\| < \infty$) to summability of the corresponding series using linear means. We generalize theorems of Kadec [41] on the connection between summability of this type and the supply of discontinuous multipliers for the family of eigenvectors.

In subsection 5° we show that a Γ-complete "non-quasi-analytic" operator T has no spectrum outside the closure of the point spectrum $\overline{\sigma_p(T)}$, while "one-sided" conditions guarantee the same property only for strongly ergodic operators T. We establish an analogue of the Wiener-Pitt theorem (see [18] and [179]) on the spectrum of a multiplier which does not contain a term which is "singular" with respect to T.

Finally, subsection 6° connects the "number" of eigenvalues of an operator T which occur on Γ with the growth of the norms of its powers $\|T^n\|$ as $n \longrightarrow \infty$; namely, the more massive the set $\sigma_p(T) \cap \Gamma$, the faster the increase in $\|T^n\|$; massiveness is measured by a γ-capacity on Γ.

2°. *Analysis.* The possibility of harmonic analysis for A-ergodic operators is based on the following assertion of Hille and Phillips [108].

LEMMA 1. *Suppose that X is a reflexive Banach space, T is a linear operator in X, and $\sigma(T) \subset \bar{\mathbf{D}}$. If*

$$K \stackrel{\text{def}}{=} \sup_{|\zeta|>1} \|R(\zeta, T)\|(|\zeta|-1) < +\infty,$$

then for each λ, $|\lambda| = 1$, the strong limit

$$\lim_{r \to 1+0} R(r\lambda, T)(r-1) \stackrel{\text{def}}{=} P_\lambda$$

exists, where P_λ is a projector onto the eigensubspace E_λ, $|\lambda| = 1$, $\|P_\lambda\| \leqslant k$ and $P_\lambda P_\mu = \mathbf{0}$ if $\lambda \neq \mu$.

COROLLARY 1. *Under the hypothesis of Lemma* 1 *the operator T admits harmonic analysis; that is, $P_\lambda x \in E^T(x)$ for each $x \in X$ and $\lambda \in \Gamma$.*

In fact, $R(\zeta, T)x \in E^T(x)$ for $|\zeta| > 1$ and $x \in X$. ●

REMARK. Lemma 1 and its corollary remain valid in the nonreflexive case if we assume completeness of T. We also note that each root vector of an A-ergodic operator corresponding to a number $\lambda \in \Gamma$ is an eigenvector. For if there exist nonzero vectors e_1 and e_2 such that $(T - \lambda I)e_2 = e_1$ and $(T - \lambda I)e_1 = \mathbf{0}$, then for the operator $T_0 = T|\mathfrak{L}(e_1, e_2)$ we have

$$T_0^n = \left(\lambda I + \begin{pmatrix} 0 & 0 \\ 1 & 0 \end{pmatrix}\right)^n = \lambda^n I + n\lambda^{n-1} \begin{pmatrix} 0 & 0 \\ 1 & 0 \end{pmatrix}$$

in the basis $\{e_1, e_2\}$ for all $n \geqslant 1$. Since T_0 is A-ergodic, we then have $|\lambda| < 1$.

COROLLARY 2. *The family of eigensubspaces $\{E_\lambda\}_{\lambda \in \Gamma}$ of an A-ergodic operator is uniformly minimal; if X is separable, the set $\sigma_p(T) \cap \Gamma$ is no bigger than countable.*

The first assertion in the corollary follows immediately from Lemma 1, and the second assertion follows from the first: if $x_\lambda \in E_\lambda$, $\lambda \in \sigma_p(T)$ and $\|x_\lambda\| = 1$, then $\|x_\lambda - x_\mu\| K \geqslant 1$ for $\lambda \neq \mu$. ●

For operators with uniformly bounded powers $\{T^k\}_{k \geqslant 0}$ we can prove much stronger assertions.

THEOREM 1. *Suppose that X is a reflexive Banach space, γ is a subgroup of the circle Γ, and T is a linear operator in X such that*

$$\sup_{n\geqslant 0}\|T^n\|=K<+\infty.$$

Then in X there exists a projector P_γ onto the subspace $E_\gamma = \mathfrak{L}(E_\lambda\colon \lambda \in \gamma)$ which commutes with all operators that commute with T. Here $|P_\gamma| \leqslant K$.

PROOF. First we let $\gamma = \{\alpha^k\colon k \in \mathbf{Z}\}$, $|\alpha| = 1$. We consider the Fejér kernel $\{K_m^\alpha\}_{m=1}^\infty$ on $\mathbf{Z}$ which corresponds to the "frequency" α:

$$K_m^\alpha(n)=\sum_{k=-m}^{m}\left(1-\frac{|k|}{m+1}\right)\alpha^{nk},\quad n\in\mathbf{Z},$$

if $\alpha^N \neq 1$ for $N \in \mathbf{Z}\setminus\{0\}$, and

$$K_m^\alpha(n)=\tau_m\sum_{k=-m}^{m}\left(1-\frac{|k|}{m+1}\right)\alpha^{nk},\quad n\in\mathbf{Z},$$

if N is the smallest natural number such that $\alpha^N = 1$ and

$$\tau_m^{-1}=\sum_{|k|\leqslant m,\ k|N}\left(1-\frac{|k|}{m+1}\right)$$

here $k|N$ means that k is divisible by N. As usual (see [54]) we have

a) $\displaystyle\int K_m^\alpha(\cdot)\overset{\text{def}}{=}\lim_{p\to\infty}\frac{1}{p+1}\sum_{n=0}^{p}K_m^\alpha(n)=1$ and $K_m^\alpha\geqslant 0$;

b) $\displaystyle\lim_{m\to\infty}\hat{K}_m^\alpha(\lambda)\overset{\text{def}}{=}\lim_{m\to\infty}\int K_m^\alpha(\cdot)\lambda^{-n}=\begin{cases}1, & \text{if } \lambda\in\gamma,\\ 0, & \text{if } \lambda\notin\gamma.\end{cases}$

We will prove, as an example, the first equality in b) for the case of a finite group $\gamma = \{1, \alpha, \ldots, \alpha^{N-1}\}$. If $\lambda = \alpha^r$, $0 \leqslant r \leqslant N-1$, then

$$\hat{K}_m^\alpha(\lambda)=\int K_m^\alpha(\cdot)\alpha^{-rn}=\lim_{p\to\infty}(p+1)^{-1}\sum_{n=0}^{p}K_m^\alpha(n)\alpha^{-rn}$$

$$=\tau_m\sum_{k=-m}^{m}\left(1-\frac{|k|}{m+1}\right)\lim_{p\to\infty}(p+1)^{-1}\sum_{n=0}^{p}\alpha^{n(k-r)}=\tau_m\sum_{\substack{|k|\leqslant m,\\ (k-r)|N}}\left(1-\frac{|k|}{m+1}\right).$$

Since $\lim_m \tau_m = 0$ and the difference $\tau_m^{-1}(1-\hat{K}_m^\alpha(\lambda))$ is bounded (it contains no more than $2r$ terms which do not exceed 1, and no more than $2m/(N-1)$ terms which do not exceed $r/(m+1)$), we have $\lim_{m\to\infty}\hat{K}_m^\alpha(\lambda) = 1$.

For $x \in X$ we set

$$\mathbf{K}_m^\alpha x=\int K_m^\alpha(-n)T^n x=\lim_{p\to\infty}\frac{1}{p+1}\sum_{n=0}^{p}K_m^\alpha(-n)T^n x=\sum_{k=-m}^{m}\left(1-\frac{|k|}{m+1}\right)P_{\alpha^k}x.$$

(The last equality holds for the case of an infinite group γ; if γ is finite, an obvious change must be made.) Then

$$\|\mathbf{K}_m^\alpha x\|\leqslant\sup_{n\geqslant 0}\|T^n x\|\cdot\int K_m^\alpha(-n)\leqslant K\|x\|.$$

Also, the sequence $\{\mathbf{K}_m^\alpha x\}_{m=0}^\infty$ is weakly convergent; in fact, if $f \in X^*$ and $x = \Sigma c_\lambda x_\lambda$ is a finite linear combination of eigenvectors $x_\lambda \in E_\lambda$, $|\lambda| = 1$, then

$$\lim_m f(\mathbf{K}_m^\alpha x) = \sum c_\lambda \lim_m f(\mathbf{K}_m^\alpha x_\lambda) = \sum c_\lambda \lim_m \int K_m^\alpha(-n)\lambda^n f(x_\lambda)$$

$$= \sum_{\lambda\in\gamma} c_\lambda f(x_\lambda) = f\Big(\sum_{\lambda\in\gamma} c_\lambda x_\lambda\Big).$$

If $x \in \mathfrak{L}(E_\lambda\colon \lambda \in \gamma)$, the sequence $\{f(\mathbf{K}_m^\alpha x)\}_{m=1}^\infty$ is the l^∞-limit of the convergent sequence $\{f(\mathbf{K}_m^\alpha x_n)\}_{m=0}^\infty$, where $x_n = \Sigma_{\lambda\in\gamma} c_\lambda^{(n)} x_\lambda$ and $x_\lambda \in E_\lambda$; and therefore it converges to $f(x)$. Consequently the system of functionals $\{P_\lambda^* X^*\colon |\lambda| = 1\}$ is total over $E_\gamma = \mathfrak{L}(E_\lambda\colon \lambda \in \gamma)$: if $x \in E_\gamma$ and $P_\lambda x = \mathbf{0}$ for all $\lambda \in \gamma$, then $\mathbf{K}_m^\alpha x = \mathbf{0}$ for $m \geqslant 0$, and $x = \mathbf{0}$. In order to conclude the proof of weak convergence, we just note that for each $f \in X^*$ and for each λ with $|\lambda| = 1$ we have existence of the limit

$$\lim_m (P_\lambda^* f)(\mathbf{K}_m^\alpha x) = \lim_m f(\mathbf{K}_m^\alpha P_\lambda x) = \begin{cases} 0, & \lambda \notin \gamma, \\ f(P_\lambda x), & \lambda \in \gamma. \end{cases}$$

Thus for each $x \in X$ we have existence of the weak limit $P_\gamma x = \lim_m \mathbf{K}_m^\alpha x$. We note also that by the Banach-Steinhaus theorem the convergence is strong:

$$\lim_m \| P_\gamma x - \mathbf{K}_m^\alpha x \| = 0$$

on elements x in the subspace E_Γ. It also follows that P_γ is a projector onto the subspace E_γ with the properties described in Theorem 1.

In the general case we consider the composite Fejér-Bochner kernel $\{B_\sigma\}_\sigma$, $\sigma = \{\alpha_1, \ldots, \alpha_N\}$:

$$B_{\sigma,m}(n) = \prod_{i=1}^N K_m^{\alpha_i}(n), \quad n \in \mathbf{Z},$$

where σ is a finite subset of γ, and we choose a "subkernel" which converges to P_γ. To do this, we consider a rational basis(22) B_γ, $B_\gamma \subset \gamma$, of the group γ. For simplicity of notation we will assume that X is a separable space; then we only have to consider countable groups γ. In this case the set B_γ is also at most countable; we let $B_\gamma = \bigcup_{r=1}^\infty \sigma_r$, where the σ_r are $\sigma_r \subset \sigma_{r+1}$. We set

$$\gamma_r = \Big\{\alpha \in \gamma : \alpha^{p_r} = \prod_{i=1}^p \lambda_i^{k_i}, \text{ where } \lambda_i \in \sigma_k, \ k_i \in \mathbf{Z}, \ p \in \mathfrak{N}\Big\};$$

p_r is the product of the first r primes; then $\gamma = \bigcup_1^\infty \gamma_r$, where γ_r is the integer hull of the rationally independent(22) set σ_r^{1/p_r}. We consider the Fejér-Bochner kernel $\{B_r\}_1^\infty$:

$$B_r \overset{\text{def}}{=} B_{\sigma_r^{1/p_r},\, m(r)}, \quad r = 1, 2, \ldots.$$

where the number $m(r)$ is chosen so large that(23) $\hat{B}_r(\lambda) = 1 - 1/r$ for all λ in a previously chosen finite subset $e_r \subset \gamma_r$; we make this choice in such a way that $\bigcup_1^\infty e_r = \gamma$. We set

$$\mathbf{B}_r x = \int B_r(-n) T^n x = \lim_{p\to\infty} \frac{1}{p+1} \sum_{n=0}^p B_r(-n) T^n x, \quad x \in X.$$

---

(22) See [5], [18], or [54], p. 18.

(23) Here and hereafter we assume that the set B_γ is infinite. The case of a finite basis makes the following arguments slightly simpler.

Since $\gamma_r \subset \gamma_{r+1}$, $r \geq 1$, and for the kernel $\{B_r\}_1^\infty$ conditions a) and b) are satisfied, so that the estimates $\|\mathbf{B}_r x\| \leqslant K\|x\|$, $r = 1, 2, \ldots$, also hold as for $\{K_m^\alpha\}_{m=1}^\infty$, we conclude that, as before, the weak limit

$$P_\gamma x = \lim_r \mathbf{B}_r x, \quad x \in X.$$

exists. Analogously we can show that P_γ is the required projector. ●

REMARK. From the proof it follows that $\lim_r \|\mathbf{B}_r x - P_\gamma x\| = 0$ for $x \in E_\Gamma = \mathfrak{L}(E_\lambda\colon |\lambda| = 1)$. For the case when $\gamma = \Gamma$ and $E_\Gamma = X$ this fact was established in [63]. It is also clear that projectors which are similar to those in Theorem 1 exist for the subspaces $E_{\lambda\gamma}$, where γ is a subgroup of the circle and λ is a number with $|\lambda| = 1$. On the other hand, there exists a (nonreflexive) separable Banach space X on which there is a unitary (that is, isometric and invertible) operator T such that the subspace E_Γ is not complemented in X (see the example below). This shows that in Theorem 1 we cannot omit the assumption of reflexivity.

COROLLARY 1. *Under the conditions in Theorem* 1 *there exists a projector* P_Γ *from* X *onto* $E_\Gamma = \mathfrak{L}(E_\lambda\colon |\lambda| = 1)$ *such that* $X = E_\Gamma + E_0$, $E_0 = (I - P_\Gamma)X$, *the operator* $T_\Gamma = T|E_\Gamma$ *is unitary (that is, isometric and invertible) after the equivalent renorming* $\|x\|_T \overset{\text{def}}{=} \sup_{n \geq 0}\|T^n x\|$, *and* $T_0 = T|E_0$ *has no eigenvalues lying on* Γ.

The only claim that is not obvious is the fact that T_Γ is an isometry. This follows from a theorem of Kronecker (see [5]) according to which $\lim_k \lambda^{n_k} = 1$ if λ runs through a finite subset σ of the circle Γ and the sequence of natural numbers $\{n_k\}_1^\infty$ is chosen in a suitable fashion; in this case $\lim_k T^{n_k}x = x$ for $x \in E_\sigma$, and the equation $\|Tx\|_T = \|x\|_T$ follows from the fact that $\|Ty\|_T \leqslant \|y\|_T$ for each $y \in X$. This argument was taken from [63]. ●

COROLLARY 2. *If* σ_1 *and* σ_2 *are rationally independent*(24) *subsets of* Γ, *then under the conditions in Theorem* 1 *the subspaces* E_{σ_1} *and* E_{σ_2} *are "almost orthogonal":* $K\|x + y\| \geqslant \max(\|x\|, \|y\|)$ *for* $x \in E_{\sigma_1}$ *and* $y \in E_{\sigma_2}$ *(orthogonal if* $K = 1$).

COROLLARY 3. *Let* T *be a complete* ($E_\Gamma = X$) *isometric operator in a separable Banach space* X. *Then there exists a sequence of projectors* $\{P_n\}_{n \geq 1}$ *such that*

1) $P_n X \subset P_{n+1}X$, $n \geqslant 1$, $\lim P_n = I$;

2) $|P_n| = 1$ *("orthogonal" projectors) and the* P_n *commute with all operators that commute with* T;

3) $T|X_n$ *is a finite orthogonal sum of rotation operators; that is, isometric operators* A *such that* $Aa_n = \alpha^n a_n$, *where* $|\alpha| = 1$ *and* n *runs through the set* $\{n \in \mathbf{Z}\colon \alpha^n \in \sigma_p(A)\}$ *and* $\{a_n\}$ *is a complete sequence in the corresponding space.*

For the proof of Corollary 3 it suffices to set $P_n = P_{\gamma_n}$, where the γ_n are the subgroups constructed in Theorem 1(25) (for the group γ which is generated by the point spectrum $\sigma_p(T)$), and then to use Corollary 2. ●

---

(24) See [5], [18], or [54].

(25) Since T is complete, all of the arguments connected with reflexivity are omitted.

Thus Corollary 3 reduces the study of complete isometric operators to the study of rotations operators. This reduction will be considered in subsection 4°.

Now we present an example which shows that some type of reflexivity assumption is necessary for the validity of Theorem 1.

EXAMPLE. Let $\alpha \in \Gamma$, where the argument of α is not commensurable with π. Let $\gamma = \{\alpha^n\colon n \in \mathbf{Z}\}$, and let X be the collection of all bounded functions on Γ which are continuous at each point of $\Gamma \backslash \gamma$ and which have discontinuities only of type I (jumps) at the points of γ. We consider the rotation operator T_α:

$$(T_\alpha f)(x) = f(\alpha x), \quad f \in X, \; x \in \Gamma.$$

If we equip X with the usual sup-norm, it is clear that T_α is an isometry from X into X. We will show that the eigenfunctions of T_α are precisely the trigonometric monomials z^n, $n \in \mathbf{Z}$. The equation $T_\alpha z^n = \alpha^n z^n$, $n \in \mathbf{Z}$, is obvious. If $T_\alpha f = \lambda f$, where $f \in X$ and $f \neq \mathbf{0}$, then $|\lambda| = 1$. It suffices to note that f is continuous on Γ. If f has a jump $s(\beta, f) \neq 0$ at a point β with $|\beta| = 1$, then the equation $T_\alpha f = \lambda f$ implies the existence of a jump $s(\alpha^n\beta, f)$ for f at points of the set $\beta\gamma$, where $|s(\alpha^n\beta, f)| = |s(\beta, f)| > 0$. This contradicts the fact that $f \in X$. Thus $E_\Gamma = \mathfrak{L}(z^n\colon n \in \mathbf{Z})$ is the space of all continuous functions on Γ. It is known that E_Γ does not have a (topological) complement in X (see [92]). ●

By means of Theorem 1 we can describe all invariant subspaces of operators of the form $T = T_1 \times T_2$ in terms of the factors T_1 and T_2, where $\sup_{n \geq 0}\|T^n\| < +\infty$ and T_1 is a complete isometry. Corollaries 4 and 5 (below) show that an isometric factor in a "cross" product is always a direct factor.

COROLLARY 4. *Suppose that X and Y are reflexive Banach spaces, $T_1\colon X \longrightarrow X$ is a Γ-complete operator, and $T_2\colon Y \longrightarrow Y$, where $\sup_{n\geq 0}\|T^n\| < +\infty$ and $T = T_1 \times T_2$. Also suppose that $\sigma_p(T_1) \cap \sigma_p(T_2) \cap \Gamma = \varnothing$. Then each T-invariant subspace $L \subset X \times Y$ decomposes into a direct product of subspaces*

$$L = L_X \times L_Y, \; L_X = \{x\colon \{x, \mathbf{0}\} \in L\}, \;\; L_Y = \{y\colon \{\mathbf{0}, y\} \in L\},$$

which are invariant with respect to T_1 and T_2 respectively. Conversely, each such subspace is clearly T-invariant.

PROOF. Let $A = T|L$. By virtue of Corollary 1, $L = L_\Gamma + L_0$, where $L_\Gamma = \mathfrak{L}(E^A_\lambda\colon |\lambda| = 1)$. Also,

$$P^A_\lambda = \lim_p \frac{1}{p+1}\sum_{k=0}^{p} (\lambda^{-1}A)^k = P^{T_1}_\lambda \times P^{T_2}_\lambda.$$

Thus, $\{x, y\} \in L_0$ implies that

$$P^A_\lambda\{x, y\} = \{P^{T_1}_\lambda x, \; P^{T_2}_\lambda y\} = \mathbf{0}, \quad |\lambda| = 1,$$

and by Theorem 1 and the remark which follows it we have $x = \mathbf{0}$. Therefore $L_0 \subset \{\mathbf{0}\} \times y$. If $T\{x, y\} = \lambda\{x, y\}$, then $T_1 x = \lambda x$ and $T_2 y = \lambda y$, and by the hypothesis of the theorem either $x = \mathbf{0}$ or $y = \mathbf{0}$. Consequently

$$L_\Gamma = (L_\Gamma \cap X \times \{0\}) \times (L_\Gamma \cap \{0\} \times Y). \bullet$$

By the same method we can obtain a more general assertion if $\sigma_p(T_1)$ and $\sigma_p(T_2)$ are disjoint. We can state this result as follows.

COROLLARY 5. *Suppose that T_1 is a Γ-complete operator in a Banach space X, and* $\dim E_\lambda^{T_1} < \infty$, $\lambda \in \Gamma$; *suppose T_2 is an operator in a reflexive space Y and $T = T_1 \times T_2$. Suppose that $\sup_{n \geqslant 0} \|T^n\| < +\infty$ and L is a subspace of the direct product $X \times Y$ which is invariant with respect to T ($TL \subset L$). Then*

$$L = (L_X \times L_Y) + G(D), \tag{1}$$

where $L_X = \{x: \{x, 0\} \in L\}$ and $L_Y = \{y: \{0, y\} \in L\}$, while $G(D) = \{\{x, Dx\}: x \in X_0\}$ is the graph of the operator $D: X_0 \longrightarrow Y$, $X_0 \subset X$, which is a) *one-to-one and closed, and* b) *diagonal in the sense that $P_\lambda^{T_1} x \in X_0$ if $x \in X_0$ and $\lambda \in \Gamma$; moreover, $D(\Sigma_\lambda x_\lambda) = \Sigma_\lambda D_\lambda x_\lambda$ for an arbitrary finite collection $\{x_\lambda: x_\lambda \in E_\lambda^{T_1}, |\lambda| = 1\}$, where the D_λ are linear operators from $E_\lambda^{T_1}$ into $E_\lambda^{T_2}$.*

The sum (1) *is direct and quasi-topological (the intersection of the summands is* $\{0\}$ *and their sum is dense in L).*

Conversely, each subspace of the form (1) *with $T_1 L_X \subset L_X$ and $T_2 L_Y \subset L_Y$ is invariant with respect to $T = T_1 \times T_2$.*

We note that the linear manifold X_0 and the operator D are not in general uniquely determined by the subspace L. Uniqueness of the representation (1) can be guaranteed only in the case $\dim E_\lambda^{T_1} \leqslant 1$, $\lambda \subset \Gamma$.

We conclude this subsection with an assertion on projections onto invariant subspaces of an ergodic operator which are different from E_γ, where γ is a subgroup of Γ. This assertion reduces the proof of the fact that these subspaces are not complemented to the proof of unboundedness of a certain "natural" projector; it is an abstract analogue of a construction of Rudin and Arens (see [23], p. 155) for the case $X = L^1(\Gamma)$, $(Tf)(\zeta) = f(\alpha\zeta)$, $\zeta \in \Gamma$, (where $|\alpha| = 1$ and $\arg \alpha$ is not commensurable with π) with the subspace $L = H^1$.

LEMMA 2. *Suppose that T is a linear operator in a separable Banach space X, $\sup_{n \in \mathbf{Z}} \|T^n\| < +\infty$ and at least one of the following two conditions is satisfied:* a) *X is reflexive;* b) *T is a Γ-complete operator. Suppose that L is a subspace of X such that $TL \subset L$ and $T^{-1}L \subset L$, and P is a bounded projector from X onto L. Then there exists a bounded projector P_L onto L which commutes with T.*

PROOF. In case a) let P_L be the weak limit

$$P_L = \lim_{i \to \infty} \frac{1}{n_i + 1} \sum_{k=0}^{n_i} T^{-k} P T^k$$

Since $PT^k x = T^k x$ for $x \in L$, we have $P_L x = x$ for these x, and therefore $P_L X = L$. From the same arguments it is clear that $P_L P = P$. Also

$$T\frac{1}{n_i+1}\sum_{k=0}^{n_i}T^{-k}PT^k=\frac{1}{n_i+1}TP+\frac{1}{n_i+1}\left(\sum_{k=1}^{n_i}T^{-(k-1)}PT^{k-1}\right)T$$

$$=\frac{1}{n_i+1}TP+\frac{1}{n_i+1}\left(\sum_{j=0}^{n_i}T^{-j}PT^j\right)T-\frac{1}{n_i+1}T^{-n_i}PT^{n_i+1}$$

Since the first and third terms in this expression are $o(1)$ as $i\longrightarrow+\infty$, we obtain $TP_L=P_LT$. Therefore

$$P_LP_L=P_L\lim_i\frac{1}{n_i+1}\sum_{k=0}^{n_i}T^{-k}PT^k=\lim_i\frac{1}{n_i+1}\sum_{k=0}^{n_i}T^{-k}P_LPT^k=P_L$$

and consequently P_L is a projector onto the subspace L.

In case b) an ergodic theorem (see [30], Chapter VIII, Corollary 5.3) implies that the limit

$$P_Lx=\lim_n\frac{1}{n+1}\sum_{k=0}^{n}T^{-k}PT^kx$$

exists for all eigenvectors x of the operator T, and by the Banach-Steinhaus theorem it exists for all other $x\in X$. The rest is proved as before. ●

REMARK. A projection P_L which commutes with T is "natural" in the following sense. Let $L=\mathfrak{L}(E_\lambda\colon\lambda\in e)$, where e is a subset of Γ. By virtue of the uniform minimality of the family $\{E_\lambda\}_{\lambda\in\Gamma}$ we have $E_\mu\cap L=\{0\}$ if $\mu\notin e$. Thus for an eigenvector $x\in E_\lambda$ we obtain $TP_Lx=P_LTx=\lambda P_Lx$ and $P_Lx\in E_\lambda$; consequently $P_Lx=x$ if $\lambda\in e$, and $P_Lx=0$ if $\lambda\notin e$. Therefore in this case we have

$$P_L\left(\sum c_\lambda x_\lambda\right)=\sum_{\lambda\in e}c_\lambda x_\lambda,$$

where $x_\lambda\in E_\lambda$ and $c_\lambda\in\mathbf{C}$. As typical examples we use the aforementioned spaces $L^1(\Gamma)$ and $H^1=\mathfrak{L}(z^n\colon n\geqslant 0)=\mathfrak{L}(E_{\alpha^n}\colon n\geqslant 0)$ with a rotation operator T_α, or the space of continuous functions $X=C$ and $L=\mathfrak{L}(E_{\alpha^n}\colon\alpha^n\in e)=\mathfrak{L}(z^n\colon\alpha^n\in e)$, where e is a given subset of Γ.

3°. *Synthesis.* As we mentioned in the preceding subsection (Lemma 1), a natural framework for considering the possibility of harmonic analysis on the unitary point spectrum is provided by the condition

$$\sup_{|\lambda|>1}\|R(\lambda,T)\|(|\lambda|-1)<+\infty.$$

Now we will show that for the synthesis problem the condition of A- or C-ergodicity is insufficient, while strong ergodicity of T gives us the possibility of both harmonic and spectral synthesis. This last remark is an immediate corollary of Theorem 1; it was also verified in [63]. To prove it we simply recall (see the remark after Theorem 1, where we must set $\gamma=\Gamma$) that for a Γ-complete, strongly ergodic operator we have $x=\lim_r\mathbf{B}_rx$ for each $x\in X$, and $\mathbf{B}_rx$ for $r\geqslant 1$ is a linear combination of the projections $P_\lambda x$, $\lambda\in\Gamma$, which by virtue of Lemma 1 belong to the subspace $E(x)=\mathfrak{L}(T^nx\colon n\geqslant 0)$.

If we restrict our attention to operators T which are A-ergodic together with their inverses T^{-1}; that is, such that

$$\sup_{\lambda \notin \Gamma} \| R(\lambda, T) \| \times \operatorname{dist}(\lambda, \Gamma) < +\infty,$$

then we can describe a natural condition which guarantees the possibility of synthesis. More precisely, we have the following result.

THEOREM 2. *Let X be a reflexive space, and let T be an invertible A-ergodic operator in X with an A-ergodic inverse; that is,*

$$\sup_{\lambda \in \mathbf{C} \setminus \Gamma} \| R(\lambda, T) \| \operatorname{dist}(\lambda, \Gamma) < +\infty. \tag{2}$$

If the spectrum $\sigma(T)$ of T is no bigger than countable, then T is complete; that is, $X = \mathfrak{L}(E_\lambda\colon |\lambda| = 1)$. In particular, for an operator of this type spectral (and harmonic) analysis is possible.

REMARK. Later examples will show that in terms of the growth of the resolvent $R(\lambda, T)$ condition (2) cannot be weakened without destroying the theorem.

The proof of Theorem 2 essentially repeats arguments in [63], where a similar theorem was established for the case $\sup_{n \in \mathbf{Z}} \| T^n \| < +\infty$. If $\lambda \in \Gamma$, then by an ergodic theorem (see [30]) we have

$$\operatorname{Ker}(T - \lambda I) + \overline{(T - \lambda I) X} = X$$

(direct sum). On the other hand, each isolated point in the spectrum of an operator which satisfies (2) is an eigenvalue. In fact, the Riesz projector $\mathcal{P}_\lambda$,

$$\mathcal{P}_\lambda = \frac{1}{2\pi i} \int_{|\lambda - \zeta| = \varepsilon} R(\zeta, T) \, d\zeta$$

projects onto a subspace X_λ such that $\sigma(T|X_\lambda) = \{\lambda\}$, and an elementary theorem of Pólya (see [94]) together with condition (2) guarantees that the resolvent $R(\zeta, T)$ has a simple pole at the point λ.

Let R denote the orthogonal complement in X^* of the subspace $\mathfrak{L} = \mathfrak{L}(E_\lambda\colon |\lambda| = 1)$; we will show that $\sigma(T^*|R) = \varnothing$. In fact, by hypothesis the spectrum $\sigma(T^*|R)$ is no bigger than countable, and it does not contain isolated points (eigenvalues), since the condition $T^*f = \lambda f$, $f \in R$, implies that $f((T - \lambda I)X) = 0$ and $f(E_\lambda) = 0$; that is, $f = \mathbf{0}$. ●

Now we construct examples which demonstrate the fact that Theorem 2 is precise and the condition of strong ergodicity in the harmonic analysis-synthesis problem is best possible.

THEOREM 3. *Let X be a Hilbert space with* $\dim X = +\infty$; *let* $\alpha_n \geqslant 1$ *for all* $n = 0, 1, \ldots,$ *with* $\lim_n \alpha_n = +\infty$; *and let μ be a positive even function on* $\mathbf{R} \setminus \{0\}$, $\mu(r) > 1$, *which is decreasing on* $(0, +\infty)$ *and is unbound near zero.*

Then there exists an operator T in X such that: a) $\mathfrak{L}(E_\lambda\colon \lambda \in \Gamma) = X$; b) $\sigma(T)$ *is countable*; c) $\|T^n\| < \alpha_{|n|}$, $n \in \mathbf{Z}$; d) $\|R(\lambda, T)\| \leqslant \mu(|\lambda| - 1)/\operatorname{dist}(\lambda, \Gamma)$, $\lambda \in \mathbf{C} \setminus \Gamma$; e) *spectral analysis is possible for T*; *and* f) *T does not admit spectral synthesis and the collection of its eigensubspaces is not minimal.*

PROOF. Let $X = \Sigma_1^\infty \oplus H_n$, $\dim H_n = n + 1$, where the H_n are Hilbert spaces. Let $\{e_k^{(n)}\}_{k=0}^n$ be a normalized basis in H_n such that

$$\lim_n \operatorname{dist}(e_0^{(n)}, \mathcal{L}(e_k^{(n)}: 1 \leqslant k \leqslant n)) = 0.$$

Then there exist vectors $x_n \in H_n$, $n \geqslant 1$, such that $\|x_n\| = 1$, $x_n = \Sigma_0^n a_{k,n} e_k^{(n)}$, $a_{0,n} \neq 0$ and $\lim_n a_{0,n} = \infty$, and therefore there exists a sequence of numbers $\{b_n\}_1^\infty$, $b_n \neq 0$, such that

$$\sum_1^\infty |b_n|^2 < +\infty, \quad \sum_1^\infty |b_n|^2 |a_{0,n}|^2 = +\infty. \tag{3}$$

We set $x = \Sigma_1^\infty \oplus b_n x_n$. On the circle Γ the numbers $\lambda_0^{(n)} = 1$ and $\lambda_1^{(n)}, \ldots, \lambda_n^{(n)}$ are chosen such that the operator T_n defined in H_n by the equations $T_n e_k^{(n)} = \lambda_k^{(n)} e_k^{(n)}$, $k = 0, \ldots, n$, satisfies conditions c) and d) and among the numbers $\lambda_k^{(n)}$, $1 \leqslant k \leqslant n$, $n = 1, 2, \ldots$, there is no repetition. It clearly suffices to do the choosing with concern only for the condition c). In order to do this, we first note that for an arbitrary $m = 0, 1, \ldots$ we have

$$\left\| T^m \sum_{k=0}^n c_{k,n} e_k^{(n)} \right\| \leqslant \sum_{k=0}^n |c_{k,n}| \leqslant \text{const} \left\| \sum_{k=0}^n c_{k,n} e_k^{(n)} \right\|, \tag{4}$$

where $\Sigma_k c_{k,n} e_k^{(n)}$ is an arbitrary vector in H_n and const depends only on n and the distance $\operatorname{dist}(e_k^{(n)}, \mathfrak{L}(e_j^{(n)}: j \neq k))$. Let the natural number m_0 be so large that $\alpha_m >$ const for $m > m_0$. Now we choose numbers $\lambda_1^{(n)}, \ldots, \lambda_n^{(n)}$ which do not equal 1 but are so close to 1 that

$$|T^k| \leqslant \min_{1 \leqslant j \leqslant m_0} \alpha_j$$

for $k = 0, 1, \ldots, m_0$ and among the numbers $\lambda_1^{(p)}, \ldots, \lambda_p^{(p)}$, $1 \leqslant p \leqslant n$, there is no repetition. This is possible since m_0 is fixed, $\alpha_j > 1$ ($j = 0, 1, \ldots$), and the norm $|T^k|$ depends continuously on $\lambda_1^{(n)}, \ldots, \lambda_n^{(n)}$. If $m > m_0$, then (4) implies that $|T^m| \leqslant \alpha_m$.

We set $T = \Sigma_1^\infty \oplus T_n$, and we show that the vector x cannot be synthesized from eigenvectors of T which belong to $E(x) \overset{\text{def}}{=} \mathfrak{L}(T^k x: k = 0, 1, \ldots)$. Here it suffices to show[26] that

$$E_1 = \mathcal{L}(e_0^{(n)}: n = 1, 2, \ldots) \cap E(x) = \{0\}.$$

Let

$$e \in E_1, \quad e = \sum_{n=1}^\infty c_n e_0^{(n)}, \quad 0 < \sum_{n=1}^\infty |c_n|^2 < +\infty,$$

and let r be the smallest number for which $c_r \neq 0$. We let f_n, $n = 1, 2, \ldots$, denote the functionals in X defined by the equations

(26) Here and hereafter the vector $\Sigma_{p=0}^\infty \oplus \delta_{n,p} e_0^{(n)}$ is identified with $e_0^{(n)}$ for the sake of simple notation; δ_{np} is the Kronecker delta.

$$f_n(e_0^{(n)})=1,\ f_n(e_k^{(n)})=0,\ 1\leqslant k\leqslant n,$$
$$f_n(e_k^{(j)})=0,\quad \text{if}\quad j\neq n.$$

Finally, let p_m be polynomials in z such that

$$e=\lim_m p_m(T)x.$$

Then

$$c_r=f_r(e)=\lim_m f_r(p_m(T)x)=\lim_m a_{0,r}b_r p_m(1);$$

that is, $\lim_m p_m(1) = M \neq 0$. Therefore

$$c_n=f_n(e)=\lim_m f_n(p_m(T)x)=b_n a_{0,n}M$$

for all $n = 1, 2, \ldots$, which together with (3) leads to a contradiction: $\Sigma_1^\infty |c_n|^2 = +\infty$.

We still have to verify condition e). Let E be an invariant subspace of T and let $y \in E$, $y \neq \mathbf{0}$. From the construction it is clear that the resolvent $R(\lambda, T)$ has first-order growth near each point of the unit circle except $\lambda = 1$. Therefore there exist spectral projectors P_λ, $\lambda \in \Gamma\backslash\{1\}$. It is also clear that if $P_\lambda y = \mathbf{0}$ for all $\lambda \in \Gamma\backslash\{1\}$, then $y = \Sigma_1^\infty \oplus y_n$, where the y_n are proportional to $e_0^{(n)}$, so that $Ty = y$. Consequently spectral analysis is possible for T. ●

THEOREM 4. *Let $\mathcal{H}_2 = L^\infty_{(1-r)^{1/2}}(H^2)$ be the space of all analytic functions f in $\mathbf{D}$ such that*

$$\|f\|_{\mathcal{H}_2}=\sup_{0\leqslant r<1}(1-r)^{1/2}\|f_r\|_{H^2}<+\infty,\quad f_r(\zeta)=f(r\zeta),\quad |\zeta|<1.$$

Suppose that $\sigma = \{\lambda_k: k = 1, \ldots\}$ is a countable subset of Γ, $X = \mathfrak{L}_{\mathcal{H}_2}/((1-\lambda z)^{-1}: \lambda \in \sigma)$, and $T = S^*|X$ is the restriction to X of the inverse shift operator S^*:

$$(S^*f)(\zeta)=\frac{f(\zeta)-f(0)}{\zeta},\quad |\zeta|<1,\quad f\in\mathcal{H}_2.$$

Then the following assertions are true:

1) *T is a complete C-ergodic operator with $\sigma_p(T) = \sigma$.*
2) *If the closure $\bar\sigma$ is not countable, then harmonic synthesis is not possible for T.*
3) *If $\bar\sigma = \Gamma$, then spectral analysis is not possible for T.*

The proof of the theorem will be based on the following two lemmas.

LEMMA 3. *The operator S^* is C-ergodic:*

$$\sup_{\substack{n\geqslant 0\\ |\lambda|=1}}\frac{1}{n+1}\left\|\sum_{k=0}^{n}S^{*k}\lambda^k\right\|_{\mathcal{H}_2}<+\infty. \tag{5}$$

PROOF. *Let $f \in \mathcal{H}_2$. Then*

$$\frac{1}{n+1}\left(\sum_{k=0}^{n}S^{*k}\lambda^k f\right)(\zeta)=\frac{1}{n+1}\sum_{m=0}^{\infty}\zeta^m\left(\sum_{k=0}^{n}\hat f(m+k)\lambda^k\right),\quad |\zeta|<1.$$

For an arbitrary r, $0 \leqslant r < 1$, we have

$$\frac{1}{(n+1)^2}\sum_{m=0}^{n} r^{2m}\left|\sum_{k=0}^{n}\hat f(m+k)\lambda^k\right|^2$$

$$\leqslant \frac{1}{(n+1)^2}\sum_{m=0}^{n} r^{2m}\left(\sum_{k=0}^{n}|\hat f(m+k)|^2\left(1-\frac{1}{n}\right)^{2k+2m}\right)\left(\sum_{k=0}^{n}\left(1-\frac{1}{n}\right)^{-2k-2m}\right)$$

$$\leqslant \frac{1}{(n+1)^2}\sum_{m=0}^{n} r^{2m}(\|f\|^2_{\mathcal{H}_2}\cdot n)(n+1)\left(1-\frac{1}{n}\right)^{-4n} \leqslant \text{const}\,\|f\|^2_{\mathcal{H}_2}(1-r)^{-1},$$

where const is an absolute constant. If $1 - 1/n \leqslant r < 1$, then

$$\frac{1}{(n+1)^2}\sum_{m=n+1}^{\infty} r^{2m}\left|\sum_{k=0}^{n}\hat f(m+k)\lambda^k\right|^2 \leqslant \frac{1}{n+1}\sum_{m=n+1}^{\infty} r^{2m}\sum_{k=0}^{n}\left|\hat f(m+k)\right|^2$$

$$\leqslant \frac{1}{n+1}\sum_{p=n+1}^{\infty}|\hat f(p)|^2\left(\sum_{k=0}^{n} r^{2p-2k}\right) = \frac{1}{n+1}\sum_{p=n+1}^{\infty}|\hat f(p)|^2 r^{2p}\left(\sum_{k=0}^{n} r^{-2k}\right)$$

$$\leqslant \sum_{p=n+1}^{\infty}|\hat f(p)|^2 r^{2p} r^{-2n} \leqslant \text{const}\,\|f\|^2_{\mathcal{H}_2}(1-r)^{-1},$$

where const is an absolute constant. If $0 \leqslant r \leqslant 1 - 1/n$, then

$$\frac{1}{(n+1)^2}\sum_{m=n+1}^{\infty} r^{2m}\left|\sum_{k=0}^{n}\hat f(m+k)\lambda^k\right|^2$$

$$\leqslant \frac{1}{(n+1)^2}\sum_{m=n+1}^{\infty} r^{2m}\left(\sum_{k=0}^{n}|\hat f(m+k)|^2\left(1-\frac{1}{m}\right)^{2m+2k}\right)\left(\sum_{k=0}^{n}\left(1-\frac{1}{m}\right)^{-2m-2k}\right)$$

$$\leqslant \frac{1}{(n+1)^2}\sum_{m=n+1}^{\infty} r^{2m}(\|f\|^2_{\mathcal{H}_2}\cdot m)(n+1)\left(1-\frac{1}{m}\right)^{-4m}$$

$$\leqslant \text{const}\,\|f\|^2_{\mathcal{H}_2}\frac{1}{n+1}\cdot\frac{2r^2}{(1-r^2)^2} \leqslant \text{const}\,\|f\|^2_{\mathcal{H}_2}(1-r)^{-1},$$

where const is an absolute constant. If we combine these estimates we obtain (5). ●

LEMMA 4. *Let $\{\mathfrak{M}_n\}_1^\infty$ be a sequence of finite measures on Γ which converges weakly to some limit measure, and suppose that*

$$\varlimsup_{n,\,k\to\infty}\ \sup_{|\Delta|\leqslant\delta}|\mathcal{M}_n(\Delta)-\mathcal{M}_k(\Delta)| \leqslant \omega(\delta),$$

where the spectrum is taken over all Borel sets $\Delta \subset \Gamma$ with $|\Delta| \leqslant \delta$, and ω is a nonnegative function on $[0, 1]$ with $\omega(+0) = 0$. Then the sequence of functions $\{f_n\}_1^\infty$

$$f_n = \int_{\Gamma}(1-tz)^{-1}\,d\mathcal{M}_n(t),$$

is norm-convergent in the spaces(27) *$A_\infty(1/(1-r))$ and $\mathcal{H}_2$.*

(27) The spaces $A_\infty(\lambda)$ were introduced in § 2.1.

The proof of the lemma consists of estimating the norm

$$\|f_k - f_m\|_{A_\infty\left(\frac{1}{1-r}\right)} = \sup_{\zeta\in\mathbf{D}} (1-|\zeta|)\,|f_k(\zeta) - f_m(\zeta)|,$$

and then the assertion follows. By weak convergence of the measures $\mathfrak{M}_n$ we have

$$V \overset{\text{def}}{=} \sup_n \operatorname{Var} \mathcal{M}_n < +\infty.$$

We will show that for each function

$$f,\ f = \int_\Gamma (1-tz)^{-1}\, d\mathcal{M}(t),\ \operatorname{Var}\mathcal{M} \leqslant V,$$

and arbitrary $n = 0, 1, \ldots$ and $\epsilon > 0$ we have

$$\|f\|_{A_\infty\left(\frac{1}{1-r}\right)} \leqslant \max_{0\leqslant k\leqslant n} |\hat{\mathcal{M}}(-k)| + \varepsilon V + V(n+1)^{-1/2} + \sup_{|\Delta|<2^{3/2}\left(1-\sqrt[n+1]{\varepsilon}\right)^{1/2}} \operatorname{Var}(\mathcal{M}\,|\,\Delta). \quad (6)$$

In fact

$$\sup_{|\zeta|<1} (1-|\zeta|)\left|\sum_{k=0}^n \hat{\mathcal{M}}(-k)\zeta^k\right| \leqslant \max_{0\leqslant k\leqslant n} |\hat{\mathcal{M}}(-k)|,$$

$$|g(\zeta)| \overset{\text{def}}{=} \left| f(\zeta) - \sum_{k=0}^n \hat{\mathcal{M}}(-k)\zeta^k\right| = |\zeta|^{n+1}\left|\int_\Gamma \frac{d\mathcal{M}_1(t)}{1-t\zeta}\right|,$$

where $d\mathfrak{M}_1(t) = t^{n+1} d\mathfrak{M}(t)$, $t \in \Gamma$. If $|\zeta|^{n+1} \leqslant \epsilon$, then $|g(\zeta)| \leqslant \epsilon V(1-|\zeta|)^{-1}$; if $|\zeta|^{n+1} > \epsilon$, then by setting

$$\Delta(\zeta) = \{t \in \Gamma : |\arg t + \arg\zeta| < \arccos|\zeta|\}$$

(see Figure 7) we obtain

$$|g(\zeta)| \leqslant |\zeta|^{n+1}\left|\int_{\Delta(\zeta)} \frac{d\mathcal{M}_1(t)}{1-t\zeta}\right| + |\zeta|^{n+1}\left|\int_{\Gamma\setminus\Delta(\zeta)} \frac{d\mathcal{M}_1(t)}{1-t\zeta}\right|$$

$$\leqslant (1-|\zeta|)^{-1}\operatorname{Var}(\mathcal{M}_1\,|\,\Delta(\zeta)) + |\zeta|^{n+1}\cdot \max_{t\notin\Delta(\zeta)} \frac{V}{|1-\zeta t|}$$

$$\leqslant (1-|\zeta|)^{-1}\operatorname{Var}(\mathcal{M}\,|\,\Delta(\zeta)) + |\zeta|^{n+1}\frac{V}{(1-|\zeta|)^{1/2}}.$$

Since

$$|\Delta(\zeta)| \leqslant 2\arccos|\zeta| < 2\arccos \varepsilon^{\frac{1}{n+1}} < 2^{3/2}\left(1-\varepsilon^{\frac{1}{n+1}}\right)^{1/2}$$

and

$$\max_{0\leqslant r\leqslant 1} r^{n+1}\times(1-r)^{1/2} \leqslant (n+1)^{-1/2},$$

for an arbitrary $\zeta \in \mathbf{D}$ we have

$$(1-|\zeta|)\,|g(\zeta)| \leqslant \varepsilon V + \sup_{|\Delta|<2^{3/2}\left(1-\varepsilon^{\frac{1}{n+1}}\right)^{1/2}} \operatorname{Var}(\mathcal{M}\,|\,\Delta) + V(n+1)^{-1/2},$$

which proves (6).

We apply this inequality to the function $f = f_k - f_m$ and choose a small $\epsilon > 0$ and a large n such that $\omega(\delta) < \epsilon$ when $\delta = 2^{3/2}(1 - \sqrt[n+1]{\epsilon})^{1/2}$ and $(n+1)^{-1} < \epsilon$.

Now we can choose a number N such that if k, $m > N$, then

$$\sup_{|\Delta|\leqslant\delta} \operatorname{Var}(\mathcal{M}_k - \mathcal{M}_m)\,|\,\Delta < \varepsilon \text{ and } \max_{0\leqslant j\leqslant n} |\hat{\mathcal{M}}_k(-j) - \hat{\mathcal{M}}_m(-j)| < \varepsilon.$$

Then

$$\|f_k - f_m\|_{A_\infty\left(\frac{1}{1-r}\right)} < \varepsilon + 2\varepsilon V + 2\varepsilon V + \varepsilon.$$

Thus $\{f_k\}_1^\infty$ converges in $A_\infty(1/(1-r))$.

For the proof of convergence in $\mathcal{H}_2$ we use an inequality of Kolmogorov (see [39]), Vol. I, Chapter IV, Theorem 3.16) to show that for a function f of the form

$$f = \int_\Gamma (1 - tz)^{-1}\, d\mathcal{M}(t)$$

we have

$$\|f\|_{\mathcal{H}_2} \leqslant \text{const}\, \|f\|^{1/2}_{A_\infty\left(\frac{1}{1-r}\right)} (\operatorname{Var} \mathcal{M})^{1/2},$$

where const is an absolute constant. In fact, if a positive function h which is defined and bounded on Γ satisfies the inequality

$$m(x) \overset{\text{def}}{=} l_1\{\zeta \in \Gamma\colon h(\zeta) > x\} \leqslant Ax^{-1}$$

for each $x > 0$, then

$$\int_\Gamma h^2 dl_1 = 2\int_0^\infty xm(x)\,dx = 2\int_0^{\sup h} xm(x)\,dx \leqslant 2A \sup h.$$

Then we just note that Kolmogorov's inequality for the function $h(\zeta) = |f(r\zeta)|$, $\zeta \in \Gamma$, $0 \leqslant r < 1$, is precisely the estimate $m(x) \leqslant \text{const}(\operatorname{Var} \mathfrak{M}) \cdot x^{-1}$, $x > 0$. ●

PROOF OF THEOREM 4. It is clear that $\sigma_p(T) \supset \sigma$; C-ergodicity of T follows from Lemma 3, and it implies continuity of the spectral projectors P_λ for each $\lambda \in \Gamma$, and therefore $\sigma_p(T) = \sigma$. If $\bar{\sigma}$ is an uncountable set, then on its principal part is concentrated a continuous measure $\mathfrak{M}$, $\hat{\mathfrak{M}}(0) \neq 0$; that is, $\mathfrak{M}(\{\lambda\}) = 0$, $\lambda \in \Gamma$ (see, for example, [146]). It is clear that there exists a sequence of point measures $\{\mathfrak{M}_n\}_1^\infty$, concentrated on σ, which is weakly convergent to $\mathfrak{M}$ and which satisfies the conditions in Lemma 4. Then $f = \int_\Gamma (1 - tz)^{-1}\, dM(t) \in X$, and $P_\lambda f = \mathbf{0}$ for each λ with $|\lambda| = 1$, since

$$(P_\lambda f)(\zeta) = \lim_{\rho\to 1+0} (\rho - 1) R(\lambda\rho, T) f(\zeta) = \lim_{\rho\to 1+0} (\rho - 1) \frac{f(\zeta) - f\left(\frac{1}{\lambda\rho}\right)}{\zeta - \frac{1}{\lambda\rho}}$$

$$= \lim_{\rho\to 1+0} (\rho - 1)\lambda\rho \frac{f\left(\frac{1}{\lambda\rho}\right)}{1 - \lambda\rho\zeta} = 0$$

for all ζ, $|\zeta| < 1$. Consequently $f \notin \{\mathbf{0}\} = \mathfrak{L}(P_\lambda f\colon \lambda \in \Gamma)$.

We now prove the last assertion in the theorem. From what we said in the previous paragraph for the case $\overline{\sigma} = \Gamma$ we have the inclusion

$$X \supset \mathcal{H}_2^0 = \{f \in A(\mathbf{D}) : \|f_r\|_{H^2} = o((1-r)^{-1/2}),\ r \to 1-0\},$$

so that it suffices to prove that analysis is not possible for $T_0 = T|\mathcal{H}_2^0$. To do this, we note that if g is a function in $\mathcal{K}(\mathcal{H}_2^0)^*$ (see §2.3 for the definition), then for its Taylor coefficients $\hat{g}(n)$ we have the inequalities

$$|\hat{g}(n)| = |\langle g, z^n\rangle| \leqslant \text{const}\,\|z^n\|_{\mathcal{H}_2^0} \leqslant \text{const}\, n^{-1/2},\ n \geqslant 1.$$

On the other hand, $\Sigma_0^\infty |\hat{g}(n)| n^{1/2} < +\infty$ implies that

$$\sum_0^\infty |\hat{f}(n)|\,|\hat{g}(n)| \leqslant \text{const} \sum_0^\infty |\hat{g}(n)|(n+1)^{1/2} < +\infty,$$

for each $f \in \mathcal{H}_2^0$, and therefore $g \in \mathcal{K}(\mathcal{H}_2^0)^*$. Thus $l_A^1(n^{1/2}) \subset \mathcal{K}(\mathcal{H}_2^0)^* \subset l_A^\infty(n^{1/2})$. As we have noted more than once (see Chapter 0 and §3.0), the possibility of individual spectral analysis for the operator $T_0 = S^*|\mathcal{H}_2^0$ would imply the weak invertibility of each function g in $\mathcal{K}(\mathcal{H}_2^0)^*$, $g(\zeta) \neq 0$, $|\zeta| < 1$, with the weak topology $\sigma(\mathcal{K}(\mathcal{H}_2^0)^*, \mathcal{H}_2^0)$. But the latter situation is not possible. In fact, the set of functions

$$\mathcal{J} = \{f \in \mathcal{K}(\mathcal{H}_2^0)^* : \overline{\lim_{r\to 1-0}}\,(1-r)\log|f(r)| \leqslant -1\}$$

does not contain the function $\mathbf{1}$ and is not empty (for instance, it is known that the function $g = (1-z)^n \exp(z-1)^{-1}$ belongs to this set for large n). It is closed in the weak topology just mentioned since it contains limits of weakly convergent (countable) sequences:[28] $f_n \in \mathcal{J}$ and $f_n \xrightarrow{w} f$ implies that $K = \sup_n \|f_n\|_{\mathcal{K}} < +\infty$, and therefore inside the circle $\gamma = \{\zeta \in \mathbf{C}: |\zeta - 1/2| = 1/2\}$ we have

$$|f_n(\zeta)| \leqslant \|f_n\|_{l_A^\infty(n^{1/2})}(1-|\zeta|)^{-1/2} \leqslant \text{const}\,|1-\zeta|^{-1}$$

for all $n = 1, 2, \ldots$. The condition $f_n \in \mathcal{J}$ means that in the canonical factorizations (see [23] or Chapter 0 and §2.1) of the functions $(1-z)f_n$ which are bounded in $\operatorname{Int}\gamma$ there is a factor $\exp((z+1)/(z-1))$. Consequently the functions

$$(1-z) f_n \exp\frac{1+z}{1-z}$$

are uniformly bounded in $\operatorname{Int}\gamma$ and converge pointwise to

$$(1-z) f \exp\frac{1+z}{1-z}.$$

Therefore $f \in \mathcal{J}$. ●

Another method of obtaining theorems like Theorems 3 and 4 consists of applying the techniques of complete extensions which we used in §3.1. We will prove one result in this area which shows that the properties of Γ-completeness and A- or C-ergodicity of an operator do not eliminate the existence of invariant subspaces when the operator is constructed in an "arbitrarily complicated" fashion.

---

(28) See Banach's theorem ([11], Chapter IV, §2, Theorem 5).

THEOREM 5. *Let X_0 be a Banach space, and let T_0 be a linear operator which can be extended to a complete operator with a norm less than* 1. *Then there exists a C-ergodic Γ-complete extension of T_0.*

REMARK. 1. By the theorem in §3.1, each completely continuous operator T_0 with $\|T_0\| < 1$ has the property stated in Theorem 5. An estimate for the norm of its complete extension can easily be detected in the constructions in §3.1. A minor modification of the proof to be given below shows that in the hypothesis of Theorem 5 we can replace the estimate on the norm of the extension by an analogous estimate on its spectral radius.

PROOF OF THE THEOREM. First we show that if the operator A is defined by means of the diagram

$$\begin{array}{ccc} H & \oplus & X_1 \\ \Lambda\downarrow & \searrow^{K} & \downarrow T_1, \\ H & \oplus & X_1 \end{array} \tag{7}$$

where $|T_1| < 1, |K| < +\infty$, and Λ is a C-ergodic operator, then A is also C-ergodic. In fact,(29)

$$A^n = \Lambda^n \oplus (T_1^n + T_1^{n-1}K + \ldots + T_1 K\Lambda^{n-2} + K\Lambda^{n-1}),$$

$$\left|\frac{1}{m}\sum_{n=1}^{m} A^n\right| = \max\left\{\left|\frac{1}{m}\sum_{n=1}^{m}\Lambda^n\right|, \left|\frac{1}{m}\sum_{n=1}^{m}\left(\sum_{j=0}^{n-1} T_1^j K\Lambda^{n-j-1} + T_1^n\right)\right|\right\}$$

$$\leqslant \text{const} + \left|\frac{1}{m}\sum_{n=1}^{m}\sum_{j=0}^{n-1} T_1^j K\Lambda^{n-j-1}\right| = \text{const} + \left|\frac{1}{m}\sum_{j=0}^{m-1} T_1^j \sum_{p=0}^{m-j-1} K\Lambda^p\right|$$

$$\leqslant \text{const} + \sum_{j=0}^{m-1} |T_1|^j \frac{1}{m}\left|\sum_{p=0}^{m-j-1} K\Lambda^p\right| \leqslant \text{const}.$$

Now let T_1 be the extension of T_0 which is mentioned in the hypothesis of Theorem 5, and let $H = \Sigma_1^\infty \oplus H_n$ with the norm $\|\Sigma_1^\infty \oplus f_n\| = \Sigma_1^\infty \|f_n\|_{H_n}$, where all of the spaces H_n coincide with the space X which was constructed in the previous theorem. We set $\Lambda = \Sigma_1^\infty \oplus \Lambda_n$, where $\Lambda_n = T$ for each $n = 1, 2, \ldots$ and T is the operator in part 3) of Theorem 4. Let $\{\nu_n\}_1^\infty$ be a complete sequence of eigenvectors in X_1 for the operator T_1, $T_1\nu_n = \mu_n\nu_n$, $n \geqslant 1$; let $\{e_n\}_1^\infty$ be a sequence of eigenfunctions of the operator T (from Theorem 4) corresponding to the unitary eigenvalues λ_n; here $Te_n = \lambda_n e_n$, $|\lambda_n| = 1$, $n \geqslant 1$. We define the operator A in the space $Y = H \oplus X_1$ by means of diagram (7), where

$$K = \sum_{n=1}^{\infty} \oplus K_n;\ K_n\colon\ H_n \to X_1,\ K_n e_m = k(n, m)\,v_n;\ n, m = 1, 2, \ldots,$$

and the numbers $k(n, m)$ will be chosen later. This choice can be made in such a way that A will be the desired extension of T_0.

---

(29) We will assume, as in §3.1, that the direct product $H \oplus X$ is equipped with one of the conventional norms; for instance, $\|h \oplus x\| = \|h\| + \|x\|$.

We will find the eigenvectors of A. Let

$$y(n, m) = \left(\sum_{p=1}^{\infty} \oplus \delta_{pn} e_m\right) \oplus a(n, m) v_n,$$

where $a(n, m) \in \mathbf{C}$, $n, m = 1, 2, \ldots$. Then

$$Ay(n, m) = \lambda_m \left(\sum_{p=1}^{\infty} \oplus \delta_{pn} e_m\right) \oplus k(n, m) v_n + a(n, m) \mu_n v_n.$$

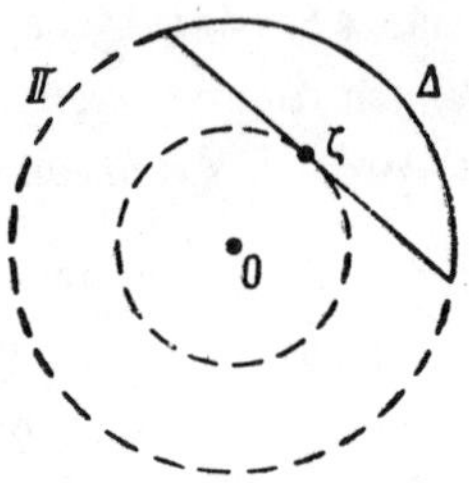

FIGURE 7
$\Delta = \overline{\Delta(\zeta)}$.

We subject the numbers k and a to the condition

$$k(n, m) + a(n, m)\mu_n = \lambda_m a(n, m), \quad n, m = 1, 2, \ldots . \tag{8}$$

Then the vectors $y(n, m)$ are eigenvectors(30) of A.

Continuity of K follows from the condition $\sup_{n \geqslant 0} \| K_n \| < +\infty$:

$$\left\| K\left(\sum_{n=1}^{\infty} \oplus f_n\right)\right\| = \left\| \sum_{n=1}^{\infty} K_n f_n \right\| \leqslant \sum_{n=1}^{\infty} \| K_n f_n \| \leqslant (\sup_{n \geqslant 1} \| K_n \|) \left\| \sum_{n=1}^{\infty} \oplus f_n \right\|.$$

The operator K_n is continuous in Y if and only if $\{k(n, m)\}_{m=1}^{\infty}$ is a moment sequence with respect to $\{e_m\}_1^{\infty}$; that is, there exists a linear functional $\Phi_n \in X^*$ such that

$$\Phi_n(e_m) = k(n, m), \quad m = 1, 2, \ldots,$$

where $\| K_n \| = \|\Phi_n\|$. On the other hand, the closure of the linear span of the vectors $\{y(n, m)\colon m = 1, 2, \ldots\}$ contains the vector $\mathbf{0} \oplus v_n$ if and only if the sequence $\{a(n, m)\}_{m=1}^{\infty}$ is not a moment sequence with respect to $\{e_m\}_1^{\infty}$ (see the argument in the corresponding part of the proof of the theorem in §3.1). Thus, by virtue of (8), the proof will be complete if we establish the existence of functionals $\Phi_n \in X^*$, $n \geqslant 1$, such that

a) $\sup_{n \geqslant 1} \|\Phi_n\| < +\infty$, and

b) the sequence $\{\Phi_n(e_m)(\lambda_m - \mu_n)^{-1}\}_{m=1}^{\infty}$ is not a moment sequence for any $n = 1, 2, \ldots$.

If to each $\Phi \in X^*$ there corresponds a functional Ψ such that $\Phi(e_m)(\lambda_m - \mu_n)^{-1} = \Psi(e_m)$, $m = 1, 2, \cdots$, then $(T^* - \mu_n I)\Psi = \Phi$ (completeness of the sequence $\{e_m\}_1^{\infty}$ in X). This would imply that $(T^* - \mu_n I)X^* = X^*$, which contradicts the fact that μ_n with $|\mu_n| < 1$ is the eigenvalue of T which was constructed in part 3) of Theorem 4. ●

REMARK 2. In subsection 5° we will construct an example of a Γ-complete C-ergodic operator T which acts in a Hilbert space and has a collection of eigenvectors for which there exists a complete biorthogonal system of functionals, and nevertheless $0 \in \sigma(T)$. This operator can also be used for constructions which are analogous to Theorem 5.

4°. *Expansions in eigenvectors of isometric operators.* Corollary 3 of Theorem 1

---

(30) In this construction of A the multiplicity of each eigenvalue λ_m, $m = 1, 2, \ldots$, is infinite. It is clear that by choosing $\Lambda_n = \nu_n T$, $|\nu_n| = 1$, where $\{\nu_n\colon n = 1, 2, \ldots\}$ is a rationally independent set on Γ, we obtain an operator with a simple unitary point spectrum, satisfying the same requirements that A satisfies. The version given in the text involves simpler notation.

enables us to approximate complete isometric operators by (finite) orthogonal sums of rotation operators. We recall that by a rotation operator we mean a complete isometry T such that $\sigma_p(T) \subset \{g^n: n \in \mathbf{Z}\}$ for some g with $|g| = 1$. We will establish some convergence (or summability) criteria for spectral decompositions of isometric operators. As is clear from the previous remark, we can restrict our attention to rotation operators, since a relation between different orthogonal "series" is known (see [40]) and it consists of estimates which are like Bessel inequalities for the norms of the corresponding projections. These inequalities essentially depend on the metric characteristics of the underlying space X (for example, the modulus of convexity). In the general case of an arbitrary Banach space X it is impossible to obtain estimates for the terms of a spectral decomposition

$$x \sim \sum_{\lambda \in \sigma_p(T)} P_\lambda x,$$

other than an analogue of the Riemann-Lebesgue lemma: the set $\{\lambda \in \Gamma: \|P_\lambda x\| \geqslant \epsilon > 0\}$ is finite for each $x \in X$ and $\epsilon > 0$. Finiteness of this set is obvious if x is a linear combination of eigenvectors, and it can be shown for other x by taking limits (uniformly with respect to λ).

More precise estimates on the "harmonics" $P_\lambda x$ can be obtained only by considering vectors x which are "smoother" with respect to T than in the random case. This method, which is often used in Fourier series, will be discussed here briefly for general rotation operators. As a basic tool we could use the "integral" $\int$ which was used in the proof of Theorem 1 (invariant mean on the group $\{T^k\}_{k\in\mathbf{Z}}$), but it is easier to make a reduction to the case of a circle. Unless we say otherwise, we will use the following notation:(31) $T = T_g$, $T_g e_n = g^n e_n$, $n \in \mathbf{Z}$, $\|e_n\| = 1$; T is an isometry in X, and $X = \mathfrak{L}(e_n: n \in \mathbf{Z})$; $P_n \overset{\text{def}}{=} P_{g^n}$, $n \in \mathbf{Z}$; $\arg g$ is not commensurable with π.

LEMMA 5. *If $T = T_g$ is a rotation operator in X, then the operator T_ζ, $T_\zeta e_n = \zeta^n e_n$, $n \in \mathbf{Z}$, is an isometry in X for each $\zeta \in \Gamma$, and the vector-valued function $\zeta \longrightarrow T_\zeta x$ is continuous for each $x \in X$.*

The proof of the lemma is obvious and consists of applying the Banach-Steinhaus theorem to the power sequence $\{T^{k_p}\}_{p=1}^\infty$ with $\lim_p g^{k_p} = \zeta$; then $\lim_p T^{k_p} x = T_\zeta x$, $x \in X$. Strong continuity of the mapping $\zeta \longrightarrow T_\zeta$, $\zeta \in \Gamma$, also follows from similar arguments.

COROLLARY 1. *For arbitrary $x \in X$ and $n \in \mathbf{Z}$ we have*

$$P_n x = \int_\Gamma \zeta^{-n} T_\zeta x \, dl_1(\zeta).$$

Now it is clear that for the spectral expansion

---

(31) We consider a rotation operator with "no gaps in the spectrum" $\sigma_p(T) = \{g^n: n \in \mathbf{Z}\}$; the general case then just requires a change of notation.

$$x \sim \sum_{n\in\mathbf{Z}} P_n x \tag{9}$$

we have the same results as in the theory of Fourier series of homogeneous (in the sense of Šilov) function spaces (see [116]). We now formulate these results for the general case of an arbitrary rotation operator, so that we can supplement them with some results which are new even for Fourier series. We will write $x \in \mathrm{Lip}(T, \omega)$ if

$$\|(T_h - I)x\| \leqslant \mathrm{const}\, \omega(|h-1|), \quad |h| = 1,$$

where ω is a convex, increasing, continuous function on $[0, +\infty)$ with $\omega(0) = 0$; the class $\mathrm{Lip}(T, \omega)$ with $\omega(t) = t^\alpha$, $0 < \alpha \leqslant 1$, will be denoted by $\mathrm{Lip}(T, \alpha)$.

COROLLARY 2. *If $x \in \mathrm{Lip}(T, \omega)$, then*

$$d_n(x) \overset{\mathrm{def}}{=} \inf_{c_k} \left\| x - \sum_{k=-n}^{n} c_k P_k x \right\| \leqslant \mathrm{const}\, \omega\left(\frac{1}{n}\right).$$

COROLLARY 3. *If $\sigma_n(x) = \Sigma_{-n}^{n} P_k x$, $x \in X$, then*

1) $\|\sigma_n\| \leqslant \mathrm{const} \log(n+2)$, $n \geqslant 0$;
2) $\lim_n \|x - \sigma_n(x)\| = 0$ *for* $d_n(x) = o(1/\log n)$, $n \to \infty$;
3) $\lim_n \|x - \sigma_n(x)\| = 0$ *for* $x \in \mathrm{Lip}(T, \omega)$ *and* $\omega(t) = o(1/\log t^{-1})$, $t \to 0$.

COROLLARY 4. *If $x \in \mathrm{Lip}(T, \omega)$ and*

$$\int_0^1 \frac{\omega(t)}{t^{3/2}}\, dt < +\infty, \tag{10}$$

then $\Sigma_{n\in\mathbf{Z}} P_n x$ is weakly unconditionally convergent to the vector x.

The proof of Corollary 2 is obtained by integrating the Jackson kernel (see, for example, Chapter X in [39]), and Corollary 3 is obtained by integrating the Dirichlet kernel (see Chapter II in [39]). Assertion 2) of this corollary is an analogue of a theorem of Lebesgue, and assertion 3) is an analogue of the Dini-Lipschitz theorem. The proof of Corollary 4 is the same as the proof of a classical theorem of Bernšteĭn, which it generalizes (see [5], Chapter IX, §2). All of these theorems correspond to the case of a rotation operator (through an angle which is not commensurable with π) in the space $X = C = C(\Gamma)$ of all functions which are continuous on Γ; here e_n, $n \in \mathbf{Z}$, is a trigonometric monomial. The case $X = L^1(\Gamma)$ is mentioned in [39] (vol. I, p. 179). In the classical situation $X = C$ we can make a stronger assertion than in Corollary 4: the series $\Sigma_{n\in\mathbf{Z}} P_n x$ converges absolutely; that is, $\Sigma_{n\in\mathbf{Z}} \|P_n x\| < +\infty$. In fact, in this case $P_n x = c_n e_n$, $n \in \mathbf{Z}$, and there exists a functional Φ (here δ is a measure concentrated at the point 1) such that $\Phi(e_n) = 1$, $n \in \mathbf{Z}$. Unconditional convergence of the numerical series $\Sigma_{n\in\mathbf{Z}} c_n \Phi(e_n)$ implies that $\Sigma_{n\in\mathbf{Z}} |c_n| = \Sigma_{n\in\mathbf{Z}} \|P_n x\| < +\infty$. In the general case of a Banach space we cannot deduce absolute convergence without additional assumptions on (9). These assumptions must be like Young-Hausdorff estimates (see [39]) involving the norms $\|x\|$ and $\|P_n x\|$, $n \in \mathbf{Z}$. Before we formulate our theorem on absolute convergence of the Fourier series (9), we recall that by a *space E which is symmetric in the sense of Semenov* we mean a normed space of two-sided sequences $\{a_n\}_{n\in\mathbf{Z}}$ such that $\|\{a_{p(n)}\}\| = \|\{a_n\}\|_E$ for an arbitrary permutation p of the group $\mathbf{Z}$ and the conditions $|b_n| \leqslant |a_n|$ and $a \in E$ imply that $b \in E$ and $\|b\|_E \leqslant \|a\|_E$; a *fundamental function* φ_F of a symmetric space F is defined by the formula $\varphi_F(n) = \|\chi_n\|_F$, where χ_n is the characteristic function of a subset of $\mathbf{Z}$ which consists of n elements, $n = 1, 2, \ldots$; the *modulus of convexity* of the space X is the function δ_X,

$$\delta_X, \quad \delta_X(t) = \inf\left(1 - \frac{1}{2}\|x + y\|\right),$$

where the infinum is taken over all pairs $x, y \in X$ such that $\|x\| = \|y\| = 1$ and $\|x - y\| = t$.

THEOREM 6. *Let $x \in \mathrm{Lip}(T, \omega)$.*

1) *If X is reflexive and* (10) *holds, then $\sum_{n\in\mathbf{Z}}\delta_X(\|P_n x\|) < +\infty$.*

2) *If $\mathfrak{M}$ is a finite measure on the unit ball B^* of the conjugate space X^*,*

$$\|y\|_{\mathfrak{M}} \overset{\text{def}}{=} \int_{B^*} |f(y)|\, d\mathfrak{M}(f)$$

for $y \in X$, and (10) *holds, then $\sum_{n\in\mathbf{Z}}\|P_n x\|_{\mathfrak{M}} < +\infty$.*

3) *If E is a symmetric sequence space and $\|Jy\|_E \leqslant \mathrm{const}\cdot\|y\|$, $y \in X$, where $Jy = \{\|P_n y\|\}_{n\in\mathbf{Z}}$, then the condition*

$$\sum_{k=1}^{\infty} \omega\left(\frac{1}{2^k}\right)\varphi_{E^*}(2^k) < +\infty \tag{11}$$

implies that $\sum_{n\in\mathbf{Z}}\|P_n x\| < +\infty$.

PROOF. Assertion 1) follows immediately from a theorem of Kadec (see [40]) and Corollary 4.

Assertion 2) follows from Corollary 4 and the estimate(32)

$$\sum_{n\in\mathbf{Z}} |f(P_n x)| < \mathrm{const}\,\|f\|_{X^*},$$

which has to be integrated with respect to the measure $\mathfrak{M}$.

We now prove assertion 3). Let χ_n denote the characteristic function of the subset $\{k\colon 2^{n-1} \leqslant |k| < 2^n\}$ of the set $\mathbf{Z}$ of all integers. Let $h \in \Gamma$ and $x \in \mathrm{Lip}(T, \omega)$. Then

$$\begin{aligned}\|(T_h - I)x\| &\geqslant \mathrm{const}\,\|J(T_h x - x)\|_E = \mathrm{const}\,\|\{\|P_k x\|\cdot|h^k - 1|\}_{k\in\mathbf{Z}}\|_E \\ &\geqslant \mathrm{const}\,\|\{\chi_n(k)\|P_k x\|\cdot|h^k - 1|\}_{k\in\mathbf{Z}}\|_E;\end{aligned}$$

that is ($|h - 1| = 2^{-n}$),

$$\omega\left(\frac{1}{2^n}\right) \geqslant \mathrm{const}\,\|\{\chi_n(k)\|P_k x\|\}_{k\in\mathbf{Z}}\|_E \geqslant \mathrm{const}\,\|\chi_n\|_{E^*}^{-1} \sum_{2^{n-1}\leqslant|k|<2^n} \|P_k x\|.$$

---

(32) This estimate follows, for example, from a theorem of Gel'fand [43] on continuity of convex functionals.

By virtue of (11) we have

$$\sum_{-\infty}^{\infty}\|P_k x\|\leqslant \text{const}\sum_{n=0}^{\infty}\omega\left(\frac{1}{2^n}\right)\varphi_{E^*}(2^n)<+\infty. \bullet$$

We now make some remarks on the theorem which we just proved. The theorem of Bernšteĭn on absolute convergence (for the case $X = C(\Gamma)$) is obtained in the way that we proved the second and third assertions in Theorem 6. In fact, in 2) it suffices to set $\|y\|_{\mathfrak{M}} = \int_\Gamma |y(t)|\,dl_1(t)$ (the measure $\mathfrak{M}$ is concentrated on the δ-measures δ_t, $t \in \Gamma$), and in 3) it suffices to choose $E = l^2$ (here J reduces to the trivial inclusion $C(\Gamma) \subset L^2(\Gamma)$). If $E = l^q$, $1 \leqslant q \leqslant \infty$, then $\varphi_{E^*}(n) = n^{1/p}$, $n \geqslant 1$, $1/p + 1/q = 1$, and condition (11) becomes

$$\sum_{k=1}^{\infty}\omega\left(\frac{1}{2^k}\right)2^{k/p}<+\infty. \tag{12}$$

Let $X = L^p(\Gamma)$ and let T be a rotation operator (through an angle which is not commensurable with π). As before, $\{e_n\}_{n\in\mathbf{Z}}$ is a trigonometric system. If $2 \leqslant p \leqslant \infty$, then $X \subset L^2(\Gamma)$, and for E we can take l^2. Since condition (12) with $p = 2$ coincides with (10), Theorem 6 implies a theorem of Szász [174] on absolute convergence of the Fourier series of a function x which has an L^p-modulus of continuity (10). If $1 \leqslant p \leqslant 2$, then by the Young-Hausdorff theorem we know that for E we can choose l^q, $1/q + 1/p = 1$. The condition (12) implies absolute convergence of the Fourier series for $x \in \mathrm{Lip}_{L^p}\omega$. In particular, this is so for $x \in \mathrm{Lip}_{L^p}\alpha$ and $\alpha > 1/p$. (Szász [174]).

On the other hand, it is well known that these cases of Theorem 6 cannot be strengthened; the condition $x \in \mathrm{Lip}_{L^p}1/p$ for $1 \leqslant p \leqslant 2$ or $x \in \mathrm{Lip}_{L^p}1/2$ for $2 \leqslant p \leqslant \infty$ does not imply absolute convergence of the Fourier series (9) for the function x.

We note that part 3) of Theorem 6 remains valid if we assume that E is not symmetric and only has a monotone norm (eliminating the assumption of invariance of the norm with respect to permutations). In this case condition (11) has to be replaced by

$$\sum_{k=1}^{\infty}\omega\left(\frac{1}{2^k}\right)\|\chi_k\|_{E^*}<+\infty,$$

where χ_k is the characteristic function of the set $\{n \in \mathbf{Z}\colon 2^{k-1} \leqslant |n| < 2^k\}$. A typical space with a monotone norm is $l^p(w_n)$ or the analogous "weighted" Orlicz space.

Now we consider an arbitrary rotation operator for which we can only establish summability of (9) by using arithmetic means. Between this case and the case of absolute convergence of (9) (and also outside this range) there is a whole scale of summation methods which we have not considered. Later we will prove a general result on these summation methods and their connection with the supply of continuous multipliers (see below for the definitions).

Let X be a Banach space and let $\mathfrak{X} = \{x_n\}_1^\infty$ be a minimal sequence in X; that is,

$$\operatorname{dist}(x_n, \mathscr{L}(x_k\colon k\neq n,\ k\geqslant 1))>0,\quad n\geqslant 1.$$

Then there exists a biorthogonal sequence of functionals $\{f_n\}_1^\infty$:

$$f_n(x_k) = (x_k, f_n) = \delta_{nk};\ n,\ k = 1,\ 2,\ \ldots .$$

Suppose that $\mathfrak{X}$ that is a complete sequence in X; to each $x \in X$ we associate its Fourier series

$$x \sim \sum_n (x,\ f_n)\, x_n. \tag{13}$$

We will say that the series (13) is *summable by the method V* if

(a) $\Sigma_{n\geqslant 1} \nu_n(t)(x, f_n)x_n$ converges for each t, $0 \leqslant t < 1$; and

(b) the limit $\lim_{t\to 1} \Sigma_{n\geqslant 1} \nu_n(t)(x, f_n)\, x_n$ exists.

Here the ν_n, $n \geqslant 1$, are piecewise continuous, bounded functions on $[0, 1)$. A natural condition for permanence* of this method is the equality

$$\lim_{t\to 1} v_n(t) = 1,\ n \geqslant 1,$$

which we will assume to be valid.

By a multiplier of a sequence $\mathfrak{X}$ we mean an operator M defined on the linear span of $\mathfrak{X}$ by the equations

$$Mx_n = \mu_n x_n,\ n \geqslant 1; \tag{14}$$

the μ_n are complex numbers, $n \geqslant 1$.

THEOREM 7. *Let X be a Banach space, and let $\mathfrak{X} = \{x_n\}_1^\infty$ be a complete sequence in X with a total biorthogonal sequence $\{f_n\}_1^\infty$. Let V be a summation method which has the property that convergence of the series* (b) *is uniform on compact subsets of* $[0, 1)$ *for an arbitrary $x \in X$. The following assertions are equivalent*:

1) $(V) - \Sigma_{n\geqslant 1}(x, f_n)x_n = x$ *for an arbitrary* $x \in X$.

2) *The multiplier* (14) *with*

$$\mu_n = \int_0^1 v_n(t)\, d\mathcal{M}(t),\ n \geqslant 1, \tag{15}$$

is continuous in X for an arbitrary finite measure $\mathfrak{M}$ on $[0, 1)$.

We postpone the proof of Theorem 7, and we deduce two theorems of Kadec (see [41]) on the description of bases and Cesàro bases in terms of multipliers. We state these theorems as follows.

A. $\{x_n\}_1^\infty$ is a basis in X if and only if each nonnegative, monotone decreasing sequence $\{\mu_n\}_1^\infty$ generates a continuous multiplier (14).

B. $\{x_n\}_1^\infty$ is a Cesàro basis in X if and only if each nonnegative convex sequence $\{\mu_n\}_1^\infty$ generates a continuous multiplier (14).

We will illustrate the connection between Theorems 7 and A. We select a sequence $\{t_n\}_1^\infty$, $0 \leqslant t_n < 1$, $t_n < t_{n+1}$, $n = 1, 2, \ldots, t_n \longrightarrow 1$, and we set

$$v_n(t) = \begin{cases} 0, & 0 \leqslant t \leqslant t_n, \\ 1, & t_n < t < 1. \end{cases}$$

***Editor's note.** I.e. regularity.

In order to obtain Theorem A from Theorem 7, it suffices to note the equivalence of the assertions i) $\mathfrak{M}_n \downarrow 0$; $\mathfrak{M}_n > 0$, and ii) there exists a measure $\mathfrak{M}$ on $[0, 1)$ with $\mathfrak{M} \geqslant 0$ such that $M_n = \mathfrak{M}([t_n, 1))$, $n \geqslant 1$.(33)

By analogy we consider the case of Theorem B. We note that one special case of Theorem B, where $X = C(\Gamma)$ or $X = L^1(\Gamma)$ and $\{x_n\}_1^\infty$ is the trigonometric system, is usually treated (the necessity part) by special methods in the theory of Fourier series (see [39], vol. I, Chapter IV, §11). An application of Theorem A to a trigonometric basis in $L^p(\Gamma)$, $1 < p < \infty$, shows that monotone decreasing sequences are multipliers from L^p into L^p.

We mention another corollary of Theorem 7.

C. A uniformly minimal(34) sequence $\{x_n\}_0^\infty$ is an Abel basis in X if and only if each positive definite sequence $\{\mu_n\}_1^\infty$ (that is, a sequence such that $\Sigma_{i=0}^m(-1)^i C_m^i \mu_{i+k+1} \geqslant 0$; $m, k = 0, 1, 2, \ldots$) generates a continuous multiplier (14).

In fact, in this case $v_n(t) \equiv t^{n-1}$, $n \geqslant 1$, and positive definiteness of $\{\mu_n\}_1^\infty$ is equivalent to solvability of the moment problem (15).

PROOF OF THEOREM 7. First we show that 1) ⇒ 2). Here we note that for each t the operator

$$V(t): x \to \sum_{n=1}^{\infty} v_n(t)(x, f_n)x_n$$

is linear and continuous, and $\sup_{0 \leqslant t < 1} \| V(t) \| < \infty$. In fact, by virtue of 1) it suffices to show that $\sup_{0 \leqslant t \leqslant 1-\delta} \| V(t) \| < \infty$ for each $\delta > 0$, and this follows from the Banach-Steinhaus theorem and the incompatibility of the following assertions:

α) $\sup_n |(V(t_n)x, F)| = \infty$, $0 \leqslant t_n \leqslant 1 - \delta$, $x \in X$, $F \in X^*$;

β) $|\Sigma_{1 \leqslant k \leqslant N} v_k(t)(x, f_k(x_k, F) - (V(t)x, F)| < \epsilon$ for all t, $0 \leqslant t \leqslant 1 - \delta$, and for sufficiently large N (see the hypothesis of the theorem). Since the function $t \longrightarrow V(t)$, $0 \leqslant t < 1$, is strongly measurable and $\sup_t \| V(t) \| < \infty$, the operator $M = \int_0^1 V(t)\,d\mathfrak{M}(t)$ is continuous in X for an arbitrary (finite) measure $\mathfrak{M}$. It is clear that $Mx_n = \mu_n x_n$, where

$$\mu_n = \int_0^1 v_n(t)\,d\mathcal{M}(t), \quad n \geqslant 1.$$

In order to show that 2) ⇒ 1), it again suffices to show that $\sup_{0 \leqslant t < 1} \| V(t) \| < \infty$. From assertion 2) and the closed graph theorem it follows that the mapping $\mathfrak{M} \longrightarrow M_{\mathfrak{M}}$, which takes a measure $\mathfrak{M}$ to a multiplier (bounded operator in X) by means of formulas (14) and (15), is continuous. Consequently $\| M_{\mathfrak{M}} \| \leqslant \text{const} \cdot \|\mathfrak{M}\|$. But $V(t) = M_{\delta(t)}$, where $\delta(t)$ is a unit loading at the point t, $0 \leqslant t < 1$. Therefore $\| V(t) \| \leqslant \text{const}$, $0 \leqslant t < 1$. ●

To conclude this subsection, we note that for a sequence $\mathfrak{X}$ of eigenvectors of a complete isometry T each continuous multiplier M is: a) the strong limit of a

(33) We note that Theorems A and B lead to "finite" methods V (that is, $v_n(t) \equiv 0$ for $n > N(t)$), so that the hypothesis of Theorem 7 is automatically satisfied.

(34) Uniform minimality (that is, $\sup_n \|x_n\| \cdot \|f_n\| < +\infty$) is used only to satisfy the condition of uniform convergence on compact subsets of $[0, 1)$ from Theorem 7 in the case of Abel summation.

sequence of finite-dimensional multipliers, and b) the strong limit of a sequence of polynomials in T. The first of these assertions follows from the fact that $Mx = \lim_{r\to\infty} M\mathbf{B}_r x$ for each $x \in X$ (here $\mathbf{B}_r$ is the Fejér-Bochner operator in the proof of Theorem 1), and the second assertion follows from what we have already proved and the formula $\mathbf{B}_r x = \int B_r(-n)T^n x$, $x \in X$, $r \geqslant 1$ (see the proof of Theorem 1). We just have to note that the strong operator topology is metrizable on each operator ball if X is a separable space, and therefore we can apply a theorem on choosing a diagonal sequence (see [43]).

5°. *Localization of the point spectrum.* Continuing our investigation of $\boldsymbol{\Gamma}$-complete operators, we consider the problem of localization of the spectrum for these operators. This problem consists of describing those $\boldsymbol{\Gamma}$-complete operators T for which

$$\overline{\sigma_p(T) \cap \boldsymbol{\Gamma}} = \sigma(T);$$

here the bar denotes the closure in $\mathbf{C}$. The first half of this subsection will be devoted to a description of general conditions for localization of the unitary point spectrum, and the second half treats the analogous problem for multipliers with respect to eigenvectors of rotation operators.

THEOREM 8. 1) *Suppose that X is a Banach space, T is an invertible linear operator in X, $X = \mathfrak{L}(E_\lambda: \lambda \in \boldsymbol{\Gamma})$, and*

$$\sum_{n\in\mathbf{Z}} \frac{|\log \|T^n\||}{n^2+1} < +\infty. \tag{16}$$

Then $\sigma(T) = \overline{\sigma_p(T) \cap \boldsymbol{\Gamma}}$.

2) *Suppose that X is a Hilbert space,* $\dim X = +\infty$, *and $\{w_n\}_{n\in\mathbf{Z}}$ is a countable positive sequence such that $1 \leqslant w_{n+k} \leqslant w_n w_k$, $n, k > 0$, $\log w_n = o(n)$ and $n^p = o(w_n)$ for arbitrary $p > 0$ as $n \longrightarrow \infty$, and*

$$\sum_{N=1} \oplus \mathscr{L}_{p_N},$$

Then there exists an operator T which is continuous in X together with its inverse such that (a) $\|T^n\| \leqslant w_n$, $n \geqslant 0$; (b) $X = \mathfrak{L}(E_\lambda: \lambda \in \boldsymbol{\Gamma})$; (c) $\sigma(T) = \boldsymbol{\Gamma}$; (d) $\sigma_p(T)$ *is a countable, nowhere dense subset of* $\boldsymbol{\Gamma}$.

PROOF. 1) It suffices to show that T is regular at each point λ_0 such that $|\lambda_0| = 1$ and $\lambda_0 \notin \overline{\sigma_p(T)}$. It is clear that for a linear combination $x = \Sigma_1^n c_k x_{\lambda_k}$ of eigenvectors x_{λ_k} of the operator T ($x_{\lambda_k} \in E_{\lambda_k}$, $\lambda_k \in \boldsymbol{\Gamma}$) the function

$$\lambda \to R(\lambda, T)x = \sum_{k=1}^{n} c_k \frac{x_{\lambda_k}}{\lambda - \lambda_k}$$

is regular in the disk B of radius $\operatorname{dist}(\lambda_0, \sigma_p(T))$ centered at λ_0, and

$$\|R(\lambda, T)x\| \leqslant \|R(\lambda, T)\| \cdot \|x\|$$

for $|\lambda| \neq 1$. If x is a limit point, $x = \lim_n x_n$, then the family of functions $\{R(\lambda, T)x_n\}_1^\infty$ is normal in B by virtue of (16) and the Levinson-Seaberg theorem, and therefore the limit function $\lambda \longrightarrow R(\lambda, T)x$ can also be analytically continued into the disk B. Since T is a complete operator, it follows that the resolvent $R(\lambda, T)$ can be continued across λ_0, and $\lambda_0 \notin \sigma(T)$.

2) We consider the Hilbert space of sequences $l^2_{\mathbf{Z}}(w_n^{-1})$; that is, sequences $x = \{x_n\}_{n\in\mathbf{Z}}$ such that $\Sigma_{n\in\mathbf{Z}}|x_n|^2 w_n^{-2} \leqslant +\infty$. From the hypothesis of the theorem it follows that $\Sigma_{n\in\mathbf{Z}} w_n^{-2} < +\infty$, and therefore $x_\lambda = \{\lambda^n\}_{n\in\mathbf{Z}} \in l^2_{\mathbf{Z}}(w_n^{-1})$ for arbitrary $\lambda \in \mathbf{\Gamma}$. We choose a countable set $\{\lambda_k: k \in \mathfrak{N}\}$ on $\mathbf{\Gamma}$ which has one limit point. We note that $\{x_{\lambda_k}: k \geqslant 1\}$ is a complete set in $l^2_{\mathbf{Z}}(w_n^{-1})$. In fact, if $f = \{\hat{f}(n)\}_{n\in\mathbf{Z}}$ is an element of the conjugate space $l^2_{\mathbf{Z}}(w_n)$ and $f \perp x_{\lambda_k}$, $k \geqslant 1$, then

$$\langle f, x_{\lambda_k}\rangle = \sum_{n\in\mathbf{Z}} \hat{f}(n)\,\bar{\lambda}_k^n \overset{\text{def}}{=} f(\bar{\lambda}_k) = 0, \quad k \geqslant 1.$$

Since the Carleman-de la Vallée-Poussin-Mandelbrojt theorem (see [60]) implies that the class of functions $\{\hat{f}: f \in l^2_{\mathbf{Z}}(w_n)\}$ is quasi-analytic on $\mathbf{\Gamma}$, we have $\hat{f} \equiv 0$ and $f = \mathbf{0}$. Let N be a given natural number. There exists a number $p = p_N$, $p_N \geqslant 1$, such that

$$\|(\lambda I - S)^{-1} | \mathcal{L}_p\| > N$$

for each λ with $\lambda^N = 1$, $|\lambda| = 1$; here S is the shift operator in $l^2_{\mathbf{Z}}(w_n^{-1})$, and $\mathfrak{L}_p = \mathfrak{L}(x_{\lambda_k}: 1 \leqslant k \leqslant p)$. In fact, if for some λ with $|\lambda| = 1$ we have

$$\overline{\lim_{p\to\infty}} \|(\lambda I - S)^{-1} | \mathcal{L}_p\| < +\infty,$$

then by the completeness of $\{x_{\lambda_k}: k \geqslant 1\}$ the point λ does not belong to the spectrum of S, while $Sx_\lambda = \bar{\lambda}x_\lambda$ and $x_\lambda \in l^2_{\mathbf{Z}}(w_n^{-1})$.

On the other hand, it is clear that $\|S^n\| = w_n$, $n \in \mathbf{Z}$. If we identify X with the orthogonal sum $\Sigma_{N=1}^\infty \oplus \mathfrak{L}_{p_N}$ and the operator T with $\Sigma_{N=1}^\infty \oplus (S|\mathfrak{L}_{p_N})$, we obtain the desired example. ●

REMARK. Condition (16) is a "condition of non-quasi-analyticity"; it was used in investigations of spectral analysis by Šilov [114], Domar [126], Gurariĭ [27] (Banach algebras), Ljubič and Macaev [57], Wermer [178], and Bishop [119] (the problem of "separation of the spectrum"). An approximate form of this condition (equivalent to (16) under minor "regularity" assumptions) is

$$\int_0^2 \log\log \max_{|\zeta|=r} \|R(\lambda, T)\|\, dr < +\infty.$$

We turn to "one-sided" conditions, which we impose only on positive powers T^n of the operator T or on its resolvent $R(\lambda, T)$ for $|\lambda| > 1$, and which guarantee the possibility of localization of the spectrum. In this case the breaking point between localizability and the lack of it for $\sigma_p(T)$ is close to the condition of strong ergodicity of T.

THEOREM 9. 1) *Let X be a Banach space and let T be a $\mathbf{\Gamma}$-complete, strongly ergodic operator in X. Then $\overline{\sigma_p(T) \cap \mathbf{\Gamma}} = \sigma(T)$.*

2) *Suppose that X is a Hilbert space, $w_n \uparrow +\infty$, and $\{\log w_n\}_0^\infty$ is a positive sequence which is convex upward. Then there exists an operator T in X such that* (a)

$\|T^n\| \leqslant w_n$, $n \geqslant 0$, *and* T *is C-ergodic*; (b) $\mathfrak{L}(E_\lambda\colon |\lambda| = 1) = X$; (c) *harmonic synthesis is possible for* T; *and* (d) $0 \in \sigma(T)$.

PROOF. The first assertion follows from Theorem 8 and Corollary 1 of Theorem 1; we will prove the second assertion, using the following lemma.

LEMMA 6. *Suppose that the (right) shift operator* S *is bounded in the space* $l^2(w_n)$, $w_n \geqslant 1$, *and the sequence* $\{\log w_n\}_0^\infty$ *is convex upward. Then* S *is C-ergodic if and only if*

$$\sup_{n\geqslant 0} \frac{w_n^2}{n+1} \sum_{k=0}^{n} \frac{1}{w_k^2} < +\infty. \tag{17}$$

PROOF OF THE LEMMA. If $x = \{x(k)\}_{k=0}^\infty \in l^2(w_n)$, then

$$\left(\frac{1}{n+1} \sum_{k=0}^{n} S^k x\right)(m) = \frac{1}{n+1} \sum_{k=0}^{n} x(m-k), \tag{18}$$

where we let $x(j) = 0$ if $j < 0$. Therefore

$$\sum_{m=0}^{\infty} w_m^2 \frac{1}{(n+1)^2} \left| \sum_{k=0}^{n} x(m-k) \right|^2 \leqslant \frac{1}{n+1} \sum_{m=0}^{\infty} w_m^2 \frac{1}{n+1} \left(\sum_{k=0}^{n} w_{m-k}^{-2} \right)$$

$$\times \left(\sum_{k=0}^{n} |x(m-k)|^2 w_{m-k}^2 \right) \leqslant \|x\|^2 \sup_{m,\, n} \frac{w_m^2}{n+1} \sum_{k=0}^{n} w_{m-k}^{-2} \overset{\text{def}}{=} \|x\|^2 \cdot A^2.$$

Conversely, if we choose m and n such that

$$w_m^2 \sum_{k=0}^{n} w_{m-k}^{-2} = w_m^2 \sum_{k=m-n}^{m} w_k^{-2} > (A^2 - \varepsilon)(n+1),$$

then without loss of generality we can assume that $m \geqslant n$. We set $x = \{0, \ldots, w_{m-n}^{-2}, \ldots, w_m^{-2}, 0, \ldots\}$ Then

$$\left\| (2n+1)^{-1} \sum_{j=0}^{2n} S^j x \right\|^2 \geqslant \sum_{p=m}^{m+n} (2n+1)^{-2} \left| \left(\sum_{j=0}^{2n} S^j x \right)(p) \right|^2 w_p^2$$

$$= (2n+1)^{-2} \left(\sum_{k=m-n}^{m} w_k^{-2} \right)^2 \left(\sum_{p=m}^{m+n} w_p^2 \right)$$

$$\geqslant n(2n+1)^{-2}(A^2 - \varepsilon)(n+1) \left(\sum_{k=m-n}^{m} w_k^{-2} \right) \geqslant \frac{A^2 - \varepsilon}{5} \|x\|^2,$$

where the second inequality follows from the monotonicity of $\{w_n\}_0^\infty$. Therefore

$$A\sqrt{5} \leqslant \sup_{n\geqslant 0} \frac{1}{n+1} \left\| \sum_{k=0}^{n} S^k \right\| \leqslant A.$$

It is clear that if $A < +\infty$, then (17) holds. Conversely, if $m \leqslant n$, then

$$\frac{w_m^2}{n+1}\sum_{k=0}^{n} w_{m-k}^{-2} = \frac{w_m^2}{n+1}\sum_{k=0}^{m} w_k^{-2} \leqslant \frac{w_m^2}{m+1}\sum_{k=0}^{m} w_k^{-2},$$

and if $m > n$, then

$$\frac{w_m^2}{n+1}\sum_{k=0}^{n} w_{m-k}^{-2} = \frac{1}{n+1}\sum_{k=0}^{n}\frac{w_m^2}{w_{m-1}^2}\cdot\frac{w_{m-1}^2}{w_{m-2}^2}\cdots\frac{w_{m-k+1}^2}{w_{m-k}^2}$$

$$\leqslant \frac{1}{n+1}\sum_{k=0}^{n}\frac{w_n^2}{w_{n-1}^2}\cdot\frac{w_{n-1}^2}{w_{n-2}^2}\cdots\frac{w_{n-k+1}^2}{w_{n-k}^2} = \frac{w_n^2}{n+1}\sum_{k=0}^{n} w_{n-k}^{-2},$$

where the inequality follows from the convexity of $\{\log w_n\}_{n=0}^{\infty}$. Consequently condition (17) is equivalent to the condition $A < +\infty$. ●

Returning to the proof of Theorem 9, we consider the "cyclic shift" operator T_N in the finite-dimensional space $l^2(w_n, N)$, which consists of finite sequences $x = \{x(k)\}_{k=1}^{\infty}$, $x(k) = 0$ for $k > N$, with the norm inherited from $l^2(w_n)$. More precisely, we define T_N on the unit vectors e_k, $0 \leqslant k \leqslant N$, by the formula

$$T_N e_k = e_{k+1}, \quad 0 \leqslant k < N, \quad T_N e_N = e_0.$$

Here we will assume that (17) is satisfied; for if this is not so, we choose a minorant $\{\underline{w}_n\}_0^{\infty}$ such that $\lim_n \underline{w}_n = +\infty$; $1 \leqslant \underline{w}_n \leqslant w_n$, $n \geqslant 1$; $\{\log \underline{w}_n\}_0^{\infty}$ is an upward convex sequence (as is $\{\log w_n\}_0^{\infty}$); and $\underline{w}_n^2 = \omega(n)$, where $\omega'(t)/\omega(t) = o(1/t)$ as $t \to \infty$. Then

$$\frac{\underline{w}_n^2}{n+1}\sum_{k=0}^{n}\frac{1}{\underline{w}_k^2} \leqslant \text{const}\,\frac{\omega(n)}{n+1}\int_1^n \frac{dt}{\omega(t)} \leqslant \text{const}$$

$$+\,\text{const}\,\frac{\omega(n)}{n+1}\int_1^n \frac{\omega'(t)\,t}{\omega(t)^2}\,dt = \text{const} + \text{const}\,\frac{\omega(n)}{n+1}\cdot o\left(\int_1^n \frac{dt}{\omega(t)}\right) \leqslant \text{const}.$$

We show that under these conditions all of the operators T_N are C-ergodic, where the corresponding constants are bounded by the number in (17). We write the mean of T_N in the form

$$\left(\frac{1}{n+1}\sum_{k=0}^{n} T_N^k x\right)(m) = \frac{1}{n+1}\sum_{k=0}^{m} x(k) + \frac{1}{n+1}\sum_{k=0}^{n-m-1} x(N-k)$$

for $m \leqslant n \leqslant N$, and in the form

$$\left(\frac{1}{n+1}\sum_{k=0}^{n} T_N^k x\right)(m) = \frac{1}{n+1}\sum_{k=0}^{n} x(m-k)$$

for $n \leqslant m \leqslant N$. Then

$$\left\|\frac{1}{n+1}\sum_{k=0}^{n} T_N^k x\right\|^2 = \sum_{m=0}^{n} w_m^2\frac{1}{(n+1)^2}\left|\sum_{k=0}^{m} x(k) + \sum_{k=0}^{n-m-1} x(N-k)\right|^2 +$$

$$+\sum_{m=n+1}^{N} w_m^2 \frac{1}{(n+1)^2}\left|\sum_{k=0}^{n} x(m-k)\right|^2 \leqslant 4\sum_{m=0}^{n} w_m^2 \frac{1}{(n+1)^2}\left|\sum_{k=0}^{m} x(k)\right|^2$$

$$+4\sum_{m=0}^{n} w_m^2 \frac{1}{(n+1)^2}\left|\sum_{k=0}^{n-m-1} x(N-k)\right|^2 + \sum_{m=n+1}^{N} w_m^2 \frac{1}{(n+1)^2}\left|\sum_{k=0}^{n} x(m-k)\right|^2 .$$

If we compare the first and third terms in this sum with (18), we can see that these terms do not exceed the corresponding terms of the arithmetic means for the shift operator S. We estimate the second term:

$$\sum_{m=0}^{n} w_m^2 \frac{1}{(n+1)^2}\left|\sum_{k=0}^{n-m-1} x(N-k)\right|^2 \leqslant \|x\|^2 \sum_{m=0}^{n} \frac{w_m^2}{(n+1)^2} \sum_{k=0}^{n-m-1} w_{N-k}^{-2}$$

$$\leqslant \|x\|^2 \sum_{m=0}^{n} \frac{w_m^2}{(n+1)^2} \cdot \frac{n-m+1}{w_{N-n+m}^2} \leqslant \|x\|^2,$$

since $N-n+m \geqslant m$. If we combine these inequalities we arrive at the estimate

$$\left\|\frac{1}{n+1}\sum_{k=0}^{n} T_N^k\right\| \leqslant \text{const}, \quad 0 \leqslant n \leqslant N,$$

where const does not depend on N. From periodicity considerations it follows that a similar estimate (with a constant no larger than 2 const) is valid for all $n = 1, 2, \ldots$. It is also clear that $\|T_N^k\| = w_k$ for $0 \leqslant k \leqslant N$ and $\|T_N^k\| \leqslant w_k$ for all $k = 0, 1, \ldots$.

We choose a sequence $\{\tau_N\}_1^\infty$ of rationally independent numbers on Γ, and we set

$$T = \sum_{N=1}^{\infty} \oplus \tau_N T_N,$$

identifying the Hilbert spaces X and $\Sigma_{N=1}^\infty \oplus l^2(w_n, N)$. Assertion (a) of Theorem 9 has already been proved, and (b) and (d) are obvious (the latter since $\|T_N^{-1}\| = \|T_N^N\| = w_N \longrightarrow +\infty$). Assertion (c) follows from the possibility of harmonic analysis (Corollary 1 of Lemma 1) and orthogonality of the subspaces $l^2(w_n, N)$. •

REMARK 1. We will list several results on C-ergodic shift operators which can be deduced from Lemma 6. They can be useful for ascertaining the critical speed of growth of the powers $\|T^n\|$, $n \longrightarrow \infty$, which does not eliminate C-ergodicity of T; a trivial upper estimate is $\|T^n\| \leqslant \text{const} \cdot (n+1)$, $n \geqslant 0$. Suppose, as in Lemma 6, that S is the shift operator in $l^2(w_n)$. Then the following assertions are true:

(1) $\Sigma_{n \geqslant 1} w_n^{-2} = +\infty$ if we have C-ergodicity (for short we will say that we have (C)).

(2) For $w_n = n^\alpha$, $\alpha \geqslant 0$, (C) is equivalent to $0 \leqslant \alpha < ½$; if $w_n = \varphi(n)$, $n \geqslant 0$, and φ is a function in the Hardy field $\mathbf{H}$, and $\log \varphi(n) \sim ½\log n$, then we do not have (C); if for each α, $\alpha < ½$, $n^{-\alpha} w_n \uparrow$ (beginning at some point), then we do not have (C).

(3) If $\{\log w_n\}_0^\infty$ is a convex sequence and we have (C), then

$$w_n \leqslant K_m \frac{n^{1/2}}{\log^m (n+1)}, \quad n \geqslant 1, \ m \geqslant 1;$$

(4) If $\sup_{n\geqslant 1} n^{-1/2} \| S^n \| = +\infty$, then we do not have (C).

(5) If

$$\sup_n \sup_{j\geqslant n} w_{j+n} w_j^{-1} = +\infty,$$

then we do not have (C) (we note that

$$\lim_n \sup_{j\geqslant 0} w_{j+n} w_j^{-1} = \lim_n \| S^n \| = +\infty$$

is compatible with (C)).

REMARK 2. An assertion which is similar to part 2) of Theorem 9 can be obtained in another way. Namely, we consider the space $L^2(Gdl_1)$ and the operator $T = T_\alpha$, $(Tf)(\zeta) = f(\alpha\zeta)$, $|\zeta| = 1$; for a special choice of the weight $G \geqslant 0$ and the number α with $|\alpha| = 1$ we can verify properties (a)–(d) of Theorem 9 for T_α. In this case $\sigma_p(T) = \{\alpha^n : n \in \mathbf{Z}\}$, and the eigenfunctions are trigonometric monomials. A problem which is similar to the synthesis problem in this situation was treated in [123].

Almost all of the difficulties in constructing examples for Theorem 9 are connected with the requirement that X be a Hilbert space. If we seek to construct an operator T in some Banach space with arbitrarily slow growth of the norms $\| T^n \|$ as $n \longrightarrow \infty$ without the property of localization of the spectrum, we can use the following simple argument. We set $X = c_0(w_n^{-1})$, where $w_n \uparrow +\infty$, $w_{n+k} \leqslant w_n w_k$, is a given sequence, and we let $T = S^*$ be the (left) shift operator in X. Then $\| T^n \| = w_n$, $n \geqslant 0$, $\mathfrak{L}(E_\lambda : |\lambda| = 1) = X$, and $0 \in \sigma(T)$. Also $\Gamma \subset \sigma_p(T)$; in the next subsection we will show that in the Hilbert space case no similar conclusion can be drawn if the powers $\| T^n \|$ grow so slowly that $\Sigma_{n\geqslant 0} \| T^n \|^{-2} = +\infty$.

Now we will give a sketchy treatment of the problem of localization of the spectrum for multipliers with respect to a system of eigenvectors of a given (isometric) rotation operator T. As we already mentioned, multipliers are closely connected with summability of formal Fourier series; an investigation of their spectral characteristics sheds some light on other properties of complete isometries. The question of localization of the spectrum for multipliers can be regarded as a question about the validity of the "spectral mapping theorem" in its traditional form; namely, if the operator A is a "function" of the operator T (that is, $A = f(T)$), is it true that $\sigma(A) = f(\sigma(T))$? By a "function of the operator T" we mean an element in the weakly closed algebra $\mathfrak{R}(T)$ which is generated by T and I. In subsection 4° we noted that $\mathfrak{R}(T) = M(\mathfrak{X})$ is the collection of all multipliers for the family $\mathfrak{X}$ of eigensubspaces of T. In the classical context where T is a rotation in $L^p(\Gamma)$ through an angle which is not commensurable with π, multipliers in $\mathfrak{R}(T)$ are multipliers of Fourier series:

$$M\Big(\sum_{n\in\mathbf{Z}} \hat{g}(n) z^n\Big) = \sum_{n\in\mathbf{Z}} \hat{g}(n) \mu_n z^n, \quad g \in L^p.$$

In particular, if $p = 2$, the sequence $\{\mu_n\}_{n\in\mathbf{Z}}$ can be an arbitrary bounded sequence, and if $p = 1$ or ∞, it must be the sequence of Fourier coefficients of a finite measure on Γ. Clearly the Fourier coefficients of a measure $\mathfrak{M}$ generate a continuous multiplier M even in the general case of an arbitrary rotation operator T:

$$M = \int_\Gamma T_\zeta d\mathcal{M}(\zeta).$$

In [179] Wiener and Pitt showed that there exists a finite measure $\mathfrak{M}$ on the circle Γ whose Fourier coefficients are bounded away from zero; that is,

$$\inf_{n \in \mathbf{Z}} |\hat{\mathcal{M}}(n)| > 0,$$

but $\{1/\hat{\mathfrak{M}}(n)\}_{n \in \mathbf{Z}}$ is not the collection of Fourier coefficients of any measure. In other words, the corresponding multiplier M in the spaces L^1, L^∞, and C does not admit localization of the spectrum. An example of a similar multiplier for an arbitrary L^p, $p \neq 2$, $1 \leqslant p \leqslant \infty$, was presented in [145]. On the other hand, in [179] it was shown (see also [18], §32) that the impossibility of localization of the spectrum for Fourier multipliers $M \sim \{\hat{\mathfrak{M}}(n)\}_{n \in \mathbf{Z}}$ is connected, roughly speaking, with the existence of a large, singular, continuous component of $\mathfrak{M}$. For measures of the form $\mathfrak{M} = \mathfrak{M}_a + \mathfrak{M}_d$, where $\mathfrak{M}_d$ is a point measure and $\mathfrak{M}_a$ is an absolutely continuous measure on Γ, the spectrum of the multiplier M in the spaces C, L^∞, and L^1 coincides with $\overline{\{\hat{\mathfrak{M}}(n)\}_{n \in \mathbf{Z}}}$.

Now we introduce an analogue of the Wiener-Pitt theorem for multipliers of the form $M = M_a + M_d$, where M_d and M_a are "pointwise" and "absolutely" continuous operators with respect to a fixed rotation T in X. Let $\mathfrak{X} = \{e_n\}_{n \in \mathbf{Z}}$ be a sequence of (simple) eigenvectors of T, $Te_n = g^n e_n$, $n \in \mathbf{Z}$, $|g| = 1$, where $\arg g$ is not commensurable with π. We let $M_a(\mathfrak{X})$ denote the collection of all completely continuous multipliers of the sequence $\mathfrak{X}$; let $\mathfrak{A}$ be the algebra which is generated by $M_a(\mathfrak{X})$ and I; let $\mathfrak{D}$ be the uniform(35) closure of the linear span of all rotations T_ζ, $\zeta \in \Gamma$. The symbol $\mathfrak{M}(A)$ will denote the space of maximal ideals of a Banach algebra A.

THEOREM 10. *Let T be a rotation in a Banach space X with a simple point spectrum $\{g^n, n \in \mathbf{Z}\}$, and let $\mathfrak{X} = \{e_n\}_{n \in \mathbf{Z}}$ be its system of eigenvectors. Let M be a multiplier of $\mathfrak{X}$: $Me_n = \mu_n e_n$, $n \in \mathbf{Z}$, where $M = M_a + M_d$ with $M_a \in M_a(\mathfrak{X})$ and $M_d \in \mathfrak{D}$. Then*

$$\sigma(M) = \overline{\{\mu_n : n \in \mathbf{Z}\}}.$$

PROOF. First we observe that $M_a(\mathfrak{X})$ is the uniform closure of the set of finite-dimensional multipliers, for if $U \in M_a(\mathfrak{X})$, then $\lim_r \mathbf{K}_r U = U$ not only in the sense of strong convergence but also in the operator norm (since U is completely continuous). Consequently the inclusion $U \in M_a(\mathfrak{X})$, $U \sim \{u_n\}_{n \in \mathbf{Z}}$ implies that $\lim_{n \to \infty} u_n = 0$.

The inclusion $\sigma(M) \supset \overline{\{\mu_n : n \in \mathbf{Z}\}}$ is obvious. For the proof of the reverse inclusion it suffices to show that under our assumptions we have

$$\inf_{n \in \mathbf{Z}} |\mu_n| > 0 \Rightarrow M^{-1} \in M(\mathcal{X}).$$

(35) That is, with respect to the operator norm.

Suppose the premise in this implication is true. Then $\mu_n = \mu_n^a + \mu_n^d$ and $\inf_{n\in\mathbf{Z}}|\mu_n^a + \mu_n^d| > 0$. Therefore $\inf_{|n|>N}|\mu_n^d| > 0$ for some N, and by almost-periodicity of the sequence $\{\mu_n^d\}_{n\in\mathbf{Z}}$ we have $\inf_{n\in\mathbf{Z}}|\mu_n^d| > 0$. Let AP denote the algebra of all almost-periodic sequences with frequencies in $\sigma_p(T)$; let $\mathfrak{D}\mathfrak{M}$ denote the algebra of sequences of Fourier coefficients of all point measures whcic are concentrated on $\sigma_p(T)$. It is clear that $\mathfrak{D}\mathfrak{M}$ is imbedded in AP as a dense subalgebra. By the Wiener-Pitt theorem an element of $\mathfrak{D}\mathfrak{M}$ which is invertible in AP is invertible in $\mathfrak{D}\mathfrak{M}$. Now we can apply a theorem of Sonis (see [182], Russian p. 244), according to which $\mathfrak{M}(AP) = \mathfrak{M}(\mathfrak{D}\mathfrak{M})$. We also have $\mathfrak{D}\mathfrak{M} \subset \mathfrak{D} \subset AP$, and therefore $\mathfrak{M}(\mathfrak{D}) = \mathfrak{M}(AP)$. From this equation and the condition $\inf_{n\in\mathbf{Z}}|\mu_n^d| > 0$ we conclude that the operator M_d is invertible in $\mathfrak{D}$. Therefore M is the sum of an isomorphism M_d and a completely continuous operator M_a, and in order to complete the proof of the theorem we have to show that M is one-to-one. But if $Mx = \mathbf{0}$, then $P_\lambda x = \mathbf{0}$ for all $\lambda \in \Gamma$, and therefore (see subsection 3°) $x = \mathbf{0}$. ●

REMARKS. 1. If $M_a \in M_a(\mathfrak{X})$ and $M_d \in \mathfrak{D}$, then $\|M_a + M_d\| \geq \|M_d\|$, so that $\overline{M_a(\mathfrak{X}) + \mathfrak{D}} = M_a(\mathfrak{X}) + \mathfrak{D}$ at least if S is an isometry which shifts eigenvectors: $Se_n = e_{n+1}$, $n \in \mathbf{Z}$. In fact, if x is a linear combination of eigenvectors and y is a linear combination of biorthogonal functionals, then by the almost-periodicity of the sequence $\{\mu_n^d\}_{n\in\mathbf{Z}}$ we obtain

$$|\langle M_d x,\ y\rangle| \leq \overline{\lim_n}\, |\langle M_d S^n x,\ S^n y\rangle|,$$

$$\overline{\lim_n}\, |\langle M_d S^n x,\ S^n y\rangle| \leq \overline{\lim_n}\, |\langle M S^n x,\ S^n y\rangle| + \overline{\lim_n}\, |\langle M_a S^n x,\ S^n y\rangle|$$
$$= \overline{\lim_n}\, |\langle M S^n x,\ S^n y\rangle| \leq \|M\| \cdot \|x\| \cdot \|y\|,$$

and therefore $\|M_d\| \leq \|M_a + M_d\|$. The operator S is isometric, for instance, if $X = L^p$ and $\{e_n\}_{n\in\mathbf{Z}}$ is a trigonometric system.

2. In the case $X = C$ or $X = L^1(\Gamma)$, there is a continuous projection onto the subspace $\mathfrak{D}$ from all of $M(\mathfrak{X})$ (which extracts the discrete component of a measure), so that each multiplier can be written in the form $M = M_a + M_d + M_s$, where M_s is generated by the singular component of the original measure $\mathfrak{M}$. It is not clear that we can find a similar representation in the general case.

3. We can establish the following assertion on "one-sided localization of the spectrum": the infinite components of the complements of the sets $\sigma(M)$ and $\overline{\{\mu_n\colon n \in \mathbf{Z}\}}$ coincide for each $M \in M(\mathfrak{X})$ if and only if

$$(M) \leq \text{const} \sup_{n\in\mathbf{Z}} |\mu_n|$$

for each $M \in M(X)$; here $r(M) = \lim_n \|M^n\|^{1/n}$ is the spectral radius of M.

6°. *Massiveness of the unitary point spectrum.* As we already mentioned, boundedness of the sequence of norms $\{\|T^n\|\}_{n\geq 0}$ implies (in a separable space) countability of the boundary point spectrum $\sigma_p(T) \cap \Gamma$. In the general case we can show that the more massive the set $\sigma_p(T) \cap \Gamma$ and the "smoother" the unit ball in the original space, the faster the sequence $\{\|T^n\|\}_{n\geq 0}$ will grow. Massiveness is measured by means of γ-capacities, which come from the theory of trigonometric series and potential theory (see [5], [39], and [146]).

THEOREM 11. 1) *Suppose that X is a separable Hilbert space, T is a linear operator in X, and γ is an integrable positive function on Γ with positive Fourier coefficients. If the γ-capacity of the set $\sigma_p(T) \cap \Gamma$ is positive, then*

$$\sum_{n\geqslant 0} \hat{\gamma}(n+N)\|T^n\|^{-2} < +\infty$$

for some $N \geqslant 0$. If the Lebesgue measure of $\sigma_p(T) \cap \Gamma$ is positive, then

$$\sum_{n\geqslant 0}\|T^n\|^{-2} < \infty.$$

2) *If $\{w_n\}_{n\geqslant 0}$ is a sequence of numbers such that $1 \leqslant w_{n+k} \leqslant w_n w_k$, $n, k \geqslant 0$, and $\Sigma_{n\geqslant 0} w_n^{-2} < +\infty$, then there exists an operator T in a separable Hilbert space X such that $\|T^n\| \leqslant w_n$ and $\sigma_p(T) \supset \Gamma$.*

PROOF. 1) First we assume that T^* is a cyclic operator, and later we eliminate this assumption. Thus we let f be a cyclic vector of T^*, and let $y_\lambda \in E_\lambda$, $y_\lambda \neq \mathbf{0}$, $\lambda \in \sigma_p(T)$. There exists a closed set $e \subset \sigma_p(T) \cap \Gamma$ for which γ-cap $e > 0$. We note that $(y_\lambda, f) \neq 0$ for each $\lambda \in e$. In fact, $(y_\lambda, T^{*n}f) = \lambda^n(y_\lambda, f)$, $n \geqslant 0$, and $(y_\lambda, f) = 0$ implies that $y_\lambda = \mathbf{0}$. We set $x(\lambda) = y_\lambda/(y_\lambda, f)$, $\lambda \in e$. Then $\lambda \longrightarrow x(\lambda)$ is a Borel mapping. We have

$$(x(\lambda), T^{*n}f) = \lambda^n, \quad \lambda \in e,$$

and for each $h \in X$ there exists a sequence $\{h_n\}_{n\geqslant 1}$ of linear combinations of the vectors $T^{*n}f$, $n \geqslant 0$, which converges to h. Therefore the function $\lambda \longrightarrow (x(\lambda), h)$ is a pointwise limit of polynomials. We will assume that the values of $x(\cdot)$ are normalized as follows: $\|x(\lambda)\| = 1$, $\lambda \in e$.

Since γ-cap $e > 0$, there exists (see [146]) a positive measure ν which is concentrated on e such that

$$0 < \int_\Gamma\int_\Gamma \gamma(\lambda\bar{\mu})\, d\nu(\lambda)\, d\nu(\mu) < +\infty.$$

We set $h_k = \int_\Gamma \bar{\lambda}^k x(\lambda)\, d\nu(\lambda)$, $k \in \mathbf{Z}$. Let $\{a_n\}_1^\infty$ be an orthonormal basis in X. Then

$$\|h_k\|^2 = \sum_{n=1}^{\infty} |a(n, k)|^2, \quad \text{where } a(n, k) = \int_\Gamma \bar{\lambda}^k (x(\lambda), a_n)\, d\nu(\lambda), \; n \geqslant 1, \; k \in \mathbf{Z}.$$

Since the $a(n, k)$ are Fourier coefficients of the measure $(x(\cdot), a_n)d\nu$, we have

$$\int_\Gamma\int_\Gamma \gamma(\lambda\bar{\mu})(x(\lambda), a_n)\overline{(x(\mu), a_n)}\, d\nu(\lambda)\, d\nu(\mu) = \sum_{k\in\mathbf{Z}} \hat{\gamma}(k)|a(n, k)|^2.$$

By Fubini's theorem we have

$$\int_\Gamma\int_\Gamma \gamma(\lambda\bar{\mu})(x(\lambda), x(\mu))\, d\nu(\lambda)\, d\nu(\mu) =$$

$$=\sum_{n=1}^{\infty}\int_{\Gamma}\int_{\Gamma}\gamma(\lambda\bar{\mu})(x(\lambda),\ a_n)\overline{(x(\mu_n),\ a_n)}\,d\nu(\lambda)\,d\nu(\mu)$$

$$=\sum_{n=1}^{\infty}\sum_{k\in\mathbf{Z}}\hat{\gamma}(k)\,|\,a(n,\ k)\,|^2=\sum_{k\in\mathbf{Z}}\hat{\gamma}(k)\,\|\,h_k\,\|^2<+\infty.$$

If $\|h_k\|^2 = 0$ for all $k \in \mathbf{Z}$, then $x(\lambda) = \mathbf{0}$ for ν-almost all λ, while $\|x(\lambda)\| = 1$ for $\lambda \in e$ and $\nu(e) > 0$. Thus there exists $N \in \mathbf{Z}$ such that $\|h_N\| > 0$. Then for $n \geqslant N$ we have

$$\|h_N\|=\|T^{n-N}h_n\|\leqslant\|T^{n-N}\|\cdot\|h_n\|,$$

and therefore

$$\sum_{n\geqslant N}\hat{\gamma}(n)\,\|T^{n-N}\|^{-2}<+\infty.$$

We still have to reduce the original problem to the case which we just considered. We note that $\sigma_p(T)$ is a Borel set for each T. This follows from the fact that the punctured ball $B\backslash\{\mathbf{0}\}$ is a countable union of weakly compact sets, and the mapping $y_\lambda \longrightarrow \lambda$, $y_\lambda \in E_\lambda \cap B$, is continuous is the weak topology on the set $B\backslash\{\mathbf{0}\}$. Therefore $\{\lambda \in \Gamma$: $(x_\lambda, a_n) \neq 0\}$ is also a Borel set for each $n = 1, 2, \ldots$ (see L. N. Nikol′skaja [181]). Consequently the inner γ-capacity of one of these sets is positive. We denote the corresponding basis vector a_n by f; we set $L = \mathfrak{L}(T^{*n}f\colon n \geqslant 0)$, and we let P be an orthogonal projector onto L and $T_1 = PT|L$. Then $T_1^n = PT^n|L$, $n > 0$, so that $\|T_1^n\| \leqslant \|T^n\|$, $n \geqslant 0$, $T_1^* = T^*|L$ is a cyclic operator and $\sigma_p(T_1) \supset \{\lambda \in \Gamma\colon (x_\lambda, f) \neq 0\}$.

2) It suffices to set $X = l^2(w_n^{-1})$ and to let T be the (left) shift operator S^* in X. ●

§3.4. Multiplicity of the spectrum and simultaneous approximation by polynomials in the weighted case

1°. The preceding sections, as well as all of Chapter 2, were devoted to problems which centered around the concepts of spectral analysis and synthesis. We described weakly invertible elements for a given operator T; that is, those elements $x \in X$ for which $X = E^T(x) \stackrel{\text{def}}{=} \mathfrak{L}(T^n x\colon n \geqslant 0)$. In the general case the existence of weakly invertible (or cyclic) vectors is nontrivial. In this section we will consider multiple shift operators; that is, operators T of the form $T = S \times \cdots \times S$, where S is a shift operator in some space of (generalized) functions.

In order to state an explicit formulation of the question, we note first of all that the problem of existence of a cyclic vector is a part of the more general problem of determining the multiplicity of the spectrum of T; that is, of determining the number

$$m_1(T)=\inf\{\operatorname{card}K : K\subset X,\ \ \mathcal{L}(T^nK : n\geqslant 0)=X\},$$

where card K is the cardinality of the set K. We will always work with separable spaces, so that $m_1(T)$ is a natural number or $\aleph_0$. In particular, simplicity of the spectrum of T (that is, the equation $m_1(T) = 1$) means that T has a cyclic vector. If the adjoint operator T^* has eigenvectors, then for T-cyclicity of an element

$x \in X$ it is necessary that $(x, f_\lambda) \neq 0$ for each $\lambda \in \sigma_p(T^*)$, and each eigenfunctional $f_\lambda \in E_\lambda^{T^*}$. Theorems which establish the converse are often called "closure theorems of Wiener type" (see [117] and [118]). If T^* is complete, the question of validity of a "closure theorem" is a dual form of the spectral analysis problem.

The multiplicity m_1 and the analogous "two-sided" multiplicity for an invertible operator T

$$m_2(T) = \inf\{\operatorname{card} K : K \subset X,\ \mathcal{L}(T^n K : n \in \mathbf{Z}) = X\}$$

became well known as spectral invariants of selfadjoint (or unitary, normal) operators in a Hilbert space. There they have many natural properties, for instance, additivity, which means that $m_i(T \oplus T) = 2m_i(T)$, $i = 1, 2$. We will consider the question of preserving additivity in the case when $T = S$, the shift operator. More precisely, let X be a space of generalized functions(36) on Γ which is invariant with respect to the shift operator S, where $Sf = zf$, $f \in X$; here, as before, z is the identity mapping on Γ. The norm $\|\cdot\|_X$ will usually be given in terms of the Fourier coefficients $\hat{f}(n)$, $n \in \mathbf{Z}$, which uniquely determine f. We will always assume that the set $\mathcal{P}$ of all polynomials in z and z^{-1} is dense in X, so that $m_2(S) = 1$. One generating element is, for instance, the function $\mathbf{1} = z^0$. When it is necessary, we will impose conditions (like "non-quasi-analyticity") on X which guarantee that $m_1(S) = 1$. The main problem in this section consists of describing the properties of X which lead to a fall in the multiplicity of the direct product to $m_i(S \times S) = 1$, $i = 1, 2$ in contrast to the expected value $m_i(S \times S) = 2$. Our fundamental result, roughly speaking, consists of the fact that preservation of the multiplicity $m_i(S \times S) = 2$, $i = 1, 2$, occurs only when X is realized in terms of (ordinary) measurable functions on Γ. For instance, $m_i(S \times S) = 2$ for $i = 1, 2$ if $X = \hat{l}^p_{\mathbf{Z}}$ and $1 \leqslant p \leqslant 2$, and $m_i(S \times S) = 1$ if $p > 2$; here $\hat{l}^p_{\mathbf{Z}}$ is the space of all generalized functions f whose sequence of Fourier coefficients $(\hat{f})$ belongs to $l^p_{\mathbf{Z}}$. Similar notation will be used in other cases.

We point out an obvious parallel between the theorems in this section and the method of applying the capacity concept in the theory of trigonometric series (see [5], [39], or [146]).

The connection between multiplicity of the spectrum and simultaneous approximations is based on the definition of the numbers $m_i(S \times S)$. The equation $m_i(S \times S) = 1$, $i = 1, 2$, means that there exist "weight" functions $f, g \in X$ such that to each pair of functions h and k in X there corresponds a sequence of polynomials $\{p_n\}_{n \geqslant 1}$ ($p_n \in \mathcal{P}$ or only $P_n \in \mathcal{P}_A$, depending on whether $i = 2$ or $i = 1$) with the simultaneous approximation property

$$\lim_n \|h - p_n f\|_X = 0, \quad \lim_n \|k - p_n g\|_X = 0. \tag{1}$$

It is clear that these functions f and g must be weakly invertible in X. Therefore the

(36) Here we have in mind what is meant in [146] by the term "hyperdistributions": elements of X are formal power series $f = \sum_{n \in \mathbf{Z}} \hat{f}(n) z^n$ for which $\hat{f}(n) = O(e^{\epsilon|n|})$ as $|n| \longrightarrow \infty$ for an arbitrary $\epsilon > 0$.

possibility of achieving the approximation (1) is equivalent to completeness of the diagonal

$$=D(\mathscr{P}\times\mathscr{P})=\{\{p,p\}:p\in\mathscr{P}\} \quad \text{or} \quad D_A=D(\mathscr{P}_A\times\mathscr{P}_A)$$

in the direct product $X_f \times X_g$, where X_f and X_g are spaces with weighted norms (for instance, $\|h\|_f = \|hf\|_X$). Later we will show that existence of f and g for which D or D_A is dense in $X_f \times X_g$ depends on completeness of the diagonal D in the space $X \times \Sigma$, where Σ is the metric space of all (classes of) measurable functions on Γ.

Finally, we mention that questions on simultaneous approximations by polynomials (that is, questions on completeness of the diagonals D or D_A or the like in spaces of the type $X \times Y$) have been treated recently by many authors (see, for example, [107], [172] and [173]). It is not hard to see that each such question leads to a uniqueness theorem "with an additional condition". As an example of this kind of theorem we cite the lemma in §3.1.

We will use the following notation: $\mathscr{P}_A$ is the collection of all polynomials in z, and $\mathscr{P}$ is the collection of all polynomials in z and z^{-1}; $D = D(\mathscr{P}\times\mathscr{P}) \stackrel{\text{def}}{=} \{\{p, p\}: p \in \mathscr{P}\}$ and analogously for D_A; $\hat{f} = \{\hat{f}(n)\}_{n\in\mathbf{Z}}$ is the sequence of Fourier coefficients of a "hyperdistribution" f; Σ is the space of all (classes of) measurable functions on Γ with convergence in measure (denoted by $f_n \Rightarrow f$ or $(\Sigma)\text{-}\lim_n f_n = f$; distance in Σ: $\rho_\Sigma(f, g) = \int_\Gamma(|f-g|/(1+|f-g|))\,dl_1$); if X is a normed space which contains $\mathscr{P}$, then $\mathscr{P}^X$ is the space of polynomials with the norm induced by X; j is the natural imbedding of $\mathscr{P}$ (or $\mathscr{P}^X$) into Σ: $j(p) = p$, $p \in \mathscr{P}$.

In this section we will encounter spaces of three types: (1) *homogeneous spaces* (h.s.) of generalized functions (in the sense of Šilov [116]), namely, spaces such that $\|p_\alpha\|_X = \|p\|_X$ for each polynomial p and each α with $|\alpha| = 1$; here $p_\alpha(\zeta) = p(\alpha\zeta)$, $\zeta \in \Gamma$; (2) *spaces with a monotone norm* (s.m.n.), namely, spaces such that the conditions $x \in X$ and $|\hat{y}(n)| \leqslant |\hat{x}(n)|$, $n \in \mathbf{Z}$, imply that $y \in X$ and $\|y\|_X \leqslant \|x\|_X$; (3) *symmetric spaces* (s.s) of generalized functions (in the sense of Semenov [97], [98]), namely, those s.m.n. in which the norm is invariant with respect to permutations, i.e. the conditions $x \in X$ and $\hat{y}(n) = \hat{x}(k(n))$, $n \in \mathbf{Z}$, imply that $y \in X$ and $\|y\|_X = \|x\|_X$ for an arbitrary permutation k of the set $\mathbf{Z}$ of all integers.

All spaces will be regarded as being in $\mathcal{K}$-duality with their conjugates (see §2.3); that is, for each polynomial p and functional $\Phi \in X^*$ we have

$$\langle p, \Phi\rangle = \sum_{n\in\mathbf{Z}} \hat{p}(n)\hat{\Phi}(n),$$

where $\hat{\Phi}(n) = \hat{\Phi}(z^n)$, $n \in \mathbf{Z}$, and for Φ "with compact support"(37) ($\hat{\Phi}(n) \neq 0$ only for finitely many n) we will identify Φ and $\sum_{n\in\mathbf{Z}}\hat{\Phi}(n)z^{-n}$, and we write $\langle p, \Phi\rangle = \int p\Phi$.

2°. *Multiplicity of the spectrum and the imbedding $X \longrightarrow \Sigma$.* We will often encounter the following condition (A) imposed on the norm of a Banach space X:

(A) $L^p \subset X$ for some p, $1 \leqslant p \leqslant \infty$; for $p = \infty$ we also assume that each

(37) Sometimes for all Φ.

functional Φ which is continuous in X is generated by some function φ in L^1:

$$\langle f, \Phi\rangle = \int_\Gamma f\varphi\, dl_1, \quad f \in \mathcal{P}.$$

Here $L^p = L^p(\Gamma, l_1)$ is the space of all (classes of) functions on Γ which are pth power integrable (essentially bounded if $p = \infty$). From the closed graph theorem and condition (A) it follows that

$$\|f\|_X \leqslant \text{const}\, \|f\|_{L^p}, \quad f \in L^p.$$

This condition is used as the basis for establishing the equations $m_i(S \times S) = 1$, $i = 1, 2$. If X is comparable to L^p but assumption (A) is not satisfied, then $X \subset L^p$ and thus $m_i(S \times S) \geqslant 2$.

LEMMA 1. *If Σ_0 is a closed subspace of Σ which is invariant with respect to the shift operator S, then $\Sigma_0 = \chi_e \Sigma$, where e is a measurable subset of Γ and χ_e is its characteristic function.*

PROOF. Since $\mathcal{P}$ is dense in Σ (Luzin's theorem; see [5]) and Σ is a topological algebra with respect to the usual algebraic operations, it suffices to show that $z^{-1} \in E^S(1)$; that is, $(\Sigma)\text{-}\lim_n p_n = z^{-1}$ for an appropriate choice of the polynomials $p_n \in \mathcal{P}_A$. But by Runge's theorem (see [100]) z^{-1} can be uniformly approximated by polynomials in z on an arbitrary closed proper subset of Γ, and therefore $z^{-1} \in E^S(1)$. ●

LEMMA 2. *If the imbedding*

$$j : \mathcal{P}^X \to \Sigma \tag{2}$$

is continuous, then $m_1(S \times S) \geqslant 2$ and $m_2(S \times S) = 2$.

PROOF. Since $\mathcal{P}$ is dense in X, continuity of (2) implies that $X \subset \Sigma$, and convergence in X implies convergence in measure. It is also clear that $m_i(S \times S) \leqslant 2$, $i = 1, 2$. If $\{f, g\}$ is a cyclic (or 2-cyclic) element of the operator $S \times S$, then $f, g \neq 0$ almost everywhere(38) on Γ. Let $\{p_n\}_{n \geqslant 1}$ be a sequence of polynomials such that $\lim_n \{p_n f, p_n g\} = \{\mathbf{0}, \mathbf{1}\}$. Then $p_n \Rightarrow \mathbf{0}$; but this contradicts convergence of the second coordinate to $\mathbf{1}$. Consequently $m_i(S \times S) > 1$, $i = 1, 2$. ●

THEOREM 1. *Let X be a h.s. of generalized functions which satisfies condition (A).*

1) *If $m_2(S \times S) = 2$, then the operator* (2) *admits a closure.*

2) *If $m_1(S \times S) \geqslant 2$ and*

$$\sum_{n \in \mathbf{Z}} (n^2 + 1)^{-1} |\log \|S^n\|| < +\infty,$$

then $j | \mathcal{P}_A^X$ admits a closure.

The proofs of assertions 1) and 2) will be given together. Suppose that the

---

(38) By "almost everywhere" we mean with respect to the normalized Lebesgue measure l_1 unless we say something to the contrary.

operator (2) (or $j|\mathcal{P}_A^X$) cannot be closed. This means that the closure $\bar{D}$ (respectively, $\bar{D}_A$) of its graph $D = D(\mathcal{P} \times \mathcal{P})$ in the space $X \times \Sigma$ contains an element of the form $\{\mathbf{0}, f\}$, $f \neq \mathbf{0}$. On the other hand, D (respectively, $\bar{D}_A$) is closed and invariant with respect to $S \times S$ in the space $X \times \Sigma$. Therefore the intersection $\bar{D} \cap (\{\mathbf{0}\} \times \Sigma)$ is closed in $\{\mathbf{0}\} \times \Sigma$ and is invariant with respect to the same operator. By Lemma 1 we have $\bar{D} \cap (\{\mathbf{0}\} \times \Sigma) = \{\mathbf{0}\} \times (\chi_e \Sigma)$, where e is a measurable subset of Γ and χ_e is its characteristic function. Since $f \in \chi_e \Sigma$, we have $|e| = l_1(e) > 0$. But X is a homogeneous space and the set D (respectively, D_A) is invariant with respect to rotations of the circle. Consequently the same is true for $\chi_e \Sigma$; that is, $e = \Gamma$.

Thus the assumption that (2) (or $j|\mathcal{P}_A^X$) cannot be closed implies the existence of a sequence of polynomials $\{p_n\}_{n \geqslant 1}$ with $p_n \in \mathcal{P}$ (respectively, $p_n \in \mathcal{P}_A$) such that

$$\lim_n \|p_n\|_X = 0, \qquad (\Sigma)\text{-}\lim_n p_n = \mathbb{1}.$$

Let $\mathfrak{S}_n = \{\zeta \in \Gamma: |p_n(\zeta) - 1| \geqslant \epsilon_n\}$, where the $\epsilon_n > 0$ are given numbers. These numbers can be chosen so that $\epsilon_n \downarrow 0$ and $\Sigma_{n \geqslant 1} |\mathfrak{S}_n| < +\infty$. We set

$$e_k = \bigcap_{n=k}^{\infty} (\Gamma \setminus \mathfrak{S}_n) = \Gamma \setminus \left(\bigcup_{n=k}^{\infty} \mathfrak{S}_n \right), \quad k = 1, 2, \ldots.$$

Then $e_k \subset \Gamma \setminus \mathfrak{S}_k$, $e_k \subset e_{k+1}$, and $\lim_k e_k = 1$.

Let $\Delta_k = e_{k+1} \setminus e_k$, $k \geqslant 1$, $\Delta_0 = e_1$, and

$$f = \sum_{k \geqslant 0} c_k \chi_{\Delta_k},$$

where we choose positive numbers c_k which converge so rapidly to zero that:

(a) $\sum_{k \geqslant 0} c_k < +\infty$; (b) $\sum_{k \geqslant 0} |\Delta_k| \log c_k = -\infty$; (c) $\lim_n \left\| (p_n - 1) \sum_{k \geqslant n} c_k \chi_{\Delta_k} \right\|_X = 0$.

In order to verify (c) it suffices to use condition (A) and to take, for example,

$$|c_k| < 2^{-k} (\max_{1 \leqslant j \leqslant k} \|p_j - 1\|_{L^\infty})^{-1}, \quad k \geqslant 1.$$

Therefore

$$\|p_n f - f\|_X \leqslant \left\| (p_n - 1) \sum_{k=0}^{n-1} c_k \chi_{\Delta_k} \right\|_X + o(1) \leqslant \text{const} \left\| (p_n - 1) \sum_{k=0}^{n-1} c_k \chi_{\Delta_k} \right\|_{L^\infty} + o(1)$$
$$\leqslant \text{const} \sum_{k=0}^{n-1} |c_k| \cdot \|(p_n - 1) \chi_{\Delta_k}\|_{L^\infty} + o(1) = o(1),$$

since $\Delta_k \subset e_n \subset \Gamma \setminus \mathfrak{S}_n$ and $\|(p_n - 1)\chi_{\Delta_k}\|_{L^\infty} \leqslant \epsilon_n$, $1 \leqslant k < n$. Consequently

$$\lim_n \{p_n f, p_n\} = \{f, \mathbf{0}\}$$

in $X \times X$.

Now we note that f is an S-cyclic element of X. This follows from conditions (A) and (b) and the Helson-Lowdenslager theorem (if $p < +\infty$) or its refinement which we proved in §2.2. (if $p = +\infty$). Therefore $X \times \{\mathbf{0}\} \subset E(\{f, \mathbf{1}\})$, and in case 1) the proof is complete since $\{\mathbf{0}, \mathbf{1}\} = \{f, \mathbf{1}\} - \{f, \mathbf{0}\} \in E(\{f, \mathbf{1}\})$.

For the proof of assertion 2) we use a theorem of Šilov (see [18] or [114]), according to which $l^1_{\mathbf{Z}}(|S^n|)$ is a regular Banach algebra. It follows that there exist infinitely differentiable (on Γ) functions μ, $\mu \neq 0$, with arbitrarily small connected supports $\gamma \subset \Gamma$ whose sequences of Fourier coefficients are contained in $l^1_{\mathbf{Z}}(|S^n|)$. We choose a sequence of open arcs $\gamma_n \subset \Gamma$ and corresponding functions μ_n such that $\overline{\gamma}_n = \operatorname{supp} \mu_n$ and $\gamma_n \cap \gamma_k = \varnothing$ for $n \neq k$, and such that $\gamma = \bigcup_1^\infty \gamma_n$ is dense in Γ, $|\gamma| = 1$. Then for sufficiently small positive coefficients b_n the function

$$\mu = \sum_{n \geqslant 1} b_n \mu_n$$

has the following properties: $\mu(\zeta) \neq 0$ for almost all $\zeta \in \Gamma$; $\int_\Gamma \log|\mu|\, dl_1 = -\infty$; $\{\hat{\mu}(n)\}_{n\in\mathbf{Z}} \in l^1_{\mathbf{Z}}(|S^n|)$. The latter inclusion guarantees continuity of the multiplication operator $p \longrightarrow p\mu$, $p \in \mathcal{P}$, since $p\mu = \Sigma_{n\in\mathbf{Z}}\hat{\mu}(n)S^n p$, and therefore

$$\|p\mu\|_X \leqslant \sum_{n\in\mathbf{Z}} |\hat{\mu}(n)| \cdot \|S^n\| \cdot \|p\|_X \leqslant \mathrm{const}\, \|p\|_X.$$

Consequently

$$\lim_n \{p_n f,\ p_n \mu\} = \{f,\ \mathbf{0}\}$$

in $X \times X$ and, as before, $X \times \{\mathbf{0}\} \subset E(\{f, \mu\})$. Also, $\{\mathbf{0}, \mu\} = \{f, \mu\} - \{f, \mathbf{0}\} \in E(\{f, u\})$. We again apply the Helson-Lowdenslager theorem and find that $\{\mathbf{0}\} \times X \subset E(\{f, \mu\})$. Thus $\{f, \mu\}$ is a cyclic element of $S \times S$. ●

REMARKS. 1. It is not hard to see that in the process of proving Theorem 1 we established the equivalence of completeness of the diagonal $D(\mathcal{P} \times \mathcal{P})$ in $X \times \Sigma$ and the lack of a closure of $j\colon \mathcal{P}^X \longrightarrow \Sigma$. An analogous remark can be made in the case of "analytic" polynomials $\mathcal{P}_A$.

2. In [157] Newman considered S-invariant subspaces in $l^p_{\mathbf{Z}}$, and for the case $X = \hat{l}^p_{\mathbf{Z}}$ he introduced the concept of an X-generating set. Adapting Newman's definition to our case, we will say that a set $e \subset \Gamma$ is *X-generating* if the set $\{f\colon f \in C, f|e = \mathbf{0}\}$ is dense in X. It is clear that the imbedding j cannot be closed (and therefore $m_2(S \times S) = 1$) if there exist X-generating sets of Lebesgue measure arbitrarily close to 1.

3. An essential part of the proof of Theorem 1 was the construction of a sequence of polynomials $\{p_n\}$ for which $\lim_n \|p_n\|_X = 0$ but $(\Sigma)\text{-}\lim_n p_n = 1$. From Egorov's theorem on almost uniform convergence, Uryson's extension theorem, and condition (A) it follows that this property is equivalent to the existence of closed sets e of Lebesgue measure arbitrarily close to 1 such that

$$0 = \inf\{\|f\|_X : f \in C,\ f(\zeta) = 1,\ \zeta \in e\} \stackrel{\text{def}}{=} \operatorname{cap}_X e.$$

The definition of the function cap_X is close to the definition of capacity in potential theory (see [146]). Later applications of Theorem 1 will confirm the similarity of the definitions.

Lemma 2 and Theorem 1 reduce the question of multiplicity of the spectrum in $X \times X$ to the possibility of establishing the inclusion $X \subset \Sigma$. We consider some properties of the norm $\|\cdot\|_X$ which are connected with this possibility of realizing X in terms of measurable functions. One such property is the following:

(B) $(\|p_n\|_X \leqslant 1,\ p_n \in \mathcal{P}, \text{ and } p_n \Rightarrow \mathbf{0}) \Rightarrow (p_n \xrightarrow{w.X} \mathbf{0})$.

LEMMA 3. *Let X be a h.s. of generalized functions having property* (A).

1) *If* (B) *holds, then* (2) *admits a closure.*

2) *Let* (2) *admit a closure, and let one of the following conditions be satisfied*: (i) *X is reflexive*; (ii) *the polynomials are dense in X^**; (iii) *weak convergence on the unit ball in X coincides with convergence of the Fourier coefficients. Then* (B) *holds.*

PROOF. 1) As we noted in the proof of Theorem 1, if (2) cannot be closed, then there exists a sequence of polynomials $\{p_n\}_{n\geqslant 1}$ such that $\|p_n^{-1}\|_X \to 0$ as $p_n \Rightarrow \mathbf{0}$. This is not compatible with (B).

2) If (B) does not hold, then there exists a sequence of polynomials $\{p_n\}_{n\geqslant 1}$ such that

$$\|p_n\|_X \leqslant 1, \quad p_n \Rightarrow \mathbf{0}, \quad p_n \overset{w.X}{\not\longrightarrow} \mathbf{0}.$$

It is not hard to see that for properties (i)–(iii) we have (iii) ⇒ (i) ⇒ (ii) ⇒ (iii),[39] so that it suffices to assume that the last one is true. Thus there exists $k \in \mathbf{Z}$ such that $\overline{\lim}_n |\hat{p}_n(k)| > 0$. If we multiply the polynomials p_n by z^{-k} and pass to a subsequence if necessary, we can see that

$$\lim_n \hat{p}_n(0) = l \neq 0; \quad \sup_n \|p_n\|_X < +\infty; \quad (\Sigma)\text{-}\lim_n p_n = \mathbf{0}.$$

We again pass to a subsequence in order to get weak convergence of $\{p_n\}_{n\geqslant 1}$ in X; that is, $p_n \xrightarrow{w.X} f$ ($f \in X$, $\hat{f}(0) = l \neq 0$); and also to get convergence of the series $\sum_{n\geqslant 1} \rho_\Sigma(p_n, \mathbf{0}) < +\infty$. By a theorem in linear analysis (see [11]) there exist convex combinations

$$q_n = \sum_{k=n}^{m(n)} c_{k,n}\, p_k; \quad c_{k,n} \geqslant 0, \quad \sum_{k=n}^{m(n)} c_{k,n} = 1,$$

such that $\lim_n \|q_n - f\|_X = 0$. At the same time

$$\rho_\Sigma(q_n, \mathbf{0}) \leqslant \sum_{k=n}^{m(n)} \rho_\Sigma(c_{k,n} p_k, \mathbf{0}) \leqslant \sum_{k=n}^{m(n)} \rho_\Sigma(p_k, \mathbf{0}) \to 0, \quad n \to \infty.$$

Now let T be a rotation operator (through an angle which is not commensurable with π) in X. By an ergodic theorem (see [30] or §3.3) the mean

$$A_n = \frac{1}{n+1} \sum_{j=0}^{n} T^j, \quad n \geqslant 1,$$

(39) (iii) ⇒ (i) since (iii) implies weak compactness of the unit ball in X; (i) ⇒ (ii) since the polynomials are weakly ($\sigma(X^*, X)$) dense in X^*; (ii) ⇒ (iii) by the Banach-Steinhaus theorem.

converges strongly to the spectral projector onto the function $\mathbf{1}$. Since this convergence is uniform on compact subsets of X, we have $\lim_n \|A_n q_n - A_n f\|_X = 0$; that is,

$$\lim_n \| A_n q_n - l\mathbf{1} \|_X = 0.$$

On the other hand, since $\Sigma_n \rho_\Sigma(q_n, \mathbf{0}) < \infty$, we obtain

$$\rho_\Sigma(A_n q_n, \mathbf{0}) \leqslant \rho_\Sigma\left(\frac{1}{n+1}\sum_{j=0}^{N} T^j q_n, \mathbf{0}\right) + \sum_{j=N+1}^{n} \rho_\Sigma\left(\frac{1}{n+1} T^j q_n, \mathbf{0}\right)$$

$$\leqslant \left\| \frac{1}{n+1}\sum_{j=0}^{N} T^j q_n \right\|_{L^\infty} + \sum_{j=N+1}^{\infty} \rho_\Sigma(q_n, \mathbf{0})$$

for each N, $1 \leqslant N \leqslant n$. Therefore

$$\overline{\lim_n}\, \rho_\Sigma(A_n q_n, \mathbf{0}) \leqslant \sum_{n=N+1}^{\infty} \rho_\Sigma(q_n, \mathbf{0})$$

for each $N \geqslant 1$; that is,

$$\lim_n \rho_\Sigma(A_n q_n, \mathbf{0}) = 0.$$

Thus

$$\lim_n \| A_n q_n - l \cdot \mathbf{1} \|_X = 0,$$

but

$$(\Sigma)\text{-}\lim_n (A_n q_n - l \cdot \mathbf{1}) = -l \cdot \mathbf{1} \neq \mathbf{0};$$

that is, j is not closable. ●

REMARK. If in assertion 2) the condition of closability of j is replaced by closability of $j|\mathcal{P}_A$, then by the same arguments we obtain an "analytic" version of property (B): $\|p_n\|_X \leqslant 1$, $p_n \in \mathcal{P}_A$, and $p_n \Rightarrow \mathbf{0}$ implies that $p_n \xrightarrow{\text{w.X}} \mathbf{0}$.

THEOREM 2. *Let X be a h.s. of generalized functions which satisfies condition* (A) *and one of the three conditions* (i)–(iii) *in Lemma* 3. *If convergence in the norm of X implies convergence in measure, then $m_i(S \times S) \geqslant 2$, $i = 1, 2$. If on the unit ball in X convergence in measure does not imply weak convergence* (*of sequences*), *then $m_2(S \times S) = 1$. The same conclusion is valid for m_1* ($m_1(S \times S) = 1$) *if the preceding assumption concerning convergence holds on the ball in $X \cap \mathcal{P}_A$ and*

$$\sum_{n\in \mathbf{Z}} (n^2+1)^{-1} |\log \|S^n\|| < +\infty.$$

The proof follows immediately from Theorem 1, Lemmas 2 and 3, and the remark after Lemma 3. ●

Lemma 4. *Let X be a s.m.n. of generalized functions. Then $X \subset \Sigma \Leftrightarrow X \subset \hat{l}^2_{\mathbf{Z}}$, where the first inclusion means continuity of the operator* (2).

PROOF. The implication $\Leftarrow$ is obvious; we will prove the implication $\Rightarrow$. Let

$$l(f,\ t)=l_1\{\zeta\in\Gamma:|f(\zeta)|>t\}\quad\text{and}\quad l(t)=\sup_{\|f\|_X\leqslant 1} l(f,\ t).$$

Continuity of the imbedding $\mathcal{P}^X \longrightarrow \Sigma$ implies that $\lim_{t\to\infty} l(t) = 0$. Consequently there exists a monotone function ω on $[0, +\infty)$, $\omega \uparrow +\infty$, such that

$$\int_\Gamma \omega(|f(\zeta)|)\,dl_1 \leqslant \text{const for}\quad \|f\|_X\leqslant 1,\quad f\in\mathcal{P}.$$

If there exists an $f \in X\backslash \hat{l}^2_{\mathbf{Z}}$, then by the Paley-Zygmund theorem (see [39], Vol. 1, Chapter V, Theorem 8.28) we have

$$\sup_{0\leqslant r<1}\int_\Gamma \omega\left(\left|\sum_{n\in\mathbf{Z}} \pm r^{|n|}\hat{f}(n)\zeta^n\right|\right)dl_1(\zeta)=+\infty$$

for almost all distributions of the symbols $\pm$. Therefore $X \subset \hat{l}^2_{\mathbf{Z}}$. ●

Lemma 4 shows that a space X with a monotone norm cannot be realized by measurable functions if it is not imbedded in $\hat{l}^2_{\mathbf{Z}}$. To conclude this subsection we mention two more facts which are connected with simultaneous weighted approximations and the equations $m_i(S \times S) = 1$.

1) Whenever we establish 2-cyclicity of the operator $S \times S$ we are dealing with simultaneous completeness of the system of functions $\{z^n f, z^n\}$, $n \in \mathbf{Z}$, where $f \in X$. If we make a suitable definition of the product of elements in X with functionals in X^* (see [146]), then completeness of this system of functions in $X \times X$ is equivalent to the relation $f\Phi \notin X^*$ for each $\Phi \in X^*$.

2) The inclusion $X \subset \Sigma$ enables us to consider restrictions of generalized functions in X to arbitrary measurable subsets of Γ. In this area we can formulate another condition which guarantees that $m_2(S \times S) = 2$; namely, let f be a weakly invertible (2-cyclic) function in X and let $\{p_n\}_{n\geqslant 1}$ be a sequence of polynomials, $p_n \in \mathcal{P}$; if $\sup_{n\geqslant 1}\|p_n f\|_X < \infty$, then the set

$$\{\Phi:\Phi\in X^*,\ \sup_{n\geqslant 1}\|p_n\Phi\|_{X^*}<\infty\}$$

is (weakly) complete in X^*. We will show that this condition implies that $m_2(S \times S) = 2$. For if $\{f, g\}$ is a 2-cyclic vector, then the set $\{pf\colon p \in \mathcal{P}\}$ is dense in X. Suppose that for some sequence of polynomials we have $\lim_n\{p_n f, p_n g\} = \{\mathbf{0}, h\}$. Let $\{\Psi, \Phi\}$ be a functional in $X \times X$, where Φ belongs to the total set described above. Suppose that the number $\epsilon > 0$ and the polynomial q are chosen so that $\|g - qf\|_X < \epsilon$. Then

$$|\langle h,\ \Phi\rangle|=\lim_n|\langle p_n g,\ \Phi\rangle|\leqslant\overline{\lim_n}|\langle g-qf,\ p_n\Phi\rangle|$$

$$+\overline{\lim_n}|\langle p_n f,\ q\Phi\rangle|=\overline{\lim_n}|\langle g-qf,\ p_n\Phi\rangle|\leqslant\epsilon\sup_{n\geqslant 0}\|p_n\Phi\|_{X^*},$$

and since $\epsilon > 0$ was arbitrary, we have $\langle h, \Phi\rangle = 0$. Consequently $h = \mathbf{0}$ and $m_2(S \times S) = 2$.

A connection between 2) and the "localness" of the space X is illustrated by the following example: if $X = L^r$ with $2 \leqslant r \leqslant \infty$, then 2-cyclicity of f is equivalent to the condition that $f(\zeta) \neq 0$ for almost all $\zeta \in \Gamma$; and then the set

$$\left\{\Phi : \Phi \in L^{r'},\ \frac{1}{r}+\frac{1}{r'}=1,\quad \sup_{n\geqslant 1}\|p_n\Phi\|_{L^{r'}}<+\infty\right\},$$

where $\sup_{n\geqslant 1}\|p_n f\|_{L^r}<\infty$, contains all of the characteristic functions χ_e such that $\operatorname{ess\,inf}_e|f|>0$. It is clear that such a set is total in L^r.

3°. *Simultaneous approximations and lacunary polynomials.* One method for showing that the imbedding $j\colon \mathcal{P}^X\longrightarrow\Sigma$ is not closable consists of considering polynomials whose Fourier coefficients are nonzero only on a sparse subset of $\mathbf{Z}$.

LEMMA 5. *Let X be a h.s. of generalized functions, and suppose that*

$$\lim_{|n|\to\infty}\|z^n\|_X=0.$$

Then the operators (2) *and $j|\mathcal{P}_A^X$ are not closable.*

PROOF. Since $\|z^n\|_X=O(1)$ as $|n|\longrightarrow\infty$, we have $\hat{l}^1_{\mathbf{Z}}\subset X$. If $f\in\hat{l}^1_{\mathbf{Z}}$ and $\hat{f}(0)=0$, then

$$\overline{\lim_n}\|f(z^n)\|_X\leqslant\overline{\lim_n}\left\|\sum_{k=-N}^{N}\hat{f}(k)z^{kn}\right\|_X+\overline{\lim_n}\left\|\sum_{|k|>N}\hat{f}(k)z^{kn}\right\|_X\leqslant\text{const}\sum_{|k|>N}|\hat{f}(k)|$$

for each $N\geqslant 1$, and therefore $\lim_n\|f(z^n)\|_X=0$. Let f_ϵ be a continuously differentiable function on Γ which equals 1 on a set e_ϵ of Lebesgue measure greater than $1-\epsilon$, with $\hat{f}(0)=0$. Then $\lim_n\|f_\epsilon(z^n)\|_X=0$. Now we note that the preimage of the set e_ϵ under the mapping z^n has the same measure as e_ϵ, and therefore $f_\epsilon(\zeta^n)=1$ on a set of points $\zeta\in\Gamma$ of measure greater than $1-\epsilon$. If we choose appropriate sequences $\{\epsilon_k\}_{k\geqslant 1}$, $\lim_k\epsilon_k=0$, and $\{n(k)\}_{k\geqslant 1}$, $\lim_k n(k)=\infty$, we obtain

$$\lim_k\|f_{\epsilon_k}(z^{n(k)})\|_X=0,\qquad(\Sigma)\text{-}\lim_k f_{\epsilon_k}(z^{n(k)})=\mathbf{1}.$$

In order to show that j is not closable, we just have to produce an arbitrarily close approximation of $f_{\epsilon_k}(z^{n(k)})$ by polynomials in $\mathcal{P}$ (in the norm $\|\cdot\|_{\hat{l}^1_{\mathbf{Z}}}$, for example).

For the operator $j|\mathcal{P}_A$ we choose an outer (in the sense of Beurling; see §§2.1 and 2.2) function g_ϵ which is differentiable on Γ and has a modulus equal to ϵ on a set of measure greater than $1-\epsilon$, with $\hat{g}_\epsilon(0)=1$. If we set $f_\epsilon=\mathbf{1}-g_\epsilon$, we can finish the proof as in the preceding case. ●

LEMMA 6. *Let X be a h.s. of generalized functions such that $\hat{l}^2_{\mathbf{Z}}\subset X\subset\hat{l}^\infty_{\mathbf{Z}}$. Let the following condition hold*:

(C) $\quad\left(\|f_n\|_{\hat{l}^2_{\mathbf{Z}}}=O(1)\ \text{ and }\ \|f_n\|_{\hat{l}^\infty_{\mathbf{Z}}}=o(1)\right)\Rightarrow\left(\|f_n\|_X=o(1)\right),$

Then the operators (2) *and $j|\mathcal{P}_A$ are not closable.*

The proof makes use of a method of Newman and Katznelson (see [148] and [157]) which they applied to the investigation of uniqueness sets for the classes $\hat{l}^p_{\mathbf{Z}}$. Let f_ϵ be the function in the proof of Lemma 5, chosen in such a way that

$$\int_{\Gamma}|f_\epsilon|\,dl_1\leqslant\text{const},\quad\max_{\Gamma}|f_\epsilon|\leqslant\text{const}\,\epsilon^{-1},$$

where const is a constant which does not depend on ϵ. Suppose that the natural numbers n_k, $k = 1, \ldots, N$, are relatively prime, and let $m = \min\{n_k - 1: 1 \leqslant k \leqslant N\}$. Then

$$\left\| \frac{1}{N} \sum_{k=1}^{N} f_\varepsilon (z^{n_k}) \right\|_{l^2_{\mathbf{Z}}} \leqslant \left\| \frac{1}{N} \sum_{k=1}^{N} \sum_{j=-m}^{m} \hat{f}_\varepsilon (j) z^{jn_k} \right\|_{l^2_{\mathbf{Z}}} + \left\| \sum_{|j|>m} \hat{f}_\varepsilon (j) z^j \right\|_{l^2_{\mathbf{Z}}}$$

$$= \left(\frac{N}{N^2} \sum_{j=-m}^{m} |\hat{f}_\varepsilon (j)|^2 \right)^{1/2} + \left\| \sum_{|j|>m} \hat{f}_\varepsilon (j) z^j \right\|_{l^2_{\mathbf{Z}}} \tag{3}$$

$$\leqslant \text{const}\, (N\varepsilon)^{-1/2} + \left\| \sum_{|j|>m} \hat{f}_\varepsilon (j) z^j \right\|_{l^2_{\mathbf{Z}}}.$$

An analogous estimate in the space $l^\infty_{\mathbf{Z}}$ gives us

$$\left\| \frac{1}{N} \sum_{k=1}^{N} f_\varepsilon (z^{n_k}) \right\|_{l^\infty_{\mathbf{Z}}} \leqslant \text{const}\, N^{-1} + \sup_{|j|>m} |\hat{f}_\varepsilon (j)|. \tag{4}$$

We choose a positive sequence $\{\delta_n\}_{n\geqslant 1}$, $\lim_n \delta_n = 0$, and for each n we find numbers N_n and ϵ_n, $N_n\epsilon_n = \delta_n$ ($\lim N_n = \infty$), and a number $m = m(n)$ which depends on them such that[40]

$$\left\| \frac{1}{N_n} \sum_{k=1}^{N_n} f_{\varepsilon_n}(z^{n_k}) \right\|_X < \frac{1}{n}.$$

Since the function

$$F_n = N_n^{-1} \sum_{k=1}^{N_n} f_{\varepsilon_n} (z^{n_k})$$

is different from 1 only on the union of the preimages of the set $\Gamma\backslash e_{\epsilon_n}$ under the mappings z^{n_k}, $k = 1, \ldots, N_n$, the total measure of this union does not exceed $N_n\epsilon_n = \delta_n$. Consequently

$$\lim_n \|F_n\|_X = 0, \quad (\Sigma)\text{-}\lim_n F_n = 1,$$

and the operator (2) is not closable.

The "analytic" case of the imbedding $j|\mathcal{P}^X_A$ is considered as in Lemma 5. ●

THEOREM 3. *Let X be a h.s. of generalized functions for which the hypothesis of Lemma 6 or Lemma 5 and also condition* (A) *are satisfied. Then $m_2(S \times S) = 1$, and if*

$$\sum_{n\in\mathbf{Z}} (n^2 + 1)^{-1} |\log \| S^n \|| < +\infty \tag{5}$$

we also have $m_1(S \times S) = 1$.

The proof of Theorem 3 follows immediately from Lemmas 5 and 6 and Theorem 1.●

---

(40) Here we use inequalities (3) and (4) and condition (C).

As we noted in subsection 1°, applications to concrete spaces for this theorem and other theorems will be presented in subsection 6°. We now introduce another simple corollary.

COROLLARY. *If X is a h.s. of generalized functions and the imbedding* $\hat{l}^2_{\mathbf{Z}} \subset X$ *is completely continuous, them* $m_2(S \times S) = 1$, *and under the non-quasi-analyticity condition* (5) *we also have* $m_1(S \times S) = 1$.

In fact, complete continuity of the imbedding implies that $\lim_n \|z^n\|_X = 0$. ●

4°. *Simultaneous approximations and singular elements of X.* For generalized functions (or "hyperdistributions"), as they were defined in subsection 1°, the concept of a (closed) support makes sense (see [146]). By the support supp f of a generalized function $f = \Sigma_{n\in\mathbf{Z}}\hat{f}(n)z^n$ we mean the smallest closed subset of Γ in the complement $\mathbf{C}\backslash\operatorname{supp} f$ of which the "Cauchy-Carleman transform" of f is singular:

$$(\mathscr{K}f)(\zeta)=\begin{cases}\sum\limits_{n\geqslant 0}\hat{f}(n)\zeta^n, & |\zeta|<1,\\ -\sum\limits_{n\leqslant -1}\hat{f}(n)\zeta^n, & |\zeta|>1.\end{cases}$$

A function f is said to be singular if $l_1(\operatorname{supp} f) = |\operatorname{supp} f| = 0$. The convolution $f * g$ of generalized functions f and g is the generalized function h for which $\hat{h}(n) = \hat{f}(n)\hat{g}(n)$, $n \in \mathbf{Z}$. It is known (see Appendix I in [146]) that $\operatorname{supp}(f * g) \subset (\operatorname{supp} f)\cdot(\operatorname{supp} g)$, and our definition of the support coincides with the usual definition if f is integrable on Γ (more precisely, if $\hat{f}(n)$, $n \in \mathbf{Z}$, are the Fourier coefficients of an integrable function).

LEMMA 7. *Let X be a h.s. of generalized functions for which there exists an increasing, logarithmically convex sequence* $\{w_n\}_{n\geqslant 0}$, $w_n \geqslant 1$, *such that*

$$\sum_{n\geqslant 0}(n^2+1)^{-1}\log w_n<+\infty \text{ and } \sup_{n\in\mathbf{Z}}|\hat{g}(n)|\,w_{|n|}^{-1}<+\infty$$

for all $g \in X$. *If X contains a nonzero singular element f, then the operator* $j\colon \mathfrak{P}^X \to \Sigma$ *is not closable.*

PROOF. First we observe that if $\mathfrak{M}$ is a finite measure on Γ and $g \in X$, then

$$\|g*\mathscr{M}\|_X=\Big\|\int_{\Gamma} g_\zeta\, d\mathscr{M}(\zeta)\Big\|_X\leqslant\|g\|_X\cdot\operatorname{Var}\mathscr{M},$$

since X is a homogeneous space; here $g_\zeta(t) = g(\zeta t)$, ζ, $t \in \Gamma$. By a theorem of Šilov (see [18] and [114]) the algebra $\hat{l}^1_{\mathbf{Z}}(w_{|n|})$ is regular, and therefore it contains a family of functions $\{\Delta_n\}_{n\geqslant 1}$ which "approximates unity":

$$\lim_n \hat{\Delta}_n(k)=1,\qquad k\in\mathbf{Z},\quad \sup_{n\geqslant 1}\int_{\Gamma}|\Delta_n|\,dl_1<+\infty,$$

$$\operatorname{supp}\Delta_n\subset\{\zeta\in\Gamma:|\arg\zeta|<1/n\},\quad n=1,2,\ldots.$$

Therefore the functions $f * \Delta_n$, $n \geqslant 1$, belong to $\hat{l}^1_{\mathbf{Z}}$, and

$$\overline{\lim_n} |\operatorname{supp}(f * \Delta_n)| \leqslant \overline{\lim_n} |(\operatorname{supp} f) \cdot (\operatorname{supp} \Delta_n)| = 0. \tag{6}$$

Since

$$\| g * \Delta_n \|_X \leqslant \text{const} \| g \|_X$$

for each $g \in X$, and

$$\lim_n \| z^k * \Delta_n - z^k \|_X = 0, \quad k \in \mathbf{Z},$$

by the Banach-Steinhaus theorem we have

$$\lim_n \| f * \Delta_n - f \|_X = 0.$$

From (6) it follows that $(\Sigma)\text{-}\lim(f * \Delta_n) = \mathbf{0}$. If we replace the functions Δ_n by the arithmetic means (of high order) of partial sums of their Fourier series, we obtain a sequence of polynomials $\{q_n\}_{n \geqslant 1}$ for which

$$\lim_n \| q_n - f \|_X = 0, \quad (\Sigma)\text{-}\lim_n q_n = \mathbf{0}.$$

If we apply the same procedure of "ergodic averaging" (with the operators A_n, $n \geqslant 1$) as in the proof of part 2) of Lemma 3, we obtain polynomials p_n, $n \geqslant 1$, with the properties

$$\lim_n \| p_n - \mathbf{1} \|_X = 0, \quad (\Sigma)\text{-}\lim_n p_n = \mathbf{0}. \bullet$$

THEOREM 4. *Let X be a h.s. of generalized functions which satisfies the hypothesis of Lemma 7 and condition* (A). *Then $m_2(S \times S) = 1$.*

The proof follows from Lemma 7 and Theorem 1. ●

REMARK. The proof implies that if there exist nonzero singular elements, the space X does not have property (B).

5°. *Multiplicity of the spectrum and Banach algebras. A countably multiple shift with a simple spectrum.* In this subsection we introduce another criterion for simplicity of the spectrum of the operator $S \times S$, and we show that under additional mild assumptions the equation $m_i(S \times S) = 1$ implies that $m_i(S \times \cdots \times S) = 1$ and even $m_i(\Pi_1^\infty \oplus S) = 1$.

THEOREM 5. *Let X be a h.s. of generalized functions which satisfies condition* (A), *and suppose that the conjugate*[41] *space X^* is a regular Banach algebra with pointwise multiplication on Γ. Then $m_2(S \times S) = 1$.*

PROOF. First we show that a function $f \in X$ is 2-cyclic if and only if $\operatorname{supp} f = \Gamma$. From the hypothesis it follows that

$$\lim_{|n| \to \infty} \| z^n \|_{X^*}^{1/|n|} = 1,$$

and therefore the space of maximal ideals in the algebra X^* can be identified with a

(41) Recall that X and X^* are in K-duality.

subset of Γ. Since X is homogeneous, this subset coincides with Γ. If $f \in X$ is not a 2-cyclic element, then the polar $\mathfrak{L}^{\perp}$, $\mathfrak{L} = \mathfrak{L}(S^n f\colon n \in \mathbf{Z})$, is a proper ideal in X^*. Suppose that γ, $\varnothing \neq \gamma \neq \Gamma$, is the closed subset of Γ consisting of the common zeros of the functions in $\mathfrak{L}^{\perp}$ (in other words, the maximal ideals containing $\mathfrak{L}^{\perp}$). The set $\mathfrak{L}^{\perp}$ contains all smooth (for instance, infinitely differentiable) functions which vanish in a neighborhood of γ (see [18]). Consequently f is orthogonal to all such functions, and therefore supp $f \subset \gamma$. Thus the equation supp $f = \Gamma$ implies that f is 2-cyclic in X. The converse is clear, for if supp $f \neq \Gamma$ and the algebra X^* is regular, then there exist functions $\Phi \neq \mathbf{0}$, $\Phi \in X^*$, with $\langle z^n f, \Phi \rangle = 0$, $n \in \mathbf{Z}$.

From this cyclicity criterion it is easy to deduce the fact that the operator j: $\mathcal{P}^X \longrightarrow \Sigma$ is not closable. Here we choose a sequence $\{e_n\}_{n \geqslant 1}$ of closed, nowhere dense subsets of Γ with $\lim_n |e_n| = 1$. Let f_n be a smooth function on Γ which vanishes precisely on the set e_n, $n \geqslant 1$. Since f_n is 2-cyclic, there exists a polynomial $q_n \in \mathcal{P}$ such that $\|q_n f_n - \mathbf{1}\|_X < 1/n$, $n \geqslant 1$. It is also clear that $(\Sigma)\text{-}\lim_n f_n q_n = \mathbf{0}$. If we select polynomials p_n, $n \geqslant 1$, where are close to $q_n f_n$ in the uniform norm $\|\cdot\|_{L^\infty}$ and use condition (A), we obtain

$$\lim_n \| p_n - \mathbf{1} \|_X = 0 \quad \text{and} \quad (\Sigma)\text{-}\lim (\mathbf{1} - p_n) = \mathbf{1}.$$

Thus j is not closable, and by Theorem 1 we have $m_2(S \times S) = 1$. ●

We note that the case $X = \hat{c}^0_{\mathbf{Z}}$, $X^* = \hat{l}^1_{\mathbf{Z}}$ is contained in [157].

The theorem on simplicity of the spectrum of a countably multiple shift operator, toward which we are proceeding, means in the language of weighted approximations that we can approximate simultaneously (with corresponding weight factors) a countable collection of functions in the original space. More precisely, let X be a space of generalized functions on Γ with a monotone norm $|\cdot|_X$ (s.m.n.), and let E be a separable normed space with a norm $|\cdot|$. We let $X(E)$ denote the collection of all E-valued generalized functions $\mathbf{f}$ on Γ, $\mathbf{f} = \Sigma_{n \in \mathbf{Z}} \hat{\mathbf{f}}(n) z^n$, $\hat{\mathbf{f}}(n) \in E$, for which $|\mathbf{f}| \overset{\text{def}}{=} \Sigma_{n \in \mathbf{Z}} |\hat{\mathbf{f}}(n)| z^n \in X$, with the norm $\|\mathbf{f}\| \overset{\text{def}}{=} \|\mathbf{f}\|_X$. The operator norm in X will be denoted by $\|\cdot\|$; the right shift operator in X will be denoted by S; and the right shift operator in $X(E)$ will be denoted by S_E.

THEOREM 6. *Let X be a s.m.n. of generalized functions on Γ, let E be a separable Banach space, and let $X(E)$ be the space of E-valued generalized functions with a finite X-norm $\|\cdot\|$. Suppose that $m_i(S \times S) = 1$ for $i = 1$ or 2. Then $m_i(S_E \times S_E) = 1$ for $i = 1$ or 2 respectively.*

Before we prove Theorem 6, we will discuss briefly what it says. Suppose that E has a basis $\{\delta_n\}_{n \geqslant 0}$ with $|\delta_n| = 1$ for $n \geqslant 1$. Then each function $\mathbf{f}$ in $X(E)$ can be written in the form $\mathbf{f} = \Sigma_{n \geqslant 1} f_n \delta_n$, where $f_n \in X$, $n \geqslant 1$. If p is a polynomial, then $p(S_E)\mathbf{f} = \Sigma_{n \geqslant 1} (pf_n)\delta_n$. Let $\mathbf{f}$ be a 1-cyclic element of $X(E)$ and let $\mathbf{g}$ be an arbitrary element of the same space. For an arbitrary $\epsilon > 0$ there exists a polynomial $p \in \mathcal{P}_A$ such that $\|p(S_E)\mathbf{f} - \mathbf{g}\| < \epsilon$. Therefore monotonicity of the norm in X and uniform minimality of the basis $\{\delta_n\}_{n \geqslant 1}$ (see §2.3) imply that $|pf_n - g_n|_X < \text{const}\ \epsilon$ for all $n \geqslant 1$, where const depends only on the choice of the basis $\{\delta_n\}_{n \geqslant 1}$. For example, we can set $E = l^r$ and $X = \hat{l}^q_{\mathbf{Z}}$, $1 \leqslant r, q \leqslant \infty$ (for $r, q = \infty$ we consider spaces of sequences which converge to zero). For $r = q > 2$ (concerning the equation $m_i(S \times S) = 1$, see subsection 6°), Theorem 6 implies that there exists a set of functions $\{f_n: n \geqslant 1\}$ in $\hat{l}^r_{\mathbf{Z}}$ such that $\Sigma_{n \geqslant 1} |f_n|^r_{\hat{l}^r_{\mathbf{Z}}} < +\infty$, and for any other analogous set of functions $\{g_n: n \geqslant 1\}$ and an arbitrary $\epsilon > 0$ we can find a polynomial $p \in \mathcal{P}_A$ such that

$$\sum_{n \geqslant 1} |pf_n - g_n|^r_{\hat{l}^r_{\mathbf{Z}}} < \varepsilon.$$

For $r = \infty$ and for an arbitrary X (with $m_i(S \times S) = 1$) we find that if

$$\sum_{k \in \mathbf{Z}} \max_n |\hat{g}_n(k)| z^k \in X,$$

then we can achieve the approximation

$$\left| \sum_{k \in \mathbf{Z}} \max_n \left| (pf_n)^\wedge (k) - \hat{g}_n(k) \right| z^k \right|_X < \varepsilon,$$

and thus $\max_n |pf_n - g_n|_X < \epsilon$. Theorem 6 also implies an analogous assertion for simultaneous weighted approximations of an arbitrary finite set of functions in X (the case $E = \mathbf{C}^n$). In [47] it is shown that Theorem 6 implies the existence of a cyclic vector for the composition operator $f \longrightarrow f \circ \varphi$, $\varphi(t) = t^2$, in the space $C_0[0, 1]$ of all continuous functions on $[0, 1]$ which equal zero at the endpoints.

LEMMA 8. *Let X be a space of generalized functions on Γ for which $m_i(S \times S) = 1$ for $i = 1$ or 2. Then there exists an element $g \in X$ such that for arbitrary δ and k (with $\delta > 0$ and $k \in \mathbf{Z}$) there exist polynomials p and q (in $\mathcal{P}_A$ or $\mathcal{P}$ respectively) such that*

$$|pg|_X < \delta, \quad |qg|_X < 2, \quad |qpg - z^k|_X < \delta.$$

PROOF OF THE LEMMA. By assumption there exist $f, g \in X$ such that the vector $\{f, g\}$ is cyclic (respectively, 2-cyclic) for the operator $S \times S$ and $|f|_X \leqslant 1$. Consequently we can find a polynomial p such that $|pg|_X < \delta$ and $|pf - z^k|_X < \delta/2$. Let $0 < \epsilon < \delta/2\|p\|$ and $\epsilon < 1$. We choose a polynomial q such that $|f - qg|_X < \epsilon$. Then

$$|pgq - z^k|_X \leqslant |pgq - pf|_X + |pf - z^k|_X < \|p\| \cdot |gq - f|_X + \delta/2 < \delta.$$

Also

$$|qg|_X \leqslant |qg - f|_X + |f|_X < \varepsilon + 1 < 2. \bullet$$

PROOF OF THEOREM 6. Let $\{e_n: n \geqslant 1\}$ be a family of vectors in E which is dense in the unit ball. First we consider the case of 2-cyclicity, $m_2(S \times S) = 1$, and then we will list the changes which must be made for the proof of the fact that $m_1(S \times S) = 1$. We will seek a cyclic vector in $X(E)$ of the form

$$\mathbf{f} = \sum_{n \geqslant 1} a_n (q_n g) e_n, \tag{7}$$

where $a_n > 0$, q_n is a polynomial, g is the function in Lemma 8, and $(q_n g)e_n$ denotes an E-valued function h_n with $\hat{h}_n(k) = e_n(q_n g)\hat{}(k)$, $k \in \mathbf{Z}$. The sequences $\{a_n\}_{n\geqslant 1}$ and $\{q_n\}_{n\geqslant 1}$ will be constructed by induction.

We set $a_1 = 1$ and $q_1 = 1$, and we suppose that $a_1, \ldots, a_n$ and $q_1, \ldots, q_n$ have already been chosen so that for some polynomial p_n we have

$$0 < \left\| \frac{p_n(S_E)}{a_n} \left(\sum_{k=1}^{n} a_k (q_k g) e_k \right) - \mathbf{1} \cdot e_n \right\| < \frac{1}{n}. \tag{8}$$

We choose $a_{n+1} > 0$ such that

$$4 \| p_n \| \frac{a_{n+1}}{a_n} < \frac{1}{n}, \quad a_{n+1} < \frac{1}{2} a_n. \tag{9}$$

Using Lemma 8, we let p_{n+1} and q_{n+1} be polynomials such that

$$| p_{n+1} g |_X < \frac{a_{n+1}}{2(n+1) \sum_{k=1}^{n} | e_k | \cdot \| q_k \|}, \quad | q_{n+1} g |_X < 2, \tag{10}$$

$$| p_{n+1} g q_{n+1} - \mathbf{1} |_X < 1/2(n+1). \tag{11}$$

Then

$$\left\| \frac{p_{n+1}(S_E)}{a_{n+1}} \left(\sum_{k=1}^{n+1} a_k (q_k g) e_k \right) - \mathbf{1} \cdot e_{n+1} \right\| \leqslant \frac{1}{a_{n+1}} \left\| p_{n+1}(S_E) \left(\sum_{k=1}^{n} a_k (q_k g) e_k \right) \right\|$$

$$+ \| (p_{n+1} q_{n+1} g) e_{n+1} - \mathbf{1} \cdot e_{n+1} \| \leqslant \frac{1}{a_{n+1}} \sum_{k=1}^{n} | e_k | \cdot \| q_k \| \cdot | p_{n+1} g |_X a_k$$

$$+ 1/2(n+1) < 1/(n+1).$$

Thus we have constructed sequences $\{a_n\}$, $\{q_n\}$ and $\{p_n\}$, $n \geqslant 1$, satisfying (8)–(11). We will show that the sum $\mathbf{f}$ of the absolutely convergent series (7) is a 2-cyclic vector for S_E. We have

$$\left\| \frac{p_n(S_E)}{a_n} \mathbf{f} - \mathbf{1} \cdot e_n \right\| \leqslant \left\| \frac{p_n(S_E)}{a_n} \left(\sum_{k=1}^{n} a_k (q_k g) e_k \right) - \mathbf{1} \cdot e_n \right\|$$

$$+ \left\| \frac{p_n(S_E)}{a_n} \left(\sum_{k \geqslant n+1} a_k (q_k g) e_k \right) \right\| \leqslant 1/n + \frac{1}{a_n} \sum_{k \geqslant n+1} \| (p_n q_k g) e_k \| a_k$$

$$\leqslant 1/n + \frac{a_{n+1}}{a_n} \| p_n \| \cdot 2 \sum_{m \geqslant 0} 2^{-m} < 1/n + 1/n = 2/n,$$

where the second inequality follows from condition (8) and monotonicity of the norm, while the third follows from conditions (9) and (10). If e is an arbitrary element in E with $|e| \leqslant 1$, then

$$\varliminf_{n} | e - e_n | = 0,$$

and therefore

$$\varliminf_n \left\| \frac{p_n(S_E)}{a_n} \mathbf{f} - \mathbf{1} \cdot e \right\| = 0.$$

Consequently all functions of the form $\mathbf{1} \cdot e$, $e \in E$, belong to the 2-invariant subspace of $X(E)$ spanned by $\mathbf{f}$, and $\mathbf{f}$ is a 2-cyclic element for S_E.

In the case $m_1(S \times S) = 1$ we enumerate the family of functions $\{z^k e_m : k \in \mathbf{Z}, m \geqslant 1\}$ as a sequence $\{\mathcal{E}_n\}_{n \geqslant 1}$ and at the nth step we make sure that (8) is satisfied with $\mathbf{1} \cdot e_n$ replaced by $\mathcal{E}_n$ at the expense of having to choose new polynomials p_n and q_n (in (11) we replace $\mathbf{1}$ by z^k). As a result we obtain

$$\varliminf_n \left\| \frac{p_n(S_E)}{a_n} \mathbf{f} - z^k \cdot e \right\| = 0$$

for each $e \in E$ with $|e| \leqslant 1$. Therefore $\mathbf{f}$ is cyclic with respect to S_E. ●

6°. *Classical spaces.* Here we present applications of the theorems in subsections 2°–5° to the classical spaces of linear analysis $\hat{l}^p_{\mathbf{Z}}$, $\hat{l}^p_{\mathbf{Z}}(w_n)$, Orlicz, Lorentz, Marcinkiewicz spaces, and others. Underlying these applications is a translation of conditions (B), (C), and so forth, into the language of symmetric spaces of generalized functions, which account for almost all of the cases which we mentioned. We will use some results in the theory of symmetric spaces which were developed by Semenov in [97] and [98]. First we consider the interrelations between conditions (B) and (C), j (that is, closability of j), j_A (closability of $j | \mathcal{P}_A$), the existence of singular elements ($\exists\, f \neq \mathbf{0}$, $|\operatorname{supp} f| = 0$), and multiplicity of the spectrum:

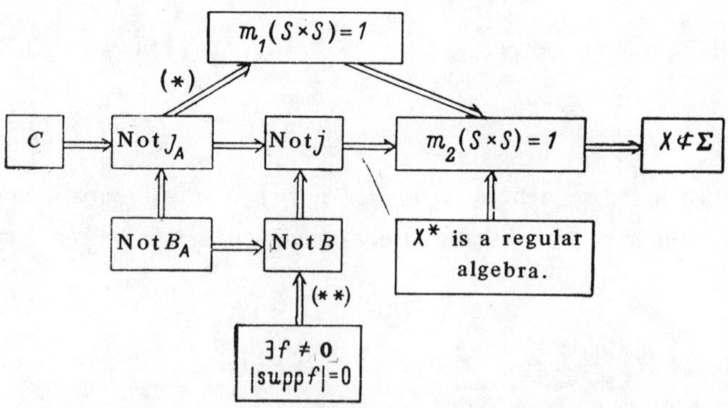

Here the symbols (*) and (**) indicate the conditions of non-quasi-analyticity in Theorems 2 and 3 and Theorem 4 respectively.

Let X be a symmetric space of generalized functions, and let φ_X be its fundamental function; that is,([42]) $\varphi_X(n) = \|\chi_n\|_X$, $n \geqslant 1$, where χ_n is a polynomial such that $\hat{\chi}_n$ is the characteristic function of a finite subset of $\mathbf{Z}$ which contains n elements. Then $1 \leqslant \varphi_X(n) \leqslant n$, and without loss of generality we can assume that φ_X is a convex (upward) function. Later we will determine conditions on φ_X which guarantee that $m_i(S \times S) = 1$ or $m_i(S \times S) = 2$. Recall that in the latter case it suffices to require that $X \subsetneq \Sigma$, and for a s.s. (or even a s.m.n.) of generalized functions this is equivalent to the inclusion $X \subset \hat{l}^2_{\mathbf{Z}}$.

([42]) Here we revert to our previous notation, which we abandoned in 5°: $\|\cdot\|_X$ is the norm in the space X.

LEMMA 9. *Let X be a s.s. of generalized functions with a convex fundamental function φ_X, and suppose that ψ_X, $\psi_X(n) = n/\varphi_X(n)$, is also convex. If*

$$\sum_{n\geqslant 1} (\psi_X(n+1) - \psi_X(n))^2 < +\infty, \tag{12}$$

then $X \subset \hat{l}^2_{\mathbf{Z}}$. If

$$\sum_{n\geqslant 1} (\varphi_X(n+1) - \varphi_X(n))^2 < +\infty, \tag{13}$$

then $\hat{l}^2_{\mathbf{Z}} \subset X$.

PROOF. It is known (see [97] and [98]) that $X \supset \Lambda(\varphi_X)$, where $\Lambda(\varphi_X)$ is the Lorentz space consisting of all generalized functions $f = \sum_{n\in\mathbf{Z}}\hat{f}(n)z^n$ such that

$$\sum_{n\geqslant 1} \hat{f}(n)^* (\varphi_X(n+1) - \varphi_X(n)) < +\infty,$$

where (henceforth) $\{\hat{f}(n)^*\}_{n\geqslant 1}$ is a nonincreasing permutation of the sequence of moduli $\{|\hat{f}(n)|\}_{n\in\mathbf{Z}}$. If (13) holds and $f \in \hat{l}^2_{\mathbf{Z}}$, then

$$\sum_{n\geqslant 1} \hat{f}(n)^* (\varphi_X(n+1) - \varphi_X(n)) \leqslant \|f\|_{\hat{l}^2_{\mathbf{Z}}} \Big(\sum_{n\geqslant 1} (\varphi_X(n+1) - \varphi_X(n))^2\Big)^{1/2} < +\infty$$

and $f \in X$. The first assertion in the lemma is obtained from the already established transition to the dual space. Namely, let

$$X' = \Big\{g : \sum_{n\in\mathbf{Z}} |\hat{g}(n)|\,|\hat{f}(n)| < +\infty, \quad \forall f \in X\Big\}$$

and $X'' = (X')'$. Then $\varphi_{X'} = \psi_X$ and $X \subset X''$. If (12) is satisfied, then the second assertion in the lemma implies that $X' \supset \hat{l}^2_{\mathbf{Z}}$, and therefore $X \subset X'' \subset \hat{l}^2_{\mathbf{Z}}$. ●

REMARK. Conditions (12) and (13) are sharp, on the class of all symmetric spaces with a given fundamental function, for the validity of the respective inclusions. Rougher (but more obvious) sufficient conditions are

$$\sum n^{-3/2}\psi_X(n) < +\infty, \quad \sum n^{-3/2}\varphi_X(n) < \infty$$

respectively. In fact, summation by parts gives us

$$\sum_{n\geqslant 1} n^{-1/2} (\psi_X(n+1) - \psi_X(n)) < +\infty,$$

and by virtue of monotonicity of the general term we have

$$\{\psi_X(n+1) - \psi_X(n)\} = o(n^{-1/2}),$$

so that the series

$$\sum_{n\geqslant 1} (\psi_X(n+1) - \psi_X(n))^2$$

converges.

LEMMA 10. 1) *Condition* (13) *implies condition* (C). 2) *If the functional p,*

$$p(f) = \Big\|\sum_{n\in\mathbf{Z}} |\hat{f}(n)|^{1/2} z^n\Big\|^2_X, \quad f \in \mathscr{P},$$

determines a norm on $\mathcal{P}$, $\hat{l}^2_{\mathbf{Z}} \subset X \subset \hat{l}^\infty_{\mathbf{Z}}$ *and* $n^{-1/2}\varphi_X(n) = o(1)$, *then condition* (C) *is satisfied.*

PROOF. 1) Let

$$\|f_n\|_{\hat{l}^2_{\mathbf{Z}}} \leqslant \text{const}, \ \lim_n \|f_n\|_{\hat{l}^\infty_{\mathbf{Z}}} = 0.$$

Then

$$\|f_n\|_{\Lambda(\varphi_X)} = \sum_{k\geqslant 1} \hat{f}_n(k)^* (\varphi_X(k+1) - \varphi_X(k))$$

$$\leqslant \sum_{k=1}^{N-1} \hat{f}_n(k)^* (\varphi_X(k+1) - \varphi_X(k)) + \|f_n\|_{\hat{l}^2_{\mathbf{Z}}} \Big(\sum_{k\geqslant N} (\varphi_X(k+1) - \varphi_X(k))^2\Big)^{1/2}$$

$$\leqslant \|f_n\|_{\hat{l}^\infty_{\mathbf{Z}}} (\varphi_X(N) - \varphi_X(1)) + \|f_n\|_{\hat{l}^2_{\mathbf{Z}}} \Big(\sum_{k\geqslant N} (\varphi_X(k+1) - \varphi_X(k))^2\Big)^{1/2}.$$

Therefore

$$\|f_n\|_X \leqslant \|f_n\|_{\Lambda(\varphi_X)} < \varepsilon \quad \text{for} \quad n > n(\varepsilon).$$

2) The norm p determines a symmetric space $X_{1/2}$ for which we have the fundamental function $\varphi_{X_{1/2}} = \varphi_X(n^2)$, $n \geqslant 1$. Therefore $\varphi_{X_{1/2}} \asymp \varphi_1$, where φ_1 is a convex (upward) function, and the hypothesis of the lemma implies that

$$\lim_n (\varphi_1(n+1) - \varphi_1(n)) = 0.$$

Let

$$\|f_n\|_{\hat{l}^2_{\mathbf{Z}}} \leqslant \text{const}, \ \lim_n \|f_n\|_{\hat{l}^\infty_{\mathbf{Z}}} = 0.$$

Then

$$\|f_n\|^2_X = p\Big(\sum_{k\in\mathbf{Z}} |\hat{f}_n(k)|^2 z^k\Big) \leqslant \Big\|\sum_{k\in\mathbf{Z}} |\hat{f}_n(k)|^2 z^k\Big\|_{\Lambda(\varphi_1)}$$

$$= \sum_{k\geqslant 1} (\hat{f}_n(k)^*)^2 (\varphi_1(k+1) - \varphi_1(k))$$

$$\leqslant \sum_{k=1}^{N-1} (\hat{f}_n(k)^*)^2 (\varphi_1(k+1) - \varphi_1(k)) + \{\varphi_1(N+1) - \varphi_1(N)\} \|f\|^2_{\hat{l}^2_{\mathbf{Z}}}.$$

Therefore $\|f_n\|_X < \epsilon$ for $n > n(\epsilon)$. ●

REMARK. From the proof of the lemma it is clear that for the Lorentz space $\Lambda(\varphi)$ condition (C) follows from just the inclusion $\hat{l}^2_{\mathbf{Z}} \subset \Lambda(\varphi)$; the analogous conclusion cannot be drawn for the Marcinkiewicz spaces $M(\psi)$, $\psi(n) = n/\varphi(n)$:

$$\|f\|_{M(\psi)} \stackrel{\text{def}}{=} \sup_{n\geqslant 0} \frac{1}{\psi(n)} \sum_{k=1}^{n} \hat{f}(k)^* < +\infty.$$

In fact, if we set $\varphi(n) = \psi(n) = n^{1/2}$, and $f_n = \Sigma_{k=1}^n z^k/\sqrt{n}$, we obtain

$$\|f_n\|_{M(\sqrt{n})} \geqslant \text{const} > 0, \ \|f_n\|_{\hat{l}^2_{\mathbf{Z}}} = 1, \ \|f_n\|_{\hat{l}^\infty_{\mathbf{Z}}} = o(1).$$

The condition $\varphi_X(n) = o(n^{1/2})$ implies condition (C) for a Marcinkiewicz space:

$$\|f_n\|_{M(\psi)} \leqslant \max_{1\leqslant m\leqslant N} \frac{1}{\psi(m)} \sum_{k=1}^{m} \hat{f}_n(k)^* + \sup_{m>N} \frac{\sqrt{m}}{\psi(m)} \to 0,$$

as well as for an Orlicz space L_M corresponding to an N-function M with the condition Δ_2 at the point 0 (see [50]). We will prove this last assertion. Since $\varphi_{L_M}(n) = nN^{-1}(1/n)$ (see [50], formula (9.11)) where N is complementary to the function M and N^{-1} is the inverse of N, we conclude that the condition $\varphi_{L_M}(n) = o(n^{1/2})$ is equivalent to $u^2 = o(N(u))$ as $u \to 0$ or $M(u) = o(u^2)$ as $u \to 0$. If we make use of the equivalence of convergence in the norm of L_M and M-mean convergence (see [50], Theorem 9.4), it suffices to show that the relations

$$\|f_n\|_{\hat{l}^2_{\mathbf{Z}}} = O(1), \ \|f_n\|_{\hat{l}^\infty_{\mathbf{Z}}} = o(1)$$

imply that

$$\sum_{k\in \mathbf{Z}} M(|\hat{f}_n(k)|) = o(1)$$

as $n \to \infty$. If $\epsilon > 0$, then $M(u) \leqslant \epsilon u^2$ for $0 \leqslant u \leqslant u_\epsilon$, and therefore

$$\sum_{k\in \mathbf{Z}} M(|\hat{f}_n(k)|) \leqslant \varepsilon_n \sum_{k\in \mathbf{Z}} |\hat{f}_n(k)|^2 = o(1) \quad \text{as} \quad n\to\infty. \bullet$$

COROLLARY 1. *If* (12) *holds, then* $m_i(S \times S) = 2$. *If* (13) *holds, then* $m_i(S \times S) = 1, i = 1, 2$.

These assertions follow immediately from Lemmas 9 and 10 and Theorems 2 and 3. It is clear that the condition of non-quasi-analyticity in these theorems is also satisfied, since in a symmetric space the operator S is isometric ($\|S^n\| = 1, n \in \mathbf{Z}$). •

COROLLARY 2. *Let* L_M *be an Orlicz space which is generated by an N-function M with condition* Δ_2. *If*

$$\lim_{u\to 0} u^{-2}M(u) = 0,$$

then $m_i(S \times S) = 1, i = 1, 2$. *If*

$$\varliminf_{u\to 0} u^{-2}M(u) > 0,$$

then $m_i(S \times S) = 2, i = 1, 2$.

The proof follows from Theorems 2 and 3 and the remark following Lemma 10. •

In particular, $m_i(S \times S) = 2$ in $\hat{l}^p_{\mathbf{Z}}$ for $1 \leqslant p \leqslant 2$, and $m_i(S \times S) = 1$ in $\hat{l}^p_{\mathbf{Z}}$ for $p > 2$.

LEMMA 11. *Let* $X = \hat{l}^p_{\mathbf{Z}}(w_n)$, *where* $w = \{w_n\}_{n\in\mathbf{Z}}$ *is a positive sequence for which* $SX = X$.

1) *If* $w \in l^\infty_{\mathbf{Z}}$ *and* $p > 2$, *or* $\lim_{|n|\to\infty} w_n = 0$ *and* $p = 2$, *or* $1 \leqslant p < 2$ *and* $\Sigma_{n\in\mathbf{Z}} w_n^{2p/(2-p)} < +\infty$, *then* $m_2(S \times S) = 1$; *and if the non-quasi-analyticity condition*

$$\sum_{n\in\mathbf{Z}} (n^2+1)^{-1} |\log \| S^n \|| < +\infty$$

is satisfied, then $m_1(S \times S) = 1$.

2) *If* $1/w \in l^\infty_{\mathbf{Z}}$ *and* $1 \leqslant p \leqslant 2$, *or* $p > 2$ *and* $\Sigma_{n\in\mathbf{Z}} w_n^{-2p/(p-2)} < +\infty$, *then* $m_1(S \times S) \geqslant 2$, $i = 1, 2$.

PROOF. Assertion 1) follows from the corollary to Theorem 3, and the condition given for $1 \leqslant p \leqslant 2$ is necessary and sufficient for complete continuity of the imbedding $\hat{l}^2_{\mathbf{Z}} \subset \hat{l}^p_{\mathbf{Z}}(w_n)$. The second part of the lemma follows from Theorem 1, since in this case we have $\hat{l}^p_{\mathbf{Z}}(w_n) \subset \hat{l}^2_{\mathbf{Z}}$. ●

We note that cases not covered by Lemma 11 (for instance, $w \in \hat{l}^\infty_{\mathbf{Z}}$, $1 \leqslant p < 2$, and $\Sigma_{n\in\mathbf{Z}} w_n^{2p/(2-p)} = \infty$) do not, as a rule, satisfy condition (A). Therefore we cannot use the theorems in subsections 2°–5°.

By analogous methods we can treat other spaces of generalized functions. For example, we can treat spaces defined by the relation

$$\sum_{k=-n}^{n} |\hat{f}(k)|^p = o(\psi(n)), \quad n \to \infty,$$

where ψ is a given nondecreasing function. In particular, if $p = 2$, then $m_i(S \times S) = 1$ if and only if $\lim_n \psi(n) = +\infty$.

This last fact, as well as a part of Lemma 1, can be deduced from Theorem 4 concerning singular elements. To do this, we just have to use known results on estimates for Fourier coefficients of singular measures (see [5], Chapter V, §12, and Chapter XIV, §§24–27).

APPENDIX I

The Carleman formula. If F is a holomorphic function in a domain $\{\zeta: 0<\lambda<|\zeta|<R, \operatorname{Im}\zeta>0\}$ which does not equal zero anywhere in this domain, then

$$\frac{1}{\pi R}\int_0^\pi \log|F(\mathrm{R}e^{i\theta})|\sin\theta d\theta+\frac{1}{2\pi}\int_\lambda^R\left(\frac{1}{x^2}-\frac{1}{R^2}\right)\log|F(x)F(-x)|\,dx$$

$$=\operatorname{Im}\frac{1}{2\pi}\int_0^\pi \log F(\lambda e^{i\theta})\left(\frac{\lambda e^{i\theta}}{R^2}-\frac{e^{-i\theta}}{\lambda}\right)d\theta.$$

See, for example, [53], Chapter V, §1, Theorem 1 and the appended Remark. There it is pointed out that the right side has a limit as $R\to\infty$ which equals

$$-\operatorname{Im}\frac{1}{2\pi}\int_0^\pi \log F(\lambda e^{i\theta})\frac{e^{-i\theta}}{\lambda}\,d\theta.$$

Warschawski's Theorem (see [15], §7). Let

$$\Omega=\{\zeta: \zeta\in\mathbf{C},\ \theta\leqslant\operatorname{Im}\zeta\leqslant\theta+s(\operatorname{Re}\zeta)\},$$

where θ is a number and s is a positive differentiable function on $\mathbf{R}$ (or only on a half-line $[a,+\infty)$, $a\in\mathbf{R}$) such that

$$\lim_{t\to+\infty}s'(t)=0,\quad \int^\infty\frac{s'^2}{s}dt<+\infty.$$

Let Z be a conformal mapping of the strip Ω onto the strip Π which takes $+\infty$ to $+\infty$. Then there exists a constant C such that

$$\operatorname{Re}Z(\sigma+i\tau)=C+\pi\int_a^\sigma\frac{dt}{s(t)}+r(\sigma,\tau),\quad \sigma+i\tau\in\Omega,$$

and

$$\lim_{\sigma\to\infty}\max_\tau|r(\sigma,\tau)|=0.$$

APPENDIX II

PROOF OF THEOREM 3 IN §1.2. The proof begins with an argument due to Warschawski [15]. Let $\Omega = \{u + iv: |v| < s(u)\}$ be the strip described in the hypothesis of Theorem 3; let Z be a conformal mapping of Ω onto Π, and let $W = Z^{-1}$. Let $u_k + iv_k \in \Omega$, $k = 1, 2$, let $\sigma_1 < \sigma_2$, and let $z_k = Z(u_k + iv_k) = x_k + iy_k$. We define the function j by the equation

$$j(u) = \frac{\pi}{2}\int_0^u \frac{dt}{s(t)}, \quad u \geqslant 0.$$

LEMMA 1. *For an arbitrary $\epsilon > 0$ we have*

$$x_2 - x_1 \leqslant \frac{\pi}{2}\int_{u_1}^{u_2} \frac{dt}{s} + \frac{\pi}{6}\int_{u_1}^{u_2} \frac{s'^2}{s}\,dt + 2\pi^2(1+\varepsilon)\left(\frac{1}{j(u_1)} + \frac{1}{j(u_2)}\right)$$

for large u_1 and u_2 with $u_1 < u_2$.

The proof refines one of the steps in the reasoning in [15], where it is shown (see formula (6.4) in [15]) that

$$x_2 - x_1 \leqslant \frac{\pi}{2}\int_{u_{1*}}^{u_2^*} \frac{dt}{s} + \frac{\pi}{6}\int_{u_{1*}}^{u_2^*} \frac{s'^2}{s}\,dt$$

with the notation

$$u_{k*} = \min_{|y| \leqslant \pi/2} \operatorname{Re} W(x_k + iy), \quad u_k^* = \max_{|y| \leqslant \pi/2} \operatorname{Re} W(x_k + iy), \quad k = 1, 2.$$

Here (see [15], §4)

$$\frac{\pi}{2}\int_{u_{1*}}^{u_1} \frac{dt}{s} \leqslant \bar{x} - \underline{x}, \tag{1}$$

$$\bar{x} = \max_v \operatorname{Re} Z(u_1 + iv), \quad \underline{x} = \min_v \operatorname{Re} Z(u_{1*} + iv),$$

$$\bar{x} - \underline{x} = (\bar{x} - x_1) + (x_1 - \underline{x}) \leqslant 2\pi(\tan m(u_1) + \tan m(u_{1*})), \tag{2}$$

where $m(u) = \max_v |\arg Z'(u + iv)|$.

LEMMA 2. *For an arbitrary $\epsilon > 0$ we have*

$$|\arg Z'(u + iv)| \leqslant \frac{\frac{\pi}{2}(1+\varepsilon)}{j(u)}, \quad |v| < s(u),$$

for large values of u.

PROOF OF LEMMA 2. Let $\operatorname{Re} \zeta > 0$ and $f(\zeta) = W(\log \zeta)$. By the Poisson formula, for a function f' which is continuous in $\operatorname{Re} \zeta \geqslant 0$ we have

$$\arg f'(\zeta) = \int_{-\infty} \frac{\xi}{(\eta - t)^2 + \xi^2} K(t)\, dt - \arg \zeta,$$

where $\zeta = \xi + i\eta$ and

$$K(t) = \arctan s'\left(\operatorname{Re} W\left(\log t + i\frac{\pi}{2}\right)\right)$$

for $t > 0$, and

$$K(t) = -\arctan s'\left(\operatorname{Re} W\left(\log |t| - i\frac{\pi}{2}\right)\right)$$

for $t < 0$. If we set $\zeta = e^z$, $z \in \Pi$, then for each a, $1 < a < |\zeta|$, we obtain

$$|\arg W'(z)| = |\arg \zeta f'(\zeta)| \leqslant \int_{-a}^{a} \frac{\xi}{(\eta - t)^2 + \xi^2}\, dt + \max_{|t| \geqslant a} |K(t)|$$

$$\leqslant 2 \int_0^a \frac{dt}{|\zeta| - t} + \max_{|t|\geqslant a} |K(t)| = 2 \log \frac{1}{1 - \frac{a}{|\zeta|}} + \max_{|t|\geqslant a} |K|.$$

Also,

$$\max_{|t|\geqslant a} |K(t)| = \max_{u \geqslant \log a} \left|\arctan s'\left(\operatorname{Re} W\left(u \pm i\frac{\pi}{2}\right)\right)\right|,$$

and if $W(u + i\pi/2) = x + iy$, then $y = s(x)$ and $u = j(x) + o(1)$ by Warschawski's theorem. Therefore

$$\max_{|t|\geqslant a} |K(t)| \leqslant \max_{j(x) \geqslant \log a + o(1)} |s'(x)|.$$

We introduce some notation: $b = \log a$, $b \geqslant 0$; $\operatorname{Re} z = \sigma$, so that $|\zeta| = e^{\operatorname{Re} z} = e^\sigma$ and $\sigma > b$. Then

$$|\arg W'(z)| \leqslant 2 \log (1 - e^{b-\sigma})^{-1} + \max_{j(x) \geqslant b + o(1)} |s'(x)|. \tag{3}$$

We find a lower estimate for the derivative $|s'|$. Let

$$\gamma = \log \frac{1}{s}, \quad \gamma' = -s'/s, \quad \gamma'' = \frac{s'^2 - ss''}{s^2}, \quad \gamma''/\gamma'^2 = 1 - ss''/s'^2.$$

By condition b) the functions γ' and γ'' are sign-definite for large values of the argument. By condition a) we have $\lim_{t\to\infty} \gamma = +\infty$, and therefore $\gamma' > 0$. If $\gamma'' < 0$, then

$$j(x) = \frac{\pi}{2} \int_0^x \frac{dt}{s} = \frac{\pi}{2} \int_0^x e^\gamma dt = \frac{\pi}{2} \cdot \frac{1}{\gamma'(x)} e^{\gamma(x)} + \frac{\pi}{2} \int_0^x e^\gamma \frac{\gamma''}{\gamma'^2}\, dt$$

$$\leqslant \frac{\pi}{2} \cdot \frac{1}{\gamma'(x)} e^{\gamma(x)} + \text{const}, \tag{4}$$

$$j(x) \leqslant \frac{\pi + o(1)}{2\gamma'(x)} e^{\gamma(x)} = \frac{\pi + o(1)}{2|s'(x)|}, \quad |s'(x)| \leqslant \frac{\pi + o(1)}{2j(x)}.$$

If $\gamma'' > 0$, then by condition b) the function γ''/γ'^2 is monotone, and consequently the limit $\lim_{t\to\infty} \gamma''/\gamma'^2 = L$ exists. As we noted in the proof of Lemma 1 in §1.1, we must have $L \leqslant 0$, and therefore $L = 0$. Then by Lemma 1 in §1.1 we have

$$j(x)=\frac{\pi}{2}\int_0^x e^{\gamma}dt\sim\frac{\pi}{2\gamma'(x)}e^{\gamma(x)}=\frac{\pi}{2|s'(x)|}.$$

Thus in both cases the inequalities in (4) are valid. If we continue the estimate in (3), we obtain

$$|\arg W'(z)|\leqslant 2\log(1-e^{b-\sigma})^{-1}+\frac{\pi+o(1)}{2b}$$

for large a, $1<a=e^b<|\zeta|=e^\sigma$, where $\zeta=e^z$, $z\in\Pi$. Therefore

$$|\arg W'(z)|\leqslant\min_{0<b<\sigma}\left\{2\log(1-e^{b-\sigma})^{-1}+\frac{\pi+\frac{\varepsilon}{3}}{2b}\right\}\leqslant\frac{\pi+\frac{\varepsilon}{3}}{2(\sigma-2\log\sigma-\text{const})}\leqslant\frac{\pi+\frac{\varepsilon}{2}}{2\sigma}$$

for arbitrary $\epsilon>0$ and $\sigma=\operatorname{Re}z>\sigma_\epsilon$. Thus for $z=Z(u+iv)$ we have

$$|\arg Z'(u+iv)|=|\arg W'(z)|\leqslant\frac{\pi+\frac{\varepsilon}{2}}{2\sigma}\leqslant\frac{\pi+\varepsilon}{2j(u)}. \bullet$$

REMARK. If we make use of the symmetry of Ω (that is, the fact that $\arg f'(\zeta)=0$, $\zeta\in\mathbf{R}$) and the Poisson formula for a quarter-plane, we also obtain $|\arg Z'(u+iv)|\leqslant\text{const}\cdot j(u)^{-2}$.

Returning to the proof of Lemma 1, we extend the inequalities (1) and (2):

$$\bar{x}-x\leqslant\frac{\pi^2+\varepsilon/2}{j(u_1)}+\frac{\pi^2+\varepsilon/2}{j(u_{1*})}\leqslant\frac{2\pi^2+\varepsilon}{j(u_1)},\quad \frac{\pi}{2}\int_{u_{1*}}^{u_1}\frac{dt}{s}\leqslant\frac{2\pi^2+\varepsilon}{j(u_1)}.\,(^1)$$

Analogously we estimate the integrals $\int_{u_2}^{u_2^*}s^{-1}\,dt$, $\int_{u_1^*}^{u_1}(s'^2/s)\,dt$, and $\int_{u_2}^{u_2^*}(s'^2/s)\,dt$ (recall that s is a convex function with $\lim s=0$ and $s'<0$, so that $\lim s'=0$; similarly the latter two integrals are $o(1/j(u_1))$ and $o(1/j(u_2))$, respectively). Finally,

$$x_2-x_1\leqslant\frac{\pi}{2}\int_{u_1}^{u_2}\frac{dt}{s}+\frac{\pi}{6}\int_{u_1}^{u_2}\frac{s'^2}{s}dt+(2\pi^2+\varepsilon)\left(\frac{1}{j(u_1)}+\frac{1}{j(u_2)}\right). \bullet$$

LEMMA 3. *Suppose that $w=w(l)$ is a natural parameterization of the curve C_-: $u=-s(v)$ (here l is arc length), $\alpha(l)$ is the slope of the tangent to C_- at the point $w(l)$, Δ_l is the segment of the normal to C_- at the point $w(l)$ contained in Ω, and $\Delta(l)$ is its length. Then*

$$x_2^*-x_{1*}\geqslant\pi\int_{l_1}^{l_2}\frac{dl}{\Delta(l)}-\pi\int_{l_1}^{l_2}|\alpha'(l)|\,dl, \tag{5}$$

where

$$x_2^*=\max_{w\in\Delta_{l_2}}\operatorname{Re}Z(w),\quad x_{1*}=\min_{w\in\Delta_{l_1}}\operatorname{Re}Z(w).$$

(1) We use the equality $j(u_1)-j(u_1^*)=o(1)$ (see [15], §6.3(b)).

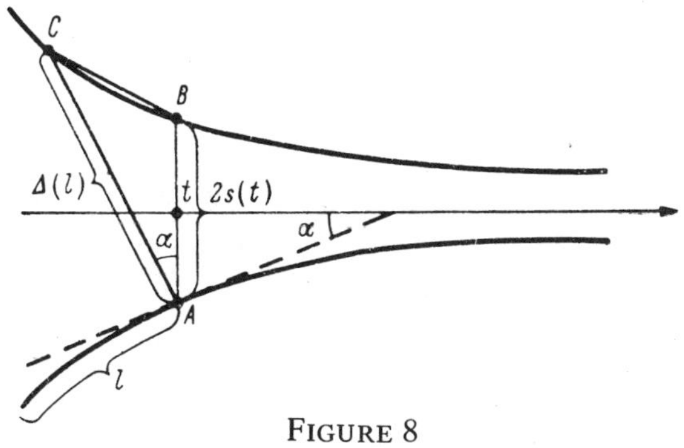

FIGURE 8

PROOF OF THE LEMMA. Following the arguments in [15] (§9, steps 1–3 in the proof of Theorem V), it suffices to note that $\Delta(l)\alpha'(l) < 1$ for all sufficiently large l. Let $t = \operatorname{Re} w(l)$. Then $\Delta(l) \sim 2s(t)$ as $l \longrightarrow \infty$ (Figure 8), since the equations $\alpha(l) = \arctan|s'(t)|$ and $\lim_{t>\infty} s' = 0$ give us $\lim_{l\to\infty}\alpha(l) = 0$, and by the same reasoning the angle B tends to $\Pi/2$ as $l \longrightarrow \infty$. Also

$$\frac{d\alpha}{dl} = [\alpha(l(t))]' \, t_l' (l(t)) = \frac{s''(t)}{1+s'^2(t)} \cdot \frac{1}{l'(t)} = \frac{s''(t)}{(1+s'^2(t))^{3/2}},$$

and therefore

$$\Delta(l)\,\alpha'(l) \sim \frac{2s(t)\,s''(t)}{(1+s'^2(t))^{3/2}} < 1$$

for large values of t by condition d). ●

LEMMA 4. *In the notation of Lemma* 1 *we have*

$$x_2 - x_1 \geqslant \frac{\pi}{2}\int_{u_1}^{u_2} \frac{dt}{s} - \frac{7}{4}\pi(1+\varepsilon)\int_{u_1}^{u_2} \frac{s'^2}{s}\,dt - \pi\,|s'(u_2) - s'(u_1)| - 2\pi^2(1+\varepsilon)\left(\frac{1}{j(u_1)} + \frac{1}{j(u_2)}\right) \tag{6}$$

for an arbitrary $\epsilon > 0$ *and for large* u_1 *and* u_2 *with* $u_1 < u_2$.

PROOF OF THE LEMMA. We find a lower estimate for the right side of (5), using the notation in Lemma 3 and Figure 8:

$$\frac{1}{2s(t)} - \frac{1}{\Delta(l)\cos\alpha} \leqslant \frac{1}{2s(t)} - \frac{1}{\Delta(l)} = \frac{\Delta^2 - 4s^2}{2s\Delta(\Delta+2s)},$$

where $\Delta^2 \leqslant |BC|^2 + 4s^2 + 2|BC|2s|\cos B|$. If we set $\delta = \Delta(l)\sin\alpha$, then $\delta \sim 2s(t)|s'(t)|$ and $|BC| \sim \delta$ (see the remark on the angle B in the proof of Lemma 3). Furthermore, by the mean-value theorem we have

$$|\cos B| = \sin\left(B - \frac{\pi}{2}\right) = \sin\arctan|s'(\xi)|, \quad t - \delta \leqslant \xi \leqslant t,$$

and by the monotonicity of $|s'|$ we have(2)

$$\begin{aligned} |\cos B| &\leqslant \sin\arctan|s'(t-\delta)| \leqslant |s'(t-\delta)| \leqslant |s'(t)| + \delta|s''(\xi)| \\ &\leqslant |s'(t)| + (2+\varepsilon)|s'(t)|\,s(t)\,|s''(\xi)| \\ &\leqslant |s'(t)|\{1 + (2+\varepsilon)s(\xi)|s''(\xi)|\} < 2|s'(t)| \end{aligned}$$

(2) For arbitrary $\epsilon > 0$ and large $t > t_\epsilon$.

by virtue of condition d) if ϵ is sufficiently small. Now we obtain

$$\frac{1}{2s(t)}-\frac{1}{\Delta(l)\cos\alpha}\leqslant\frac{|BC|^2+2|BC|\,2s\,|\cos B|}{(16-\varepsilon)[s(t)]^3}$$

$$\leqslant\frac{s'(t)^2}{\left(4-\frac{\varepsilon}{3}\right)s(t)}+\frac{s'(t)^2}{\left(1-\frac{\varepsilon}{3}\right)s(t)}\leqslant\left(\frac{7}{4}+\varepsilon\right)\frac{s'(t)^2}{s(t)}$$

for small $\epsilon > 0$ and large t. Consequently

$$\pi\int_{l_1}^{l_2}\frac{dl}{\Delta(l)}=\pi\int_{u_1}^{u_2}\frac{dt}{\cos\alpha}\cdot\frac{1}{\Delta(l)}\geqslant\frac{\pi}{2}\int_{u_1}^{u_2}\frac{dt}{s}-\left(\frac{7}{4}+\varepsilon\right)\pi\int_{u_1}^{u_2}\frac{s'(t)^2}{s(t)}\,dt,$$

and by Lemma 3

$$x_2^*-x_{1*}\geqslant\frac{\pi}{2}\int_{u_1}^{u_2}\frac{dt}{s}-\left(\frac{7}{4}+\varepsilon\right)\pi\int_{u_1}^{u_2}\frac{s'(t)^2}{s(t)}\,dt-\pi\int_{u_1}^{u_2}\frac{s''}{1+s'^2}\,dt$$

$$\geqslant\frac{\pi}{2}\int_{u_1}^{u_2}\frac{dt}{s}-\left(\frac{7}{4}+\varepsilon\right)\pi\int_{u_1}^{u_2}\frac{s'^2}{s}\,dt-\pi(s'(u_2)-s'(u_1)).$$

By Lemma 1 of [15] we have

$$x_2^*-x_2\leqslant 2\pi\tan\{m(u_2)+\alpha(l(u_2))\},\text{ where } m(u_2)=\max_{w\in\Delta_{l_2}}|\arg Z'(w)|.$$

The first term within the tan braces can be estimated by means of Lemma 2:

$$m(u_2)\leqslant\frac{\pi}{2}(1+\varepsilon)/j(u_2-\delta);$$

the second term can be estimated by means of (4):

$$\alpha(l(u_2))=\arctan|s'(u_2)|\leqslant\left(\frac{\pi}{2}+\varepsilon\right)\Big/j(u_2)$$

for an arbitrary $\epsilon > 0$ and sufficiently large u_2. Also, for some ξ, $u_2-\delta\leqslant\xi\leqslant u_2$, we can write

$$j(u_2-\delta)=j(u_2)-\delta\frac{\pi}{2}\cdot\frac{1}{s(\xi)}\geqslant j(u_2)-\frac{(2+\varepsilon)\pi}{2}\cdot\frac{s(u_2)|s'(u_2)|}{s(\xi)}$$

$$\geqslant j(u_2)\left\{1-\frac{(2+\varepsilon)\pi}{2j(u_2)}|s'(u_2)|\right\}\geqslant\left(1-\frac{\varepsilon}{2}\right)j(u_2).$$

Therefore

$$x_2^*-x_2\geqslant 2\pi^2(1+\varepsilon)/j(u_2)$$

for an arbitrary $\epsilon > 0$ and $u_2 > u_\epsilon$. Analogously we can estimate the difference x_1-x_{1*}. If we add these estimates we arrive at the inequality (6). •

From Lemmas 1 and 4 and inequality (4) we immediately obtain Theorem 3 in §1.2. •

APPENDIX III

We introduce a simple example which illustrates once again (see §2.7) the unsuitability of estimates "with a shift in the argument" for finding precise criteria for weak invertibility. We agree that we are considering only methods of inversion which are close to those which we used in Chapter 2. Let $X = A(\lambda)$, where λ is a positive increasing function on $[0, 1)$. A necessary step in all of the constructions in §§2.1–2.7 was the proof of the inclusion $f(\log f) \in A(\Lambda)$, where $f \in A(\Lambda)$ and $f(\zeta) \neq 0$ for $|\zeta| < 1$ and $A(\Lambda)$ is the space in which we seek weak invertibility of f. Even the coarse criteria in §2.1 together with the estimates in Chapter 1 give us weak invertibility in the power scale $\bigcup_{c<\infty} A(\lambda^c)$, or even $\bigcup_{\epsilon<1} A(\lambda^{1+\epsilon})$ (a theorem of Beurling type; see §2.7). If we use an estimate with a shift in the argument of the type

$$|\log f(\zeta)| \leqslant \text{const} \log \lambda(\alpha|\zeta| + (1-\alpha)), \quad |\zeta| < 1,$$

where $0 < \alpha < 1$, then it is impossible to conclude that $\log f$ belongs to $\bigcup_{c<\infty} A(\lambda^c)$ or to any space of the type $\bigcup_{n=1}^{\infty} A(h_n(\lambda))$ with preassigned (arbitrarily rapidly increasing) functions h_n. This is implied by the following simple example.

EXAMPLE. Let h be an arbitrary positive continuous function on $(0, +\infty)$, and let α be a number such that $0 < \alpha < 1$. Then there exists a convex continuous function λ, defined on $[0, 1)$, such that for $0 \leqslant r < 1$ we have

$$\lambda(\alpha r + (1-\alpha)) \geqslant h(\lambda(r)). \tag{1}$$

In fact, we define λ on $\Delta_0 = [0, 1-\alpha]$ to be any function (as long as it is convex) such that $h(\lambda(0)) < \lambda(1-\alpha)$. Let $\Gamma r = \alpha r + (1-\alpha)$, $0 \leqslant r < 1$. On the interval $\Gamma\Delta_0 = \Delta_1$ we define λ as a linear function such that (1) is satisfied for $r = \Delta_0$ and the function $\lambda|(\Delta_0 \cup \Delta_1)$ is convex (downward). We continue this inductive construction and arrive at the required function λ.

It is clear that in a similar fashion we can also establish inequalities of the type

$$\lambda(\alpha r + (1-\alpha)) \geqslant h(\lambda(r), \lambda'(r), \ldots, \lambda^{(n)}(r)),$$

where h is a positive continuous function on $\mathbf{R}^n$.

BASIC NOTATION

$\mathbf{C}$ is the complex plane; $\mathbf{R}$ is the real line.

$\mathbf{D} = \{\zeta: \zeta \in \mathbf{C}, |\zeta| < 1\}$ is the unit disk in $\mathbf{C}$; $\partial\mathbf{D}$ is its boundary.

Int γ is the interior of a closed contour $\gamma \subset \mathbf{C}$.

$f\downarrow$ ($f\uparrow$) denotes decrease (increase) of a real-valued function f.

∂A is the topological boundary of a set $A \subset \mathbf{C}$.

$f \sim g$, $f = O(g)$ and $f = o(g)$ denote the standard comparisons of functions f and g.

$f \asymp g$ means that $f = O(g)$ and $g = O(f)$ simultaneously.

$f \circ g$ denotes composition of the mappings f and g, defined where $f(g(x))$ makes sense.

$\hat{f}(n)$ are the Taylor (Fourier) coefficients of a function (distribution) f.

z is the identity mapping of the disk $\mathbf{D}$ (or its boundary $\partial\mathbf{D}$) onto itself: $z(\zeta) = \zeta$, $\zeta \in \mathbf{D}$.

X^* is the space of all continuous linear functionals on a linear space X (the conjugate space).

$\mathfrak{L}(A)$ is the closed linear hull of a set A in a linear space.

dist(a, A) is the distance from a point a to a set A.

$\|T\|_X$ is the norm of an operator T in a space X.

$\mathcal{P}_A$ and $\mathcal{P}$ denote the sets of all polynomials in z and all polynomials in z and z^{-1}, respectively.

$\mathbf{Z}$ is the set (group) of all integers.

$R(\lambda, T) = (\lambda I - T)^{-1}$ is the resolvent of an operator T.

$\sigma(T)$ and $\sigma_p(T)$ denote the spectrum and point spectrum respectively of an operator T.

$\mathfrak{N}$ is the set of all natural numbers.

$\delta_{nk} = 0$ if $n \neq k$ and $\delta_{nk} = 1$ if $n = k$.

$A(\mathbf{D})$ is the space of all functions which are analytic in $\mathbf{D}$.

$x_n \xrightarrow{\text{w.}X} x$ denotes weak convergence in a space X.

$|e|$ is the Lebesgue measure of a set e (here $e \subset \mathbf{R}$ or $e \subset \partial\mathbf{D}$).

BIBLIOGRAPHY

1. N. I. Ahiezer, *Lectures on the theory of approximation*, 2nd rev. ed., "Nauka", Moscow, 1965; English transl. of 1st ed., Ungar, New York, 1956. MR 20 #1872; 32 #6108.

2. N. I. Ahiezer and I. M. Glazman, *The theory of linear operators in Hilbert space*, 2nd rev. ed., "Nauka", Moscow, 1966; English transl. of 1st ed., Vols. I, II, Ungar, New York, 1961. MR 34 #6527; 41 #9015a, b.

3. K. I. Babenko, *On conjugate functions*, Dokl. Akad. Nauk SSSR 62 (1948), 157–160. (Russian) MR 10, 249.

4. ———, *On certain classes of spaces of infinitely differentiable functions*, Dokl. Akad. Nauk SSSR 132 (1960), 1231–1234 = Soviet Math. Dokl. 1 (1960), 738–741. MR 23 #A280.

5. N. K. Bari, *Trigonometric series*, Fizmatgiz, Moscow, 1961; English transl., Macmillan, New York; Pergamon Press, Oxford, 1964. MR 23 #A3411; 30 #1347.

6. S. N. Bernšteĭn, *Sur un théorème de M. Szegö*, Prace Mat.-Fiz. 44 (1937), 9–14.

7. ———, *Le problème de l'approximation des fonctions continues sur tout l'axe réel et l'une de ses applications*, Bull. Soc. Math. France 52 (1924), 399–410.

8. M. Š. Birman and S. B. Èntina, *The stationary method in the abstract theory of scattering*, Izv. Akad. Nauk SSSR Ser. Mat. 31 (1967), 401–430 = Math. USSR Izv. 1 (1967), 391–420. MR 35 #790.

9. M. S. Brodskiĭ, *Triangular and Jordan representations of linear operators*, "Nauka", Moscow, 1969; English transl., Transl. Math. Monographs, vol. 32, Amer. Math. Soc., Providence, R.I., 1971. MR 41 #4283; 48 #904.

10. ———, *Operators with nuclear imaginary components*, Acta Sci. Math. (Szeged) 27 (1966), 147–155. (Russian) MR 34 #1843.

11. N. Bourbaki, *Espaces vectoriels topologiques*. Chaps. 1–5 and fascicule de résultats, Actualités Sci. Indust., nos. 1189, 1229, 1230, Hermann, Paris, 1953, 1955. MR 14, 880; 17, 1109.

12. ———, *Fonctions d'une variable réelle (théorie élémentaire)*, 2nd ed., Chaps. 1–7, Actualités Sci. Indust., nos. 1074, 1132, Hermann, Paris, 1958, 1961. MR 13, 631; 27 #1340.

13. B. M. Byčkov, *On complete systems and bases in the spaces H'_p*, Candidate's Dissertation, Rostov-on-Don, 1967. (Russian)

14. G. Valiron, *Fonctions analytiques*, Presses Univ. France, Paris, 1954. MR 15, 861.

15. S. E. Warschawski, *On conformal mapping of infinite strips*, Trans. Amer. Math. Soc. 51 (1942), 280–335. MR 4, 9.

16. V. S. Videnskiĭ, *On normally increasing functions*, Uspehi Mat. Nauk 9 (1954), no. 2 (60), 212–213. (Russian)

17. V. F. Gapoškin, *A generalization of the theorem of M. Riesz on conjugate functions*, Mat. Sb. 46 (88) (1958), 359–372. (Russian) MR 20 #6000.

18. I. M. Gel'fand, D. A. Raĭkov and G. E. Šilov, *Commutative normed rings*, Fizmatgiz, Moscow, 1960; English transl., Chelsea, New York, 1964. MR 23 #A1242; 34 #4940.

19. I. M. Glazman and Ju. I. Ljubič, *Finite-dimensional linear analysis: A systematic presentation in problem form*, "Nauka", Moscow, 1969; English transl., M. I. T. Press, Cambridge, Mass., 1974.

20. G. M. Goluzin, *Geometric theory of functions of a complex variable*, 2nd ed., "Nauka", Moscow, 1966; English transl., Transl. Math. Monographs, vol. 26, Amer. Math. Soc., Providence, R.I., 1969. MR 36 #2793; 40 #308.

21. V. V. Golubev, *Single-valued analytic functions. Automorphic functions*, Fizmatgiz, Moscow, 1961. (Russian) MR 26 #294.

22. A. A. Gol'dberg, *On an inequality connected with logarithmic convex functions*, Dopovidī Akad. Nauk Ukraïn. RSR 1957, 227–230; English transl., Amer. Math. Soc. Transl. (2) 34 (1963), 165–167. MR 19, 950.

23. K. Hoffman, *Banach spaces of analytic functions*, Prentice-Hall Ser. in Modern Analysis, Prentice-Hall, Englewood Cliffs, N.J., 1962. MR **24** #A2844.

24. I. C. Gohberg and M. G. Kreĭn, *Introduction to the theory of linear nonselfadjoint operators in Hilbert space*, "Nauka", Moscow, 1965; English transl., Transl. Math. Monographs, vol. 18, Amer. Math. Soc., Providence, R.I., 1969. MR **36** #3137; **39** #7447.

25. ———, *Theory and application of Volterra operators in Hilbert space*, "Nauka", Moscow, 1967; English transl., Transl. Math. Monographs, vol. 24, Amer. Math. Soc., Providence, R.I., 1970. MR **36** #2007; **41** #9041.

26. I. C. Gohberg and O. I. Soĭbel'man, *Some remarks on similarity of operators*, Mat. Issled. 2 (1967), vyp. 3, 166–170. (Russian) MR **37** #3387.

27. V. P. Gurariĭ, *The structure of primary ideals in the rings of functions integrable with increasing weight on the half-axis*, Dokl. Akad. Nauk SSSR **178** (1968), 991–994 = Soviet Math. Dokl. 9 (1968), 209–213. MR **37** #3361.

28. ———, *Harmonic analysis of bounded and increasing functions*, Doctoral Dissertation, Kharkov, 1969. (Russian)

29. D. I. Gurevič, *Research on a problem of L. Schwartz*, Candidate's Dissertation, Moscow, 1972. (Russian)

30. N. Dunford and J. T. Schwartz, *Linear operators*. I, II, Interscience, New York, 1958, 1963. MR **22** #8302; **32** #6181.

31. M. M. Džrbašjan, *On metric criteria for the completeness of the system of polynomials, with respect to weighted approximation*, Dokl. Akad. Nauk SSSR **66** (1949), 1037–1040. (Russian) MR **11**, 94.

32. ———, *Some questions of the theory of weighted polynomial approximations in the complex domain*, Mat. Sb. **36** (78) (1955), 353–440. (Russian) MR **17**, 31.

33. ———, *Uniqueness theorems for Fourier transforms and for infinitely differentiable functions*, Izv. Akad. Nauk Armjan. SSR Ser. Fiz.-Mat. Nauk **10** (1957), no. 6, 7–24. (Russian) MR **21** #261.

34. ———, *Theory of factorization of functions meromorphic in the disk*, Mat. Sb. 79 (121) (1969), 517–615 = Math. USSR Sb. 8 (1969), 493–592. MR **41** #3772.

35. M. A. Evgrafov, *The method of near systems in the space of analytic functions and its application to interpolation*, Trudy Moskov. Mat. Obšč. 5 (1956), 89–201; English transl., Amer. Math. Soc. Transl. (2) 16 (1960), 195–314. MR **18**, 472; **22** #8133.

36. ———, *Asymptotic estimates and entire functions*, 2nd rev. ed., Fizmatgiz, Moscow, 1962; English transl. of 1st ed., Gordon and Breach, New York, 1961. MR **27** #4924; **31** #1376.

37. V. P. Zaharjuta and V. I. Judovič, *The general form of a linear functional in* H_p', Uspehi Mat. Nauk 19 (1964), no. 2 (116), 139–142. (Russian) MR **32** #2891.

38. V. S. Zaharjan, *Uniqueness theorems for certain classes of functions holomorphic in the circle*, Mat. Sb. **63** (105) (1964), 3–22. (Russian) MR **28** #2217.

39. A. Zygmund, *Trigonometric series*. Vols. 1, 2, 2nd rev. ed., Cambridge Univ. Press, New York, 1959. MR **21** #6498.

40. M. I. Kadec, *Unconditionally convergent series in a uniformly convex space*, Uspehi Mat. Nauk 11 (1956), no. 5 (71), 185–190. (Russian) MR **18**, 733.

41. ———, *Bases and their spaces of coefficients*, Dopovidi Akad. Nauk Ukraïn. RSR **1964**, 1139–1141. (Ukrainian) MR **30** #385.

42. Ju. A. Kaz'min, *The completeness of certain types of sequences of analytic functions*, Sibirsk. Mat. Ž. 7 (1966), 70–82 = Siberian Math. J. 7 (1966), 55–64. MR **33** #7834.

43. L. V. Kantorovič and G. P. Akilov, *Functional analysis in normed spaces*, Fizmatgiz, Moscow, 1959; English transl., Internat. Ser. Monographs Pure Appl. Math., vol. 46, Macmillan, New York, 1964. MR **22** #9837; **35** #4699.

44. È. M. Kegejan, *Simultaneous approximation in a disc*, Izv. Akad. Nauk Armjan. SSR Ser. Mat. 1 (1966), 317–330. (Russian) MR **35** #385.

45. M. V. Keldyš, *Sur l'approximation en moyenne par polynomes des fonctions d'une variable complexe*, Mat. Sb. **16** (58) (1945), 1–20. MR **7**, 64.

46. ———, *On the characteristic values and characteristic functions of certain classes of non-self-adjoint equations*, Dokl. Akad. Nauk SSSR 77 (1951), 11–14. (Russian) MR **12**, 835.

47. A. K. Kitover, *The spectral properties of unitary operators in C*, Zap. Naučn. Sem. Leningrad. Otdel. Mat. Inst. Steklov. (LOMI) 22 (1971), 47–64 = J. Soviet Math. 2 (1974), 99–111. MR **44** #7335.

48. I. F. Krasičkov, *Closed ideals in locally convex algebras of entire functions*. II, Izv. Akad. Nauk SSSR Ser. Mat. 32 (1968), 1024–1032 = Math. USSR Izv. 2 (1968), 979–986. MR **38** #2605.

49. I. F. Krasičkov, *Closed ideals in locally convex algebras of entire functions. Algebras of minimal type*, Sibirsk. Mat. Ž. 9 (1968), 77–96 = Siberian Math. J. 9 (1968), 59–71. MR 37 #5685.

50. M. A. Krasnosel'skiĭ and Ja. B. Rutickiĭ, *Convex functions and Orlicz spaces*, GITTL, Moscow, 1958; English transl., Noordhoff, Groningen, 1961. MR 21 #5144; 23 #A4016.

51. B. I. Korenbljum, *Invariant subspaces of the shift operator in a weighted Hilbert space*, Mat. Sb. 89 (131) (1972), 110–137 = Math. USSR Sb. 18 (1972), 111–138. MR 47 #5621.

52. ———, *Closed ideals of the ring* A^n, Funkcional. Anal. i Priložen. 6 (1972), no. 3, 38–52 = Functional Anal. Appl. 6 (1972), 203–214 (1973). MR 48 #2776.

53. B. Ja. Levin, *Distribution of zeros of entire functions*, GITTL, Moscow, 1956; English transl., Transl. Math. Monographs, vol. 5, Amer. Math. Soc., Providence, R.I., 1964. MR 19, 402; 28 #217.

54. B. M. Levitan, *Almost-periodic functions*, GITTL, Moscow, 1953. (Russian) MR 15, 700.

55. V. B. Lidskiĭ, *Conditions for completeness of a system of root subspaces for nonself-adjoint operators with discrete spectra*, Trudy Moskov. Mat. Obšč. 8 (1959), 83–120; English transl., Amer. Math. Soc. Transl. (2) 34 (1963), 241–281. MR 21 #6539.

56. F. S. Lisin, *A problem of mean-square approximation related to the investigation of generators of the algebra* l^1, Mat. Zametki 3 (1968), 703–706 = Math. Notes 3 (1968), 449–451. MR 37 #5402.

57. Ju. I. Ljubič and V. I. Macaev, *Operators with separable spectrum*, Mat. Sb. 56 (98) (1962), 433–468; English transl., Amer. Math. Soc. Transl. (2) 47 (1965), 89–129. MR 25 #2450.

58. Ju. I. Ljubič, *Completeness conditions for a system of eigenvectors of a correct operator*, Uspehi Mat. Nauk 18 (1963), no. 1 (109), 165–171. (Russian) MR 27 #4081.

59. ———, *Conservative operators*, Uspehi Mat. Nauk 20 (1965), no. 5 (125), 221–225. (Russian) MR 34 #4925.

60. S. Mandelbrojt, *Séries adhérentes, régularisation des suites, applications*, Gauthier-Villars, Paris, 1952. MR 14, 542.

61. A. S. Markus, *Some criteria for the completeness of a system of root vectors of a linear operator in a Banach space*, Mat. Sb. 70 (112) (1966), 526–561; English transl., Amer. Math. Soc. Transl. (2) 85 (1969), 51–91. MR 35 #7151.

62. ———, *The problem of spectral synthesis for operators with point spectrum*, Izv. Akad. Nauk SSSR Ser. Mat. 34 (1970), 662–688 =Math. USSR Izv. 4 (1970), 670–696. MR 41 #9030.

63. A. S. Markus, L. N. Nikol'skaja and N. K. Nikol'skiĭ, *The unitary spectrum of a contraction in a Banach space*, Zap. Naučn. Sem. Leningrad. Otdel. Mat. Inst. Steklov. (LOMI) 22 (1971), 65–74 = J. Soviet Math. 2 (1974), 112–119. MR 44 #7337.

64. A. I. Markuševič, *Conformal mapping of domains with variable boundaries, with applications to the approximation of analytic functions by polynomials*, Candidate's Dissertation, Moscow State Univ., Moscow, 1934. (Russian)

65. L. A. Ljusternik, O. A. Červonenkis and A. R. Janpol'skiĭ, *Mathematical analysis. Computation of the elementary functions*, Fizmatgiz, Moscow, 1963; English transl., *Handbook for computing elementary functions*, Pergamon Press, Oxford and New York, 1965. MR 28 #1733; 32 #584.

66. V. I. Macaev, *Some theorems on the completeness of root subspaces of completely continuous operators*, Dokl. Akad. Nauk SSSR 155 (1964), 273–276 = Soviet Math. Dokl. 5 (1964), 396–399. MR 28 #2444.

67. ———, *On the growth of entire functions that admit a certain estimate from below*, Dokl. Akad. Nauk SSSR 132 (1960), 283–286 = Soviet Math. Dokl. 1 (1960), 548–552. MR 23 #A328.

68. ———, *A method for the estimation of the resolvents of non-self-adjoint operators*, Dokl. Akad. Nauk SSSR 154 (1964), 1034–1037 = Soviet Math. Dokl. 5 (1964), 236–240. MR 28 #1495.

69. ———, *Factorization in some classes of functions*, Abstracts of Reports, All-Union Conf. Theory of Functions of a Complex Variable, Kharkov, 1971, pp. 138–139. (Russian)

70. S. N. Mergeljan, *On the completeness of systems of analytic functions*, Uspehi Mat. Nauk 8 (1953), no. 4 (56), 3–63; English transl., Amer. Math. Soc. Transl. (2) 19 (1962), 109–166. MR 15, 411; 24 #A1410.

71. ———, *Weighted approximations by polynomials*, Uspehi Mat. Nauk 11 (1956), no. 5 (71), 107–152; English transl., Amer. Math. Soc. Transl. (2) 10 (1958), 59–106. MR 18, 734.

72. N. K. Nikol'skiĭ, *Spaces and algebras of Toeplitz matrices operating in* l^p, Sibirsk. Mat. Ž. 7 (1966), 146–158 = Siberian Math. J. 7 (1966), 118–126. MR 33 #1733.

73. ———, *Invariant subspaces of unitary operators*, Vestnik Leningrad. Univ. 21 (1966), no. 19, 36–43. (Russian) MR 34 #4910.

74. ———, *Invariant subspaces of weighted shift operators*, Mat. Sb. 74 (116) (1967), 171–190 = Math. USSR Sb. 3 (1967), 159–176. MR 37 #4659.

75. ———, *Closed ideals in certain algebras of entire functions*, Sibirsk. Mat. Ž. 9 (1968), 211–215 = Siberian Math. J. 9 (1968), 160–163. MR 37 #446.

76. ———, *Invariant subspaces of the shift operator in some sequence spaces*, Candidate's Dissertation, Leningrad, 1966. (Russian)

77. ———, *Invariant subspaces of the shift operator and bounded approximation almost everywhere*, Trudy Mat. Inst. Steklov. 96 (1968), 243–257 = Proc. Steklov Inst. Math. 96 (1968), 307–323. MR 42 #1629; #3555.

78. ———, *Basicity and unicellularity of weighted shift operators*, Izv. Akad. Nauk SSSR Ser. Mat. 32 (1968), 1123–1137 = Math. USSR Izv. 2 (1968), 1077–1090. MR 38 #6374.

79. ———, *Invariant subspaces and bases of generalized primitives in the sense of A. O. Gel'fond and A. F. Leont'ev*, Mat. Issled. 3 (1968), vyp. 3 (9), 101–116. (Russian) MR 41 #2434.

80. ———, *Complete extensions of Volterra operators*, Izv. Akad. Nauk SSSR Ser. Mat. 33 (1969), 1349–1355 = Math. USSR Izv. 3 (1969), 1271–1276. MR 49 #1188.

81. ———, *On perturbations of the spectrum of unitary operators*, Mat. Zametki 5 (1969), 341–349 = Math. Notes 5 (1969), 207–211.

82. ———, *Spectral synthesis for a shift operator and zeros in certain classes of analytic functions smooth up to the boundary*, Dokl. Akad. Nauk SSSR 190 (1970), 780–783 = Soviet Math. Dokl. 11 (1970), 206–209. MR 41 #5999.

83. ———, *Nonstandard ideals, unicellularity and algebras that are connected with the shift operator*, Zap. Naučn. Sem. Leningrad. Otdel. Mat. Inst. Steklov. (LOMI) 19 (1970), 156–195 = Sem. Math. V. A. Steklov Math. Inst. Leningrad. 19 (1970), 91–111. MR 45 #9179.

84. ———, *The multiple shift with a simple spectrum*, Zap. Naučn. Sem. Leningrad. Otdel. Mat. Inst. Steklov. (LOMI) 19 (1970), 227–236 = Sem. Math. V. A. Steklov Math. Inst. Leningrad 19 (1970), 132–136. MR 45 #948.

85. ———, *Spectral analysis on the unitary spectrum. The point spectrum*, Dokl. Akad. Nauk SSSR 199 (1971), 544–547 = Soviet Math. Dokl. 12 (1971), 1123–1126. MR 44 #5805.

86. ———, *On spectral analysis on the unitary spectrum. The point spectrum*. II, Zap. Naučn. Sem. Leningrad. Otdel. Mat. Inst. Steklov. (LOMI) 22 (1971), 75–88 = J. Soviet Math. 2 (1974), 120–130. MR 45 #5792.

87. ———, *Spectral synthesis and the problem of weighted approximation in certain spaces of analytic functions that grow near the boundary*, Izv. Akad. Nauk Armjan. SSR Ser. Mat. 6 (1971), 345–367. (Russian) MR 45 #7504.

88. ———, *The problem of localization of ideals in spaces and algebras of analytic functions*, Abstracts of Reports, All-Union Conf. Theory of Functions of a Complex Variable, Kharkov, 1971, pp. 160–161. (Russian)

89. ———, *Unilateral estimates of the modulus of harmonic functions in a disk and in a strip*, Dokl. Akad. Nauk SSSR 205 (1972), 522–525 = Soviet Math. Dokl. 13 (1972), 957–961. MR 47 #5261.

90. ———, *A criterion for weak invertibility in spaces of analytic functions that are defined by growth conditions*, Zap. Naučn. Sem. Leningrad. Otdel. Mat. Inst. Steklov. (LOMI) 30 (1972), 106–129 = J. Soviet Math. 4 (1975), 393–412. MR 49 #3515.

91. V. P. Palamodov, *Linear differential operators with constant coefficients*, "Nauka", Moscow, 1967; English transl., Die Grundlehren der math. Wissenschaften, Band 168, Springer-Verlag, Berlin and New York, 1970. MR 39 #4517; 41 #8793.

92. A. Pełczyński, *Linear extensions, linear averagings, and their applications to linear topological classifications of spaces of continuous functions*, Rozprawy Mat. 58 (1968). MR 37 #3335.

93. A. I. Plotkin, *On the behavior of Taylor series in spaces of summable analytic functions*, Vestnik Leningrad. Univ. 25 (1970), no. 1, 52–59 = Vestnik Leningrad Univ. Math. 3 (to appear).

94. G. Pólya and G. Szegö, *Aufgaben und Lehrsatze aus der Analysis*. Vols. 1, 2, Die Grundlehren der math. Wissenschaften, Bände 19, 20, Springer-Verlag, Berlin, 1925; English transl., Die Grundlehren der math. Wissenschaften, Bände 193, 216, Springer-Verlag, Berlin and New York, 1972, 1975.

95. I. I. Privalov, *Boundary properties of analytic functions*, 2nd ed., GITTL, Moscow, 1950; German transl., Hochschulbücher für Math., Band 25, VEB Deutscher Verlag, Berlin, 1956. MR 13, 926; 18, 727.

96. B. Sz.-Nagy and C. Foiaş, *Analyse harmonique des opérateurs de l'espace de Hilbert*, Masson, Paris; Akad. Kiadó, Budapest, 1967; English rev. transl., North-Holland, Amsterdam; American Elsevier, New York; Akad. Kiadó, Budapest, 1970. MR 37 #778; 43 #947.

97. E. M. Semenov, *Interpolation of linear operators in symmetric spaces*, Dokl. Akad. Nauk SSSR **164** (1965), 746–749 = Soviet Math. Dokl. 6 (1965), 1294–1298. MR 32 #8139.

98. ———, *A method for establishing interpolation theorems in symmetric spaces*, Dokl. Akad. Nauk SSSR **185** (1969), 1243–1246 = Soviet Math. Dokl. **10** (1969), 507–511. MR 40 #1799.

99. V. I. Smirnov, *Über die Ränderzuordnung bei konformer Abbildung*, Math. Ann. **107** (1932), 313–323.

100. V. I. Smirnov and N. A. Lebedev, *The constructive theory of functions of a complex variable*, "Nauka", Moscow, 1964; English transl., *Functions of a complex variable: Constructive theory*, M.I.T. Press, Cambridge, Mass., 1968. MR 30 #2152; 37 #5369.

101. P. K. Suetin, *Polynomials orthogonal over a region and Bieberbach polynomials*, Trudy Mat. Inst. Steklov. **100** (1971) = Proc. Steklov Inst. Math. **100** (1971).

102. G. C. Tumarkin, *Approximation in the mean to functions on rectifiable curves*, Mat. Sb. **42** (84) (1957), 79–128. (Russian) MR 19, 545.

103. K. M. Fišman and G. M. Sas'ko, *On certain systems of functions which form quasipower bases in spaces of functions analytic in a disk*, Dokl. Akad. Nauk SSSR **146** (1962), 314–317 = Soviet Math. Dokl. 3 (1962), 1311–1314. MR 25 #5406.

104. V. P. Havin, *Analytic representations of linear functionals in spaces of harmonic and analytic functions continuous in a closed region*, Dokl. Akad. Nauk SSSR **151** (1963), 505–508 = Soviet Math. Dokl. 4 (1963), 1039–1042. MR 27 #2636.

105. S. Ja. Havinson, *The theory of extremal problems for bounded analytic functions satisfying additional conditions inside the domain*, Uspehi Mat. Nauk 18 (1963), no. 2 (110), 25–98 = Russian Math. Surveys 18 (1963), no. 2, 23–96. MR 27 #300.

106. ———, *Analytic functions of bounded type*, Mathematical Analysis 1963, Akad. Nauk SSSR Inst. Naučn. Informacii, Moscow, 1965, pp. 5–80. (Russian) MR 33 #4288.

107. ———, *A notion of completeness that takes into account the magnitude of the coefficients of the approximating polynomials*, Izv. Akad. Nauk Armjan. SSR Ser. Mat. 6 (1971), 221–234. (Russian) MR 45 #5641.

108. E. Hille and R. S. Phillips, *Functional analysis and semi-groups*, rev. ed., Amer. Math. Soc. Colloq. Publ., vol. 31, Amer. Math. Soc., Providence, R.I., 1957. MR 19, 664.

109. S. V. Hruščev, *The removal of singularities for Cauchy integrals and an analog of the Hinčin-Ostrowski theorem for sequences of increasing functions*, Dokl. Akad. Nauk SSSR **214** (1974), 524–527 = Soviet Math. Dokl. 15 (1974), 198–202.

110. A. L. Šaginjan, *On an incompleteness test for a system of analytic functions*, Akad. Nauk Armjan. SSR Dokl. 5 (1946), 97–100. (Russian) MR 8, 455.

111. ———, *A fundamental inequality in the theory of functions and its applications*, Izv. Akad. Nauk Armjan. SSR Ser. Fiz.-Mat. Nauk 12 (1959), 3–25. (Russian) MR 21 #5025.

112. F. A. Šamojan, *Description of closed ideals and certain factorization questions in algebras of growing functions that are analytic in the disc*, Izv. Akad. Nauk Armjan. SSR Ser. Mat. 5 (1970), 419–433. (Russian) MR 45 #905.

113. H. S. Shapiro, *Some remarks on weighted polynomial approximations of holomorphic functions*, Mat. Sb. **73** (115) (1967), 320–330 = Math. USSR Sb. 2 (1967), 285–294. MR 36 #395.

114. G. E. Šilov, *On regular normed rings*, Trudy Mat. Inst. Steklov. 21 (1947). (Russian) MR 9, 596.

115. ———, *On rings of functions with uniform convergence*, Ukrain. Mat. Ž. 3 (1951), 404–411. (Russian) MR 14, 884.

116. ———, *Mathematical analysis: Special course*, 2nd ed., GITTL, Moscow, 1961; English transl., Pergamon Press, New York, 1965. MR 24 #A1180; 32 #2519.

117. Arne Beurling, *On two problems concerning linear transformations in Hilbert space*, Acta Math. **81** (1948), 239–255. MR 10, 381.

118. ———, *A critical topology in harmonic analysis on semigroups*, Acta Math. **112** (1964), 215–228. MR 29 #6255.

119. Errett Bishop, *A duality theorem for an arbitrary operator*, Pacific J. Math. 9 (1959), 379–397. MR 22 #8339.

120. Louis de Branges, *Perturbations of self-adjoint transformations*, Amer. J. Math. **84** (1962), 543–560. MR 27 #4083.

121. A. Brown and P. R. Halmos, *Algebraic properties of Toeplitz operators*, J. Reine Angew. Math. 213 (1963/64), 89–102. MR 28 #3350; errata, 30, p. 1205.

122. L. Brown, A. Shields and K. Zeller, *On absolutely convergent exponential sums*, Trans. Amer. Math. Soc. 96 (1960), 162–183. MR 26 #332.

123. J. S. Byrnes and D. J. Newman, *Completeness preserving multipliers*, Proc. Amer. Math. Soc. 21 (1969), 445–450. MR 39 #1893.

124. M. L. Cartwright, *On analytic functions regular in the unit circle*. I, II, Quart. J. Math. Oxford Ser. 4 (1933), 246–257; 6 (1935), 94–105.

125. Paul Civin, *Inequalities for trigonometric integrals*, Duke Math. J. 8 (1941), 656–665. MR 3, 108.

126. Yngve Domar, *Closed primary ideals in a class of Banach algebras*, Math. Scand. 7 (1959), 109–125. MR 22 #5909.

127. W. F. Donoghue, Jr., *The lattice of invariant subspaces of a completely continuous quasi-nilpotent transformation*, Pacific J. Math. 7 (1957), 1031–1035. MR 19, 1066.

128. Peter L. Duren, *Extension of a result of Beurling on invariant subspaces*, Trans. Amer. Math. Soc. 99 (1961), 320–324. MR 22 #9851.

129. Leon Ehrenpreis, *Mean periodic functions*. I. *Varieties whose annihilator ideals are principal*, Amer. J. Math. 77 (1955), 293–328; Appendix, 731–733. MR 16, 1122; 17, 878.

130. Frank Forelli, *Invariant subspaces in* L^1, Proc. Amer. Math. Soc. 14 (1963), 76–79. MR 26 #1743.

131. ———, *The Marcel Riesz theorem on conjugate functions*, Trans. Amer. Math. Soc. 106 (1963), 369–390. MR 26 #5340.

132. J. M. Freeman, *Perturbations of the shift operator*, Trans. Amer. Math. Soc. 114 (1965), 251–260. MR 30 #2342.

133. Ralph Gellar, *Cyclic vectors and parts of the spectrum of a weighted shift*, Trans. Amer. Math. Soc. 146 (1969), 69–85. MR 41 #4277b.

134. H. L. Hamburger, *Über die Zerlegung des Hilbertschen Raumes durch vollstetige lineare Transformationen*, Math. Nachr. 4 (1951), 56–69. MR 12, 718.

135. G. H. Hardy, *Orders of infinity*, Cambridge Univ. Press, New York, 1924.

136. G. H. Hardy and J. E. Littlewood, *Some properties of conjugate functions*, J. Reine Angew. Math. 167 (1932), 405–423.

137. ———, *Some more theorems concerning Fourier series and Fourier power series*, Duke Math. J. 2 (1936), 354–382.

138. W. K. Hayman, *The minimum modulus of large integral functions*, Proc. London Math. Soc. (3) 2 (1952), 469–512. MR 15, 22.

139. L. I. Hedberg, *Weighted mean approximation in Carathéodory regions*, Math. Scand. 23 (1968), 113–122 (1969). MR 41 #2028.

140. Henry Helson, *Lectures on invariant subspaces*, Academic Press, New York and London, 1964. MR 30 #1409.

141. Henry Helson and David Lowdenslager, *Invariant subspaces*, Proc. Internat. Sympos. Linear Spaces (Jerusalem, 1960), Jerusalem Academic Press, Jerusalem; Pergamon Press, Oxford, 1961, pp. 251–262. MR 28 #487.

142. Henry Helson and Gabor Szegö, *A problem in prediction theory*, Ann. Mat. Pura Appl. (4) 51 (1960), 107–138. MR 22 #12343.

143. Lars Hörmander, *Generators for some rings of analytic functions*, Bull. Amer. Math. Soc. 73 (1967), 943–949. MR 37 #1977.

144. R. Hunt, B. Muckenhoupt and R. Wheeden, *Weighted norm inequalities for the conjugate function and Hilbert transform*, Trans. Amer. Math. Soc. 176 (1973), 227–251. MR 47 #701.

145. Satoru Igari, *Functions of* L^p*-multipliers*, Tôhoku Math. J. (2) 21 (1969), 304–320.

146. J.-P. Kahane and R. Salem, *Ensembles parfaits et séries trigonométriques*, Actualités Sci. Indust., no. 1301, Hermann, Paris, 1963. MR 28 #3279.

147. Yitzhak Katznelson, *An introduction to harmonic analysis*, Wiley, New York and London, 1968. MR 40 #1734.

148. ———, *Sets of uniqueness for some classes of trigonometrical series*, Bull. Amer. Math. Soc. 70 (1964), 722–723. MR 29 #2592.

149. J. J. Kelleher and B. A. Taylor, *Closed ideals in locally convex algebras of analytic functions*, J. Reine Angew. Math. 255 (1972), 190–209. MR 46 #6046.

150. Paul J. Koosis, *Sur l'approximation pondérée par des polynomes et par des sommes d'exponentielles imaginaires*, Ann. Sci. École Norm. Sup. (3) 81 (1964), 387–408. MR 31 #1505.

151. S. T. Kuroda, *On a theorem of Weyl-von Neumann*, Proc. Japan Acad. **34** (1958), 11–15. MR 21 #1537.

152. K. de Leeuw and I. Glicksberg, *Applications of almost periodic compactifications*, Acta Math. **105** (1961), 63–97. MR 24 #A1632.

153. C. N. Linden, *Functions regular in the unit circle*, Proc. Cambridge Philos. Soc. **52** (1956), 49–60. MR 17, 472.

154. ———, *The minimum modulus of functions regular and of finite order in the unit circle*, Quart. J. Math. Oxford Ser. (2) **7** (1956), 196–216. MR 20 #3982.

155. Edgar R. Lorch, *The integral representation of weakly almost-periodic transformations in reflexive vector spaces*, Trans. Amer. Math. Soc. **49** (1941), 18–40. MR 2, 224.

156. John von Neumann, *Charakterisierung des Spektrum eines Integraloperators*, Actualités Sci. Indust., no. 229, Hermann, Paris, 1935.

157. D. J. Newman, *The closure of translates in l^p*, Amer. J. Math. **86** (1964), 651–667. MR 29 #1492.

158. D. J. Newman and H. S. Shapiro, *The Taylor coefficients of inner functions*, Michigan Math. J. **9** (1962), 249–255. MR 26 #6371.

159. D. J. Newman, J. T. Schwartz and H. S. Shapiro, *On generators of the Banach algebras l_1 and $L_1(0, \infty)$*, Trans. Amer. Math. Soc. **107** (1963), 466–484. MR 27 #575.

160. Eric A. Nordgren, *Invariant subspaces of a direct sum of weighted shifts*, Pacific J. Math. **27** (1968), 587–598. MR 39 #2003.

161. E. A. Nordgren, H. Radjavi and P. Rosenthal, *On density of transitive algebras*, Acta Sci. Math. (Szeged) **30** (1969), 175–179. MR 40 #6276.

162. Marvin Rosenblum, *Summability of Fourier series in $L^p(d\mu)$*, Trans. Amer. Math. Soc. **105** (1962), 32–42. MR 28 #3287.

163. L. A. Rubel and A. L. Shields, *The space of bounded analytic functions on a region*, Ann. Inst. Fourier (Grenoble) **16** (1966), fasc. 1, 235–277. MR 33 #6440.

164. Walter Rudin, *The closed ideals in an algebra of analytic functions*, Canad. J. Math. **9** (1957), 426–434. MR 19, 641.

165. ———, *Projections on invariant subspaces*, Proc. Amer. Math. Soc. **13** (1962), 429–432. MR 25 #1460.

166. Baltasar Rodriguez-Salinas, *Functions with null moments*, Rev. Acad. Ci. Madrid **49** (1955), 331–368. (Spanish) MR 18, 204.

167. Stephen A. Scheinberg, *A convergence question in H^p*, Proc. Amer. Math. Soc. **30** (1971), 120–124. MR 44 #439.

168. O. F. G. Schilling, *Ideal theory on open Riemann surfaces*, Bull. Amer. Math. Soc. **52** (1946), 945–963. MR 8, 454.

169. Laurent Schwartz, *Théorie générale des fonctions moyenne-périodiques*, Ann. of Math. (2) **48** (1947), 857–929. MR 9, 428.

170. Harold S. Shapiro, *Weakly invertible elements in certain function spaces, and generators in ℓ_1*, Michigan Math. J. **11** (1964), 161–165. MR 29 #3620.

171. ———, *Weighted polynomial approximation and boundary behaviour of analytic functions*, Contemporary Problems in Theory Anal. Functions (Internat. Conf., Erevan, 1965), "Nauka", Moscow, 1966, pp. 326–335. MR 35 #383.

172. James D. Stafney, *A permissible restriction on the coefficients in uniform polynomial approximation to $C[0, 1]$*, Duke Math. J. **34** (1967), 392–396. MR 35 #3329.

173. M. V. Subbarao, *Some classes of complete sequences and approximations in normed linear spaces*, Publ. Ramanujan Inst. No. 1 (1968/69), 189–197. MR 42 #2237.

174. Otto Szász, *Über die Fourierschen Reihen gewisser Funktionenklassen*, Math. Ann. **100** (1928), 530–536.

175. Gabor Szegö, *Über einen Satz des Herrn Serge Bernstein*, Schr. Königsberger Gelehrten Ges. **5** (1928), 59–70.

176. B. A. Taylor and D. L. Williams, *Ideals in rings of analytic functions with smooth boundary values*, Canad. J. Math. **22** (1970), 1266–1283. MR 42 #7905.

177. John Wermer, *On invariant subspaces of normal operators*, Proc. Amer. Math. Soc. **3** (1952), 270–277. MR 14, 55.

178. ———, *The existence of invariant subspaces*, Duke Math. J. **19** (1952), 615–622. MR 14, 384.

179. N. Wiener and H. R. Pitt, *On absolutely convergent Fourier-Stieltjes transforms*, Duke Math. J. **4** (1938), 420–436.

180. Edward Kerlin and Alan Lambert, *Strictly cyclic shifts on l_p*, Acta Sci. Math. (Szeged) 35 (1973), 87–94. MR 48 #7000.

181. L. N. Nikol'skaja, *Structure of the point spectrum of a linear operator*, Mat. Zametki 15 (1974), 149–158 = Math. Notes 15 (1974), 83–87.

182. M. A. Naĭmark, *Normed rings*, 2nd rev. ed., "Nauka", Moscow, 1968; English transl., *Normed algebras*, Wolters-Noordhoff, Groningen, 1972.